*Now available in a lower priced paperback edition in the Wiley Classics Library.

Continued on back end papers

Statistical Methods for Survival Data Analysis

Statistical Methods for Survival Data Analysis

Second Edition

ELISA T. LEE
Center for Epidemiologic Research and
Department of Biostatistics and Epidemiology
College of Public Health
University of Oklahoma Health Sciences Center
Oklahoma City, Oklahoma

A Wiley-Interscience Publication
JOHN WILEY & SONS, INC.
New York • Chichester • Brisbane • Toronto • Singapore

Copyright © 1992 by John Wiley & Sons, Inc.

All rights reserved. Published simultaneously in Canada.

Reproduction or translation of any part of this work
beyond that permitted by Section 107 or 108 of the
1976 United States Copyright Act without the permission
of the copyright owner is unlawful. Requests for
permission or further information should be addressed to
the Permissions Department, John Wiley & Sons, Inc.

Library of Congress Cataloging in Publication Data:
Lee, Elisa T.
  Statistical methods for survival data analysis / Elisa T. Lee. --
2nd ed.
      p.   cm. -- (Wiley series in probability and mathematical
statistics. Applied probability and statistics)
    "A Wiley-Interscience publication."
    Includes bibliographical references and index.
    ISBN 0-471-61592-7
    1. Medicine -- Research -- Statistical methods.   2. Failure time data
analysis. 3. Prognosis -- Statistical methods.   I. Title.
    II. Series.
    R853.S7L43   1992
    610'.72--dc20   91-27926
                                                                 CIP

Printed and bound in the United States of America by Braun-Brumfield, Inc.

10 9 8 7 6 5 4 3 2 1

To the memory of my parents
Mr. Chi-Lan Tan and
Mrs. Hwei-Chi Lee Tan

Contents

Preface

The term *survival data* has been used in a broad sense for data involving time to a certain event such as to death, to relapse, and to onset of a disease. In the past decade, applications of the statistical methods for survival data analysis have been extended beyond biomedical and reliability research to other fields, for example, criminology, sociology, marketing, and health insurance practice. The second edition of *Statistical Methods for Survival Data Analysis* is intended to meet the need for a single volume covering the methodologies appropriate for the analysis of survival data. The book has been written for biomedical investigators, statisticians, epidemiologists, and researchers in other disciplines who are involved or interested in analyzing survival data. It covers the most commonly used methods, parametric and nonparametric, in survival data analysis and can be used as a reference resource or textbook. In addition, it provides guidelines for the planning and design of clinical trials. Some of the guidelines are also applicable to other types of research, for example, epidemiologic studies. Most of the statistical methods described in the book are applicable to clinical investigations, epidemiologic studies, social science research, and studies in other areas. The present edition remains application oriented and the mathematical level has been kept to a minimum. Some knowledge of calculus and matrix algebra is needed in a few sections. However, readers with only college algebra will find most of the book understandable. In addition to a large number of real-life examples in the text several large data sets are provided as exercises for the reader to use.

The following improvements have been incorporated into the second edition:

1. In addition to clinical life-table analysis, population life tables are discussed.

2. The concepts of standardized mortality ratio (SMR) and standardized incidence ratio (SIR) are introduced.

3. A goodness-of-fit test for modeling survival data involving censored observations has been added.

4. The section on Cox's proportional hazards model has been expanded to cover stratification and the verification of the underlying assumption of proportional hazards.

5. A discussion on the relationship between odds ratios and coefficients of the linear logistic regression model is presented. This regression model is extended to case–control studies. A test of the goodness-of-fit for the logistic regression method is also included.

6. Methods for determining sample sizes in clinical trials now include survival time as the endpoint.

7. The issues of repeated significance testing and group sequential design are introduced.

8. The updated reference list includes a large number of recently published papers.

In the past 10 years, many computer programs for survival data analysis have been written for mainframe computers and microcomputers. The current edition refers the reader to these computer programs in various chapters. Because of the wide availability of these programs, most of the FORTRAN programs listed in the first edition have been removed. Only the computer program that generates the gamma probability plot is retained. This program has received the most requests from readers of the first edition in the past 10 years.

I would like to thank the many researchers, teachers, and students who have used the first edition of the book. I apologize to and also appreciate the patience of those who have been trying to get a copy of the first edition but could not in the past several years because it was unavailable. They have motivated me to work on this second edition, and I appreciate very much their support of the book. Many colleagues and readers have provided helpful suggestions for the second edition; it is impossible to list them all. Special thanks go to Dr. Min Lu, Wansu Chen, and Dr. J. L. Yeh, who have assisted me in working out examples and searching for papers in the literature. Prior to her retirement, Ms. Beatrice Shube of John Wiley & Sons was enthusiastic about publishing a second edition of the book. Since then, Ms. Kate Roach has provided continuing support of the project. I highly value and appreciate their enthusiasm and advice.

Finally, I am grateful for the patience and support of my family, my husband Sam and children Vivian and Jennifer. Without their constant and dependable help this book could not have been completed.

Elisa T. Lee

Statistical Methods for Survival Data Analysis

CHAPTER 1

Introduction

1.1 PRELIMINARIES

This book is written for biomedical researchers, epidemiologists, consulting statisticians, students taking a first course on survival data analysis, and others interested in survival time study. It deals with the statistical methods for analyzing survival data derived from laboratory studies of animals or from clinical and epidemiologic studies of humans who have acute or chronic diseases.

Survival time can be broadly defined as the time to the occurrence of a given event. This event can be the development of a disease, response to a treatment, relapse, or death. Therefore, survival time can be tumor-free time, the time from the start of treatment to response, length of remission, and time to death. Survival data can include survival time, response to a given treatment, and patient characteristics related to response, survival, and the development of a disease. In the past, the study of survival data has focused on predicting the probability of response, survival, or mean lifetime, and comparing the survival distributions of experimental animals or of human patients. In recent years, the identification of risk and/or prognostic factors related to response, survival, and the development of a disease has become equally important. In this book, special consideration is given to the study of survival data in biomedical sciences, though all the methods are suitable for applications in industrial reliability, social sciences, and business. Examples of survival data in these fields are lifetime of electronic devices, components or systems (reliability engineering), felons' time to parole (criminology), duration of first marriage (sociology), length of newspaper or magazine subscription (marketing), and workmen's compensation claims (insurance) and their various influencing *risk* or *prognostic* factors.

1.2 CENSORED DATA

Many researchers consider survival data analysis to be merely the application of two conventional statistical methods to a special type of problem:

1

parametric if the distribution of survival times is known to be normal and nonparametric if the distribution is unknown. This assumption would be true if the survival times of all the subjects were exact and known. However, some survival times are not. Further, the survival distribution is often skewed or far from being normal. Thus there is a need for new statistical techniques. One of the most important developments is due to a special feature of survival data in the life sciences that occurs when some subjects in the study have not experienced the event of interest at the end of the study or time of analysis. For example, some patients may still be alive or in remission at the end of the study period. The exact survival times of these subjects are unknown. These are called *censored observations* or *censored times* and can also occur when individuals are lost to follow-up after a period of study. There are three types of censoring.

1. Type I Censoring

Animal studies usually start with a fixed number of animals, to which the treatment or treatments is given. Because of time and/or cost limitations, the researcher often cannot wait for the death of all the animals. One option is to observe for a fixed period of time, say six months, after which the surviving animals are sacrificed. Survival times recorded for the animals that died during the study period are the times from the start of the experiment to their death. These are called *exact* or *uncensored* observations. The survival times of the sacrificed animals are not exactly known but are recorded as at least the length of the study period. These are called *censored* observations. Some animals could be lost or die accidentally. Their survival times, from the start of an experiment to loss or death, are also censored observations. In Type I censoring, if there are no accidental losses, all censored observations equal the length of the study period.

For example, suppose that six rats have been exposed to carcinogens by injecting tumor cells into their foot-pads. The times to develop a tumor of a given size are observed. The investigator decides to terminate the experiment after 30 weeks. Figure 1.1 plots the development times of the tumors Rats A, B, and D develop tumors after 10, 15, and 25 weeks, respectively. Rats C and E do not develop tumors by the end of the study; their tumor-free times are thus 30-plus weeks. Rat F died accidentally without any tumors after 19 weeks of observation. The survival data (tumor-free times) are 10, 15, 30+, 25, 30+, and 19+ weeks. (The plus sign indicates a censored observation.)

2. Type II Censoring

Another option in animal studies is to wait until a fixed portion of the animals have died, say, 80 of 100, after which the surviving animals are sacrificed. In this case, if there are no accidental losses, the censored observations equal the largest uncensored observation. For example, in an experiment of six rats (Figure 1.2), the investigator may decide to terminate

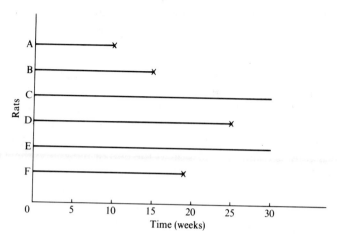

Figure 1.1 An example of Type I censored data.

the study after four of the six rats have developed tumors. The survival or tumor-free times are then 10, 15, 35+, 25, 35, and 19+ weeks.

3. *Type III Censoring*

In most clinical studies the period of study is fixed and patients enter the study at different times during that period. Some may die before the end of the study; their exact survival times are known. Others may withdraw before the end of the study and are lost to follow-up. Still others may be alive at the end of the study. For "lost" patients, survival times are at least from their entrances to the last contact. For patients still alive, survival times are at least from entry to the end of the study. These last two kinds of observations are censored observations. Since the entry times are not simultaneous, the

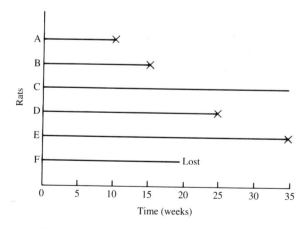

Figure 1.2 An example of Type II censored data.

censored times are also different. For example, suppose that six patients with acute leukemia enter a clinical study during a total study period of one year. Suppose also that all six respond to treatment and achieve remission. The remission times are plotted in Figure 1.3. Patients A, C, and E achieve remission at the beginning of the second, fourth, and ninth months and relapse after four, six, and three months, respectively. Patient B achieves remission at the beginning of the third month but is lost to follow-up four months later; the remission duration is thus at least four months. Patients D and F achieve remission at the beginning of the fifth and tenth month, respectively, and are still in remission at the end of the study; their remission times are thus at least eight and three months. The respective remission times of the six patients are 4, 4+, 6, 8+, 3, and 3+ months.

Type I and Type II censored observations are also called *singly censored data* and Type III *progressively censored data* by Cohen (1965). Another commonly used name for Type III censoring is *random censoring*. All of these types of censoring are *right censoring* or *censoring to the right*. When there are no censored observations, the set of survival times is *complete*.

We will study descriptive and analytic methods for complete, singly censored, and progressively censored survival data using numerical and graphical techniques. Analytic methods discussed include *parametric* and *nonparametric*. Parametric approaches are used either when a suitable model or distribution is fitted to the data or when a distribution can be assumed for the population from which the sample is drawn. Commonly used survival distributions are the exponential, Weibull, lognormal, and gamma. If a survival distribution is found to fit the data properly, the survival pattern can then be described by the parameters in a compact way. Statistical inference can be based on the distribution chosen. If the search for an appropriate model of distribution is too time consuming or not economical or no theoretical distribution adequately fits the data, nonparametric methods, which are generally easy to apply, should be considered.

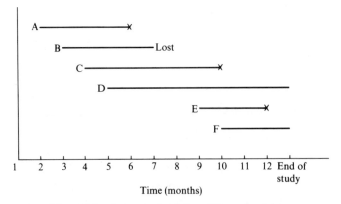

Figure 1.3 An example of Type III censored data.

1.3 SCOPE OF THE BOOK

This book is divided into five parts.

Part I (Chapters 1–3) defines survival functions and gives examples of survival data analysis. Survival distribution is most commonly described by three functions: *survivorship function* (also called cumulative survival rate or survival function), *probability density function*, and *hazard function* (hazard rate or age-specific rate). Chapter 2 defines these three functions and their equivalence relationships. Chapter 3 illustrates survival data analysis with five examples taken from actual research situations. Clinical and laboratory data are systematically analyzed in progressive steps and the results are interpreted. Section and chapter numbers are given for quick reference. The actual calculations are given as example or left as exercises in the chapters where the methods are discussed. Four sets of data are provided in the exercise section for the reader to analyze. These data are referred to in the various chapters.

Part II (Chapters 4 and 5) introduces some of the most widely used nonparametric methods for estimating and comparing survival distributions. Chapter 4 deals with the nonparametric methods for estimating the three survival functions: for small or moderate samples, the Kaplan and Meier product-limit (PL) estimate, and for large samples, the life-table technique (population life tables and clinical life tables). Also covered is standardization of rates by direct and indirect methods including the standardized mortality ratio. Chapter 5 is devoted to nonparametric techniques for comparing survival distributions. A common practice is to compare the survival experiences of two or more groups differing in their treatment or in a given characteristic. Several nonparametric tests are described.

Part III (Chapters 6–9) introduces the parametric approach to survival data analysis. Although nonparametric methods play an important role in survival studies, parametric techniques cannot be ignored. Chapter 6 introduces and discusses the exponential, Weibull, lognormal, and gamma survival distributions. Practical applications of these distributions taken from the literature are included. Chapter 7 introduces two kinds of graphical methods, probability plotting and hazard plotting, along with four computational tests for goodness-of-fit.

An important part of survival data analysis is model or distribution fitting. Once an appropriate statistical model for survival time has been constructed and its parameters estimated, its information can help predict survival, develop optimal treatment regimens, plan future clinical or laboratory studies, and so on. The graphical technique is a simple informal way to select a statistical model and estimate its parameters. When a statistical distribution is found to fit the data well, the parameters can be estimated by analytical methods. Chapter 8 discusses the analytical estimation procedures for survival distributions. Most of the estimation procedures are based on the maximum likelihood method. Mathematical derivations are omitted;

only formulas for the estimates and examples are given. Chapter 9 describes three parametric methods for comparing survival distributions.

A topic that has received increasing attention recently is the identification of prognostic factors related to survival time. For example, who is likely to survive longest after mastectomy and what are the most important factors that influence that survival? Another subject important to biomedical researchers and epidemiologists alike is the identification of the risk factors related to the development of a given disease and the response to a given treatment. What are the factors most closely related to the development of a given disease? Who is more likely to develop lung cancer, diabetes, or coronary disease? In many diseases such as cancer, patients who respond to treatment have a better prognosis than patients who do not. The question then is what the factors are that influence response. Who is more likely to respond to treatment and thus perhaps survive longer?

Part IV (Chapters 10 and 11) deals with prognostic/risk factors and survival. Chapter 10 introduces methods for identifying important prognostic factors. Chapter 11 discusses methods for identifying risk factors.

Part V (Chapters 12 and 13) attempts to provide guidelines for the planning and design of clinical trials. If a study is not well planned and designed, its results can be misleading. The two chapters give a brief discussion of the types of clinical trials and some important considerations in their design.

Appendix A describes a numerical procedure for solving nonlinear equations, namely the Newton–Raphson method. This method is suggested in Chapters 8, 10, and 11. Appendix B gives a FORTRAN computer program for the gamma probability plot. Programs in other packages, such as the BMDP, SPSS, and SAS, useful in survival data analysis, are referred to in the various chapters.

Most nonparametric techniques discussed here are easy to understand and simple to apply. Parametric methods require an understanding of the survival distributions. Unfortunately most of the survival distributions are not simple. Readers without calculus may find it difficult to apply them on their own. However, if the main purpose is not model fitting, most parametric techniques can be substituted for by their nonparametric competitors. In fact, a large percentage of the survival studies in clinical or epidemiologic journals are analyzed by nonparametric methods. Researchers not interested in survival model fitting should read the chapters and sections on nonparametric methods.

BIBLIOGRAPHICAL REMARKS

The book by Gross and Clark (1975) was the first to discuss parametric models and nonparametric and graphical techniques for both complete and censored survival data. Since then, several other books have been published in addition to the first edition of this book (Lee, 1980). Elandt–Johnson and Johnson (1980) give an extensive discussion on the construction of life tables, model fitting, competing risk, and mathematical models of biological processes of disease progression and aging. Kalbfleisch and Prentice (1980) focus on regression problems with survival data, particularly Cox's proportional hazards model. Miller (1981) covers a number of parametric and nonparametric methods for survival analysis. Cox and Oakes (1984) also give a concise coverage of the topic with an emphasis on the examination of explanatory variables.

Nelson (1982) gives a good discussion on parametric, nonparametric, and graphical methods. The book is more suited for industrial reliability engineers than for biomedical researchers and so are Hahn and Shapiro (1967), Mann, Schafer, and Singpurwalla (1974), and Gertsbakh (1989). In addition, Lawless (1982) gives a broad coverage of the area with applications in engineering and biomedical sciences.

Most of these books take a more rigorous mathematical approach and require a knowledge of mathematical statistics.

CHAPTER 2

Functions of Survival Time

Survival times are data that measure the time to a certain event such as failure, death, response, relapse, the development of a given disease, parole, or divorce. These times are subject to random variations, and like any random variables, form a distribution. The distribution of survival times is usually described or characterized by three functions: (1) the *survivorship function*, (2) the *probability density functions*, and (3) the *hazard function*. These three functions are mathematically equivalent—if one of them is given, the other two can be derived.

In practice, the three survival functions can be used to illustrate different aspects of the data. A basic problem in survival data analysis is to estimate from the sampled data one or more of these three functions and to draw inferences about the survival pattern in the population.

In this chapter, Section 2.1 defines three *survival functions*. Section 2.2 discusses the equivalence relationship among the three functions.

2.1 DEFINITIONS

Let T denote the survival time. The distribution of T can be characterized by the following three equivalent functions.

1. *Survivorship Function* (*or Survival Function*). This function, denoted by $S(t)$, is defined as the probability that an individual survives longer than t:

$$S(t) = P \text{ (an individual survives longer than } t)$$
$$= P(T > t) \tag{2.1}$$

From the definition of the cumulative distribution function $F(t)$ of T,

$$S(t) = 1 - P \text{ (an individual fails before time } t)$$
$$= 1 - F(t) \tag{2.2}$$

Here $S(t)$ is a nonincreasing function of time t with the properties

$$S(t) = 1 \qquad \text{for } t = 0$$

and

$$S(t) = 0 \qquad \text{for } t = \infty$$

that is, the probability of surviving at least at the time zero is 1 and that of surviving an infinite time is zero.

The function $S(t)$ is also known as the *cumulative survival rate*. To depict the course of survival, Berkson (1942) recommended a graphic presentation of $S(t)$. The graph of $S(t)$ is called the *survival curve*. A steep survival curve, such as the one in Figure 2.1(*a*), represents low survival rate or short survival time. A gradual or flat survival curve such as in Figure 2.1(*b*) represents high survival rate or longer survival.

The survivorship function or the survival curve is used to find the 50th percentile (the median) and other percentiles (e.g., 25th and 75th) of survival time and to compare survival distributions of two or more groups. The median survival times in Figures 2.1(*a*) and (*b*) are approximately 5 and 36 units of time, respectively. The mean is usually used to describe the central tendency of a distribution, but in survival distributions the median is often better because a small number of individuals with exceptionally long or short lifetimes will cause the mean survival time to be disproportionately large or small.

In practice, if there are no censored observations, the survivorship function is estimated as the proportion of patients surviving longer than t:

$$\hat{S}(t) = \frac{\text{number of patients surviving longer than } t}{\text{total number of patients}} \qquad (2.3)$$

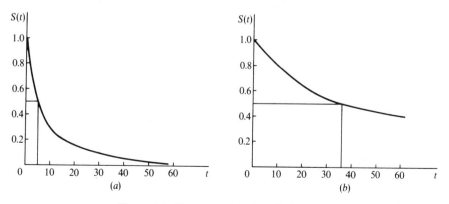

Figure 2.1 Two examples of survival curves.

where the circumflex denotes an *estimate* of the function. When censored observations are present, the numerator of (2.3) cannot always be determined. For example, consider the following set of survival data, 4, 6, 6+, 10+, 15, 20. Using (2.3), we can compute $\hat{S}(5) = \frac{5}{6} = 0.833$. However, we cannot obtain $\hat{S}(11)$ since the exact number of patients surviving longer than 11 is unknown. Either the third or the fourth patient (6+ and 10+) could survive longer than or less than 11. Thus, when censored observations are present, (2.3) is no longer appropriate for estimating $S(t)$. Nonparametric methods of estimating $S(t)$ for censored data will be discussed in Chapter 4.

   2. *Probability Density Function (or Density Function)*. Like any other continuous random variable, the survival time T has a probability density function defined as the limit of the probability that an individual fails in the short interval t to $t + \Delta t$ per unit width Δt, or simply the probability of failure in a small interval per unit time. It can be expressed as

$$f(t) = \lim_{\Delta t \to 0} \frac{P\{\text{an individual dying in the interval } (t, t + \Delta t)\}}{\Delta t} \qquad (2.4)$$

   The graph of $f(t)$ is called the *density curve*. Figures 2.2(a) and (b) give two examples of the density curve. The density function has the following two properties:

   1. $f(t)$ is a nonnegative function:

$$f(t) \geq 0 \qquad \text{for all } t \geq 0$$
$$= 0 \qquad \text{for } t < 0$$

   2. The area between the density curve and the t axis is equal to 1.

   In practice, if there are no censored observations, the probability density function $f(t)$ is estimated as the proportion of patients dying in an interval per unit width:

$$\hat{f}(t) = \frac{\text{number of patients dying in the interval beginning at time } t}{(\text{total number of patients})(\text{interval width})}$$

$$(2.5)$$

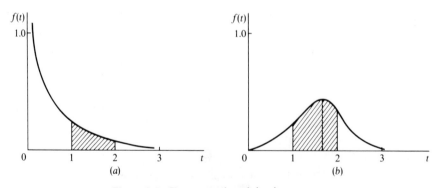

Figure 2.2   Two examples of density curves.

Similar to the estimation of $S(t)$, when censored observations are present, (2.5) is not applicable. We will discuss an appropriate method in Chapter 4.

The proportion of individuals that fail in any time interval and the peaks of high frequency of failure can be found from the density function. The density curve in Figure 2.2(a) gives a pattern of high failure rate at the beginning of the study and decreasing failure rate as time increases. In Figure 2.2(b), the peak of high failure frequency occurs at approximately 1.7 units of time. The proportion of individuals that fail between 1 and 2 units of time is equal to the shaded area between the density curve and the t axis. The density function is also known as the *unconditional failure rate*.

3. *Hazard Function*. The hazard function $h(t)$ of survival time T gives the *conditional failure rate*. This is defined as the probability of failure during a very small time interval, assuming that the individual has survived to the beginning of the interval, or as the limit of the probability that an individual fails in a very short interval, t to $t + \Delta t$ per unit time, given that the individual has survived to time t:

$$h(t) = \lim_{\Delta t \to 0} \frac{P\{\text{an individual of age } t \text{ fails in the time interval } (t, t + \Delta t)\}}{\Delta t}$$

(2.6)

The hazard function can also be defined in terms of the cumulative distribution function $F(t)$ and the probability density function $f(t)$:

$$h(t) = f(t) / \{1 - F(t)\}$$
(2.7)

The hazard function is also known as the *instantaneous failure rate, force of mortality, conditional mortality rate*, and *age-specific failure rate*. It is a measure of the proneness to failure as a function of the age of the individual in the sense that the quantity $\Delta t h(t)$ is the expected proportion of age t individuals who will fail in the short time interval t to $t + \Delta t$. The hazard function thus gives the risk of failure per unit time during the aging process. It plays an important role in survival data analysis.

In practice, when there are no censored observations, the hazard function is estimated as the proportion of patients dying in an interval per unit time, given that they have survived to the beginning of the interval:

$$\hat{h}(t) = \frac{\text{number of patients dying in the interval beginning at time } t}{(\text{number of patients surviving at } t)(\text{interval width})}$$

$$= \frac{\text{number of patients dying per unit time in the interval}}{\text{number of patients surviving at } t}$$
(2.8)

Actuaries usually use the average hazard rate of the interval in which the number of patients dying per unit time in the interval is divided by the average number of survivors at the midpoint of the interval:

$$\hat{h}(t) =$$

$$\frac{\text{number of patients dying per unit time in the interval}}{(\text{number of patients surviving at } t) - \frac{1}{2}(\text{number of deaths in the interval})}$$

$$(2.9)$$

The actuarial estimate in (2.9) gives a higher hazard rate then (2.8) and thus a more conservative estimate.

The hazard function may increase, decrease, remain constant, or indicate a more complicated process. Figure 2.3 plots several kinds of hazard function. For example, patients with acute leukemia who do not respond to treatment have an increasing hazard rate, $h_1(t)$, $h_2(t)$ is a decreasing hazard function that, for example, indicates the risk of soldiers wounded by bullets who undergo surgery. The main danger is the operation itself, and this danger decreases if the surgery is successful. An example of a constant hazard function, $h_3(t)$, is the risk of healthy individuals between 18 and 40 years of age whose main risks of death are accidents. The so-called bathtub curve, $h_4(t)$, describes the process of human life. During an initial period, the risk is high (high infant mortality). Subsequently, $h(t)$ stays approximately constant until a certain time, after which it increases because of wear-out failures. Finally, patients with tuberculosis have risks that increase initially, then decrease after treatment. Such an increasing then decreasing hazard function is described by $h_5(t)$.

The *cumulative hazard function* is defined as

$$H(t) = \int_0^t h(x)\, dx \qquad (2.10)$$

It will be shown in Section 2.2 that

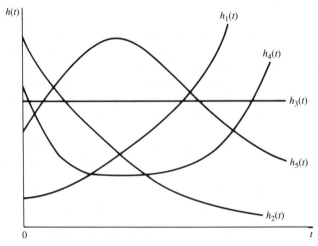

Figure 2.3   Examples of the hazard function.

$$H(t) = -\log_e S(t) \qquad (2.11)$$

Thus, at $t = 0$, $S(t) = 1$, $H(t) = 0$, and at $t = \infty$, $S(t) = 0$, $H(t) = \infty$. The cumulative hazard function can be any value between zero and infinity.

The following examples illustrates how these functions can be estimated from a complete sample of grouped survival times without censored observations.

Example 2.1

The first three columns of Table 2.1 give the survival data of 40 patients with myeloma. The survival times are grouped into intervals of five months. The estimated survivorship function, density function, and hazard function are also given, with the corresponding graphs plotted in Figures 2.4(a)–(c).

The estimated survivorship function, $\hat{S}(t)$, is calculated following (2.3) at the beginning or the end of each interval. For example, at the beginning of the first interval, all of the 40 patients are alive, $\hat{S}(0) = 1$, and at the beginning of the second interval, 35 of the 40 patients are still alive, $\hat{S}(5) = \frac{35}{40} = 0.875$. Similarly, $\hat{S}(10) = \frac{28}{40} = 0.700$. The estimated density function $\hat{f}(t)$ is computed following (2.5). For example, the density function of the first interval (0–5) is $5/(40 \times 5) = 0.025$ and that of the second interval (5–10) is $7/(40 \times 5) = 0.035$. The estimated density function is plotted at the midpoint of each interval [Figure 2.4(b)]. The estimated hazard function, $\hat{h}(t)$, is computed following the actuarial method given in (2.9). For example, the hazard function of the first interval is $5/[5(40 - \frac{5}{2})] = 0.027$ and that of the second interval is $7/[5(35 - \frac{7}{2})] = 0.044$. The estimated hazard function is also plotted at the midpoint of each interval [Figure 2.4(c)].

Table 2.1 Survival Data and Estimated Survival Functions of 40 Myeloma Patients

Survival Time t (Months)	Number of Patients Surviving at Beginning of Interval	Number of Patients Dying in Interval	$\hat{S}(t)$	$\hat{f}(t)$	$\hat{h}(t)$
0–5	40	5	1.000	0.025	0.027
5–10	35	7	0.875	0.035	0.044
10–15	28	6	0.700	0.030	0.048
15–20	22	4	0.550	0.020	0.040
20–25	18	5	0.450	0.025	0.065
25–30	13	4	0.325	0.020	0.072
30–35	9	4	0.225	0.020	0.114
35–40	5	0	0.125	0.000	0.000
40–45	5	2	0.125	0.010	0.100
45–50	3	1	0.075	0.005	0.080
≥50	2	2	0.050	—	—

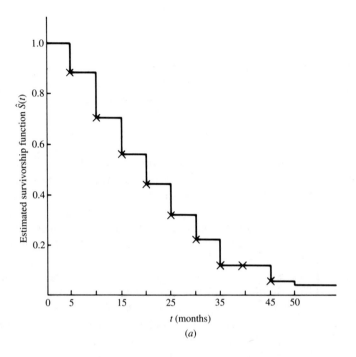

t (months)

(*a*)

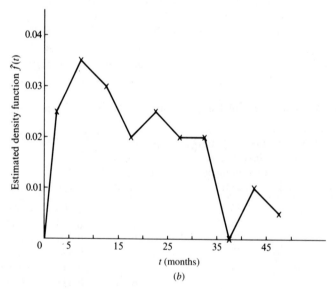

t (months)

(*b*)

Figure 2.4   Estimated survival functions of myeloma patients.

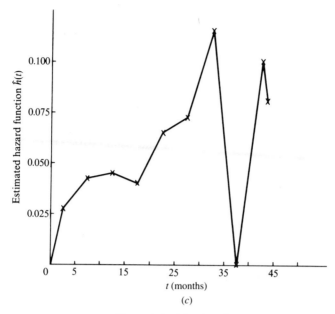

Figure 2.4 (*Continued*)

From Table 2.1 or Figure 2.4, the median survival time of myeloma patients is approximately 17.5 months and the peak of the high frequency of death occurs in 5–10 months. In addition, the hazard function shows an increasing trend and reaches its peak at approximately 32.5 months and then fluctuates.

2.2 RELATIONSHIPS OF THE SURVIVAL FUNCTIONS

The three functions defined in Section 2.1 are mathematically equivalent. Given any one of them, the other two can be be derived. Readers not interested in the mathematical relationship among the three survival functions can skip this section without loss of continuity.

1. From (2.2) and (2.7),

$$h(t) = f(t)/S(t) \tag{2.12}$$

This relationships can also be derived from (2.6) using basic definitions of conditional probabilities.

2. Since the probability density function is the derivative of the cumulative distribution function,

$$f(t) = \frac{d}{dt}[1 - S(t)] = -S'(t) \tag{2.13}$$

3. Substituting (2.13) into (2.12) yields

$$h(t) = -\frac{S'(t)}{S(t)} = -\frac{d}{dt}\log_e S(t) \tag{2.14}$$

4. Integrating (2.14) from zero to t and using $S(0) = 1$, we have

$$-\int_0^t h(x)\,dx = \log_e S(t)$$

or

$$H(t) = -\log_e S(t)$$

or

$$S(t) = \exp[-H(t)] = \exp\left[-\int_0^t h(x)\,dx\right] \tag{2.15}$$

5. From (2.12) and (2.15) we obtain

$$f(t) = h(t)\exp[-H(t)] \tag{2.16}$$

Hence, if $f(t)$ is known, the survivorship function $S(t)$ can be obtained from the basic relationship between $f(t)$, $F(t)$, and (2.2). The hazard function can then be determined from (2.12). If $S(t)$ is known, $f(t)$ and $h(t)$ can be determined from (2.13) and (2.12), respectively or first $h(t)$ can be derived from (2.14) and then $f(t)$ from (2.12). If $h(t)$ is given, $S(t)$ and $f(t)$ can be obtained, respectively, from (2.15) and (2.16). Thus, given any one of the three survival functions, the other two can easily be derived. The following example illustrates these equivalence relationships.

Example 2.2
Suppose that the survival time of a population has the following density function:

$$f(t) = e^{-t} \qquad t \geq 0$$

Using the definition of the cumulative distribution function,

$$F(t) = \int_0^t f(x)\,dx = \int_0^t e^{-x}\,dx = -e^{-x}\big|_0^t = 1 - e^{-t}$$

From (2.2) we obtain the survivorship function

$$S(t) = e^{-t}$$

The hazard function can then be obtained from (2.12):

$$h(t) = e^{-t}/e^{-t} = 1$$

Section 6.1 gives a complete treatment of this distribution.

BIBLIOGRAPHICAL REMARKS

The three survival functions and their equivalents are discussed in every text that is cited in the Bibliographical Remarks in Chapter 1.

EXERCISES

2.1 Consider the survival data given in Exercise Table 2.1 and compute and plot the estimated survivorship function, the probability density function, and the hazard function.

Exercise Table 2.1

Year of Follow-up	Number Alive at Beginning of Interval	Number Dying in Interval
0–1	1100	240
1–2	860	180
2–3	680	184
3–4	496	138
4–5	358	118
5–6	240	60
6–7	180	52
7–8	128	44
8–9	84	32
≥ 9	52	28

2.2 Exercise Table 2.2 is a life table for the total population (of 100,000 live births) in the United States, 1959–1961. Compute and plot the estimated survivorship function, the probability density function, and the hazard function.

Exercise Table 2.2

Age Interval	Number Living at Beginning of Age Interval	Number Dying in Age Interval
0–1	100,000	2,593
1–5	97,407	409
5–10	96,998	233
10–15	96,765	214
15–20	96,551	440
20–25	96,111	594
25–30	95,517	612
30–35	94,905	761
35–40	94,144	1,080
40–45	93,064	1,686
45–50	91,378	2,622
50–55	88,756	4.045
55–60	84,711	5,644
60–65	79,067	7,920
65–70	71,147	10,290
70–75	60,857	12,687
75–80	48,170	14,594
80–85	33,576	15,034
85 and over	18,542	18,542

Source: U.S. National Center for Health Statistics, Life Tables 1959–1961, Vol. 1, No. 1, "United States Life Tables 1959–61," December, 1964, pp. 8–9.

2.3 Derive (2.12) using (2.6) and basic definitions of conditional probability.

2.4 Given the hazard function

$$h(t) = c$$

derive the survivorship function and the probability density function.

2.5 Given the survivorship function

$$S(t) = \exp(-t^{\gamma})$$

derive the probability density function and the hazard function.

CHAPTER 3

Examples of Survival Data Analysis

The investigator who has assembled a large amount of data must decide what to do with it and what it indicates. In this chapter, we take several sets of survival data from actual research situations and analyze them.

Example 3.1 analyzes two sets of data obtained, respectively, from two and three treatment groups to compare the treatment's abilities to prolong life. Example 3.2 is an example of the life-table technique for large samples. Example 3.3 gives remission data from two treatments; the investigator seeks a well-known distribution for the remission patterns to compare the two groups. Example 3.4 studies survival data and several other patient characteristics to identify important prognostic factors; the patient characteristics are analyzed individually and simultaneously for their prognostic values. Example 3.5 introduces a case in which the interest is to identify risk factors in the development of a given disease. The Exercises present four sets of real data for the reader to use.

EXAMPLE 3.1 COMPARISON OF TWO TREATMENTS AND THREE DIETS

A. Comparison of Two Treatments

Thirty melanoma patients (Stages 2–4) were studied to compare the immunotherapies BCG (*Bacillus* Calmette–Guérin) and *Corynebacterium parvum* for their abilities to prolong remission and survival time. The age, sex, disease stage, treatment received, remission duration, and survival time are given in Table 3.1. All the patients were resected before treatment began and thus had no evidence of melanoma at the time of first treatment.

The usual objective of this type of data is to determine the length of remission and survival and to compare the distributions of remission and survival time in each group. Before comparing the remission and survival

Table 3.1   Data for 30 Resected Melanoma Patients

Patient Number	Age	Sex[a]	Initial Stage	Treatment Received[b]	Remission Duration[c]	Survival Time[c]
1	59	2	3B	1	33.7+	33.7+
2	50	2	3B	1	3.8	3.9
3	76	1	3B	1	6.3	10.5
4	66	2	3B	1	2.3	5.4
5	33	1	3B	1	6.4	19.5
6	23	2	3B	1	23.8+	23.8+
7	40	2	3B	1	1.8	7.9
8	34	1	3B	1	5.5	16.9+
9	34	1	3B	1	16.6+	16.6+
10	38	2	2	1	33.7+	33.7+
11	54	2	2	1	17.1+	17.1+
12	49	1	3B	2	4.3	8.0
13	35	1	3B	2	26.9+	26.9+
14	22	1	3B	2	21.4+	21.4+
15	30	1	3B	2	18.1+	18.1+
16	26	2	3B	2	5.8	16.0+
17	27	1	3B	2	3.0	6.9
18	45	2	3B	2	11.0+	11.0+
19	76	2	3A	2	22.1	24.8+
20	48	1	3A	2	23.0+	23.0+
21	91	1	4A	2	6.8	8.3
22	82	2	4A	2	10.8+	10.8+
23	50	2	4A	2	2.8	12.2+
24	40	1	4A	2	9.2	12.5+
25	34	1	3A	2	15.9	24.4
26	38	1	4A	2	4.5	7.7
27	50	1	2	2	9.2	14.8+
28	53	2	2	2	8.2+	8.2+
29	48	2	2	2	8.2+	8.2+
30	40	2	2	2	7.8+	7.8+

[a]1 male; 2, female.
[b]1, BCG; 2, C. parvum.
[c]Remission and survival times are in months.

Source: Kindly provided by Dr. Richard Ishmael.

distributions, we must examine if the two treatment groups are comparable with respect to prognostic factors. Let us use the survival time to illustrate the steps. (The remission time could be similarly analyzed.)

1. *Estimate and Plot the Survival Function of the Two Treatment Groups.* The resulting curves are called *survival curves*. Points on the curve estimate the proportion of patients who will survive at least a given period of time.

Table 3.2   **Kaplan–Meier Product-Limit Estimate of Survival Function** $S(t)$

	BCG Patients				
Death time (t)	3.9	5.4	7.9	10.5	19.5
$\hat{S}(t)$	0.909	0.818	0.727	0.636	0.477

	C. parvum patients				
Death time (t)	6.9	7.7	8.0	8.3	24.4
$\hat{S}(t)$	0.947	0.895	0.839	0.774	0.516

For such samples with progressively censored observations, the Kaplan–Meier product-limit (PL) method is appropriate for estimating the survival function. It does not require any assumptions about the form of the function that is being estimated. Section 4.1 discusses this method in detail. Computer programs for the method can be found in BMDP (Dixon et al., 1988), SAS (1990), and SPSS (1988).

Table 3.2 gives the PL estimate of the survival function $\hat{S}(t)$ for the two treatment groups. Note that $S(t)$ is estimated only at death times; however, the censored observations are used to estimate $S(t)$. The *median survival time* can be estimated by linear interpolation. For BCG patients the median survival time was about 18.2 months. The median survival time for the *C. parvum* group cannot be calculated since 15 of the 19 patients were still alive. Most computer programs give not only $\hat{S}(t)$ but also the standard error of $\hat{S}(t)$ and the 75-, 50-, and 25-percentile points.

Figure 3.1 plots the estimated survival function $\hat{S}(t)$ for patients receiving

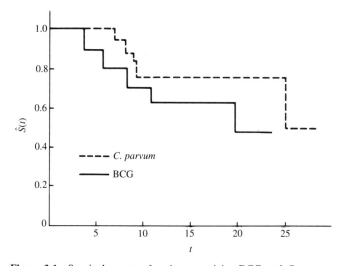

Figure 3.1 Survival curves of patients receiving BCG and *C. parvum*.

the two treatments. The median survival time (50-percentile point) for the BCG group can also be determined graphically. The survival curves clearly show that *C. parvum* patients had slightly better survival experience than BCG patients. For example, 50% of the BCG patients survived at least 18.2 months while about 61% of the *C. parvum* patients survived that long.

2. *Examine the Prognostic Homogeneity of the Two Groups.* The next question to ask is whether the difference in survival between the two treatment groups is statistically significant. Is the difference shown by the data significant or simply random variation in the sample? A statistical test of significance is needed. However, a statistical test without considering patient characteristics makes sense only if the two groups of patients are homogeneous with respect to prognostic factors. It has been assumed thus far that the patients in the two groups are comparable and that the only difference between them is treatment. Thus, before performing a statistical test, it is necessary to examine the homogeneity between the two groups.

Although prognostic factors for melanoma patients are not well established, it has been reported that the female and the young have a better survival experience than the male and the old. Also, the disease stage plays an important role in survival. Let us check the homogeneity of the two treatment groups with respect to age, sex, and disease stage.

The age distributions are estimated and plotted in Figure 3.2. The median age is 39 for the BCG group and 43 for the *C. parvum* patients. To test the significance of the difference between the two age distributions, the two-sample *t*-test (Armitage 1971, Daniel 1987) or nonparametric tests such as the Mann–Whitney *U*-test and the Kolmogorov–Smirnov test (Marascuilo and McSweeney 1977) are appropriate. However, the generalized Wilcoxon tests given in Sections 5.1.1 and 5.1.4 can also be used, since they reduce to the Mann–Whitney *U*-test when the data are complete. There is no need to acquire another computer program for the Mann–Whitney *U*-test or the Kolmogorov–Smirnov test. Using Gehan's generalized Wilcoxon test, the difference between the two age distributions is not found to be statistically significant. More about the test is given in Section 5.1.

The number of male and female patients in the two treatment groups is given in Table 3.3. Sixty-four percent of the BCG patients and 42% of the *C. parvum* patients are female. A chi-square test can be used to compare the two proportions (see Section 11.1.2). It can be used only for $r \times c$ tables in which the entries are frequencies and not for tables in which the entries are mean values or medians of a certain variable. For a 2×2 table, the chi-square value can be computed by hand. Computer programs for the test can be found in many computer program packages, such as BMDP (Dixon et al., 1988), SPSS (1988), and SAS (1990).

The chi-square value for treatment by sex in Table 3.3 is 1.29 with one degree of freedom, which is not significant at the 0.05 or 0.10 level. Therefore the difference between the two proportions is not statistically significant.

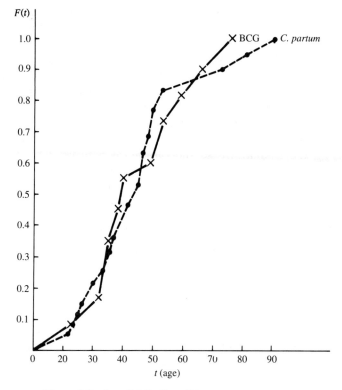

Figure 3.2 Age distribution of two treatment groups.

The number of Stage 2 patients and the number of patients with more advanced disease in the two treatment groups are also given in Table 3.3. Eighteen percent of the BCG patients are at Stage 2 against 21% of the *C. parvum* patients. However, a chi-square test result shows that the difference is not significant.

Thus we can say that the data do not show heterogeneity between the two treatment groups. If heterogeneity is found, the groups can be divided into subgroups of members who are similar in their prognoses (details are discussed in Section 13.3).

Table 3.3 Treatment by Sex and Disease Stage

Sex	BCG Number	%	C. parvum Number	%	Total	Disease Stage	BCG Number	%	C. parvum Number	%	Total
Male	4	36	11	58	15	2	2	18	4	21	6
Female	7	64	8	42	15	3 and 4	9	82	15	79	24
	11		19		30		11		19		30

3. *Compare the Two Survival Distributions.* There are several parametric and nonparametric tests to compare two survival distributions. Chapters 5 and 9 describe them. Since we have no information of the survival distribution that the data follow, we would continue to use nonparametric methods to compare the two survival distributions. The four tests described in Sections 5.1.1–5.1.4 are suitable. The performance of these tests is discussed at the end of Section 5.1. We choose Gehan's generalized Wilcoxon test here to demonstrate the analysis procedure only because of its simplicity of calculation.

In testing the significance of the difference between two survival distributions, the hypothesis is that the survival distribution of the BCG patients is the same as that of the *C. parvum* patients. Let $S_1(t)$ and $S_2(t)$ be the survival function of the BCG and *C. parvum* groups, respectively. The null hypothesis is

$$H_0: S_1(t) = S_2(t)$$

The alternative hypothesis chosen is two-sided:

$$H_1: S_1(t) \neq S_2(t)$$

since we have no prior information concerning the superiority of either of the two treatments. The slight difference between the two estimated survival curves could be due to random variation. The one-sided alternative $H_1: S_1(t) < S_2(t)$ should be considered inappropriate.

Using Gehan's generalized Wilcoxon test, the difference in survival distribution of the two treatment groups is found to be insignificant ($p = 0.33$). Therefore, we do not reject the null hypothesis that the two survival distributions are equal. Although our conclusion is that the data do not provide enough evidence to reject the hypothesis, "not to reject the null hypothesis" does not automatically mean "to accept the null hypothesis." The difference between the two statements is that the error probability of the latter statement is usually much larger than that of the former.

B. Comparison of Three Diets

A laboratory investigator interested in the relationship between diet and the development of tumors divided 90 rats into three groups and fed them with low-fat, saturated, and unsaturated diets, respectively (King et al. 1979). The rats were of the same age and species and were in similar physical condition. An identical amount of tumor cells were injected into a foot pad of each rat. The rats were observed for 200 days. Many developed a recognizable tumor early in the study period. Some were tumor free at the end of the 200 days. Rat 16 in the low-fat groups and rat 24 in the saturated group died accidentally after 140 days and 170 days, respectively, with no

Table 3.4 Tumor-Free Time (Days) of 90 Rats on Three Different Diets

Rat	Low-Fat	Rat	Saturated	Rat	Unsaturated
1	140	1	124	1	112
2	177	2	58	2	68
3	50	3	56	3	84
4	65	4	68	4	109
5	86	5	79	5	153
6	153	6	89	6	143
7	181	7	107	7	60
8	191	8	86	8	70
9	77	9	142	9	98
10	84	10	110	10	164
11	87	11	96	11	63
12	56	12	142	12	63
13	66	13	86	13	77
14	73	14	75	14	91
15	119	15	117	15	91
16	140+	16	98	16	66
17	200+	17	105	17	70
18	200+	18	126	18	77
19	200+	19	43	19	63
20	200+	20	46	20	66
21	200+	21	81	21	66
22	200+	22	133	22	94
23	200+	23	165	23	101
24	200+	24	170+	24	105
25	200+	25	200+	25	108
26	200+	26	200+	26	112
27	200+	27	200+	27	115
28	200+	28	200+	28	126
29	200+	29	200+	29	161
30	200+	30	200+	30	178

Source: King et al. 1979. Data are used by permission of the author.

evidence of tumor. Table 3.4 gives the tumor-free time—the time from injection to the time that a tumor develops or to the end of the study. Fifteen of the 30 rats on the low-fat diet developed a tumor before the experiment was terminated. The rat that died had a tumor-free time of at least 140 days. The other 14 rats did not develop any tumor by the end of the experiment; their tumor-free times were at least 200 days. Among the 30 rats in the saturated diet group, 23 developed a tumor, one died tumor free after 170 days, and six were tumor free at the end of the experiment. All 30 rats in the unsaturated diet group developed tumors within 200 days. The two early deaths can be considered losses to follow-up. The data are singly censored if the two early deaths are excluded.

The investigator's main interest here is to compare the three diets' abilities to keep the rats tumor free. To obtain information about the distribution of the tumor-free time, we can first estimate the *survival* (tumor-free) function of the three diet groups. The three survival functions are estimated using the Kaplan–Meier PL method and are plotted in Figure 3.3. The median tumor-free times for the low-fat, saturated, and unsaturated groups are 188, 107, and 91 days, respectively. Since the three groups are homogeneous, we can skip the step that checks for homogeneity and compare the three distributions of tumor-free time.

The *K*-sample test described in Section 5.3.3 can be used to test the significance of the differences among the three diet groups. Using this test, the investigator finds that the differences among the three groups are highly significant ($p = 0.002$). Note that the *K*-sample test can tell the investigator only that the differences among the groups are statistically significant. It cannot tell which two groups contribute the most to the differences— whether the low-fat diet produces a significantly different tumor-free time from the saturated diet or the saturated diet is significantly different from the unsaturated diet. All one can conclude is that the data show a significant difference among the tumor-free times produced by the three diets.

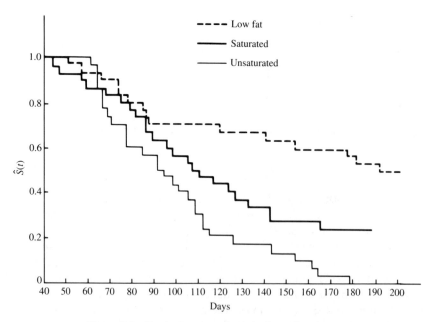

Figure 3.3 Survival curves of rats in three diet groups.

EXAMPLE 3.2 COMPARISON OF TWO SURVIVAL PATTERNS USING LIFE TABLES

When the sample of patients is so large that their groups is meaningful, the life-table technique can be used to estimate the survival distribution. A method developed by Mantel and Haenszel (1959) and applied to life tables by Mantel (1966) can be used to compare two survival patterns in the life-table analysis.

Consider the data of male patients with localized cancer of the rectum diagnosed in Connecticut from 1935 to 1954 (Myers 1969). A total of 388 patients were diagnosed between 1935 and 1944, and 749 patients were diagnosed between 1945 and 1954. For such large sample sizes the data can be grouped and tabulated as shown in Table 3.5. The 10 intervals indicate the number of years after diagnosis. For the tabulated life tables the survival function $S(t_i)$ can be estimated for each interval t_i. Section 4.2 discusses the estimation procedures of $S(t_i)$ and density and hazard functions. The survival, density, and hazard functions are the three most important functions that characterize a survival distribution.

The $\hat{S}(t_i)$ column in Table 3.5 gives the estimated survival function for the two time periods: these are plotted in Figure 3.4. Patients diagnosed in the 1945–1954 period had considerably longer survival times (median 3.87 years) than patients diagnosed in the 1935–1944 period (median 1.58 years). The five-year survival rate is frequently used by cancer researchers and can be easily determined from the life table. Patients diagnosed in 1935–1944 had a

Table 3.5 Life Table for Male Patients with Localized Cancer of Rectum Diagnosed in Connecticut, 1935–1944 and 1945–1954[a]

Interval	1935–1944					1945–1954				
(t_i)	n_i'	d_i	$w_i + l_i$	n_i	$\hat{S}(t_i)$	n_i'	d_i	$w_i + l_i$	n_i	$\hat{S}(t_i)$
1	388	167	2	387.0	0.5685	749	185	10	744.0	0.7513
2	219	45	1	218.5	0.4514	554	88	10	549.0	0.6309
3	173	45	1	172.5	0.3336	456	55	10	451.0	0.5539
4	127	19	0	127.0	0.2837	391	43	10	386.0	0.4922
5	108	17	0	108.0	0.2390	338	32	14	331.0	0.4446
6	91	11	1	90.5	0.2100	292	31	52	266.0	0.3928
7	79	8	0	79.0	0.1887	209	20	38	190.0	0.3514
8	71	5	0	71.0	0.1754	151	7	24	139.0	0.3337
9	66	6	1	65.5	0.1593	120	6	25	107.5	0.3151
10	59	7	0	59.0	0.1404	89	6	24	77.0	0.2905

[a]Symbols: n_i', number of patients alive at beginning of interval t_i; d_i, number of patients dying during interval t_i; $w_i + l_i$, number of patients withdrawn alive or lost to follow-up during interval t_i; $n_i = n_i' - \frac{1}{2}(w_i - l_i)$; $\hat{S}(t_i)$ cumulative proportion surviving from beginning of study to end of interval t_i.

Source: Myers (1969).

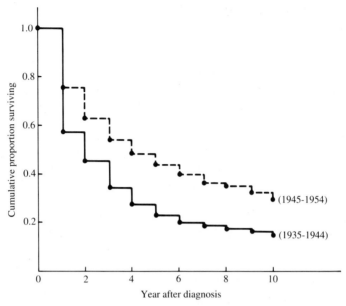

Figure 3.4 Survival curves for male patients with localized cancer of the rectum, diagnosed in Connecticut, 1935–1944 versus 1945–1954.

five-year survival rate of 0.2390, or 23.9%. The patients diagnosed in 1945–1954 had a rate of 0.4446, or 44.5%. The five-year survival rate is only a convention. As medical science advances, a five-year survival for patients with certain kinds of cancer is no longer a miracle and cannot be considered a cure. In comparing two sets of survival data, one can compare the proportion of patients surviving some stated periods such as five years, or the five-year survival rates. However, one cannot anticipate that two survival patterns will always stand in a superior–inferior relationship. It is more desirable to make a whole-pattern comparison (see Sections 4.3 and 5.2).

The Mantel–Haenszel method described in Section 5.2 is a whole-pattern comparison and can be used to compare two survival patterns in life tables. Application of this method to the data in Table 3.5 results in a chi-square value of 51.996 with one degree of freedom. We can conclude that the difference between the two survival patterns is highly significant ($p < 0.001$).

Estimates of the survival function or survival rate depend on the life-table interval used. if each interval is very short, resulting in a large number of intervals, the computation becomes very tedious and the advantage of the life table is not fully taken. One assumption underlying the life table is that the population has the same survival probability in each interval. If the interval length is long, this assumption may be violated and the estimates inaccurate; this should be avoided except for rough calculations. Although the length of each interval and the total number of intervals are important, they will not cause trouble in most clinical studies since the study periods normally cover a short period of time such as one, two, or three years. Life tables with about 10–20 intervals of several months to one year each are reasonable. The investigator should also consider the disease under study. If the variation in survival is large in a short period of time, the interval length should be short. However, in some demographic or other studies it is often of interest to cover a life span from birth to age 85 or more. The number of intervals would be very large if short intervals were used. In this case five-year intervals are sufficient to take into account the important variations in survival rate estimates (Shryock et al. 1971).

EXAMPLE 3.3   FITTING SURVIVAL DISTRIBUTIONS TO REMISSION DATA

The remission times of 42 patients with acute leukemia were reported by Freireich et al. (1963) in a clinical trail undertaken to assess the ability of 6-mercaptopurine (6-MP) to maintain remission.[1] Each patient was random-ized to receive 6-MP or a placebo. The study was terminated after one year. The following remission times, in weeks, were recorded:

6-MP (21 patients): 6, 6, 6, 7, 10, 13, 16, 22, 23, 6+, 9+, 10+, 11+, 17+, 19+, 20+, 25+, 32+, 32+, 34+, 35+

Placebo (21 patients): 1, 1, 2, 2, 3, 4, 4, 5, 5, 8, 8, 8, 8, 11, 11, 12, 12, 15, 17, 22, 23

Suppose that we are interested in a distribution to describe the remission times of these patients but there is no information available as to which distribution will fit. We need to find a distribution that fits the data well. If we can find one, the remission experience can then be described by the properties of the distribution and the remission time of new patients can be predicted. Parametric tests can be used to compare the effectiveness of the two treatments, but since there is a large number of well-known functions and distributions to choose from, the search becomes an art as much as a scientific task.

The simplest and most efficient tool is the graph. Probability plotting can be done for complete data; for data that include censored observations, hazard plotting is more appropriate. Commercially available *probability* and *hazard plotting papers* are designed for this purpose. It would not be difficult to use the computer to generate these plots if probability or hazard plotting papers are unavailable. Detailed discussions of probability plotting and hazard plotting are presented in Sections 7.1–7.3. In both systems, a linear configuration indicates that the distribution can be estimated from the graph.

[1]Data are used by permission of the publisher.

Let us begin by trying to fit a distribution to the remission duration of 6-MP patients. Since the data consist of both censored and uncensored observations, we use the technique of hazard plotting. Figure 3.5 is an exponential hazard plot of the data. The configuration is fairly linear. The straight line is fitted to the points by eye. It shows that the exponential distribution fits the remission data reasonably well. For progressively censored data, hazard plotting papers are also available for the Weibull and lognormal distributions (see Figures 3.6 and 3.7, respectively). Clearly, these two plots are not straight. If we compare the three fits, the exponential distribution is by far the best.

The mean remission time estimated from the exponential plot is approximately 37 weeks. An estimate of the hazard rate (relapse rate in this case) is then $\frac{1}{37}$, or 0.027 per week, independent of time. A more accurate estimate may be obtained by the maximum likelihood method (Chapter 8) that refines the estimate $\hat{\lambda}_1$ to 0.025 per week.

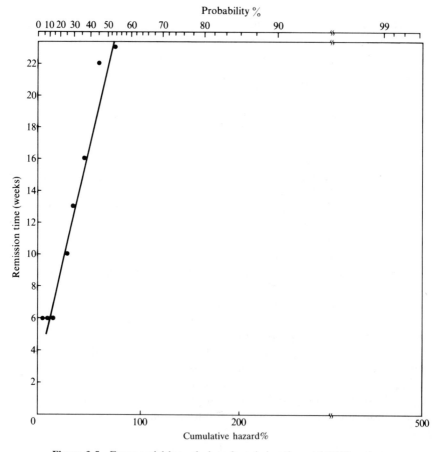

Figure 3.5 Exponential hazard plot of remission times of 6-MP patients.

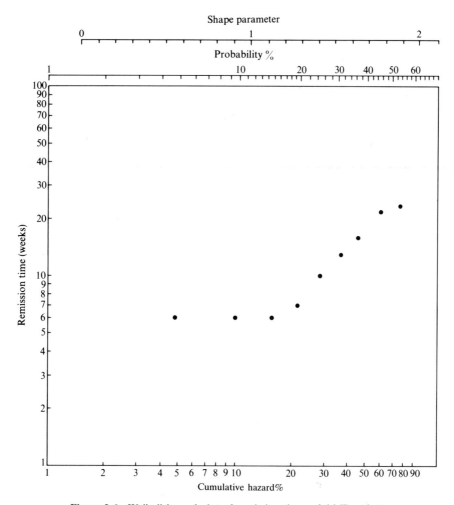

Figure 3.6   Weibull hazard plot of remission times of 6-MP patients.

An objective technique is available to test whether the exponential distribution with $\lambda = 0.025$ gives an adequate fit to the data. We can use the test proposed by Hollander and Proschan (1979) described in Section 7.4.4.

A unique property of the exponential distribution is that it shows the probability of relapse during a specified week to be a constant (0.025), irrespective of the time the patient has been in remission. In other words, the probability of future relapse for patients in remission is independent of their history. The probability for a 10-week remission, for example, can be predicted as

$$P(\text{10-week remission}) = e^{-10\lambda_1}$$

Thus, an estimated probability of a 10-week remission is $e^{-10(0.025)} = 0.78$.

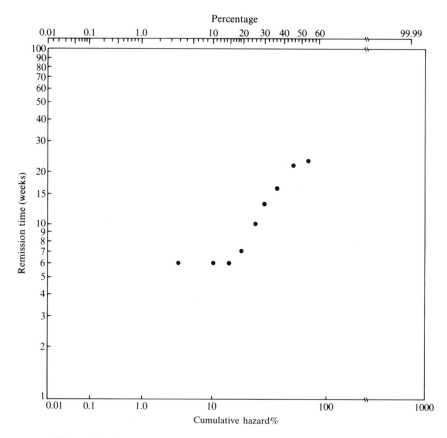

Figure 3.7 Lognormal hazard plot of remission times of 6-MP patients.

For the placebo group remission time, we can use the probability plotting technique since the data are complete (see Figure 3.8). The configuration is nearly linear except for the upper tail. Figure 3.9 shows the data plotted on Weibull probability paper. The points fall very close to a straight line. Figure 3.10 is a normal probability plot of the natural logarithm of the remission time. If the data are from a lognormal distribution, the plot should be linear. In this case, the configuration is not, and the lognormal distribution is thus inadequate for the data. Gamma probability paper has different scales for different values of the shape parameter but is available only for a few integral values of the shape parameter. It is difficult to choose an appropriate gamma probability paper without any information about the shape parameter. For this set of data, it seems unnecessary to use the gamma distribution since the Weibull and exponential plots give reasonably straight lines. If none of these plots showed a straight line and the gamma distribution was thought appropriate for practical reasons, a method sugges-

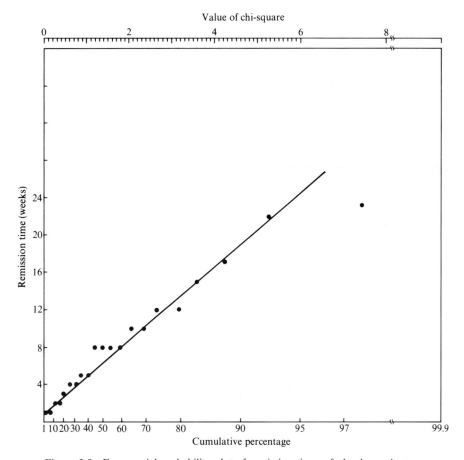

Figure 3.8   Exponential probability plot of remission times of placebo patients.

ted by Wilk et al. (1962a) can be used (details and a computer program are described in Sections 7.2, 7.5, and 8.4).

Let us examine the exponential and the Weibull plots in Figures 3.8 and 3.9. The straight line in the exponential plot fits the data points reasonably well except the largest observation. The straight line in the Weibull plot seems to give slightly better fit—the points are close to the line, again except the largest observation. We may prefer to choose the Weibull distribution. However, when we obtain estimates of the parameters, we find that there is little difference between the two fits. From the exponential plot we find that the mean remission time of this group is approximately 8.6 weeks; hence the hazard rate is approximately 1/8.6, or 0.12 per week. On the Weibull plot, a line is drawn parallel to the fitted line from the circle marked "origin" at the top of the paper. The intersection of this line and the left auxiliary scale gives an estimate of the shape parameter γ, which is approximately 1.12.

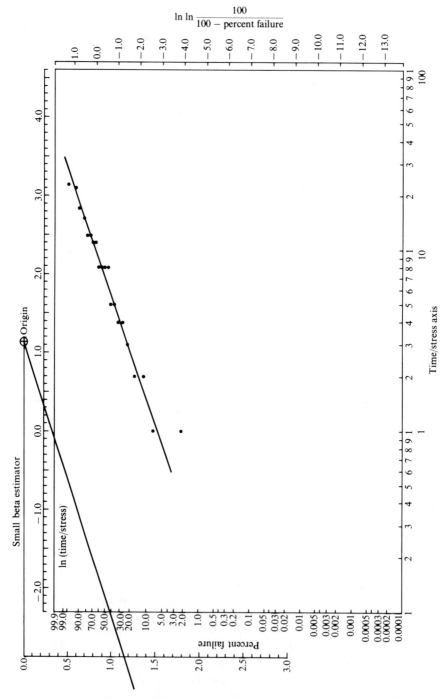

Figure 3.9 Weibull probability plot of remission times of placebo patients.

34

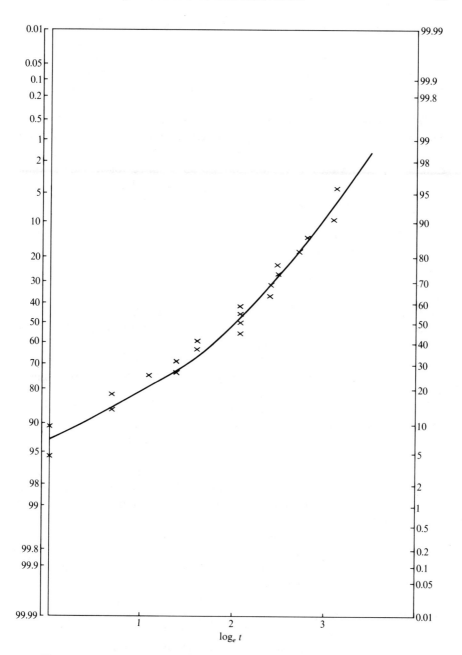

Figure 3.10 Lognormal probability plot of remission times of placebo patients.

When $\gamma = 1$, the Weibull distribution reduces to the exponential distribution. Therefore, the $\gamma = 1.12$ Weibull distribution is close to the exponential. The mean remission time obtained from the Weibull fit is approximately 9.1 months and the hazard rate is approximately 0.11 per month. Thus, for this set of data, there is no substantial difference between the exponential fit and the Weibull fit.

To test whether the exponential distribution gives an adequate fit to the data, we use the WE_1-test developed by Bartholomew (1957) and Shapiro and Wilk (Hahn and Shapiro 1967). This test is described in Section 7.4.1. Equation (7.22) yields $WE_1 = 0.025$. The 95% range for $n = 21$ is found from Table C-8 to be 0.020–0.085. Since the calculated value of WE_1 is inside this range, we do not reject the hypothesis that the data could have come from an exponential distribution.

If the physician would like to predict the probability of a 10-week remission, its value could be calculated as

$$P(\text{10-week remission}) = e^{-10\lambda_2}$$

An estimate from the data is used for λ_2 to obtain an estimated probability. Graphically, we have an estimate $\hat{\lambda}_2 = 0.12$. The maximum likelihood estimate $\hat{\lambda}_2$ obtained from (8.9) is $\frac{21}{182}$, or 0.116 (0.12 rounded off). Thus an estimated probability of a 10-week remission of a patient receiving placebo is $e^{-10(0.12)} = 0.30$, which is much smaller than that of a patient receiving 6-MP.

These graphical methods are subjective. The judgment whether the assumed distribution fits the data is based on a visual examination rather than an objective statistical test. However, the methods are very simple and do provide a great deal of information. Even in a case where none of the distributions discussed in this book fit well, graphs can help find the reasons and thus help modify the model. Therefore, graphical methods are usually recommended as the first thing to try. It must be pointed out that, although statistical tests permit us to reject a distribution as inadequate, they do not prove that the distribution is *correct*. The outcome of a statistical test depends highly on the amount (the sample size) of data available. The larger the sample size, the better the chances are of rejecting an inadequate distribution. If too few data points are available, the statistical test may lack sensitivity in detecting inadequate distributions. On the other hand, an assumed distribution rejected by an appropriate statistical test frequently cannot be established as inadequate if the sample size is too small.

When it can be assumed that times to failure are exponentially distributed in both treatment groups, two parametric tests are available to test the equality of the two parameters: the *likelihood ratio test* and Cox's F-test (see Chapter 9). Cox's F-test is the most powerful test for comparing exponential distributions and is preferred to the likelihood ratio test. The null hypothesis to be tested is

H_0: $\lambda_1 = \lambda_2$ (no difference in relapse rates between patients receiving 6-MP and patients receiving placebo)

and the alternative hypothesis is

H_1: $\lambda \neq \lambda_2$ (relapse rates in the two treatment groups are different)

Applying Cox's F-test, the computed value of the test statistic that follows the F distribution is 3.24, with 18 and 42 degrees of freedom. The upper percentage points F_α for $\alpha = 0.025$ and 0.005 for the nearest degrees for freedom available in Table C-3 are $F_{20,40,0.025} = 2.0677$ and $F_{20,40,0.005} = 2.5984$. Since the computed value of the test statistic exceeds the upper 0.5 percentage point, we reject the null hypothesis at the 0.005 level and conclude that the data provide strong evidence that the rates of relapse from remission differ in the two groups.

EXAMPLE 3.4   RELATIVE MORTALITY AND IDENTIFICATION OF PROGNOSTIC FACTORS

One thousand and twelve Oklahoma Indians (379 men and 633 women) with non-insulin-dependent diabetes mellitus (NIDDM) were examined in 1972–1980 and a mortality follow-up study was conducted in 1986–1989 (Lee et al. in press-b). The mean [standard deviation (SD)] age and duration of diabetes at baseline examination were 52 (11) and 7 (6) years, respectively. The average duration of follow-up was 10 (SD 4) years. As of December 31, 1989, 548 patients were alive, 452 (187 men and 265 women) were dead, and 12 could not be traced. Table 3.6 gives the survival time in years (T) of the first 40 male patients along with 12 potential prognostic factors: age, duration of diabetes (DUR) in years, family history of diabetes (FAM), use of insulin within one year of diagnosis (INS), use of diuretics (DIU), hypertension (HBP), retinopathy (EVD), proteinuria (PRO), fasting plasma glucose (GLU) in milligrams per deciliter, cholesterol (TC) in milligrams per deciliter, triglyceride (TG) in milligrams per deciliter, and body mass index (BMI), which is defined as weight in kilograms divided by height in meters squared.

Among other things, the authors compared the mortality experience of the diabetic patients with that of the general population in Oklahoma over the follow-up period. Taking changes in age distribution into consideration, the patients were divided into five groups according to their age at baseline examination, <35, 35–44, 45–54, 55–64, and ≥65. The expected survival rates were calculated on a yearly basis following the methods described in Section 4.3 and using the death rates given in the 1970 and 1980 Oklahoma population life tables. Death rates for the years between 1970 and 1980 and between 1980 and 1989 were estimated based on the 1970 and 1980 statistics

Table 3.6 Data of First 40 Male Patients Enrolled in Mortality Study of Oklahoma Diabetic Indians

OBS	Status[a]	T	Age	DUR	FAM[b]	INS[b]	DIU[b]	HBP[b]	EVD[b]	PRO[b]	GLU	TC	TG	BMI
1	1	12.4	44.0	3	1	0	0	1	0	1	242	392	538	34.2
2	1	14.1	43.5	1	0	0	0	0	0	0	94	158	94	42.2
3	0	14.4	47.6	4	1	0	1	1	0	0	100	195	405	33.1
4	0	14.2	36.3	3	1	0	0	0	0	0	171	212	218	38.5
5	0	14.4	54.4	7	0	0	0	0	0	0	112	204	77	31.7
6	0	12.4	50.8	4	1	0	1	0	0	0	83	206	178	41.5
7	0	12.4	50.0	2	1	0	0	0	0	0	104	178	100	39.5
8	1	7.0	66.9	8	0	0	0	0	0	1	161	189	99	29.7
9	1	13.6	40.2	14	1	0	0	1	1	0	262	241	301	27.5
10	0	14.4	54.1	4	1	0	1	1	1	0	115	183	392	24.4
11	0	12.4	38.9	3	1	0	0	1	0	1	108	237	49	32.4
12	1	9.8	51.2	4	1	0	0	0	0	1	184	114	118	26.5
13	1	0.0	53.0	6	0	0	1	0	0	1	126	206	480	34.5
14	1	12.1	45.0	5	0	0	1	1	0	0	115	238	177	18.9
15	0	14.4	38.3	2	1	0	1	0	0	0	110	204	180	31.2
16	0	12.4	40.0	5	1	0	1	1	1	1	227	159	337	39.2
17	0	14.4	44.4	9	1	0	0	0	0	0	182	193	332	32.7
18	0	14.2	48.2	5	1	0	1	1	1	1	184	171	238	33.5
19	0	12.4	36.3	6	1	0	0	0	0	0	238	196	162	24.2
20	0	13.7	41.9	3	1	0	0	0	1	0	284	170	125	30.7

	a												
21	1	13.4	49.7	15	0	0	0	0	0	248	217	234	28.0
22	1	12.6	51.5	9	1	1	1	1	1	96	174	144	24.2
23	0	14.0	44.1	1	1	0	0	0	0	116	251	153	33.3
24	0	12.4	35.9	3	1	0	0	0	0	120	129	76	30.1
25	0	12.9	50.4	1	0	1	0	0	0	128	190	123	27.7
26	0	12.4	48.0	3	0	1	1	0	0	95	207	59	28.1
27	0	14.5	40.1	0	0	0	0	1	0	89	252	112	31.7
28	0	13.4	54.5	6	0	1	0	0	0	104	490	540	30.8
29	1	13.9	69.3	0	1	1	1	0	1	122	162	209	24.2
30	1	12.0	38.5	18	1	1	1	0	1	225	183	343	43.1
31	1	3.6	63.7	12	0	0	0	0	0	211	217	124	25.1
32	1	15.4	71.8	3	1	1	0	0	0	150	227	137	26.0
33	0	10.3	59.5	1	1	0	1	0	0	180	188	308	28.1
34	1	5.8	50.1	18	1	0	0	0	0	400	200	166	26.1
35	1	2.5	75.4	1	1	1	0	1	1	281	692	364	49.7
36	0	15.0	29.6	18	0	0	0	1	1	167	154	157	60.2
37	1	5.5	60.2	1	1	0	0	0	0	351	206	141	26.0
38	1	4.5	63.9	4	1	1	1	0	1	127	180	131	21.8
39	1	6.8	57.4	17	1	0	0	1	0	153	786	466	34.1
40	1	3.6	71.3	18	1	1	1	1	1	179	488	364	25.6

[a] 1, Dead; 0, alive.
[b] 1, yes; 0, no.

and the assumption that changes in death rates between 1970 and 1980 and after 1980 follow a linear trend. The observed and expected survival curves for the groups were plotted (Figure 3.11), and ratios of the observed and expected number of deaths (O/E ratios) by age tabulated (Table 3.7).

Figure 3.11 shows that the diabetic patients had a much lower survivorship than the general Oklahoma population for this age–sex distribution. At the beginning of the 15th year after baseline examination, the relative survival for the diabetic Oklahoma Indians was only 60%. The overall O/E ratios in Table 3.7 are 2.92 [or standardized mortality ratio (SMR) 292] for men and 4.09 (or SMR 409) for women, which indicates a significantly higher mortality in the diabetic Oklahoma Indians than in the general population. Although patients in every group experienced excessive mortality, the younger patients suffered the worst.

The relationship between the 12 potential prognostic variables and the

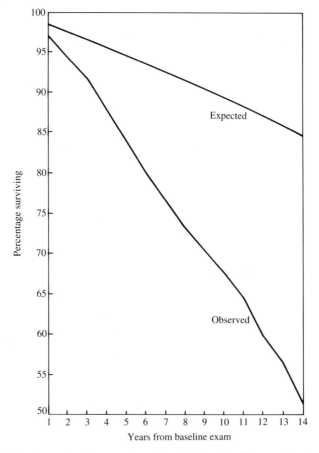

Figure 3.11 The observed and expected survivorship from baseline examination for diabetic Oklahoma indians.

Table 3.7 Observed and Expected Number of Deaths and O/E Ratios during Follow-up Period by Sex and Age at Baseline

Age at Baseline (y)	Males			Females		
	Observed	Expected	O/E	Observed	Expected	O/E
≤44	34	6.48	5.25	31	5.51	5.63
45–54	79	25.64	3.08	85	18.40	4.62
55–64	32	11.67	2.74	69	15.03	4.59
65+	42	20.32	2.06	80	25.86	3.09
Total	187	64.11	2.92	265	64.80	4.09

survival time of men was examined using univariate and multivariate methods. The procedures are summarized below.

1. *Examine the Individual Relationship of Each Variable to Survival.* One way to analyze the data is to first determine which of the 12 variables could be of significant prognostic importance. In addition to correlation analysis of these variables, the survival times in subcategories are compared (Table 3.8). Patients are grouped into subgroups in a meaningful way or in a way that maximizes the observed difference in survival time between the subgroups (subject to the constraint that each subgroup contains at least 10% of the total number of patients).

The survivorship function for every subgroup of each variable was estimated using the Kaplan–Meier method (discussed in Chapter 4) and plotted. Figure 3.12 gives an example. Survival functions among the subgroups were compared by the logrank test (one of the available tests discussed in Chapter 5). Table 3.8 shows that except cholesterol and triglyceride, every one of the 12 variables is significant. The median survival time decreases as age and duration of diabetes increase. Patients without family history of diabetes, elevated fasting plasma glucose, hypertension, or retinopathy have significantly shorter survival than patients without these characteristics. Patients with baseline BMI value greater than or equal to 30 had much better survivorship than patients with a lower BMI.

2. *Examine the Simultaneous Relationship of the Variables to Survival.* Examination of each individual variable can give only a preliminary idea of which variables might be of prognostic importance. The simultaneous effect of the variables must be analyzed by an appropriate multivariate method to determine the relative importance of each. Cox's (1972) proportional hazards model can be applied. This model, presented in Chapter 10, is a regression model that directly relates patient characteristics to the risk of failure and thus, indirectly, to survival. The assumption of this model is that the hazards for different strata of each independent (or prognostic) variable are proportional over time. This assumption was verified by a graphical method (discussed in Chapter 10) using each of the variables. Figure 3.13

Table 3.8 Survival Time by Potential Prognostic Variable for Male Diabetic Patients

Variable	Number of Patients	Number of Deaths	Median Survival Time (yr)	P value
Age (yr)				
<45	102	34	15.2	
45–54	173	79	14.2	<0.001
55–64	53	31	9.1	
≥65	46	43	5.1	
Family history of diabetes				
No	104	62	11.4	<0.01
Yes	238	104	14.8	
Duration of diabetes (yr)				
<7	207	84	15.2	
7–13	91	49	12.2	<0.001
≥14	58	48	7.9	
Use of diuretics				
No	254	117	14.5	<0.001
Yes	102	64	10.0	
Use of insulin <1 year of diagnosis				
No	317	157	13.9	<0.05
Yes	39	24	11.9	
Hypertension				
No	211	84	15.3	< 0.001
Yes	151	99	9.8	
Retinopathy				
No	332	163	13.9	
Yes	24	18	6.5	<0.001
Proteinuria				
Negative	250	112	14.5	
Slight	54	27	12.4	<0.001
Heavy	57	44	8.0	
Fasting plasma glucose				
<200	235	106	11.9	
≥200	139	81	8.4	<0.05
Cholesterol				
<240	300	144	14.8	
≥240	64	38	12.2	0.12
Triglyceride				
<220	223	105	14.1	
≥220	141	77	13.2	0.44
BMI				
<30	189	114	11.8	<0.001
≥30	184	73	15.4	

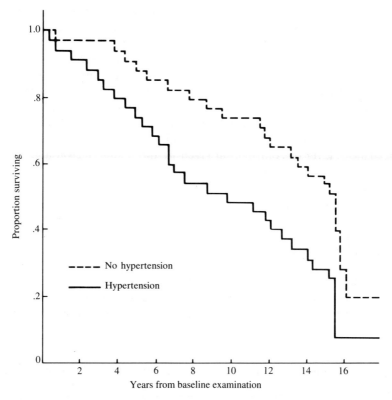

Figure 3.12 Survival curves of diabetic patients by hypertension status at baseline.

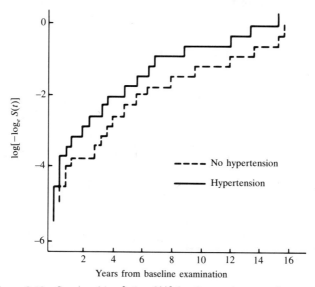

Figure 3.13 Curves of $\log_e[-\log_e S(t)]$ for the two hypertension groups.

gives an example of the graph of $\log_e[-\log_e S(t)]$ versus t for the two hypertension groups. The two almost parallel curves indicate that the hazards of dying are proportional. Therefore, the assumption of the proportional hazards model is satisfied and the method appropriate.

This model can be fitted by using a stepwise procedure that results in a ranking of the prognostic variables. The first variable selected to enter the model is the most important single variable in predicting the risk of dying. The second variable is the second most important, and so on. A significance level can be obtained from a likelihood ratio test at each step, which indicates the level of contribution given by the additional variable.

Using the proportional hazards model and a stepwise procedure, 7 of the 12 variables were identified as significant at the 0.05 level based on the likelihood ratio test at each step. These variables, the regression coefficients, and the significance levels based on the Ward test, which uses the regression coefficient and its standard error (SE) are given in Table 3.9. The sign of the coefficient indicates whether the variable is positively or negatively related to the hazard of dying. For example, age and duration of diabetes are both positively related to the risk of dying and, therefore, negatively related to the survival time. Table 3.9 also gives the ratio of risk (or hazard) for values of each variable unfavorable to survival to values of

Table 3.9 Significant Variables (at 0.05 Level) Identified by Proportional Hazards Model

Variable[a]	Regression Coefficient	P Value (Ward Test)	Relative Risk[b] Favorable	Relative Risk[b] Unfavorable	Ratio of Risk
Age	0.0558	<0.001	9.32	28.45	3.05
Hypertension: 1—yes, 0—no	0.6360	<0.001	1.00	1.89	1.89
Duration of diabetes	0.0559	<0.001	1.32	2.19	1.66
Fasting plasma glucose	0.0023	<0.010	1.35	1.58	1.17
BMI	−0.0330	0.035	0.32	0.44	0.72
Proteinuria: 1—yes, 0—no	0.3744	0.025	1.00	1.45	1.45
Use of diuretics: 1—yes, 0—no	0.4191	0.030	1.00	1.52	1.52

[a]Variables are listed in order of entry into model with a p value limit for entry of 0.05.
[b]Favorable categories are 40 years of age, no hypertension, duration of diabetes 5 years, fasting plasma glucose 130 mg/dl, BMI 35, no proteinuria and no diuretics use. Unfavorable categories are 60 years of age, hypertensive, duration of diabetes 14 years, fasting plasma glucose 200 mg/dl, BMI 25, having proteinuria and diuretics use.

that variable favorable to survival. For example, patients who were 60 years of age at baseline had a 3.05 times higher risk of dying during the follow-up period (10–16 years, average 13 years) than patients who were only 40 years old at baseline. For dichotomous variables, the ratio of risk is equal to exp(coefficient), which is also interpreted as the relative risk of the variable adjusting for the other variables. Consequently, the confidence interval for the relative risk can be calculated (not shown in Table 3.9). Based on this set of data, the authors conclude that age, hypertension, duration of diabetes, fasting plasma glucose, BMI, proteinuria, and use of diuretics are significantly related to survival. The multivariate method also showed that high values of BMI might be protective.

EXAMPLE 3.5 IDENTIFICATION OF RISK FACTORS

A study of the incidence of retinopathy in Oklahoma Indians with NIDDM was conducted in 1987–1990 as part of a prospective study of diabetic complications (Lee et al. in press-a). Among the 312 patients who were free of retinopathy at initial examination in the 1970s, 228 were found to have developed the eye disease during the 10–16-year follow-up period (average follow-up time 12.7 years). Twelve potential risk factors (assessed at time of baseline examination) were examined by univariate and multivariate methods for their relationship to retinopathy (RET): age, sex, duration of diabetes (DUR), fasting plasma glucose (GLU), initial treatment (TRT), systolic (SBP) and diastolic blood pressure (DBP), body mass index (BMI), plasma cholesterol (TC), plasma triglyceride (TG), and presence of macrovascular disease (LVD) or renal disease (RD). Table 3.10 gives the data of the first 40 patients. Among other things the authors related these variables to the development of retinopathy.

1. *Examine the Individual Relationship of Each Variable to the Development of Diabetic Retinopathy.* Table 3.11 gives some summary statistics of the eight continuous variables for patients who have developed retinopathy and for those who have not. Notice that patients who have developed the disease were younger at baseline and had much higher fasting plasma glucose, systolic and diastolic blood pressure, and plasma triglyceride than patients who have not. Table 3.12 summarizes the contingency table analysis of retinopathy incidence rates. The number of patients at risk of developing retinopathy and the number of patients who developed the disease (and rate) are given by subcategory of each potential risk factor. Using the chi-square test, it is found that there was a significant difference in the retinopathy rate among the subcategories of several variables using a significance level of 0.05: duration of diabetes, fasting plasma glucose,

Table 3.10 First 40 Patients Involved in Study of Risk Factors in Development of Diabetic Retinopathy

OBS	RET[a]	Age	Sex	DUR	TRT[b]	GLU	SBP	DBP	TC	TG	BMI	LVD[a]	RD[a]
1	1	47.6	M	4	2	100	156	98	195	405	33.1	0	0
2	0	54.4	M	7	1	112	112	78	204	77	31.7	0	0
3	0	50.8	M	4	2	83	134	80	206	178	41.5	1	0
4	1	49.4	F	5	2	276	102	70	190	222	24.1	0	0
5	1	50.0	M	2	2	104	142	86	178	100	39.5	0	0
6	1	50.7	F	7	2	242	142	78	217	268	31.6	0	1
7	1	35.3	F	2	2	130	134	80	390	564	47.0	0	0
8	0	50.2	F	5	2	130	100	70	174	128	29.8	1	0
9	1	45.0	M	5	2	115	134	100	238	177	18.9	1	0
10	0	38.3	M	2	2	110	132	80	204	180	31.2	1	0
11	0	45.8	F	5	3	130	118	68	185	316	26.6	0	1
12	1	51.6	F	3	2	141	112	78	152	77	31.2	0	0
13	1	36.3	M	6	2	238	142	94	194	162	24.2	0	0
14	1	44.7	F	16	2	190	152	90	132	161	32.0	0	0
15	1	37.2	F	1	2	126	136	90	133	211	32.7	0	0
16	1	52.6	F	9	2	159	140	76	151	132	26.7	0	0
17	1	44.1	M	1	1	116	126	76	251	153	33.3	0	0
18	1	35.9	M	3	1	120	132	80	129	76	30.1	0	0
19	1	50.4	M	1	1	128	144	90	190	123	27.7	0	0
20	1	48.0	M	1	2	95	128	74	207	59	28.1	1	0

21	0	47.5	F	1	2	85	124	82	161	190	31.6	1	0
22	1	50.1	F	6	2	138	106	72	181	135	30.5	0	1
23	0	43.3	F	1	1	104	128	86	204	198	26.1	0	0
24	1	54.5	M	0	3	104	142	84	490	540	30.8	1	0
25	1	52.2	F	3	3	304	132	84	192	119	36.9	1	0
26	1	53.3	F	3	2	249	128	72	120	85	35.5	0	0
27	1	64.3	F	6	2	297	138	80	145	64	30.1	1	0
28	1	44.6	F	4	1	139	112	80	156	111	36.1	0	1
29	1	47.1	F	2	2	169	130	84	198	99	39.0	0	1
30	1	46.5	F	7	1	159	128	78	238	157	34.5	0	0
31	0	51.5	F	3	1	147	128	78	185	182	32.0	0	0
32	1	59.5	M	3	2	180	132	78	188	308	28.1	1	0
33	1	52.0	F	6	2	183	142	84	175	68	37.6	0	0
34	1	45.7	F	6	2	180	138	80	179	189	44.1	1	0
35	1	48.2	F	18	2	267	158	100	195	112	25.2	1	0
36	1	57.4	M	0	1	159	172	108	219	294	33.7	0	1
37	1	42.0	F	4	2	158	106	68	224	157	33.6	0	0
38	1	50.7	F	1	2	211	142	84	390	645	37.1	0	1
39	1	53.8	F	0	1	177	154	80	175	208	35.3	0	1
40	0	56.9	M	1	1	98	116	70	146	97	26.5	0	1

[a] 1, yes; 0, no.
[b] 1, Diet only; 2, oral agent; 3, insulin.

Table 3.11 Summary Statistics for Eight Variables by Retinopathy Status at Follow-up

| | Retinopathy Status | | | | |
| | No | | Yes | | |
Variable	Mean	SD	Mean	SD	P Value
Age	50.0	9.0	47.2	7.4	0.01
Duration of diabetes	4.2	4.5	4.8	4.4	0.34
Fasting plasma glucose	141.8	65.6	196.3	76.6	<0.0001
Systolic blood pressure	128.0	15.7	132.6	17.3	0.04
Diastolic blood pressure	80.3	10.8	84.9	10.1	<0.001
Body mass index	32.3	6.3	32.5	5.9	0.76
Cholesterol	204.4	66.0	206.8	58.7	0.76
Triglyceride	180.5	111.1	234.4	273.3	0.01

systolic and diastolic blood pressure, and treatment. It appears that patients with poor glucose control or high blood pressure or treated with oral agents or insulin have a higher incidence of retinopathy. In addition, patients with high triglyceride levels tend to have higher incidence of retinopathy ($p = 0.064$). However, patients who had developed macrovascular disease at the time of baseline examination had a lower retinopathy incidence. The authors state that this may be due to the fact that 68% of the patients who had macrovascular disease either died (54%) during the follow-up period or were lost to follow-up (14%). Many of these patients may have developed retinopathy, particularly the patients who have died, but were not included. Therefore, the lower incidence of retinopathy in patients who had macrovascular disease at baseline is likely the result of a selection bias. Similarly, the large number of death plus the losses to follow-up may also contribute to the drop in retinopathy rate in patients who had had diabetes for more than 12 years at baseline. Among the 80 patients in this duration of diabetes category, 56% have died and 10% did not participate in the follow-up examination. The large number of deaths may also be responsible for the finding that patients who survived long enough to develop retinopathy were younger at baseline. The deceased patients were significantly older (mean 57 years) than the survivors who participated in the follow-up examination (mean 48 years).

2. *Examine the Simultaneous Relationship of the Variables to the Development of Retinopathy.* Univariate analysis of each variable using the contingency table or the chi-square test gives a preliminary idea of which individual variables might be of prognostic importance. The simultaneous effect of all the variables can be analyzed by the linear logistic regression model (discussed in Section 11.3) to determine the relative importance of each.

Table 3.12 Cumulative Incidence Rates of Retinopathy by Baseline Variables

Variable	Number of Persons at Risk	Developed Retinopathy Number	Developed Retinopathy Percent	P-Value
Sex				
Female	211	151	71.6	0.384
Male	101	77	76.2	
Age (yr)				
<35	13	10	76.9	
35–44	101	77	76.2	0.242
45–54	155	115	74.2	
≥55	43	26	60.5	
Duration of diabetes (yr)				
<4	153	105	68.6	
4–7	113	86	76.1	0.033
8–11	23	22	95.7	
≥12	23	15	65.2	
Fasting plasma glucose (mg/dl)				
<140	117	62	53.0	
140–199	90	74	82.2	<0.001
≥200	105	92	87.6	
Systolic blood pressure (mm Hg)				
<130	145	95	65.5	
130–159	149	115	78.8	0.016
≥160	20	18	85.7	
Diastolic blood pressure (mm Hg)				
<85	179	118	65.9	
85–94	87	73	83.9	0.004
≥95	46	37	80.4	
Plasma cholesterol (mg/dl)				
<240	267	193	72.3	0.442
≥240	45	35	77.8	
Plasma triglyceride (mg/dl)				
<250	237	167	70.5	0.064
≥250	75	61	81.3	
Body mass index (kg/m^2)				
<28	73	49	67.1	
28–33	121	94	77.7	0.261
≥34	118	85	72.0	
Renal disease				
No	251	179	71.3	0.155
Yes	61	49	80.3	
Macrovascular disease				
No	205	157	76.6	0.053
Yes	107	71	66.4	
Treatment (initial)				
Diet alone	115	62	53.9	<0.001
Oral agent	158	136	86.1	
Insulin	37	29	78.4	

The 12 variables were fitted to the linear logistic regression model using a stepwise procedure. The variables most significantly related to the development of retinopathy were found to be initial treatment, fasting plasma glucose, age, and diastolic blood pressure ($p \leq 0.001$). Table 3.13 gives the regression coefficients of the four most significant variables ($p \leq 0.05$), the standard errors, and adjusted odds ratios [exp(coefficient)]. The p values used here are the significance levels based on the likelihood ratio test or the improvement in the maximum likelihood due to the addition of the variable in the stepwise procedure. This method is more powerful than the Wald test, which is based on the standardized regression coefficients (Chapter 11). The results are consistent with those in the univariate analysis.

On the basis of the regression coefficients, the probability of developing retinopathy during a 10–16-year follow-up can be estimated by substituting values of the risk factors into the regression equation,

$$\log_e \frac{P}{1-P} = -2.373 + 1.495(\text{oral agent}) + 0.882(\text{insulin}) + 0.014(\text{GLU})$$
$$- 0.074(\text{age}) + 0.048(\text{DBP})$$

For example, for a 50-year-old patient who is on oral agents and whose fasting plasma glucose and diastolic blood pressure are 170 mg/dl and 95 mmHg, respectively, the chance of developing retinopathy in the next 10–16 years in 91%.

The linear logistic regression model is useful in identifying important risk factors. However, complete measurements of all the variables are needed; missing data are a problem. In this example, complete data are available on most of the patients. This may not always be the case. Though there are methods of coping with missing data (discussed in Section 10.1), none is perfect. Thus it is extremely important for investigators to make every effort to obtain complete data on every subject.

Table 3.13 Results of Logistic Regression Analysis

Variable	Coefficient	Standard Error	exp(coefficient)	Coefficient/SE
Constant	−2.373	1.557		
Initial treatment				
Oral agent	1.495	0.330	4.459	4.53
Insulin	0.882	0.488	2.416	1.81
Fasting plasma				
glucose	0.014	0.003	1.014	4.67
Age	−0.074	0.019	0.929	−3.89
Diastolic blood				
pressure	0.048	0.015	1.049	3.20

BIBLIOGRAPHICAL REMARKS

It is impossible to cite all the published examples of survival data analysis similar to those in this chapter. Others can be found in the literature, for example, *Biometrics, Biometrika, Cancer, Journal of Chronic Disease, Journal of the National Cancer Institute, American Journal of Epidemiology, The Journal of the American Medical Association*, and *The New England Journal of Medicine*. Below are listed a few in three general classifications.

1. *Survival Patterns and Comparison of Survival Patterns.* Armitage (1959), Bodey et al. (1971), Bonadonna et al. (1976), Carbone et al. (1968), Cutler and Ederer (1958), Cutler et al. (1957, 1959, 1960a,b, 1967), Frei et al. (1973), Freireich et al. (1963), Parker et al. (1946), Sullivan et al. (1975), Kirk et al. (1980), Kashiwagi et al. (1985), Oliver et al. (1988), Geller et al. (1989), and Aaby et al. (1990).

2. *Survival Model Fitting.* Birnbaum and Saunders (1958), Byar et al. (1974), Feinleib and MacMahon (1960), Hokanson et al. (1977), Myers et al. (1973), Pike (1966), Zelen (1966), Zippin and Armitage (1966), DeWals and Bouckaert (1985), Bolin and Greene (1986), Shulz et al. (1986), Walle et al. (1987), Horner (1987), Eggermont (1988), and Juckett and Rosenberg (1990).

3. *Risk or Prognostic Factor Identification.* Freireich et al. (1974), George et al. (1973), Hart et al. (1977), Minow et al. (1977), Hammond et al. (1984), Baumgartner et al. (1987), Farchi et al. (1987), Asal et al. (1988b), Winkleby et al. (1988), Kaplan et al. (1989), Hanson et al. (1989), Broderick et al. (1990), Teitelman et al. (1990), Mulders et al. (1990), and Chiasson et al. (1990).

EXERCISES

The four set of data below are taken from actual research situations. Although the data can be used for various analyses throughout the book, the reader is asked here only to describe in detail how the data can be analyzed. The data appear in examples and other exercises in subsequent chapters.

3.1 Thirty-three patients with hypernephroma were treated with combined chemotherapy, immunotherapy, and hormonal therapy. Exercise Table 3.1 gives the age, sex, date treatment began, response status, date of death or last follow-up, survival status, and results of five pretreatment skin tests. The investigator is interested in the response and survival of the patients and in identifying prognostic factors. How would you analyze the data?

Exercise Table 3.1 Data for 33 patients with Hypernephroma

Patient Number	Age	Sex	Date Treatment Started	Response[a]	Date of Death or Last Follow-up	Status[b]	Skin Test Results[c]				
							Monilia	Mumps	PPD	PHA	SK-SD
1	53	F	3/31/77	1	10/1/77	0	7 × 7	23 × 23	0 × 0	25 × 25	0 × 0
2	61	M	6/18/76	0	8/21/76	1	10 × 10	15 × 20	0 × 0	13 × 13	9 × 9
3	53	F	2/1/77	3	10/1/77	0	0 × 0	7 × 7	0 × 0	25 × 25	0 × 0
4	48	M	12/19/74	2	1/15/76	1	0 × 0	0 × 0	0 × 0	0 × 0	0 × 0
5	55	M	11/10/75	0	1/15/76	1	12 × 12	ND	10 × 10	8 × 8	5 × 5
6	62	F	10/7/74	2	4/5/75	1	10 × 10	5 × 5	0 × 0	7 × 7	5 × 5
7	57	M	10/28/74	0	1/6/75	1	15 × 15	15 × 15	0 × 0	0 × 0	10 × 10
8	53	M	10/6/75	2	6/18/77	1	0 × 0	ND	0 × 0	12 × 12	0 × 0
9	45	M	4/11/77	0	10/1/77	0	6 × 4	4 × 4	0 × 0	0 × 0	0 × 0
10	58	M	8/4/76	3	2/11/77	1	13 × 13	13 × 13	22 × 22	23 × 23	0 × 0
11	61	F	1/1/77	3	10/1/77	0	0 × 0	8 × 8	17 × 17	11 × 11	0 × 0
12	61	M	7/25/76	1	10/1/77	0	9 × 9	12 × 12	0 × 0	20 × 20	0 × 0
13	77	M	5/8/75	0	9/26/75	1	0 × 0	0 × 0	0 × 0	0 × 0	0 × 0
14	55	M	4/27/77	2	10/1/77	0	0 × 0	0 × 0	15 × 15	10 × 10	0 × 0
15	50	M	4/20/77	3	10/1/77	0	0 × 0	14 × 14	5 × 5	32 × 32	21 × 21
16	42	M	8/24/76	0	10/1/77	0	11 × 11	7 × 7	0 × 0	12 × 12	0 × 0

#	Age	Sex	Date	Code	Date	Code					
17	50	F	1/8/75	0	6/30/75	1	0×0	0×0	0×0	0×0	0×0
18	66	F	9/8/76	3	10/1/77	0	9×9	10×10	6×6	15×15	11×11
19	58	M	2/18/75	0	10/1/77	0	0×0	0×0	0×0	0×0	ND
20	62	M	5/12/76	0	10/17/76	1	2×2	ND	ND	3×3	2×2
21	71	F	10/22/76	3	12/12/76	1	10×10	6×6	0×0	12×12	0×0
22	44	M	6/6/77	3	10/1/77	0	10×10	10×10	0×0	20×20	0×0
23	69	M	6/21/76	0	10/13/76	1	0×0	15×15	25×25	25×25	0×0
24	56	M	6/7/77	2	10/1/77	0	0×0	7×7	0×0	0×0	0×0
25	57	M	11/16/76	0	12/10/76	1	11×11	5×5	0×0	20×20	0×0
26	69	M	5/10/77	0	7/25/77	1	0×0	0×0	0×0	15×15	0×0
27	60	M	6/29/77	0	7/7/77	1	0×0	0×0	0×0	26×26	0×0
28	60	M	7/21/75	3	10/1/77	0	11×11	20×20	10×10	18×18	0×0
29	72	M	7/19/75	0	10/18/75	1	10×10	0×0	7×7	10×10	0×0
30	42	F	3/3/75	0	4/23/75	1	0×0	ND	0×0	0×0	0×0
31	57	M	2/24/77	2	10/1/77	0	5×5	8×8	0×0	25×15	0×0
32	66	M	6/15/77	3	10/1/77	0	0×0	15×15	0×0	10×10	0×0
33	59	M	3/4/77	0	4/2/77	1	0×0	0×0	0×0	16×16	0×0

[a] 0, No response; 1, complete response; 2, partial response; 3, stable.
[b] 0, Alive; 1, dead.
[c] ND, not done.

Source: Kindly provided by Dr. Richard Ishmael.

Exercise Table 3.2 Data of 58 Patients with Hypernephroma

Patient Number	Sex[a]	Age	Nephrectomy[b]	Time of Nephrectomy[c]	Treatment[a]	Response[e]	Survival Time	Status[f]	Lung Metastasis[b]	Bone Metastasis[b]
1	2	53	1	0.0	1	1	77	0	1	0
2	1	69	1	4.0	1	2	18	1	0	1
3	1	61	0	−9.0	1	0	8	1	1	0
4	2	52	1	2.0	1	2	68	1	1	0
5	1	46	1	2.0	1	2	35	1	0	1
6	1	55	1	0.0	1	0	8	1	1	0
7	2	62	1	0.0	1	2	26	1	1	0
8	1	53	1	0.0	1	2	84	1	0	1
9	1	70	0	−9.0	1	0	17	1	1	0
10	1	48	1	0.0	1	3	52	1	1	0
11	1	58	1	1.5	1	0	26	1	1	1
12	1	61	1	5.0	1	1	108	0	1	0
13	1	77	1	4.0	1	0	18	1	1	0
14	1	56	1	0.0	1	3	72	1	0	1
15	1	55	1	0.0	1	2	38	1	1	1
16	1	50	1	4.0	1	3	−99	9	1	0
17	1	75	1	0.0	1	0	9	1	1	0
18	1	43	1	2.0	1	3	56	1	0	0
19	1	69	1	1.0	1	2	36	1	1	1
20	2	59	1	1.5	1	2	108	1	0	1
21	2	71	1	0.0	1	2	10	1	1	0
22	1	56	1	0.0	1	2	36	1	1	0
23	1	57	0	−9.0	1	0	6	1	1	0
24	1	69	1	8.0	1	0	9	1	1	0
25	1	72	0	−9.0	1	0	12	1	1	0
26	1	67	1	0.0	1	9	5	0	1	0
27	1	41	1	2.0	1	2	104	0	1	0
28	1	77	1	10.0	1	0	6	1	1	0
29	2	63	1	2.0	1	3	115	1	0	1
30	2	42	1	12.0	1	0	9	1	0	0
31	1	59	0	−9.0	1	0	21	1	0	0

	a		b	c	d	e					f
32	1	62	1	5.0	1	0	14	1	1	1	0
33	1	65	1	0.0	1	0	52	1	1	1	0
34	2	53	0	-9.0	1	0	9	1	0	0	1
35	1	57	1	0.0	1	2	48	1	1	1	0
36	2	60	0	-9.0	1	0	15	1	1	1	1
37	1	59	1	0.0	1	0	5	1	1	1	0
38	1	75	1	0.0	2	3	28	1	1	0	0
39	2	53	1	0.0	2	2	25	0	1	0	0
40	2	67	1	5.0	2	3	25	0	0	1	1
41	1	58	1	8.0	2	3	40	1	1	1	1
42	1	62	1	8.0	2	0	16	1	1	1	1
43	1	69	0	-9.0	2	0	8	1	1	0	0
44	1	44	1	0.0	2	2	70	1	1	1	1
45	1	60	1	1.0	2	0	6	1	0	1	1
46	1	57	0	-9.0	2	0	8	1	1	1	1
47	1	45	1	2.0	2	4	12	1	1	1	0
48	2	50	1	1.0	2	4	20	1	1	1	1
49	1	58	0	-9.0	2	4	8	1	1	1	0
50	1	51	1	0.0	2	3	-99	0	0	0	0
51	1	59	1	3.0	2	3	12	1	1	1	1
52	1	53	1	0.0	2	1	181	0	1	0	0
53	1	70	1	0.5	2	0	20	1	1	1	0
54	1	69	1	3.0	2	0	14	1	1	1	1
55	1	62	1	0.0	2	3	26	1	0	1	0
56	1	52	1	2.0	2	0	16	1	1	1	0
57	1	77	1	2.0	2	2	30	1	1	1	0
58	1	61	1	8.0	2	0	20	1	1	1	0

[a]1, Male; 2, female.
[b]1, Yes; 0, No.
[c]Number of years prior to treatment, negative value—no nephrectomy.
[d]1, Combined chemotherapy and immunotherapy; 2, others.
[e]0, No response; 1, complete response; 2, partial response; 3, stable; 4, increasing disease; 9, unknown.
[f]1, Dead; 0, alive; 9, unknown.

Source: Kindly provided by Dr. Richard Ishmael.

Exercise Table 3.3   Data of the 102 Patients with Stages 3 and 4 Melanoma

Patient Number	Age	Sex[a]	Family History of Melanoma[b]	Remission Time (months)[c]	Remission Status[d]	Survival Time (months)	Survival Status[e]	Stage	Skin tests[f]					
									Monilia	Mumps	PPD	PHA	SK-SD	Tricophyton
1	58	2	9	42.0	0	42.0	0	3B	18	16	0	20	14	99
2	50	2	0	3.3	1	3.9	1	3B	7	8	0	0	7	0
3	76	1	0	6.1	1	10.5	1	3B	0	15	0	0	99	0
4	66	2	9	2.3	1	6.0	0	3B	8	0	0	10	0	0
5	33	1	9	5.1	1	20.6	1	3B	17	10	0	5	18	99
6	55	2	0	11.1	0	21.8	0	4B	99	99	99	99	99	99
7	25	2	0	36.5	0	36.5	0	3B	7	5	0	8	90	99
8	23	1	0	24.3	0	24.3	0	3B	10	20	0	7	30	0
9	30	1	0	28.7	0	28.7	0	3B	8	99	0	6	10	17
10	34	1	9	7.7	0	7.7	0	3B	15	99	0	0	0	0
11	34	1	0	29.3	0	29.3	0	3B	15	99	0	7	10	0
12	26	2	0	5.9	1	19.3	1	3AB	0	99	0	42	25	0
13	27	1	9	2.6	1	6.9	1	3AB	15	99	0	10	10	0
14	72	2	9	16.7	0	18.0	0	4B	10	99	99	15	0	7
15	70	2	0	14.6	0	14.6	0	3B	0	99	0	16	0	20
16	82	2	0	-99.0	9	23.6	0	4B	9	99	99	17	10	0
17	43	1	9	-99.0	9	3.9	1	4B	25	0	0	7	20	5
18	52	1	9	-99.0	9	7.3	1	4B	12	99	0	5	15	0
19	34	1	9	-99.0	9	9.8	1	4B	13	20	0	15	30	30
20	48	1	0	26.5	0	26.5	0	4A	10	99	0	88	5	0
21	62	1	9	18.0	1	25.4	0	3AB	5	99	0	10	0	24
22	49	1	9	4.3	1	8.0	1	3B	0	99	8	3	0	0
23	46	1	0	0.3	1	13.8	1	3B	0	5	0	7	0	0
24	53	2	1	21.5	0	21.5	0	3A	16	20	14	14	18	12
25	21	2	9	-99.0	9	9.3	1	4B	10	5	0	10	15	0

Row														
26	25	1	9	−99.0	9	1.2	1	4B	15	10	5	18	10	5
27	35	2	0	−99.0	9	20.0	1	4B	11	7	0	10	0	2
28	66	2	9	−99.0	9	12.5	0	4B	99	99	99	99	99	99
29	54	2	0	−99.0	9	7.4	1	4B	99	99	99	99	99	99
30	43	2	0	−99.0	9	4.7	0	4B	13	19	9	11	30	0
31	40	1	0	13.3	0	13.3	0	3B	0	5	25	12	10	0
32	16	1	0	0.0	0	0.0	0	3B	7	10	14	10	0	0
33	59	1	9	−99.0	9	25.8	0	4B	0	99	0	18	0	35
34	64	1	9	16.5	0	16.5	0	3B	8	7	0	0	5	0
35	52	1	8	−99.0	9	2.5	1	4B	9	14	0	12	0	0
36	−99	2	0	−99.0	9	13.8	1	4B	30	75	10	35	0	12
37	27	1	0	−99.0	9	4.2	1	4B	10	12	0	10	0	3
38	60	2	9	5.4	1	11.4	1	4B	20	10	0	10	5	0
39	73	1	0	−99.0	9	5.8	0	3B	0	9	0	40	0	30
40	50	2	0	13.5	0	13.5	1	4B	0	8	0	10	0	0
41	63	2	0	−99.0	9	2.7	1	4B	6	10	0	15	0	10
42	56	1	0	−99.0	9	0.9	1	3A	0	6	0	15	0	0
43	62	2	9	2.1	1	8.0	0	3AB	0	8	11	32	6	20
44	57	1	0	−99.0	9	0.0	0	3B	99	24	0	20	99	0
45	56	2	1	12.1	1	16.1	1	4B	0	5	99	25	17	30
46	41	2	0	−99.0	9	13.3	0	4A	99	99	0	99	99	99
47	40	2	0	10.1	0	10.1	0	4B	0	15	99	15	0	20
48	81	2	0	−99.0	9	0.0	0	3B	0	99	0	99	23	99
49	61	1	0	8.4	0	8.4	0	3AB	8	4	15	10	20	8
50	62	2	0	7.7	0	7.7	0	3B	0	11	0	16	0	0
51	34	1	0	15.1	1	24.4	1	4A	0	9	0	5	3	0
52	62	2	9	1.1	1	10.5	1	4B	0	99	0	99	0	0
53	63	2	9	−99.0	9	22.2	1	4B		0	0	8		0
54	56	1	9	−99.0	9	7.4	1	4B		99		6		0
55	66	2	9	−99.0	9	1.3	1	4B		0		22		0

Exercise Table 3.3 (*Continued*)

Patient Number	Age	Sex[a]	Family History of Melanoma[b]	Remission Time (months)[c]	Remission Status[d]	Survival Time (months)	Survival Status[e]	Stage	Skin tests[f]					
									Monilia	Mumps	PPD	PHA	SK-SD	Tricophyton
56	62	1	0	11.1	0	20.5	1	4B	25	20	0	10	15	18
57	68	2	0	−99.0	9	13.8	1	4B	20	99	0	17	15	15
58	45	1	0	−99.0	9	6.3	1	4B	28	15	0	10	17	50
59	58	1	9	−99.0	9	8.5	0	3B	10	17	0	25	30	25
60	55	1	9	−99.0	9	5.8	0	4B	99	99	99	99	99	99
61	63	2	1	7.3	1	8.7	0	3B	0	9	0	23	12	15
62	53	1	9	36.4	0	36.4	0	3B	5	35	20	17	6	0
63	45	1	0	−99.0	9	5.9	1	4B	0	0	0	0	0	0
64	41	1	0	−99.0	9	1.7	1	4B	0	99	0	5	0	6
65	43	1	9	−99.0	9	3.9	1	4B	25	0	0	7	20	5
66	80	1	0	5.8	0	11.0	1	4B	99	99	99	99	99	99
67	75	2	9	−99.0	9	3.8	1	4B	0	99	0	6	0	0
68	47	2	9	−99.0	1	15.9	0	3B	0	0	0	20	15	0
69	64	2	9	6.7	0	6.7	0	3AB	0	5	0	18	0	0
70	38	1	9	−99.0	9	1.6	1	4B	99	99	99	99	99	99
71	27	1	0	6.0	0	6.0	0	3B	8	15	20	27	20	10
72	56	1	9	−99.0	9	4.1	0	4B	0	0	0	0	0	0
73	60	2	9	−99.0	9	2.8	0	3A	99	99	99	99	99	99
74	80	2	9	−99.0	9	0.2	0	4B	0	20	99	40	0	20
75	38	1	9	−99.0	9	7.0	0	4B	0	0	0	15	12	12
76	71	1	9	6.2	0	6.2	0	4A	99	99	99	99	99	99
77	57	2	9	6.1	0	6.1	0	4B	28	20	0	19	20	20
78	69	1	0	−99.0	9	2.1	0	4B	15	15	15	10	0	0
79	17	2	9	4.9	0	4.9	0	3B	99	99	99	99	99	99
80	64	2	0	−99.0	9	1.6	0	4B	99	99	99	99	99	99

81	91	1	0	6.5	0	8.3	1	4A	99	99	99	99	99
82	40	2	0	1.7	1	4.6	0	3B	99	99	99	99	99
83	63	1	0	7.3	0	28.0	1	3A	99	99	99	99	99
84	40	1	9	−99.0	9	16.1	1	4B	99	99	99	99	99
85	53	1	9	−99.0	9	4.5	1	4B	99	99	99	99	99
86	41	1	0	21.2	0	21.2	0	3A	99	99	99	99	99
87	27	1	9	−99.0	9	4.0	1	4B	99	99	99	99	99
88	−99	−9	0	−99.0	0	7.8	1	4B	99	99	99	99	99
89	45	2	9	−99.0	9	4.4	0	4B	99	99	99	99	99
90	50	2	9	−99.0	9	4.2	1	4B	99	99	99	99	99
91	47	1	9	−99.0	9	1.5	1	4B	99	99	99	99	99
92	63	1	9	−99.0	9	3.5	0	4B	99	99	99	99	99
93	52	1	9	−99.0	9	0.4	1	4B	99	99	99	99	99
94	53	1	9	−99.0	9	2.5	1	4B	99	99	99	99	99
95	60	2	9	−99.0	9	1.1	0	4B	99	99	99	99	99
96	35	1	9	−99.0	9	11.1	1	4B	99	99	99	99	99
97	24	2	9	1.2	0	1.2	0	3B	99	99	99	99	99
98	80	2	0	−99.0	9	1.9	0	3A	99	99	99	99	99
99	−99	−9	0	4.6	1	6.7	0	3B	99	99	99	99	99
100	60	2	9	0.9	0	0.9	0	3AB	99	99	99	99	99
101	60	2	9	−99.0	9	4.3	0	4B	99	99	99	99	99
102	35	1	0	5.2	0	5.2	0	3B	99	99	99	99	99

[a] 1, Male; 2, female; −9, unknown.
[b] 1, Yes; 0, No; 9, unknown.
[c] −99, Never in remission during study period.
[d] 1, Relapsed; 0, still in remission; 9, never in remission during study period.
[e] 1, Dead; 0, still alive.
[f] In millimeters; 99, unknown.

Source: Lee et al. (1979)

Exercise Table 3.4 Data of 149 Diabetic Patients

						Variable at Baseline				
Patient Number	Status[a]	Survival Time (yr)	Age (yr)	BMI	Age at Diagnosis (yr)	Smoking Status[b]	SBP (mm Hg)	DBP (mm Hg)	ECG[c]	CHD[d]
1	1	12.4	44	34.2	41	0	132	96	1	0
2	1	12.4	49	32.6	48	2	130	72	1	0
3	1	9.6	49	22.0	35	2	108	58	1	1
4	1	7.2	47	37.9	45	0	128	76	2	1
5	1	14.1	43	42.2	42	2	142	80	1	0
6	1	14.1	47	33.1	44	0	156	94	1	0
7	1	12.4	50	36.5	48	0	140	86	2	1
8	1	14.2	36	38.5	33	2	144	88	1	0
9	1	12.4	50	41.5	47	1	134	78	1	1
10	1	14.5	49	34.1	45	0	102	68	1	0
11	1	12.4	50	39.5	48	2	142	84	1	0
12	1	10.8	54	42.9	43	0	128	74	1	0
13	0	10.9	42	29.8	36	2	156	86	1	0
14	1	10.3	44	33.2	43	2	102	58	1	0
15	0	13.6	40	27.5	26	2	146	98	1	0
16	1	11.9	48	25.3	48	0	120	68	2	1
17	1	12.5	50	31.6	44	1	142	76	1	0
18	1	5.9	47	26.3	38	1	144	82	1	0
19	1	12.4	38	32.4	36	2	150	98	2	1
20	1	14.1	35	47.0	33	1	134	78	1	0
21	0	9.8	51	26.5	47	2	130	76	1	0
22	1	7.2	40	43.9	34	0	122	92	1	0
23	1	3.5	54	32.3	52	1	132	80	1	0
24	1	0.0	53	34.5	47	2	150	88	1	1
25	0	12.1	45	18.9	40	1	134	98	3	0
26	1	1.9	41	32.0	31	1	142	90	1	1
27	1	8.6	34	33.9	30	2	124	66	2	0
28	1	14.0	38	23.7	28	0	102	60	1	0

29	1	14.3	43	24.8	43	0	134	80	1	0
30	1	12.4	45	26.6	41	2	118	66	2	1
31	1	12.4	40	39.2	35	2	192	108	1	0
32	1	14.4	44	32.7	36	1	122	78	1	0
33	1	14.2	48	33.5	43	2	122	92	1	0
34	1	14.5	51	31.2	49	2	112	74	1	0
35	1	12.4	36	24.2	30	1	142	90	1	0
36	1	14.3	52	31.6	48	2	152	96	1	0
37	0	13.7	41	30.7	39	2	112	74	1	0
38	1	13.4	49	28.0	35	0	118	84	1	0
39	1	12.5	44	32.0	29	2	152	88	1	0
40	1	14.4	37	32.7	36	2	136	88	1	0
41	1	12.6	51	24.2	42	0	134	90	2	0
42	1	13.8	47	18.7	42	0	130	78	1	0
43	1	14.0	45	25.6	36	2	108	72	2	1
44	1	6.8	38	22.8	27	0	126	66	1	0
45	1	12.4	35	30.1	33	1	132	78	1	1
46	1	12.9	50	27.7	49	2	144	88	1	0
47	1	8.9	53	27.6	49	1	126	68	1	1
48	1	12.4	48	28.1	47	2	128	70	1	0
49	1	14.5	40	31.7	37	2	132	82	2	0
50	1	13.0	43	26.1	42	1	128	80	2	0
51	1	13.4	54	30.8	54	1	142	80	1	0
52	1	10.6	52	36.9	50	1	132	80	1	0
53	1	13.9	69	24.2	63	2	148	78	1	0
54	1	16.9	38	27.5	26	1	170	100	2	0
55	1	3.6	50	27.3	44	0	140	90	1	1
56	1	10.2	64	30.1	58	0	138	76	1	1
57	1	15.7	44	36.1	41	2	112	78	3	0
58	1	12.0	38	43.1	39	0	140	78	1	0
59	0	6.7	62	34.6	58	0	138	78	2	0
60	1	11.6	47	39.0	45	0	130	82	1	1
61	0	2.0	78	28.7	77	0	178	86	2	0
62	1	10.2	49	28.2	43	2	158	80	1	0

Exercise Table 3.4 (*Continued*)

Patient Number	Status[a]	Survival Time (yr)	Age (yr)	BMI	Age at Diagnosis (yr)	Smoking Status[b]	SBP (mm Hg)	DBP (mm Hg)	ECG[c]	CHD[d]
							Variable at Baseline			
63	1	3.6	63	25.1	46	1	168	88	3	1
64	1	15.4	71	26.0	59	0	146	88	1	0
65	1	11.3	51	32.0	49	2	128	76	1	0
66	1	10.3	59	28.1	57	1	132	76	1	1
67	1	5.8	50	26.1	49	1	154	80	1	0
68	0	8.0	66	45.3	49	0	154	92	1	0
69	1	14.6	42	30.0	41	1	122	80	1	0
70	1	11.4	40	35.7	36	2	144	76	2	1
71	1	7.2	67	28.1	61	0	178	96	1	0
72	1	5.5	86	32.9	61	0	162	60	1	0
73	1	11.1	52	37.6	46	1	142	80	1	0
74	1	16.5	42	43.4	37	0	120	76	1	0
75	1	10.9	60	25.4	60	0	124	64	1	0
76	1	2.5	75	49.7	57	1	174	82	2	1
77	0	10.8	81	35.2	81	0	142	88	1	0
78	1	4.7	60	37.3	39	0	160	78	1	0
79	0	5.5	60	26.0	42	0	122	68	3	1
80	1	4.5	63	21.8	60	2	162	98	1	1
81	1	9.0	62	18.2	43	0	132	72	2	1
82	1	6.8	57	34.1	41	2	116	60	3	1
83	0	3.6	71	25.6	54	1	152	84	3	1
84	1	12.1	58	35.1	45	0	144	68	2	1
85	1	8.1	42	32.5	28	1	98	68	3	1
86	1	11.1	45	44.1	40	0	138	76	1	1
87	0	7.0	66	29.7	59	1	138	78	1	0
88	1	1.5	61	29.2	54	0	184	80	2	1
89	1	11.7	48	25.2	30	2	158	98	1	0
90	1	0.3	82	25.3	50	0	176	96	1	1

91	1	13.6	35	25.8	34	1	118	72	1	0
92	1	15.0	57	33.7	57	2	172	98	1	0
93	1	11.2	56	39.5	55	1	182	100	1	1
94	1	3.0	49	32.9	48	0	144	90	2	1
95	1	13.7	50	37.1	50	0	142	80	1	0
96	1	10.2	53	35.3	53	2	154	76	1	0
97	1	12.4	71	29.3	70	0	122	60	1	0
98	1	1.1	55	22.1	33	2	222	102	2	1
99	1	16.3	69	23.6	43	0	150	80	1	1
100	1	6.7	59	26.1	55	2	142	66	1	0
101	0	15.4	47	32.5	45	2	128	82	1	0
102	0	7.6	75	29.8	67	0	122	76	3	1
103	1	3.6	80	24.4	80	1	162	88	2	1
104	1	11.5	57	26.3	54	0	172	82	2	1
105	0	13.5	52	30.8	46	2	132	70	1	0
106	0	10.6	48	29.4	46	0	112	68	1	1
107	1	6.5	57	29.1	47	1	138	92	2	0
108	1	14.3	58	30.1	56	0	128	74	1	1
109	1	11.6	51	31.0	37	2	132	78	1	0
110	0	15.4	33	34.0	33	2	120	78	1	1
111	0	11.0	36	38.1	33	1	122	70	1	0
112	1	11.0	52	37.0	46	0	140	98	3	0
113	1	4.8	64	31.2	57	2	172	88	1	0
114	1	14.8	31	38.8	29	1	136	76	3	1
115	1	1.8	69	22.3	56	0	152	74	1	0
116	1	15.8	59	25.0	58	0	126	80	1	1
117	1	14.1	38	31.3	38	2	104	58	1	0
118	1	4.6	49	59.7	49	1	142	82	1	0
119	0	15.5	49	34.0	41	0	128	76	1	0
120	1	7.2	68	29.4	66	1	122	58	3	0
121	1	14.5	40	43.2	41	1	122	70	1	1
122	1	10.5	36	35.1	32	2	122	68	1	0
123	1	14.3	60	37.0	54	0	122	70	1	0
124	0	2.2	74	27.1	54	1	168	84	2	1

Exercise Table 3.4 (*Continued*)

							Variable at Baseline			
Patient Number	Status[a]	Survival Time (yr)	Age (yr)	BMI	Age at Diagnosis (yr)	Smoking Status[b]	SBP (mm Hg)	DBP (mm Hg)	ECG[c]	CHD[d]
125	1	5.0	61	27.6	51	0	162	82	1	0
126	1	12.4	54	25.2	51	0	116	76	1	0
127	1	1.1	35	25.8	34	2	126	82	1	0
128	1	15.4	46	32.2	42	2	180	98	1	0
129	1	14.3	40	41.6	41	2	132	98	1	0
130	1	15.6	53	39.8	52	0	150	88	1	0
131	0	12.5	66	26.6	54	1	106	70	1	1
132	1	12.3	61	33.3	55	0	154	88	1	0
133	1	14.8	41	27.7	38	1	122	76	1	0
134	1	10.2	64	26.6	51	2	130	68	1	0
135	1	12.3	41	25.0	38	2	120	58	1	0
136	1	10.3	46	54.3	45	1	144	86	1	0
137	1	8.5	80	29.4	79	1	134	60	1	1
138	1	10.2	63	33.1	60	1	148	80	2	1
139	0	10.0	72	27.3	68	1	170	78	3	1
140	1	7.3	41	36.9	33	0	160	92	2	1
141	0	15.3	52	40.2	36	0	154	96	1	0
142	1	14.0	53	32.7	48	2	124	76	2	1
143	1	15.8	61	33.2	57	1	130	70	1	0
144	1	11.4	53	41.4	47	1	156	78	1	0
145	0	5.5	75	35.8	66	0	162	78	1	0
146	1	11.0	40	34.0	38	2	132	76	1	0
147	1	7.3	61	19.9	37	0	120	60	2	1
148	0	10.6	62	30.6	49	0	160	86	2	1
149	1	10.5	49	30.8	47	1	146	86	1	0

[a]Status: 0, dead; 1, alive.
[b]0, No; 1, ex-smoker; 2, current.
[c]1, Normal; 2, Borderline; 3, abnormal. [d]0, No; 1, yes.

64

3.2 In a study undertaken to compare the treatments given to hypernephroma patients and to relate response and survival to surgery, metastasis, and treatment time, data from 58 patients were collected (Exercise Table 3.2). How would you analyze the data to answer these questions?

   a. Do patients who had nephrectomy have a higher response rate?

   b. Is the time of nephrectomy related to response and survival?

   c. Are there significant differences between the treatments?

   d. What are the most important variables related to response and survival?

3.3 Exercise Table 3.3 gives the age, sex, family history of melanoma, remission duration, survival time, stage, and results of six pretreatment skin tests (the larger diameter is given) of 102 Stage 3 and 4 melanoma patients (Lee et al. 1982).

   a. Study the immunocompetence of melanoma patients by investigating skin test results.

   b. Determine if age, sex, or pretreatment skin test results are predictive to remission and survival time.

   c. Find theoretical distributions that describe the survival and remission patterns.

3.4 One hundred and forty-nine diabetic patients were followed for 17 years (a subset of data from Lee et al. 1988). Exercise Table 3.4 gives the survival time from baseline examination, survival status, and several potential prognostic factors at baseline: age, body mass index (BMI), age at diagnosis of diabetes, smoking status, systolic blood pressure (SBP), diastolic blood pressure (DBP), electrocardiogram reading (ECG), and whether the patient had any coronary heart disease (CHD). Identify the important prognostic factors that are associated with survival.

CHAPTER 4

Nonparametric Methods of Estimating Survival Functions

In this chapter we discuss methods of estimating the three survival (survivorship, density, and hazard) functions for censored data. Unfortunately, the simple method of Example 2.1 cannot be applied if some of the patients are alive at the time of analysis and therefore their exact survival times unknown. Nonparametric or distribution-free methods are quite easy to understand and apply. They are less efficient than parametric methods when survival times follow a theoretical distribution and more efficient when no suitable theoretical distributions are known. Therefore, we suggest using nonparametric methods to analyze survival data before attempting to fit a theoretical distribution. If your main objective is to find a model for the data, estimates obtained by nonparametric methods and graphs can be helpful in choosing a distribution.

Of the three survival functions, survivorship or its graphical presentation, the survival curve, is the most widely used. Section 4.1 introduces the product-limit (PL) method of estimating the survivorship function developed by Kaplan and Meier (1958). With the increased availability of computers, this method is applicable to small, moderate, and large samples. However, if the data have already been grouped into intervals or the sample size is very large, say in the thousands, or the interest is in a large population, it may be more convenient to perform a life-table analysis. Section 4.2 is devoted to the discussion of population and clinical life tables. The PL estimates and life-table estimates of the survivorship function are essentially the same. Many authors use the term *life-table estimates* for the PL estimates. The only difference is that the PL estimate is based on individual survival times while in the life-table method survival times are grouped into intervals. The PL estimate can be considered as a special case of the life-table estimate where each interval contains only one observation.

Section 4.3 discusses three other measures that describe the survival experience: the relative survival rate, the five-year survival rate, and the corrected survival rate. Section 4.4 describes two methods, direct and indirect standardization, to adjust rates to eliminate the effect of differences in population composition with respect to age and other variables. In addition, it introduces the standardized mortality rate and standardized incidence rate.

4.1 PRODUCT-LIMIT ESTIMATES OF SURVIVORSHIP FUNCTION

Let us first consider the simple case where all of the patients are observed to death so that the survival times are exact and known. Let $t_1, t_2, \ldots, t_n$ be the exact survival times of the n individuals under study. Conceptually, we consider this group of patients as a random sample from a much larger population of similar patients. We relabel the n survival times $t_1, t_2, \ldots, t_n$ in ascending order such that

$$t_{(1)} \le t_{(2)} \le \cdots \le t_{(n)}$$

Following (2.2) and (2.3), the survivorship function at $t_{(i)}$ can be estimated as

$$\hat{S}(t_{(i)}) = \frac{n-i}{n} = 1 - \frac{i}{n} \tag{4.1}$$

where $n - i$ is the number of individuals in the sample surviving longer than $t_{(i)}$. If two or more $t_{(i)}$ are equal (tied observations), the largest i value is used. For example, if $t_{(2)} = t_{(3)} = t_{(4)}$, then

$$\hat{S}(t_{(2)}) = \hat{S}(t_{(3)}) = \hat{S}(t_{(4)}) = \frac{n-4}{n}$$

This gives a conservative estimate for the tied observations.

Since every individual is alive at the beginning of the study and no one survives longer than $t_{(n)}$,

$$\hat{S}(t_{(0)}) = 1 \quad \text{and} \quad \hat{S}(t_{(n)}) = 0 \tag{4.2}$$

In practice, $\hat{S}(t)$ is computed at every *distinct* survival time. We doe not have to worry about the intervals between the distinct survival times in which no one dies and $\hat{S}(t)$ remains constant. Equations (4.1) and (4.2) show that $\hat{S}(t)$ is a step function starting at 1.0 and decreasing in steps of $1/n$ (if there are no ties) to zero. When $\hat{S}(t)$ is plotted versus t, the various percentiles of survival time can be read from the graph or calculated from $\hat{S}(t)$. The following example illustrates the method.

Example 4.1
Consider a clinical trial in which 10 lung cancer patients are followed to death. Table 4.1 lists the survival times t in months. The function $\hat{S}(t)$ is

Table 4.1   Computation of $\hat{S}(t)$ for 10 Lung Cancer Patients

t	i	$\hat{S}(t)$
4	1	$\frac{9}{10} = 0.9$
5	2	$\frac{8}{10} = 0.8$
6	3	$\frac{7}{10} = 0.7$
8	4	$\frac{4}{10} = 0.4$
8	5	$\frac{4}{10} = 0.4$
8	6	$\frac{4}{10} = 0.4$
10	7	$\frac{2}{10} = 0.2$
10	8	$\frac{2}{10} = 0.2$
11	9	$\frac{1}{10} = 0.1$
12	10	$\frac{0}{10} = 0.0$

computed following (4.1) and plotted as a step function in Figure 4.1(*a*) and as a smooth curve in Figure 4.1(*b*). The estimated median survival time is 8 months from Figure 4.1(*a*) or 7.6 from Figure 4.1(*b*). A more accurate estimate can be obtained using linear interpolation:

t	$\hat{S}(t)$
6	0.7
m	0.5
8	0.4

$$\frac{8-6}{0.4-0.7} = \frac{8-m}{0.4-0.5}$$

$$m = 8 - \frac{2(0.1)}{0.3} = 7.3 \text{ (months)}$$

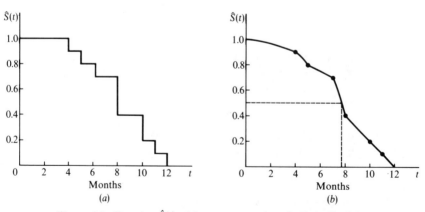

Figure 4.1   Function $\hat{S}(t)$ of lung cancer patients in Example 4.1.

Theoretically, $\hat{S}(t)$ should be plotted as a step function since it remains constant between two observed exact survival times. However, when the median survival time must be estimated from a survival curve, the smooth curve [such as Figure 4.1(b)] may give a much better estimate than the step function, as indicated in the example.

This method can only be applied if all the patients are followed to death. If some of the patients are still alive at the end of the study, a different method of estimating $S(t)$, such as the PL estimate given by Kaplan and Meier (1958), is required. The rationale can be illustrated by the following simple example.

Suppose that 10 patients join a clinical study at the beginning of 1988; during that year 6 patients die and 4 survive. At the end of the year, 20 additional patients join the study. In 1989, 3 patients who entered in the beginning of 1988 and 15 patients who entered later die, leaving one and five survivors, respectively. Suppose that the study terminates at the end of 1989 and you want to estimate the proportion of patients in the population surviving for two years or more, that is, $S(2)$.

The first group of patients in this example is followed for two years while the second group is followed only for one year. One possible estimate, the *reduced-sample estimate*, is $\hat{S}(2) = \frac{1}{10} = 0.1$, which ignores the 20 patients who are followed only for one year. Kaplan and Meier believe that the second sample, under observation for only one year, can contribute to the estimate of $S(2)$.

Patients who survived two years may be considered as surviving the first year and then surviving one more year. Thus, the probability of surviving two years or more is equal to the probability of surviving the first year and then surviving one more year. That is,

$$S(2) = P(\text{surviving first year and then surviving one more year})$$

which can be written as

$$S(2) = P(\text{surviving two years given patient has survived first year})$$

$$\times P(\text{surviving first year}) \tag{4.3}$$

The Kaplan–Meier estimate of $S(2)$ following (4.3) is

$$\hat{S}(2) = (\text{proportion of patients surviving two years given they survive for one year})$$

$$\times (\text{proportion of patients surviving one year}) \tag{4.4}$$

For the data given above, one of the four patients who survived the first year survived two years, so the first proportion in (4.4) is $\frac{1}{4}$. Four of the 10 patients who entered at the beginning of 1988 and 5 of the 20 patients who entered at the end of 1988 survived one year. Therefore, the second proportion in (4.4) is $(4+5)/(10+20)$. The PL estimate of $S(2)$ is

$$\hat{S}(2) = \tfrac{1}{4} \times \frac{4+5}{10+20} = 0.25 \times 0.3 = 0.075$$

This simple rule may be generalized as follows: The probability of surviving $k(\geq 2)$ or more years from the beginning of the study is a product of k observed survival rates

$$\hat{S}(k) = p_1 \times p_2 \times p_3 \times \cdots \times p_k \tag{4.5}$$

where p_1 denotes the proportion of patients surviving at least one year,

$\quad\quad p_2$ denotes the proportion of patients surviving the second year after they have survived one year,

$\quad\quad p_3$ denotes the proportion of patients surviving the third year after they have survived two years, and

$\quad\quad p_k$ denotes the proportion of patients surviving the kth year after they have survived $k-1$ years.

Therefore, the PL estimate of the probability of surviving any particular number of years from the beginning of the study is the product of the same estimate up to the previous year, and the observed survival rate for the particular year, that is,

$$\hat{S}(t) = \hat{S}(t-1)p_t \tag{4.6}$$

The PL estimates are maximum likelihood estimates.

In practice the PL estimates can be calculated by constructing a table with five columns following the outline below.

1. Column 1 contains all the survival times, both censored and un-censored in the order from smallest to largest. Affix a plus sign to the censored observation. If a censored observation has the same value as an uncensored, the latter should appear first.
2. The second column, labeled i consists of the corresponding rank of each observation in column 1.
3. The third column, labeled r, pertains to uncensored observations only. Let $r = i$.
4. Compute $(n-r)/(n-r+1)$ or p_i for every uncensored observation $t_{(i)}$ in column 4 to give the proportion of patients surviving up to and then through $t_{(i)}$.
5. In column 5, $\hat{S}(t)$ is the product of all values of $(n-r)/(n-r+1)$ up to and including t. If some uncensored observations are ties, the smallest $\hat{S}(t)$ should be used.

To summarize this procedure, let n be the total number of individuals whose survival times, censored or not, are available. Relabel the n survival times in order of increasing magnitude such that $t_{(1)} \leq t_{(2)} \leq \cdots \leq t_{(n)}$. Then

$$\hat{S}(t) = \prod_{t_{(r)} \leq t} \frac{n-r}{n-r+1} \qquad (4.7)$$

where r runs through those positive integers for which $t_{(r)} \leq t$ and $t_{(r)}$ is uncensored. The values of r are consecutive integers $1, 2, \ldots, n$ if there are no censored observations. If there are censored observations, they are not.

The estimated median survival time is the 50th percentile, which is the value of t at $\hat{S}(t) = 0.50$. The following example illustrates the calculation procedures.

Example 4.2

Suppose the following remission durations are observed from 10 patients ($n = 10$) with solid tumors. Six patients relapse at 3.0, 6.5, 6.5, 10, 12, and 15 months; 1 patient is lost to follow-up at 8.4 months; and 3 patients are still in remission at the end of the study after 4.0, 5.7, and 10 months. The calculation of $\hat{S}(t)$ is shown in Table 4.2.

The survivorship function $\hat{S}(t)$ is plotted in Figure 4.2; the estimated median remission time is $m = 9.8$ months.

From the calculation we notice that $\hat{S}(t)$ at $t = t_{(i)}$ is related to $\hat{S}(t)$ at $t = t_{(i-1)}$ and (4.6) can be rewritten as

$$\hat{S}(t_{(i-1)}) = \hat{S}(t_{(i-1)}) \frac{n-i}{n-i+1} \qquad (4.8)$$

where $t_{(i)}$ and $t_{(i-1)}$ are uncensored observations. For example, $\hat{S}(12) = \hat{S}(10) \times \frac{1}{2} = 0.482 \times \frac{1}{2} = 0.241$.

Table 4.2 Calculation of the PL Estimate of $\hat{S}(t)$ for Data in Example 4.2

Remission Time t	Rank i	r	$(n-r)/(n-r+1)$	$\hat{S}(t)$
3.0	1	1	$\frac{9}{10}$	$\frac{9}{10} = 0.900$
4.0+	2	—	—	—
5.7+	3	—	—	—
6.5	4	4	$\frac{6}{7}$	$\frac{9}{10} \times \frac{6}{7} = 0.771^a$
6.5	5	5	$\frac{5}{6}$	$\frac{9}{10} \times \frac{6}{7} \times \frac{5}{6} = 0.643^a$
8.4+	6	—	—	—
10.0	7	7	$\frac{3}{4}$	$\frac{9}{10} \times \frac{6}{7} \times \frac{5}{6} \times \frac{3}{4} = 0.482$
10.0+	8	—	—	—
12.0	9	9	$\frac{1}{2}$	$\frac{9}{10} \times \frac{6}{7} \times \frac{5}{6} \times \frac{3}{4} \times \frac{1}{2} = 0.241$
15.0	10	10	0	0

^{a}0.643 is used as $\hat{S}(6.5)$. It is a conservative estimate.

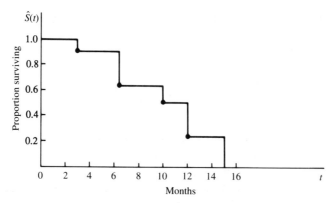

Figure 4.2   Function $\hat{S}(t)$ of Example 4.2.

If there are no censored observations or losses before t, (4.7) is equivalent to (4.1).

The variance of the PL estimate of $\hat{S}(t)$ is approximated by

$$\operatorname{Var}[\hat{S}(t)] \cong [\hat{S}(t)]^2 \sum_r \frac{1}{(n-r)(n-r+1)} \qquad (4.9)$$

where r includes those positive integers for which $t_{(r)} \leq t$ and $t_{(r)}$ corresponds to a death. For the data in Example 4.2, for example,

$$\operatorname{Var}[\hat{S}(10)] = (0.482)^2 \left(\frac{1}{9 \times 10} + \frac{1}{6 \times 7} + \frac{1}{5 \times 6} + \frac{1}{3 \times 4} \right)$$
$$= 0.0352$$

and the estimated standard error is 0.1876. In Example 4.1,

$$\operatorname{Var}[\hat{S}(6)] = (0.7)^2 \left(\frac{1}{9 \times 10} + \frac{1}{8 \times 9} + \frac{1}{7 \times 8} \right) = 0.0210$$

and the estimated standard error is 0.145. The variance may be used to obtain confidence intervals for $S(t)$.

Example **4.3**
Consider the tumor-free time in days of the 30 rats on a low-fat diet in Table 3.4. Table 4.3 gives the calculations of the PL estimates of $S(t)$ and the standard error of $\hat{S}(t)$.

Estimated $S(t)$ is plotted in Figure 3.4. The median tumor-free time is approximately 189 days.

The mean survival time μ can be shown to equal the area under the survival curve. To estimate μ, we can use

$$\hat{\mu} = \int_0^\infty \hat{S}(t) \, dt$$

that is, $\hat{\mu}$ is equal to the area under the estimated survivorship function. Thus, if the times to death are ordered as $t^{(1)} \le t^{(2)} \le \cdots \le t^{(m)}$ (if there are m uncensored observations) and $t^{(m)}$ is the largest observation of all n observations (i.e., $t^{(m)} = t_{(n)}$ when $t_{(n)}$ is an uncensored observation), then μ can be estimated as

$$\hat{\mu} = 1.000 t^{(1)} + \hat{S}(t^{(1)})(t^{(2)} - t^{(1)}) + \hat{S}(t^{(2)})(t^{(3)} - t^{(2)}) + \cdots$$
$$+ \hat{S}(t^{(m-1)})(t^{(m)} - t^{(m-1)}) \tag{4.10}$$

which is the sum of the areas of the rectangles under the survival curve formed by the uncensored observations. Consider the data in Example 4.2: $m = 6$, $t^{(1)} = 3.0$, $t^{(2)} = 6.5$, $t^{(3)} = 6.5$, $t^{(4)} = 10$, $t^{(5)} = 12$, and $t^{(6)} = 15$. The mean survival time is estimated using (4.10) as

$$\hat{\mu} = 1.000 \times 3.0 + 0.900(6.5 - 3.0) + 0.643(10 - 6.5) + 0.482(12 - 10)$$
$$+ 0.241(15 - 12)$$
$$= 3.000 + 3.150 + 2.251 + 0.964 + 0.723$$
$$= 10.088 \text{ months}$$

However, if the largest observation in the data is censored and is used as $t^{(m)}$ in (4.10), μ so obtained may be a low estimate. In such cases Irwin (1949) suggests that instead of estimating the mean survival time, one should choose a time limit L and estimate the "mean survival time limited to a time L," say $\mu_{[L]}$, by using L for $t^{(m)}$ in (4.10). For example, if in Example 4.2 the largest observation is censored, that is, $15+$, and if we let $L = 16$, then

$$\hat{\mu}_{[16]} = 3.000 + 3.150 + 2.251 + 0.964 + 0.241(16 - 12)$$
$$= 10.329 \text{ months}$$

which is the mean survival time limited to 16 months.

The variance of $\hat{\mu}$ is estimated by

$$\text{Var}(\hat{\mu}) = \sum_r \frac{A_r^2}{(n-r)(n-r+1)} \tag{4.11}$$

where r runs through those integers for which t_r corresponds to a death and A_r is the area under the curve $\hat{S}(t)$ to the right of $t_{(r)}$. The kth A_r in terms of the m uncensored observations is

$$\hat{S}(t^{(k)})(t^{(k+1)} - t^{(k)}) + \hat{S}(t^{(k+1)})(t^{(k+2)} - t^{(k+1)}) + \cdots + \hat{S}(t^{(m-1)})(t^{(m)} - t^{(m-1)})$$
$$\tag{4.12}$$

Table 4.3 Calculation of $\hat{S}(t)$ and Standard Error of $\hat{S}(t)$ for 30 Rats on a Low-Fat Diet in Table 3.4

Rat Number	Tumor-free Time t	i	r	$\dfrac{n-r}{n-r+1}$	$\hat{S}(t)$	Standard Error of $\hat{S}(t)$
3	50	1	1	$\frac{29}{30}$	$\frac{29}{30}=0.967$	$\left[(0.967)^2\left(\dfrac{1}{29\times30}\right)\right]^{1/2}=0.033$
12	56	2	2	$\frac{28}{29}$	$0.967\times\frac{28}{29}=0.933$	$\left[(0.933)^2\left(\dfrac{1}{29\times30}+\dfrac{1}{28\times29}\right)\right]^{1/2}=0.046$
4	65	3	3	$\frac{27}{28}$	$0.933\times\frac{27}{28}=0.900$	$\left[(0.900)^2\left(\dfrac{1}{29\times30}+\dfrac{1}{28\times29}+\dfrac{1}{27\times28}\right)\right]^{1/2}=0.055$
13	66	4	4	$\frac{26}{27}$	$0.900\times\frac{26}{27}=0.867$	0.062
14	73	5	5	$\frac{25}{26}$	0.833	0.068
9	77	6	6	$\frac{24}{25}$	0.800	0.073
10	84	7	7	$\frac{23}{24}$	0.767	0.077
5	86	8	8	$\frac{22}{23}$	0.733	0.081
11	87	9	9	$\frac{21}{22}$	0.700	0.084
15	119	10	10	$\frac{20}{21}$	0.667	0.086
1	140	11	11	$\frac{19}{20}$	0.633	0.088

16	140+	12	—	—	—	0.090
6	153	13	13	$\frac{17}{18}$	$0.633 \times \frac{17}{18} = 0.598$	0.091
2	177	14	14	$\frac{16}{17}$	$0.598 \times \frac{16}{17} = 0.563$	0.092
7	181	15	15	$\frac{15}{16}$	0.528	0.092
8	191	16	16	$\frac{14}{15}$	0.493	—
17	200+	17	—	—	—	—
18	200+	18	—	—	—	—
19	200+	19	—	—	—	—
20	200+	20	—	—	—	—
21	200+	21	—	—	—	—
22	200+	22	—	—	—	—
23	200+	23	—	—	—	—
24	200+	24	—	—	—	—
25	200+	25	—	—	—	—
26	200+	26	—	—	—	—
27	200+	27	—	—	—	—
28	200+	28	—	—	—	—
29	200+	29	—	—	—	—
30	200+	30	—	—	—	—

If there are no censored observations, (4.10) reduces to the sample mean $\bar{t} = \Sigma\, t_i/n$, and (4.11) reduces to

$$\mathrm{Var}(\hat{\mu}) = \mathrm{Var}(\bar{t}) = \frac{\Sigma\,(t_i - \bar{t})^2}{n^2} \tag{4.13}$$

which is not an unbiased estimate. Kaplan and Meier suggest that (4.11) and (4.13) be multiplied by $m/(m-1)$ and $n/(n-1)$, respectively, to correct the bias.

Consider the survival times in Example 4.1: the sample mean is $\bar{t} = \hat{\mu} = 8.2$ months and the estimated variance of $\hat{\mu}$, by (4.13), is 0.616. If the factor $n/(n-1) = \frac{10}{9}$ is multiplied, the estimated variance of μ becomes 0.684.

To compute the variance of $\hat{\mu}$ in Example 4.2, we first compute the five A_r's: A_1, A_4, A_5, A_7, and A_9. The first A_r is

$$A_1 = \hat{S}(t^{(1)})(t^{(2)} - t^{(1)}) + \hat{S}(t^{(2)})(t^{(3)} - t^{(2)}) + \cdots + \hat{S}(t^{(5)})(t^{(6)} - t^{(5)})$$
$$= 3.150 + 2.251 + 0.964 + 0.723 = 7.088$$

The second A_r is

$$A_4 = \hat{S}(t^{(2)})(t^{(3)} - t^{(2)}) + \cdots + \hat{S}(t^{(5)})(t^{(6)} - t^{(5)})$$
$$= 2.251 + 0.964 + 0.723 = 3.938$$

The third, fourth, and fifth A_r's are, respectively,

$$A_5 = 2.251 + 0.964 + 0.723 = 3.938$$
$$A_7 = 0.964 + 0.723 = 1.687$$
$$A_9 = 0.723$$

Thus

$$\widehat{\mathrm{Var}}(\hat{\mu}) = \frac{(7.088)^2}{9 \times 10} + \frac{(3.938)^2}{6 \times 7} + \frac{(3.938)^2}{5 \times 6} + \frac{(1.687)^2}{3 \times 4} + \frac{(0.723)^2}{1 \times 2} = 1.942$$

The estimated standard error of $\hat{\mu}$ is 1.394. If the factor $m/(m-1) = \frac{6}{5}$ is included, these results become 2.330 and 1.526, respectively.

The Kaplan–Meier method provides very useful estimates of survival probabilities and graphical presentation of survival distribution. It is the most widely used method in survival data analysis. Breslow and Crowley (1974) and Meier (1975b) have shown that under certain conditions, the estimate is consistent and asymptomatically normal. However, a few critical features should be mentioned.

1. The Kaplan–Meier estimates are limited to the time interval in which the observations fall. If the largest observation is uncensored, the PL estimate at that time equals zero. Although the estimate may not be welcomed by physicians, it is correct since no one in the sample lives longer. If the largest observation is censored, the PL estimate can never equal zero and is undefined beyond the largest observation.

2. The most commonly used summary statistic in survival analysis is the median survival time. A simple estimate of the median can be read from survival curves estimated by the PL method as the time t at which $\hat{S}(t) = 0.5$. However, the solution may not be unique. Consider Figure 4.3(a) where the survival curve is horizontal at $\hat{S}(t) = 0.5$; any t value in the interval t_1 to t_2 is a reasonable estimate of the median. A practical solution is to take the midpoint of the interval as the PL estimate of the median. Figure 4.3(b) presents a different case in which the straightforward estimate (t_1) tends to overestimate the median. A practical way to handle this problem is to connect the points and then locate the median.

3. If less than 50% of the observations are uncensored and the largest observation is censored, the median survival time cannot be estimated. A practical way to handle the situation is to use probability of surviving a given length of time, say one, three, or five years, or the mean survival time limited to a given time t.

4. The PL method assumes that censoring is independent of the survival times. In other words, the reason an observation is censored is unrelated to the cause of death. This assumption is true if the patient is still alive at the end of the study period. However, the assumption is violated if the patient develops severe adverse effects from the treatment and is forced to leave the study before death or if the patient died of a cause other than the one under study (e.g., death due to automobile accidents in a cancer survival study). When there is inappropriate censoring, the PL method is not appropriate. In prac-

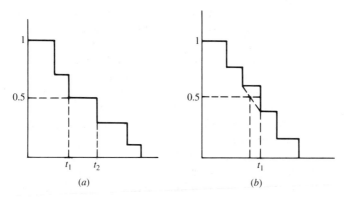

Figure 4.3 Kaplan–Meier estimate of median survival time.

tice, one way to alleviate the problem is to avoid it or to reduce it to a minimum.

5. Similar to other estimators, the standard error (SE) of the Kaplan–Meier estimator of $S(t)$ gives an indication of the potential error of $\hat{S}(t)$. The confidence interval deserves more attention than the point estimate $\hat{S}(t)$. A 95% confidence interval for $S(t)$ is $\hat{S}(t) \pm 1.96$ SE $[\hat{S}(t)]$.

4.2 LIFE-TABLE ANALYSIS

The life-table method is one of the oldest techniques for measuring mortality and describing the survival experience of a population. It has been used by actuaries, demographers, governmental agencies, and medical researchers in studies of survival, population growth, fertility, migration, length of married life, length of working life, and so on. There has been a decennial series of life tables on the entire U.S. population since 1900. States and local governments also publish life tables. These life tables, summarizing the mortality experience of a specific population for a specific period of time, are called *population life tables*. As clinical and epidemiologic research becomes more common, the life-table method has been applied to patients with a given disease who have been followed for a period of time. Life tables constructed for patients are called *clinical life tables*. Although population and clinical life tables are similar in calculation, the sources of required data are different.

4.2.1 Population Life Tables

There are two kinds of population life tables: the *cohort life table* and *current life table*. The cohort life table describes the survival or mortality experience from birth to death of a specific cohort of persons who were born at about the same time, for example, all persons born in 1900. The cohort has to be followed from 1900 until all of them die. The proportion of death (survivor) is then used to construct life tables for successive calendar years. This type of table, useful in population projection and prospective studies, is not often constructed since it requires a long follow-up period.

The current life table is constructed by applying the age-specific mortality rates of a population in a given period of time to a hypothetical cohort of 100,000 or 1,000,000 persons. The starting point is birth at year 0. Two sources of data are required for constructing a population life table: (1) census data on the number of living persons at each age for a given year at midyear and (2) vital statistics on the number of deaths in the given year for each age. For example, a current U.S. life table assumes a hypothetical cohort of 100,000 persons that is subject to the age-specific death rates based on the observed data for the United States in the 1980 census. The current life table, based on the life experience of an actual population over a short period of time, gives a good summary of current mortality. This type of life table is regularly published by government agencies of different levels. One of the most often reported statistics from current life tables in the life expectancy. The term *population life table* is often used to refer to the current life table.

In the United States, the National Center for Health Statistics publishes detailed decennial life tables after each decennial census. These complete life tables use one-year age groups. Between censuses, annual life tables are also published. The annual life tables are often seen in five-year age intervals and are called abridged life tables. Tables 4.4 and 4.5 are, respectively, a complete decennial life table for the total U.S. Population for 1979–1981 and an abridged life table for the same population for 1985. The abridged table in Table 4.5 was constructed based on a complete life table.

Current life tables usually have the following columns:

1. *Age Interval* (x *to* $x + t$). This is the time interval between two exact ages x and $x + t$; t is the length of the interval. For example, the interval 20–21 includes the time interval from the 20th birthday up to the 21st birthday (but not including the 21st birthday).

2. *Proportion of Persons Alive at Beginning of Age Interval but Dying during Interval* ($_tq_x$). The information is obtained from census data. For example, $_tq_x$ for age interval 20–21 is the proportion of persons who died on or after their 20th birthday and before their 21st birthday. It is an estimate of the conditional probability of dying in the interval given the person is alive at age x. This column is usually calculated from data of the decennial census of population and deaths occurring in the given time interval. For example, the mortality rates in Table 4.4 are calculated from the data of the 1980 Census of Population and deaths occurring in the United States in the three years 1979–1981. This column is the foundation of the life table from which all of the other columns are derived.

3. *Number Living at Beginning of Age Interval* (l_x). The initial value of l_x, the size of the hypothetical population, is usually 100,000 or 1,000,000. The successive values are computed using the formula

$$l_x = l_{x-t}(1 - _tq_{x-t})$$ (4.14)

where $1 - _tq_{x-t}$ is the proportion of persons who survived the previous age interval. For example, in Table 4.4, $t = 1$, $l_{20} = l_{19}(1 - _1q_{19}) = 97,851(1 - 0.00112) = 97,741$, which is the number of persons living at the beginning of age 20.

4. *Number Dying during Age Interval* ($_td_x$)

$$_td_x = l_x(_tq_x) = l_x - l_{x+t}$$ (4.15)

Table 4.4 Life Table for Total Population: United States, 1979–1981

Age Interval, Period of Life Between Two ages, x to $x + t$ (1)	Proportion of Persons Alive at Beginning of Age Interval Dying during Interval, $_tq_x$ (2)	Of 100,000 Born Alive		Stationary Population		Average Number of Years of Life Remaining at Beginning of Age Interval $\overset{0}{e}_x$ (7)
		Number Living at Beginning of Age Interval, l_x (3)	Number Dying During Age Interval, $_td_x$ (4)	In Age interval, L_x (5)	In This and All Subsequent Age Intervals, T_x (6)	
Days						
0–1	0.00463	100,000	463	273	7,387,758	73.88
1–7	0.00246	99,537	245	1,635	7,387,485	74.22
7–28	0.00139	99,292	138	5,708	7,385,850	74.38
28–365	0.00418	99,154	414	91,357	7,380,142	74.43
Years						
0–1	0.01260	100,000	1,260	98,973	7,387,758	73.88
1–2	0.00093	98,740	92	98,694	7,288,785	73.82
2–3	0.00065	98,648	64	98,617	7,190,091	72.89
3–4	0.00050	98,584	49	98,560	7,091,474	71.93
4–5	0.00040	98,535	40	98,515	6,992,914	70.97
5–6	0.00037	98,495	36	98,477	6,894,399	70.00
6–7	0.00033	98,459	33	98,442	6,795,922	69.02
7–8	0.00030	98,426	30	98,412	6,697,480	68.05
8–9	0.00027	98,396	26	98,383	6,599,068	67.07
9–10	0.00023	98,370	23	98,358	6,500,685	66.08
10–11	0.00020	98,347	19	98,338	6,402,327	65.10
11–12	0.00019	98,328	19	98,319	6,303,989	64.11
12–13	0.00025	98,309	24	98,297	6,205,670	63.12
13–14	0.00037	98,285	37	98,266	6,107,373	62.14
14–15	0.00053	98,248	52	98,222	6,009,107	61.16
15–16	0.00069	98,196	67	98,163	5,910,885	60.19
16–17	0.00083	98,129	82	98,087	5,812,722	59.24
17–18	0.00095	98,047	94	98,000	5,714,635	58.28

18–19	0.00105	97,953	102	97,902	5,616,635	57.34
19–20	0.00112	97,851	110	97,796	5,518,733	56.40
20–21	0.00120	97,741	118	97,682	5,420,937	55.46
21–22	0.00127	97,623	124	97,561	5,323,255	54.53
22–23	0.00132	97,499	129	97,435	5,225,694	53.60
23–24	0.00134	97,370	130	97,306	5,128,259	52.67
24–25	0.00133	97,240	130	97,175	5,030,953	51.74
25–26	0.00132	97,110	128	97,046	4,933,778	50.81
26–27	0.00131	96,982	126	96,919	4,836,732	49.67
27–28	0.00130	96,856	126	96,793	4,739,813	48.94
28–29	0.00130	96,730	126	96,667	4,643,020	48.00
29–30	0.00131	96,604	127	96,541	4,546,353	47.06
30–31	0.00133	96,477	127	96,414	4,449,812	46.12
31–32	0.00134	96,350	130	96,284	4,353,398	45.18
32–33	0.00137	96,220	132	96,155	4,257,114	44.24
33–34	0.00142	96,088	137	96,019	4,160,959	43.30
34–35	0.00150	95,951	143	95,880	4,064,940	42.36
35–36	0.00159	95,808	153	95,731	3,969,060	41.43
36–37	0.00170	95,655	163	95,574	3,873,329	40.49
37–38	0.00183	95,492	175	95,404	3,777,755	39.56
38–39	0.00197	95,317	188	95,224	3,682,351	38.63
39–40	0.00213	95,129	203	95,027	3,587,127	37.71
40–41	0.00232	94,926	220	94,817	3,492,100	36.79
41–42	0.00254	94,706	241	94,585	3,397,283	35.87
42–43	0.00274	94,465	264	94,334	3,302,698	34.96
43–44	0.00306	94,201	288	94,057	3,208,364	34.06
44–45	0.00335	93,913	314	93,756	3,114,307	33.16
45–46	0.00356	93,599	343	93,427	3,020,551	32.27
46–47	0.00401	93,256	374	93,069	2,927,124	31.39
47–48	0.00442	92,882	410	92,677	2,834,055	30.51
48–49	0.00488	92,472	451	92,246	2,741,378	29.65
49–50	0.00538	92,021	495	91,773	2,649,132	28.79
50–51	0.00589	91,526	540	91,256	2,557,359	27.94
51–52	0.00642	90,986	584	90,695	2,466,103	27.10

Table 4.4 *(Continued)*

Age Interval, Period of Life Between Two ages, x to $x + t$ (1)	Proportion of Persons Alive at Beginning of Age Interval Dying during Interval, $_tq_x$ (2)	Of 100,000 Born Alive		Stationary Population		Average Number of Years of Life Remaining at Beginning of Age Interval $\overset{\circ}{e}_x$ (7)
		Number Livint at Beginning of Age Interval, l_x (3)	Number Dying During Age Interval, $_td_x$ (4)	In Age interval, $_tL_x$ (5)	In This and All Subsequent Age Intervals, T_x (6)	
Years						
52–53	0.00699	90,402	631	90,086	2,375,408	26.28
53–54	0.00761	89,771	684	89,430	2,285,322	25.46
54–55	0.00830	89,087	739	88,717	2,195,892	24.65
55–56	0.00902	88,348	797	87,950	2,107,175	23.85
56–57	0.00978	87,551	856	87,122	2,019,225	23.06
57–58	0.01059	86,695	919	86,236	1,932,103	22.29
58–59	0.01151	85,776	987	85,283	1,845,867	21.52
59–60	0.01254	84,789	1,063	84,258	1,760,584	20.76
60–61	0.01368	83,726	1,145	83,153	1,676,326	20.02
61–62	0.01493	82,581	1,233	81,965	1,593,173	19.29
62–63	0.01628	81,348	1,324	80,686	1,511,208	18.58
63–64	0.01767	80,024	1,415	79,316	1,430,522	17.88
64–65	0.01911	78,609	1,502	77,859	1,351,206	17.19
65–66	0.02059	77,107	1,587	76,314	1,273,347	16.51
66–67	0.02216	75,520	1,674	74,683	1,197,033	15.85
67–68	0.02389	73,846	1,764	72,964	1,122,350	15.20
68–69	0.02585	72,082	1,864	71,150	1,049,386	14.56
69–70	0.02806	70,218	1,970	69,233	978,236	13.93
70–71	0.03052	68,248	2,083	67,206	909,003	13.32
71–72	0.03315	66,165	2,193	65,069	841,797	12.72
72–73	0.03593	63,972	2,299	62,823	776,728	12.14
73–74	0.03882	61,673	2,394	60,476	713,905	11.58
74–75	0.04184	59,279	2,480	58,039	653,429	11.02

Age					
75–76	0.04507	56,799	2,560	55,520	595,390	10.48
76–77	0.04867	54,239	2,640	52,919	539,870	9.95
77–78	0.05274	51,599	2,721	50,238	486,951	9.44
78–79	0.05742	48,878	2,807	47,475	436,713	8.93
79–80	0.06277	46,071	2,891	44,626	389,238	8.45
80–81	0.06882	43,180	2,972	41,694	344,612	7.98
81–82	0.07552	40,208	3,036	38,689	302,918	7.53
82–83	0.08278	37,172	3,077	35,634	264,229	7.11
83–84	0.09041	34,095	3,083	32,553	228,595	6.70
84–85	0.09842	31,012	3,052	29,486	196,042	6.32
85–86	0.10725	27,960	2,999	26,461	166,556	5.96
86–87	0.11712	24,961	2,923	23,500	140,095	5.61
87–88	0.12717	22,038	2,803	20,636	116,595	5.29
88–89	0.13708	19,235	2,637	17,917	95,959	4.99
89–90	0.14728	16,598	2,444	15,376	78,042	4.70
90–91	0.15868	14,154	2,246	13,031	62,666	4.43
91–92	0.17169	11,908	2,045	10,886	49,635	4.17
92–93	0.18570	9,863	1,831	8,948	38,749	3.93
93–94	0.20023	8,032	1,608	7,228	29,801	3.71
94–95	0.21495	6,424	1,381	5,733	22,573	3.51
95–96	0.22976	5,043	1,159	4,463	16,840	3.34
96–97	0.24338	3,884	945	3,412	12,377	3.15
97–98	0.25637	2,939	754	2,562	8,965	3.05
98–99	0.26868	2,185	587	1,892	6,403	2.93
99–100	0.28030	1,598	448	1,374	4,511	2.82
100–101	0.29120	1,150	335	983	3,137	2.73
101–102	0.30139	815	245	692	2,154	2.64
102–103	0.31089	570	177	481	1,462	2.57
103–104	0.31970	393	126	330	981	2.50
104–105	0.32786	267	88	223	651	2.44
105–106	0.33539	179	60	150	428	2.38
106–107	0.34233	119	41	99	378	2.33
107–108	0.34870	78	27	64	179	2.29
108–109	0.35453	51	18	42	115	2.25
109–110	0.35988	33	12	27	73	2.20

Source: U.S. Decennial Life Tables for 1979–81, Vol. 1, No. 1, U.S. Life Tables, DHHS Publication No. (PHS)85-1150-1, National Center for Health Statistics, 1985.

Table 4.5 Abridged Life Table for Total U.S. Population, 1985

Age Interval, Period of Life Between Two Exact Ages Stated in Years, Race, and Sex, x to $x + t$ (1)	Proportion of Persons Alive at Beginning of Age Interval Dying during Interval, $_tq_x$ (2)	Of 100,000 Born Alive		Stationary Population		Average Number of Years of Life Remaining at Beginning of Age Interval $\overset{\circ}{e}_x$ (7)
		Number Living at Beginning of Age Interval, l_x (3)	Number Dying During Age Interval, d_x (4)	In Age interval, $_tL_x$ (5)	In This and All Subsequent Age Intervals, T_x (6)	
All Races						
0–1	0.0107	100,000	1,069	99,079	7,472,607	74.7
1–5	0.0020	98,931	200	395,255	7,373,528	74.5
5–10	0.0012	98,731	123	493,320	6,978,273	70.7
10–15	0.0014	98,608	135	492,778	6,484,953	65.8
15–20	0.0040	98,473	397	491,465	5,992,175	60.9
20–25	0.0054	98,076	534	489,072	5,500,710	56.1
25–30	0.0057	97,542	553	486,326	5,011,638	51.4
30–35	0.0067	96,989	648	483,363	4,525,312	46.7
35–40	0.0086	96,341	831	479,745	4,041,949	42.0
40–45	0.0126	95,510	1,199	474,766	3,562,204	37.3
45–50	0.0197	94,311	1,854	467,259	3,087,438	32.7
50–55	0.0316	92,457	2,924	455,434	2,620,179	28.3
55–60	0.0497	89,533	4,447	437,186	2,164,745	24.2
60–65	0.0753	85,086	6,408	410,278	1,727,559	20.3
65–70	0.1092	78,678	8,588	372,847	1,317,281	16.7
70–75	0.1625	70,090	11,391	322,963	944,434	13.5
75–80	0.2349	58,699	13,788	259,866	621,471	10.6
80–85	0.3480	44,911	15,628	185,697	361,605	8.1
85 and over	1.0000	29,283	29,283	175,908	175,908	6.0

Source: U.S. Life Tables, DHHS Publication No. (PHS)85-1150-1, National Center for Health Statistics, 1985.

For example, the number of persons dying during age interval 20–21, $_1d_{20} = 97,741 \ (0.00120) = 118$ (or $_1d_{20} = 97,741 - 97,623 = 118$).

5. *Stationary Population* ($_tL_x$ *and* T_x). Here $_tL_x$ is the total number of years lived in the ith age interval or the number of person-years that l_x persons, aged x exactly, live through the interval. For those who survive the interval, their contribution to $_tL_x$ is the length of the interval t. For those who die during the interval, we may not know exactly the time of death and the survival time must be estimated. The conventional assumption is that they live one-half of the interval and contribute $\frac{1}{2}t$ to the calculation of $_tL_x$. Thus

$$_tL_x = t(l_{x+t} + \tfrac{1}{2}_td_x) \qquad (4.16)$$

For example, in Table 4.4, $_1L_{20} = 97,623 + \frac{118}{2} = 97,682$. If we do know the exact survival time of those who die in the interval, $_tL_x$ should be computed accordingly.

The symbol T_x is the total number of person-years lived beyond age t by persons alive at that age, that is,

$$T_x = \sum_{j \geq x} {}_tL_j \qquad (4.17)$$

and

$$T_x = {}_tL_x + T_{x+t} \qquad (4.18)$$

For example, in Table 4.4, $T_0 = 7,387,758$ which is the sum of all $_tL_x$ values in column 5 and $T_1 = 7,288,785$, which is $T_0 - {}_1L_0 = 7,387,758 - 98,973$.

6. *Average Remaining Lifetime or Average Number of Years of Life Remaining at Beginning of Age Interval* ($\overset{0}{e}_i$). This is also known as the life expectancy at a given age, defined as the number of years *remaining* to be lived by persons at age x:

$$\overset{0}{e}_x = \frac{T_x}{l_x} \qquad (4.19)$$

The expected age at death of a person aged x is $x + \overset{0}{e}_x$. The $\overset{0}{e}_x$ at $x = 0$ is the life expectancy at birth. For example, according to the U.S. life table for 1979–1981 the life expectancy at birth is 73.88 years and that at age 40 is 36.79 years. This means that according to the mortality rates of 1979–1981, newborns are expected to live 73.88 years and those at age 40 are expected to live another 36.79 years. The life expectancy of a population is a general indication of the capability of prolonging life. It is used to identify trends and to compare longevity. Table 4.5 shows that according to the mortality rates of 1985, the newborns and those at age 40 are expected to live 74.7 and 37.3 years, respectively. The overall increase in life expectancy indicates an improvement in longevity in the United States over the time period.

Population life tables can be constructed for various subgroups. For example, there are published life tables by sex, race, cause of death, as well as those that eliminate certain causes of death.

4.2.2 Clinical Life Tables

The actuarial life table method has been applied to clinical data for many decades. Berkson and Gage (1950) and Cutler and Ederer (1958) give a life-table method for estimating the survivorship function; Gehan (1969) provides methods for estimating all three functions (survivorship, density, and hazard).

The life-table method requires a fairly large number of observations so that survival times can be grouped into intervals. Similar to the PL estimate, the life-table method incorporates all survival information accumulated up to the termination of the study. For example, in computing a five-year survival rate of breast cancer patients, one need not restrict oneself only to those patients who have entered the study for five or more years. Patients who have entered for four, three, two, and even one year contribute useful information to the evaluation of five-year survival. In this way, the life-table technique uses incomplete data such as losses to follow-up and individuals withdrawn alive as well as complete death data.

Table 4.6 shows the format of the clinical life table. The columns are described below:

1. *Interval* $[t_i - t_{i+1})$. The first column gives the intervals into which the survival times and times to loss or withdrawal are distributed. The interval is from t_i up to but *not* including t_{i+1}, $i = 1, \ldots, s$. The last interval has an infinite length. These intervals are assumed to be fixed.

2. *Midpoint* (t_{mi}). The midpoint of each interval, designated t_{mi}, $i = 1, \ldots, s - 1$, is included for convenience in plotting the hazard and probability density functions. Both functions are plotted as t_{mi}.

3. *Width* (b_i). The width of each interval, $b_i = t_{i+1} - t_i$, $i = 1, \ldots, s - 1$, is needed for the calculation of the hazard and density functions. The width of the last interval, b_s, is theoretically infinite; no estimate of the hazard or density function can be obtained for this interval.

4. *Number Lost to Follow-up* (l_i). This is the number of individuals who are lost to observation and whose survival status is thus unknown in the ith interval ($i = 1, \ldots, s$).

Table 4.6 Format of a Life Table

Interval	Midpoint	Width	Number Lost to Follow-up	Number Withdrawn Alive	Number Dying	Number Entering Interval	Number Exposed To Risk	Conditional Proportion Dying	Conditional Proportion Surviving	Cumulative Proportion Surviving	Probability Density $f(t_{mi})$	Hazard $\hat{h}(t_{mi})$
$t-t_2$	t_{m1}	b_1	l_1	w_1	d_1	n'_1	n_1	$\hat{q}_1$	$\hat{p}_1$	$\hat{S}(t_1)=1.00$	$\hat{f}(t_{m1})$	$\hat{h}(t_{m1})$
t_2-t_3	t_{m2}	b_2	l_2	w_2	d_2	n'_2	n_2	$\hat{q}_2$	$\hat{p}_2$	$\hat{S}(t_2)$	$\hat{f}(t_{m2})$	$\hat{h}(t_{m2})$
$\ldots$	$\ldots$	$\ldots$	$\ldots$	$\ldots$	$\ldots$	$\ldots$	$\ldots$	$\ldots$	$\ldots$	$\ldots$	$\ldots$	$\ldots$
t_i-t_{i+1}	t_{mi}	b_i	l_i	w_i	d_i	n'_i	n_i	$\hat{q}_i$	$\hat{p}_i$	$\hat{S}(t_i)$	$\hat{f}(t_{mi})$	$\hat{h}(t_{mi})$
$\ldots$	$\ldots$	$\ldots$	$\ldots$	$\ldots$	$\ldots$	$\ldots$	$\ldots$	$\ldots$	$\ldots$	$\ldots$	$\ldots$	$\ldots$
$t_{s-1}-t_s$	$t_{m,s-1}$	b_{s-1}	l_{s-1}	w_{s-1}	d_{s-1}	n'_{s-1}	n_{s-1}	$\hat{q}_{s-1}$	$\hat{p}_{s-1}$	$\hat{S}(t_{s-1})$	$\hat{f}(t_{m,s-1})$	$\hat{h}(t_{m,s-1})$
$t_s-\infty$	—	—	l_s	w_s	d_s	n'_s	n_s	1	0	$\hat{S}(t_s)$	—	—

5. *Number Withdrawn Alive* (w_i). Individuals withdrawn alive in the ith interval are those known to be alive at the closing date of the study. The survival time recorded for such individuals is the length of time from entrance to the closing date of the study.

6. *Number Dying* (d_i). This is the number of individuals who die in the ith interval. The survival time of these individuals is the time from entrance to death.

7. *Number Entering the ith Interval* (n_i'). The number of individuals entering the first interval n_1' is the total sample size. Other entries are determined from $n_i' = n_{i-1}' - l_{i-1} - w_{i-1} - d_{i-1}$. That is, the number of individuals entering the ith interval is equal to the number of individuals studied at the beginning of the previous interval minus those who are lost to follow-up, are withdrawn alive, or have died in the previous interval.

8. *Number Exposed to Risk* (n_i). This is the number of individuals who are exposed to risk in the ith interval and is defined as $n_i = n_i' - \frac{1}{2}(l_i + w_i)$. It is assumed that the times to loss or withdrawal are approximately uniformly distributed in the interval. Therefore, individuals lost or withdrawn in the interval are exposed to risk of death for one-half the interval. If there are no losses or withdrawals, $n_i = n_i'$.

9. *Conditional Proportion Dying* $(\hat{q}_i)$. This is defined as $\hat{q}_i = d_i/n_i$ for $i = 1, \ldots, s-1$, and $\hat{q}_s = 1$. It is an estimate of the conditional probability of death in the ith interval given exposure to the risk of death in the ith interval.

10. *Conditional Proportion Surviving* $(\hat{p}_i)$. This is given by $\hat{p}_i = 1 - \hat{q}_i$, which is an estimate of the conditional probability of surviving in the ith interval.

11. *Cumulative Proportion Surviving* $[\hat{S}(t_i)]$. This is an estimate of the survivorship function at time t_i; it is often referred to as the cumulative survival rate. For $i = 1$, $\hat{S}(t_1) = 1$ and for $i = 2, \ldots, s$, $\hat{S}(t_i) = \hat{p}_{i-1}\hat{S}(t_{i-1})$. It is the usual life-table estimate and is based on the fact that surviving to the start of the ith interval means surviving to the start of and then through the $(i-1)$th interval.

12. *Estimated Probability Density Function* $[\hat{f}(t_{mi})]$. This is defined as the probability of dying in the ith interval per unit width. Thus, a natural estimate at the midpoint of the interval is

$$\hat{f}(t_{mi}) = \frac{\hat{S}(t_i) - \hat{S}(t_{i+1})}{b_i} = \frac{\hat{S}(t_i)\hat{q}_i}{b_i} \qquad i = 1, \ldots, s-1$$

$$(4.20)$$

13. *Hazard Function* $[\hat{h}(t_{mi})]$. The hazard function for the ith interval, estimated at the midpoint, is

$$\hat{h}(t_{mi}) = \frac{d_i}{b_i(n_i - \frac{1}{2}d_i)} = \frac{2\hat{q}_i}{b_i(1 + \hat{p}_i)} \qquad i = 1, \ldots, s-1$$

(4.21)

It is the number of deaths per unit time in the interval divided by the average number of survivors at the midpoint of the interval. That is, $\hat{h}(t_{mi})$ is derived from $\hat{f}(t_{mi})/\hat{S}(t_{mi})$ and $\hat{S}(t_{mi}) = \frac{1}{2}[\hat{S}(t_{i+1}) + \hat{S}(t_i)]$ since $S(t_i)$ is defined as the probability of surviving at the beginning, not the midpoint, of the ith interval:

$$\hat{h}(t_{mi}) = \frac{\hat{f}(t_{mi})}{\hat{S}(t_{mi})} = \frac{\hat{S}(t_i)\hat{q}_i/b_i}{\frac{1}{2}\hat{S}(t_i)(\hat{p}_i + 1)}$$

(4.22)

which reduces to (4.21).

Sacher (1956) derives an estimate of the hazard function by assuming that hazard is constant within an interval but varies among intervals. His estimate is

$$\hat{h}(t_{mi}) = \frac{(-\log_e \hat{p}_i)}{b_i}$$

(4.23)

In a Monte Carlo study, Gehan and Siddiqui (1973) show that (4.22) is less biased than (4.23).

The large-sample approximate variances of the estimated survival functions $\hat{S}(t_i)$, $\hat{f}(t_{mi})$, and $\hat{h}(t_{mi})$ in the ith interval are

$$\text{Var}[\hat{S}(t_i)] \cong [\hat{S}(t_i)]^2 \sum_{j=1}^{i-1} \frac{\hat{q}_j}{n_j \hat{p}_j}$$

(4.24)

$$\text{Var}[\hat{f}(t_{mi})] \cong \frac{[\hat{S}(t_i)\hat{q}_i]^2}{b_i} \left[\sum_{j=1}^{i-1} \frac{\hat{q}_j}{n_j \hat{p}_j} + \frac{\hat{p}_i}{n_i \hat{q}_i} \right]$$

(4.25)

and

$$\text{Var}[\hat{h}(t_{mi})] \cong \frac{[\hat{h}(t_{mi})]^2}{n_i \hat{q}_i} \{1 - [\frac{1}{2}\hat{h}(t_{mi})b_i]^2\}$$

(4.26)

Equation (4.24) is given by Greenwood (1926); Gehan (1969) derived (4.25) and (4.26). These may be used to obtain approximate confidence intervals for the various survival functions.

The graph of $\hat{S}(t_i)$ can be used to find an estimate of the median. Or let (t_j, t_{j+1}) be the interval such that $\hat{S}(t_j) \geq 0.5$ and $\hat{S}(t_{j+1}) < 0.5$. Then the median survival time t_m can be estimated by linear interpolation:

$$\hat{t}_m = t_j + \frac{[\hat{S}(t_j) - 0.5]b_j}{\hat{S}(t_j) - \hat{S}(t_{j+1})} = t_j + \frac{\hat{S}(t_j) - 0.5}{\hat{f}(t_{mj})} \tag{4.27}$$

where $\hat{f}(t_{mj})$ is defined in (4.20).

Another interesting measure that can be obtained from the life table is the *median remaining lifetime* at time t_i, denoted by $t_{mr}(i)$, $i = 1, \ldots, s - 1$. If at t_i the proportion of individual survival is $\hat{S}(t_i)$, then the proportion of individual survival at $t_{mr}(i)$ is $\frac{1}{2}\hat{S}(t_i)$. That is, one-half of the individuals who are alive at time t_i are expected to be alive at time $t_{mr}(i)$. Let (t_j, t_{j+1}) be the interval in which $\frac{1}{2}\hat{S}(t_i)$ falls, that is, $\hat{S}(t_j) \geq \frac{1}{2}\hat{S}(t_i)$ and $\hat{S}(t_{j+1}) < \frac{1}{2}\hat{S}(t_i)$. Then an estimate of $t_{mr}(i)$ is

$$\hat{t}_{mr}(i) = (t_j - t_i) + \frac{b_j[\hat{S}(t_j) - \frac{1}{2}\hat{S}(t_i)]}{\hat{S}(t_j) - \hat{S}(t_{j+1})} \tag{4.28}$$

Here $\hat{S}(t_j)$ is the estimated proportion surviving beyond the lower limit of the interval containing the median.

The variance of $\hat{t}_{mr}(i)$ is approximately

$$\text{Var}[\hat{t}_{mr}(i)] = \frac{[\hat{S}(t_i)]^2}{4n_i[\hat{f}(t_{mj})]^2} \tag{4.29}$$

Example 4.4
The following survival data for 2418 males with angina pectoris, originally reported by Parker et al. (1946), were also included in Gehan's (1969) paper. Survival time is computed from time of diagnosis in years. The life table uses 16 intervals of one year. Table 4.7 gives estimates of the various survival functions, the median remaining lifetime, and their standard errors. The survivorship function $\hat{S}(t)$ is plotted at t and the hazard and density functions $\hat{h}(t)$ and $\hat{f}(t)$ are plotted at the midpoint of the interval (Figure 4.4).

The graph of the estimated hazard function shows that the death rate is highest in the first year after diagnosis. From the end of the first year to the beginning of the tenth year, the death rate remains relatively constant, fluctuating between 0.09 and 0.12. The hazard rate is generally higher after the tenth year. Hence, the prognosis for a patient who has survived one year is better than that for a newly diagnosed patient if factors such as age, sex, and race are not considered. A similar interpretation is reached by examining the estimated median remaining lifetimes. Initially, the estimated median remaining lifetime is 5.33 years. It reaches a peak of 6.34 years at the beginning of the second year after diagnosis and then decreases. The median survival time, either read from the survival curve or using (4.27), is 5.33 years, and the five-year survival rate is 0.5193 with a standard error of 0.0103.

Table 4.7 A Life-Table Analysis of 2418 Males with Angina Pectoris

Year after Diagnosis	Midpoint	Width	Number Lost to Follow-up	Number Withdrawn Alive	Number Dying	Number Entering Interval	Number Exposed to Risk	Conditional Proportion Dying	Conditional Proportion Surviving	$\hat{S}(t_i)$	$\hat{f}(t_{mi})$	$\hat{h}(t_{mi})$	$\sqrt{Var[\hat{S}(t_i)]}$	$\sqrt{Var[\hat{f}(t_{mi})]}$	$\sqrt{Var[\hat{h}(t_{mi})]}$	$\hat{i}_{mr}(i)$	$\sqrt{Var[\hat{i}_{mr}(i)]}$
0	0.5	1.0	0	0	456	2418	2418.0	0.1886	0.8114	1.0000	0.1886	0.2082	—	0.0080	0.0097	5.33	0.17
1	1.5	1.0	39	0	226	1962	1942.5	0.1163	0.8837	0.8114	0.0944	0.1235	0.0080	0.0060	0.0082	6.35	0.20
2	2.5	1.0	22	0	152	1697	1686.0	0.0902	0.9098	0.7170	0.0646	0.0944	0.0092	0.0051	0.0076	6.34	0.24
3	3.5	1.0	23	0	171	1523	1511.5	0.1131	0.8869	0.6524	0.0738	0.1199	0.0097	0.0054	0.0092	6.23	0.24
4	4.5	1.0	24	0	135	1329	1317.0	0.1025	0.8975	0.5786	0.0593	0.1080	0.0101	0.0049	0.0093	6.22	0.19
5	5.5	1.0	107	0	125	1170	1116.5	0.1120	0.8880	0.5193	0.0581	0.1186	0.0103	0.0050	0.0106	5.91	0.18
6	6.5	1.0	133	0	83	938	871.5	0.0952	0.9048	0.4611	0.0439	0.1000	0.0104	0.0047	0.0110	5.60	0.19
7	7.5	1.0	102	0	74	722	671.0	0.1103	0.8897	0.4172	0.0460	0.1167	0.0105	0.0052	0.0135	5.17	0.27
8	8.5	1.0	68	0	51	546	512.0	0.0996	0.9004	0.3712	0.0370	0.1048	0.0106	0.0050	0.0147	4.94	0.28
9	9.5	1.0	64	0	42	427	395.0	0.1063	0.8937	0.3342	0.0355	0.1123	0.0107	0.0053	0.0173	4.83	0.41
10	10.5	1.0	45	0	43	321	298.5	0.1441	0.8559	0.2987	0.0430	0.1552	0.0109	0.0063	0.0236	4.69	0.42
11	11.5	1.0	53	0	34	233	206.5	0.1646	0.8354	0.2557	0.0421	0.1794	0.0111	0.0068	0.0306	4.00+	—
12	12.5	1.0	33	0	18	146	129.5	0.1390	0.8610	0.2136	0.0297	0.1494	0.0114	0.0067	0.0351	3.00+	—
13	13.5	1.0	27	0	9	95	81.5	0.1104	0.8896	0.1839	0.0203	0.1169	0.0118	0.0065	0.0389	2.00+	—
14	14.5	1.0	23	0	6	59	47.5	0.1263	0.8737	0.1636	0.0207	0.1348	0.0123	0.0080	0.0549	1.00+	—
15	—	—	0	0	0	30	30.0	1.0000	0.0000	0.1429	—	—	0.0133	—	—	—	—

Source: Gehan (1969).

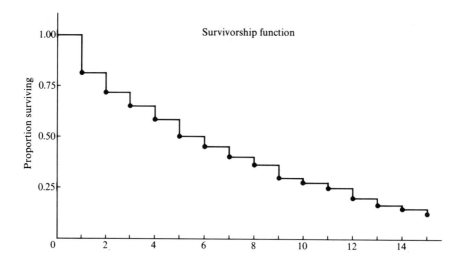

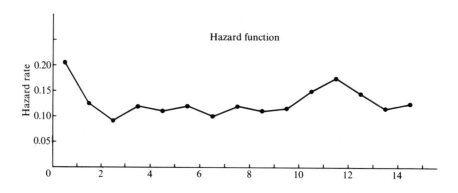

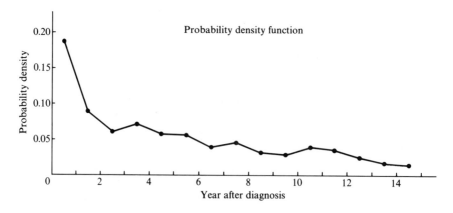

Figure 4.4 Survival functions of male patients with angina pectoris.

4.3   RELATIVE, FIVE-YEAR, AND CORRECTED SURVIVAL RATES

Another approach to large-scale survival data is the calculation of the *relative survival rate* or *annual survival ratio*. The relative survival rate evaluates the survival experience of patients in terms of the general population. Greenwood (1926) first suggested this approach for evaluating the efficacy of cancer treatment: If average survival time of the treated patients equals that of a random sample of persons of the same age, sex, and occupation, the patients could be considered "cure." Cutler et al. (1958, 1959, 1960a,b, 1967) adopted Greenwood's idea of comparing the survival experience of cancer patients with that of the general population in order to ascertain (1) the ratio of observed to expected survival rates and (2) whether, in time, the mortality rate declines to a "normal" level.

The relative survival rate is defined as the ratio of the survival rate (probability of surviving one year) for a patient under study (*observed rate*) to someone in the general population of the same age, sex, and race (*expected rate*) over a specified period of time. To provide a more precise measure of the relationship of the observed and expected survival rates, Cutler et al. suggest computing the ratio for each individual follow-up year. A relative rate of 100% means that during a specific follow-up year the mortality rates in the patient and in the general population are equal. A relative rate of less than 100% means that the mortality rate in the patients is higher than that in the general population. Cutler et al. use the survival rates in the Connecticut and U.S. life tables for the general population.

Using the notation in Table 4.6, the observed survival rate at time t_i is $\hat{p}_i$. The expected survival rate can be computed as follows: Suppose that at time t_i there are n'_i individuals alive for whom age, sex, race, and time of observation are known. Let p^*_{ij} be the survival rate of the jth individual from general population life tables (with corresponding age, sex, and race). The expected survival rate is

$$p^*_i = \frac{1}{n'_i} \sum_{j=1}^{n'_i} p^*_{ij} \qquad (4.30)$$

Then the relative survival rate at time t_i is defined by

$$r_i = \hat{p}_i / p^*_i \qquad (4.31)$$

Example 4.5 taken from Cutler et al. (1957) illustrates the interpretation of relative survival rates.

Example 4.5

A total of 9121 breast cancer cases were diagnosed in Connecticut hospitals from 1935 to 1953. The Connecticut life table for white females, 1939–1941, is used in the calculation of the expected survival rate. Table 4.8 gives the

Table 4.8 Relative Survival Rates of Breast Cancer Patients in Connecticut, 1935–1953

Years after Diagnosis	Survival Rates (%)		Relative Survival Rate (%)
	Observed	Expected	
0–1	82.9	97.2	85
1–2	83.3	97.1	86
2–3	85.9	96.9	89
3–4	86.8	96.7	90
4–5	89.2	96.6	92
5–6	90.0	96.4	93
6–7	89.9	96.4	93
7–8	91.6	96.2	95
8–9	92.0	96.1	96
9–10	92.7	96.1	96
10–11	92.9	95.9	97
11–12	94.0	95.8	98
12–13	94.1	95.3	99
13–14	91.5	95.3	96
14–15	90.6	94.9	95

Source: Cutler et al. (1957).

observed and expected survival rates as well as the relative survival rates. Figure 4.5(*a*) graphically shows these data: the survival curves for the breast cancer patients and the general population. The relative survival rates are plotted in Figure 4.5(*b*). For this group of patients, the relative survival rates, though increasing during 13 successive years, are less than 100% throughout the 15 years of follow-up. During each of the 15 years, breast cancer patient mortality rate is greater than that of the general population.

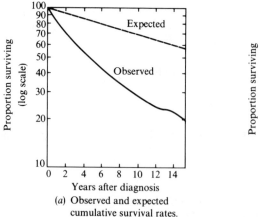

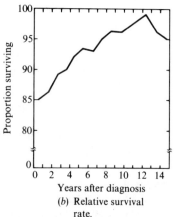

(*a*) Observed and expected cumulative survival rates.

(*b*) Relative survival rate.

Figure 4.5 Survival rates of breast cancer patients in Connecticut, 1935–1953.

Other measures of describing survival experience of cancer patients are the *five-year survival rate* and the *corrected rate*. The five-year survival rate is simply the cumulative proportion surviving at the end of the fifth year. For example, the five-year survival rate for the males with angina pectoris in Example 4.4 is 0.5193. The five-year survival rate is no longer a measure of treatment success for patients with many types of cancer since the survival of cancer patients has improved considerably in the last three decades.

Berkson (1942) suggests using a *corrected survival rate*. This is the survival rate if the disease under study alone is the cause of death. In most survival studies, the proportion of surviving patients is usually determined without considering the cause of death, which might be unrelated to the specific illness. If p_c denotes the survival rate when cancer alone is the cause of death, Berkson proposes that

$$p_c = p/p_0 \tag{4.32}$$

where p is the observed total survival rate in a group of cancer patients and p_0 is the survival rate for a group of the same age and sex in the general population. Rate p_c may be computed at any time after the initiation of follow-up; it provides a measure of the proportion of patients that escaped death from cancer up to that point. If a five-year survival rate is 0.5 and it is corrected for noncancer deaths and if we find that the five-year survival rate of the general population is 0.9, then the corrected survival rate is 0.5/0.9, or 0.56.

4.4   STANDARDIZED RATES AND RATIOS

Rates and ratios are often used in demography and epidemiology to describe the occurrence of a health-related event. For example, the standardized mortality (or morbidity) ratio (SMR) is frequently used in occupational epidemiology as a measure of risk, and the standardized death rate is commonly used in comparing mortality experiences of different populations or the same population at different times.

The concept of the SMR is very similar to that of the relative survival rate described above. It is defined as the ratio of the observed and the expected number of deaths and can be expressed as

$$\text{SMR} = \frac{\text{observed number of deaths in study population}}{\text{expected number of deaths in study population}} \times 100 \tag{4.33}$$

where the expected number of deaths is the sum of the expected deaths from the same age, sex, and race groups in the general population. The standardized morbidity ratio can similarly be calculated by simply replacing the word *deaths* by *disease cases* in (4.33). If only new cases are of interest, we call the ratio standardized incidence ratio (SIR).

The standardized death rate is only one of the many rates used to describe the health status of a population or to compare the health status of different populations. If the populations are similar with respect to demographic variables such as age, sex, or race, the *crude rate*, or the ratio of the number of persons to whom the event under study occurred to the total number of persons in the population, can safely be used for comparison.

The level of the crude rate is affected by demographic characteristics of the population for which the rate is computed. If populations have different demographic compositions, a comparison of the crude rates may be misleading. As an example consider the two hypothetical populations, Sunny City and Happy City, in Table 4.9. The crude death rate of Sunny City is 1000(1475/100,000), or 14.7 per 1000. The crude death rate of Happy City is 1000(1125/100,000), or 11.25 per 1000, which is lower than that of Sunny City even though all age-specific rates in Happy City are higher. This is mainly because there is a large proportion of older people in Sunny City. A crude death rate of a population may be relatively high merely because the population has a high proportion of older people; it may be relatively low because the population has a high proportion of younger people. Thus, one should adjust the rate to eliminate the effects of age, sex, or other differences. The procedure of adjustment is called *standardization* and the rate obtained after standardization is called the *standardized rate*.

The most frequently used methods for standardization are the *direct method* and the *indirect method*.

1. *Direct Method.* In this method a standard population is selected. The distribution across the groups with different values of the demographic characteristic (e.g., different age groups) must be known. Let $r_1, \ldots, r_k$, where k is the number of groups, be the specific rates of the different groups for the population under study. Let $p_1, \ldots, p_k$ be the proportions of people in the k groups for the standard population. The direct standardized rate is

Table 4.9 Population and Deaths of Sunny City and Happy City by Age

	Sunny City			Happy City		
Age	Population	Deaths	Age-specific Rates (per 1000)	Population	Deaths	Age-specific Rates (per 1000)
<25	25,000	25	1.00	55,000	110	2.0
25–44	40,000	50	1.25	20,000	50	2.5
45–46	20,000	200	10.00	21,000	315	15.0
≥65	15,000	1,200	80.00	4,000	650	162.5
Total	100,000	1,475		100,000	1,125	

obtained by multiplying the specific rates r_i by p_i in each group. The formula for the direct standardized rate is

$$R_{\text{direct}} = \sum_{i=1}^{k} r_i p_i \qquad (4.34)$$

As an example consider the data in Table 4.9. If we choose a standard population whose distribution is shown in the second column of Table 4.10, the direct standardized death rates for Sunny City and Happy City are, respectively, 9.37 and 17.84 per thousand. These standardized rates are more reliable than the crude rates for comparison purposes.

2. *Indirect Method.* If the specific rates r_i of the population being studied are unknown, the direct method cannot be applied. In this case, it is possible to standardize the rate by an indirect method if the following are available:

(a) The number of persons to whom the event being studied occurred (D) in the population. For example, if the death rate is being standardized, D is the number of deaths.

(b) The distribution across the various groups for the population being studied denoted by $n_1, \ldots, n_k$.

(c) The specific rates of the selected standard population denoted by $s_1, \ldots, s_k$.

(d) The crude rate of the standard population denoted by r.

The formula for indirect standardization is

$$R_{\text{indirect}} = \frac{D}{\sum_{i=1}^{k} n_i s_i} \, r \qquad (4.35)$$

The summation in (4.35) is the expected number of persons to whom the event occurred on the basis of the specific rates of the standard population. Thus, the indirect method adjusts the crude rate of the standard population by the ratio of the observed to expected number of persons to whom the event occurred in the population under study.

Table 4.11 presents an example for the death rate in the states of Oklahoma and Arizona in 1960 (data are from Grove and Hetzel 1963). The U.S. population in 1960 is used as the standard population. The crude death rate of Oklahoma (9.7 per thousand) is higher than that of Arizona (7.8 per thousand). However, the indirect standardized rates show a reverse relationship (8.6 for Oklahoma and 9.6 for Arizona), This, again, is because of the

Table 4.10 Standardized Death Rates by Direct Method for Sunny City and Happy City

Age	Standard Population	Proportion p_i	Sunny City		Happy City	
			Age-specific Death Rates r_i	Age-standardized Death Rates $p_i r_i$	Age-specific Death Rates r_i	Age-standardized Death Rates $p_i r_i$
<25	420,000	0.42	1.00	0.42	2.0	0.84
25–44	280,000	0.28	1.25	0.35	2.5	0.70
45–64	220,000	0.22	10.00	2.20	15.0	3.30
≥65	80,000	0.08	80.00	6.40	162.5	13.00
Total	1,000,000			9.37 (R_{direct})		17.84 (R_{direct})

Table 4.11 Standardized Death Rates by Indirect Method for Oklahoma and Arizona, 1960[a]

Age	Standard Population (U.S. population, 1960) Age-specific Death Rates s_i	Oklahoma Population n_i	Oklahoma Expected Deaths $n_i s_i$	Arizona Population n_i	Arizona Expected Deaths $n_i s_i$
<1	0.0270	49,103	1,325.78	34,599	934.17
1–4	0.0011	193,644	213.01	132,367	145.60
5–14	0.0005	454,972	227.49	285,830	142.92
15–24	0.0011	329,230	362.15	186,789	205.47
25–34	0.0015	279,327	418.99	169,873	254.81
35–44	0.0030	287,994	863.98	173,029	519.09
45–54	0.0076	269,147	2,045.52	136,573	1,037.95
55–64	0.0174	216,036	3,759.03	92,871	1,615.96
65–74	0.0382	157,385	6,012.11	63,634	2,430.82
75–84	0.0875	74,848	6,549.20	22,499	1,968.66
85 +	0.1986	16,598	3,296.36	4,092	812.67
Total		2,328,284	25,074	1,302,161	10,068
Crude rates (per thousand)	9.5	9.7		7.8	
Observed deaths		22,584		10,157	
Expected deaths[b]		25,074		10,068	
Standardized rate (per thousand)		$\left(\dfrac{22,584}{25,074}\right)9.5 = 8.6$		$\left(\dfrac{10,157}{10,068}\right)9.5 = 9.6$	

[a]Data are from Grove and Hetzel (1963).
[b]$\sum n_i s_i$.

differences in age distribution. There is a higher proportion of people below the age of 25 in Arizona and a higher proportion of people above the age of 54 in Oklahoma.

Results for the adjusted rates depend on the standard population selected. Hence, this selection should be done carefully. Shryock et al. (1971), when discussing death rate by age, suggest that a population with similar age distribution to the various populations under study be selected as a standard. If the death rate of two populations is being compared, it is best to use the average of the two distributions as a standard.

It should be remembered that specific rates are still the most accurate and essential indicators of the variations among populations.. Standardized rates, no matter which method is used, are meaningful only when compared with other similarly computed rates. Kitagawa (1964) also criticizes the standardized rate because if the specific rates vary in different ways between the two populations being compared, standardization will not indicate the differences and sometimes will even mask the differences.

Nevertheless, if the specific rates are not available, if a single rate for a population is desired, or if the demographic composition of the population being compared is different, the standardized rate is useful.

BIBLIOGRAPHICAL REMARKS

Kaplan and Meier's (1958) PL method is the most commonly used technique for estimating the survivorship function for samples of small and moderate sizes. However, with the aid of a computer, it is not difficult to use the method for large sample sizes.

Berkson (1942), Berkson and Gage (1950), Cutler and Ederer (1958), and Gehan (1969) have written classic reports on life-table analysis. Peto et al. (1976) published an excellent review of some statistical methods related to clinical trials. The term *life-table analysis* that they use includes the PL method. Other references on life tables are, for example, Armitage (1971), Shryock et al. (1971), Kuzma (1967), Chiang (1968), Gross and Clark (1975), and Elandt-Johnson and Johnson (1980).

Relative survival rates and corrected survival rates have been used by Cutler and co-workers in a series of survival studies on cancer patients in Connecticut in the 1950s and 1960s (Cutler et al. 1957, 1959, 1960a,b, 1967; Ederer et al. 1961). Discussions of SMR, standardized rates, and related topics can be found in many standard epidemiology textbooks, for example, Mausner and Kramer (1985), Kahn (1983), Kelsey et al. (1986), Shryock et al. (1971), Chiang (1961), and Mantel and Stark (1968).

Many computer programs are available for the PL and life-table methods. Among them are P1L in BMDP (Dixon et al. 1988) and SURVIVAL in SPSS (1987).

EXERCISES

4.1 Consider the survival time of the 30 melanoma patients in Table 3.1.

    a. Compute and plot the PL estimates of the survivorship functions $\hat{S}(t)$ of the two treatment groups and check your results with Table 3.2 and Figure 3.1.

    b. Compute the variance of $\hat{S}(t)$ for every uncensored observation.

    c. Estimate the median survival times of the two groups.

4.2 Do the same as in Exercise 4.1 for the remission durations of the two treatment groups in Table 3.1.

4.3 Compute and plot the PL estimates of the tumor-free time distributions for the saturated fat and unsaturated fat diet groups in Table 3.4. Compare your results with Figure 3.3.

4.4 Consider the remission data of 42 patients with acute leukemia in Example 3.3.

    a. Compute and plot the PL estimates of $S(t)$ at every time to relapse for the 6-MP and placebo groups.

    b. Compute the variances of $\hat{S}(10)$ in the 6-MP group and of $\hat{S}(3)$ in the placebo group.

    c. Estimate the median remission times of the two treatment groups.

4.5   **a.** Compute the survival time for each patient in Exercise Table 3.1.

    b. Estimate and plot the overall survivorship function using the PL method. What is the median survival time?

    c. Divide the patients into two groups by sex. Compute and plot the PL estimates of the survivorship functions for each group. What is the median survival time for each?

4.6 Consider the skin test results in Exercise Table 3.1. For each of the five skin tests:

    a. Divide patients into two groups according to whether they had a positive reaction. Measurements less than 10×10 (5×5 for mumps) are considered negative.

    b. Estimate and plot the survivorship functions of the two groups.

    c. Can you tell from the plots if any skin tests might predict survival time?

4.7 Consider the data of patients with cancer of the ovary diagnosed in Connecticut from 1935 to 1944 (Cutler et al. 1960b). Exercise Table 4.1 reproduces the data in life-table format. Provide a life table like Table 4.6. Interpret the results.

Exercise Table 4.1

Time from Diagnosis (yr)	Number Lost to Follow-up l_i	Number Withdrawn Alive w_i	Number Dying d_i	Number Entering n'_i
0–5	18	0	731	949
5–10	16	0	52	200
10–15	8	67	14	132
15–20	0	33	10	43

4.8 Do a complete life-table analysis for the two sets of data given in Table 3.5. Plot the three survival functions.

4.9 Do a complete life-table analysis of the data given in Exercise Table 4.2. Plot the three survival functions.

Exercise Table 4.2   Survival Data of Female Patients with Angina Pectoris

Year after Diagnosis	Number Entering Interval	Number Lost to Follow-up	Number Dying
0–1	555	0	82
1–2	473	8	30
2–3	435	8	27
3–4	400	7	22
4–5	371	7	26
5–6	338	28	25
6–7	285	31	20
7–8	234	32	11
8–9	191	24	14
9–10	153	27	13
10–11	113	22	5
11–12	86	23	5
12–13	58	18	5
13–14	35	9	2
14–15	24	7	3
15+	14	11	3

Source: R.L. Parker et al., *JAMA* 131(2): 95–100 (1946). Copyright 1946. American Medical Association.

4.10 Consider the survival times of the diabetes patients in Exercise Table 3.4. Do a complete life-table analysis of the survival time. Plot the three survival functions.

4.11 Consider the data given in Exercise Table 4.3. Compute the direct standardized death rate for the states of Oklahoma and Montana using the U.S. population of 1960 as the standard.

Exercise Table 4.3

Age	U.S. Population, 1960 (in thousands)	Proportion p_i	Oklahoma Average Death Rate (per thousand) r_i	Montana Average Death Rate (per thousand) r_i
<1	4,112	0.023	25.5	25.8
1–4	16,209	0.091	1.2	1.2
5–14	35,465	0.198	0.5	0.5
15–24	24,020	0.134	1.2	1.6
25–34	22,818	0.127	1.6	1.8
35–44	24,081	0.134	2.9	3.1
45–54	20,486	0.114	6.9	7.5
55–64	15,572	0.087	14.8	16.3
65–74	10,997	0.061	32.4	37.3
75–84	4,634	0.026	79.0	87.3
85+	929	0.005	190.4	202.8
Total	179,323	1.000		

Source: Grove and Hetzel (1963).

4.12 Given the population of Japan and Chile (Exercise Table 4.4), compute the indirect standarized death rate for the two countries using the U.S. death rate of 1960 in Table 4.11 as the standard.

Exercise Table 4.4

Age	Population (in thousands) 1960	
	Japan	Chile
<1	1,577	228
1–4	6,268	876
5–14	20,223	1,817
15–24	17,627	1,323
25–34	15,727	1,034
35–44	11,057	779
45–54	9,018	603
55–64	6,573	395
65–74	3,724	212
75–84	1,438	83
≤85	188	22
Total	93,419	7,374
Observed deaths	706,599	95,486

Source: Shryock et al. (1971).

CHAPTER 5

Nonparametric Methods for Comparing Survival Distributions

The problem of comparing survival distributions arises often in biomedical research. A laboratory researcher may want to compare the tumor-free times of two or more groups of rats exposed to carcinogens. A diabetologist may wish to compare the retinopathy-free times of two groups of diabetic patients. A clinical oncologist may be interested in comparing the ability of two or more treatments to prolong life or maintain health. Almost invariably, the disease-free or survival times of the different groups vary. These differences can be illustrated by drawing graphs of the estimated survivorship function, but that only gives a rough idea of the difference between the distributions. It does not reveal whether the differences are significant or merely chance variations. A statistical test is necessary.

Section 5.1 introduces five nonparametric tests that can be used for data with and without censored observations. Section 5.2 is devoted to the Mantel–Haenszel test, which is particularly useful in stratified analysis, a method commonly used to take account of possible confounding variables. Section 5.3 discusses the problem of comparing three or more survival distributions with or without censoring.

5.1 COMPARISON OF TWO SURVIVAL DISTRIBUTIONS

Suppose that there are n_1 and n_2 individuals who receive treatments 1 and 2, respectively. Let $x_1, \ldots, x_{r_1}$ be the r_1 failure observations and $x_{r_1+1}^{+}, \ldots, x_{n_1}^{+}$ the $n_1 - r_1$ censored observations in group 1. In group 2, let $y_1, \ldots, y_{r_2}$ be the r_2 failure observations and $y_{r_2+1}^{+}, \ldots, y_{n_2}^{+}$ the $n_2 - r_2$ censored observations. That is, at the end of the study $n_1 - r_1$ individuals who received treatment 1 and $n_2 - r_2$ individuals who received treatment 2 are still alive. Suppose that the observations in group 1 are samples from a distribution with survivorship function $S_1(t)$ and the observations in group 2 are samples from a distribution with survivorship function $S_2(t)$. Then the null hypothesis to consider is

$$H_0: S_1(t) = S_2(t) \quad \text{(treatments 1 and 2 are equally effective)}$$

against the alternative

$$H_1: S_1(t) > S_2(t) \quad \text{(treatment 1 is more effective than 2)}$$

or

$$H_2: S_1(t) < S_2(t) \quad \text{(treatment 2 is more effective than 1)}$$

or

$$H_3: S_1(t) \neq S_2(t) \quad \text{(treatments 1 and 2 are not equally effective)}$$

When there are no censored observations, standard nonparametric tests can be used to compare two survival distributions. For example, the Wilcoxon (1945) test or the Mann–Whitney (1947) U-test can test the equality of two independent populations, and the sign test can be used for paired (or dependent) samples (Marascuilo and McSweeney 1977).

In the following we introduce five nonparametric tests: Gehan's generalized Wilcoxon test (Gehan 1965a,b), the Cox–Mantel test (Cox 1959, 1972; Mantel 1966), the logrank test (Peto and Peto 1972), Peto and Peto's generalized Wilcoxon test (1972), and Cox's F-test (1964). All the tests are designed to handle censored data; data without censored observations can be considered a special case.

5.1.1 Gehan's Generalized Wilcoxon Test

In Gehan's generalized Wilcoxon test every observation x_i or x_i^+ in group 1 is compared with every observation y_j or y_j^+ in group 2 and a score U_{ij} is given to the result of every comparison. For the purpose of illustration, let us assume that the alternative hypothesis is $H_1: S_1(t) > S_2(t)$, that is, treatment 1 is more effective than treatment 2.

Define

$$U_{ij} = \begin{cases} +1 & \text{if } x_i > y_j \text{ or } x_i^+ \geq y_j \\ 0 & \text{if } x_i = y_j \text{ or } x_i^+ < y_j \text{ or } y_j^+ < x_i \text{ or } (x_i^+, y_j^+) \\ -1 & \text{if } x_i < y_j \text{ or } x_i \leq y_j^+ \end{cases}$$

and calculate the test statistic

$$W = \sum_{i=1}^{n_1} \sum_{j=1}^{n_2} U_{ij} \tag{5.1}$$

where the sum is over all $n_1 n_2$ comparisons. Hence, there is a contribution to the test statistic W for every comparison where both observations are failures (except for ties) and for every comparison where a censored observation is equal to or larger than a failure.

The calculation of W is laborious when n_1 and n_2 are large. Mantel (1967) shows that it can be calculated in an alternative way by assigning a score to each observation based on its relative ranking. In Gehan's computation each observation in sample 1 is compared with each one in sample 2. If the two samples are combined into a single pooled sample of $n_1 + n_2$ observations, it is the same as comparing each observation with the remaining $n_1 + n_2 - 1$. Let U_i, $i = 1, \ldots, n_1 + n_2$, be the number of remaining $n_1 + n_2 - 1$ observations that the ith observation is definitely greater than *minus* the number that it is definitely less than. The $n_1 + n_2$ U_i's define a finite population with mean zero and it is true that Gehan's

$$W = \sum_{i=1}^{n_1} U_i \tag{5.2}$$

where the summation is over the U_i of sample 1 only. From either (5.1) or (5.2), it is clear that W would be a large positive number if H_1 is true. Mantel also suggests that the permutational variance of W be used instead of the more complicated variance formula derived by Gehan. The permutational distribution of W can be obtained by considering all[1]

$$\binom{n_1 + n_2}{n_2} = \frac{(n_1 + n_2)!}{n_1! \, n_2!}$$

ways of selecting n_1 of the U_i at random. The test statistic W under H_0 can be considered approximately normally distributed with mean zero and variance[2]

$$\mathrm{Var}(W) = \frac{n_1 n_2 \sum_{i=1}^{n_1+n_2} U_i^2}{(n_1 + n_2)(n_1 + n_2 - 1)} \tag{5.3}$$

Since W is discrete, an appropriate continuity correction of 1 is ordinarily used when there are neither ties nor censored observations. Otherwise a continuity correction of 0.5 would probably be appropriate.

Since W has an asymptotically normal distribution with mean zero and variance in (5.3), $Z = W/\sqrt{\mathrm{Var}(W)}$ has standard normal distribution. The rejection regions are $Z > Z_\alpha$ for H_1, $Z < -Z_\alpha$ for H_2, and $|Z| > Z_{\alpha/2}$ for H_3, where $P(Z > Z_\alpha | H_0) = \alpha$.

---

[1] $n!$ reads n factorial; $n! = n(n-1)(n-2) \cdots 3 \cdot 2 \cdot 1$.
[2] This is called the permutational variance because it is obtained by considering the permutational distribution of all $(n_1 + n_2)!/n_1! n_2!$ W's.

The number U_i can be computed in two stages. The first stage yields, for each observation, unity plus the number of remaining observations that it is definitely larger than, that is, R_{1i}. The second stage yields R_{2i}, which is unity plus the number of remaining observations that the particular observation is definitely less than. Then $U_i = R_{1i} - R_{2i}$. The computations of R_{1i} and R_{2i} can be accomplished systematically in steps as illustrated in the following hypothetical example.

Example 5.1

Ten female patients with breast cancer are randomized to receive either CMF (cyclic administration of cyclophosphamide, methatrexate, and fluorouracil) or no treatment after a radical mastectomy. At the end of two years, the following times to relapse (or remission times) in months are recorded:

CMF (group 1) 23, 16+, 18+, 20+, 24+

Control (group 2) 15, 18, 19, 19, 20

The null hypothesis and the alternatives are

$H_0: S_1 = S_2$ (two treatments equally effective)

$H_1: S_1 > S_2$ (CMF more efficient than no treatment)

The computations of R_{1i}, R_{2i}, and U_i are given in Table 5.1. Thus, $W = 1 + 2 + 5 + 4 + 6 = 18$, $\text{Var}(W) = (5)(5)(208)/[(10)(9)] = 57.78$, and $Z = 18/\sqrt{57.78} = 2.368$. Suppose that the significance level used is $\alpha = 0.05$, $Z_{0.05} = 1.64$; then the computed Z value is in the rejection region. Therefore, we reject H_0 at the 0.05 level and conclude that the data show that CMF is more effective than no treatment. In fact, the approximate p value corresponding to $Z = 2.368$ is 0.009.

Note that the sum of all $n_1 + n_2$ U_i's equals zero. This fact can be used to check the computation.

5.1.2 The Cox–Mantel Test

Let $t_{(1)} < \cdots < t_{(k)}$ be the distinct failure times in the two groups together and $m_{(i)}$ the number of failure times equal to t_i, or the multiplicity of t_i, so that

$$\sum_{i=1}^{k} m_{(i)} = r_1 + r_2 \tag{5.4}$$

Further, let $R(t)$ be the set of individuals still exposed to risk of failure at time t, whose failure or censoring times are at least t. Here $R(t)$ is called the risk set at time t. Let n_{1t} and n_{2t} be the number of patients in $R(t)$ that

Table 5.1   Mantel's Procedure of Calculating U_i for Gehan's Generalized Wilcoxon Test

Observations of two samples in ascending order	15	16$^+$	18	18$^+$	19	19	20	20$^+$	23	24$^+$
Computation of R_{1i}										
Step 1 Rank from left to right omitting censored observations	1		2		3	4	5		6	
Step 2 Assign next higher rank to censored observations		2		3				6		7
Step 3 Reduce the rank of tied observations to the lower rank for the value						3				
Step 4 R_{1i}	1	2	2	3	3	3	5	6	6	7
Computation of R_{2i}										
Step 5 Rank from right to left	10	9	8	7	6	5	4	3	2	1
Step 6 Reduce the rank of tied observations to the lowest rank for the value						5				
Step 7 Reduce the rank of censored observations to 1		1		1				1		1
Step 8 R_{2i}	10	1	8	1	5	5	4	1	2	1
$U_i = R_{1i} - R_{2i}$	-9	1^a	-6	2^a	-2	-2	1	5^a	4^a	6^a

^{a} From group 1.

belong to treatment groups 1 and 2, respectively. The total number of observations, failure or censored in $R(t_{(i)})$ is $r_{(i)} = n_{1i} + n_{2i}$. Define

$$U = r_2 - \sum_{i=1}^{k} m_{(i)} A_{(i)} \tag{5.5}$$

$$I = \sum_{i=1}^{k} \frac{m_{(i)}(r_{(i)} - m_{(i)})}{r_{(i)} - 1} A_{(i)}(1 - A_{(i)}) \tag{5.6}$$

where $r_{(i)}$ is the number of observations, failure or censored, in $R(t_{(i)})$ and $A_{(i)}$ is the proportion of $r_{(i)}$ that belong to group 2. An asymptotic two-sample test is thus obtained by treating the statistic $C = U/\sqrt{I}$ as a standard normal variate under the null hypothesis (Cox 1972). The follow-ing example illustrates the procedure.

Example 5.2

Consider the remission data and the hypothesis in Example 5.1. There are $k = 5$ distinct failure times in the two groups, $r_1 = 1$ and $r_2 = 5$. To perform the Cox–Mantel test, Table 5.2 is prepared for convenience:

$$U = 5 - (0.5 + 0.5 + 2 \times 0.5 + 0.25)$$

$$= 5 - 2.25$$

$$= 2.75$$

$$I = \frac{1 \times 9}{9}(0.5 \times 0.5) + \frac{1 \times 7}{7}(0.5 \times 0.5) + \frac{2 \times 4}{5}(0.5 \times 0.5)$$

$$+ \frac{1 \times 3}{3}(0.25 \times 0.75)$$

$$= 0.25 + 0.25 + 0.4 + 0.1875$$

$$= 1.0875$$

Therefore, $C = 2.75/\sqrt{1.0875} = 2.637 > Z_{0.05} = 1.64$ and we reject H_0 at the 0.05 level and reach the same conclusion as in Example 5.1. The p value corresponding to $Z = 2.637$ is approximately 0.004.

5.1.3   The Logrank Test

Mantel's (1966) generalization of the Savage (1956) test, often refered to as the logrank test (Peto and Peto 1972), is based on a set of scores w_i assigned to the observations. The scores are functions of the logarithm of the survival function. Altshuler (1970) estimates the log survival function at $t_{(i)}$ using

$$-e(t_{(i)}) = -\sum_{j \le t_{(i)}} \frac{m_{(j)}}{r_{(j)}} \tag{5.7}$$

Table 5.2   Computations of Cox–Mantel Test

		Number in Risk Set of			
		Sample 1	Sample 2		
Distinct Failure Time t_i	$m_{(i)}$	n_{1t}	n_{2t}	$r_{(i)}$	$A_{(i)}$
15	1	5	5	10	0.5
18	1	4	4	8	0.5
19	2	3	3	6	0.5
20	1	3	1	4	0.25
23	1	2	0	2	0

where $m_{(j)}$ and $r_{(j)}$ are defined in Section 5.1.2. The scores suggested by Peto and Peto are $w_i = 1 - e(t_{(i)})$ for an uncensored observation $t_{(i)}$ and $-e(T)$ for an observation censored at T. In practice, for a censored observation $t_{i,}^{+}$, $w_i = -e(t_{(j)})$, where $t_{(j)}$ is the largest uncensored observation such that $t_{(j)} \leq t_i^{+}$. Thus, the larger the uncensored observation, the smaller its score. Censored observations receive negative scores. The w scores sum identically to zero for the two groups together. The logrank test is based on the sum S of the w scores in one of the two groups. The permutational variance of S is given by

$$\text{Var}(S) = \frac{n_1 n_2 \sum_{i=1}^{n_1+n_2} w_i^2}{(n_1 + n_2)(n_1 + n_2 - 1)} \qquad (5.8)$$

which can be rewritten as

$$V = \left\{ \sum_{j=1}^{k} \frac{m_{(j)}(r_{(j)} - m_{(j)})}{r_{(j)}} \right\} \frac{n_1 n_2}{(n_1 + n_2)(n_1 + n_2 - 1)} \qquad (5.9)$$

The test statistic $L = S/\sqrt{\text{Var}(S)}$ has an asymptotically standard normal distribution under the null hypothesis. If S is obtained from group 1, the critical region is $L < -Z_\alpha$, and if S is obtained from group 2, the critical region is $L > Z_\alpha$, where α is the significance level for testing $H_0: S_1 = S_2$ against $H_1: S_1 > S_2$. The following example illustrates the computational procedures.

Example 5.3
Consider the data and hypotheses in Example 5.1. The test statistic of the logrank test can be computed by tabulating $m_{(i)}$, $r_{(i)}$, $m_{(i)}/r_{(i)}$, and $e(t_{(i)})$ as in Table 5.3. Since every observation in the two samples, censored or not, is

Table 5.3   Computations of Logrank Test

Remission Times in Both Samples t_i	$m_{(i)}$	$r_{(i)}$	$m_{(i)}/r_{(i)}$	$e(t_{(i)})$	w_i
15	1	10	0.100	0.100	0.990[a]
16+	—	—	—	—	−0.100
18	1	8	0.125	0.225	0.775[a]
18+	—	—	—	—	−0.225
19	2	6	0.333	0.558	0.442[a]
20	1	4	0.250	0.808	0.192[a]
20+	—	—	—	—	−0.808
23	1	2	0.500	1.308	−1.308
24+	—	—	—	—	−1.308

[a]From sample 2.

assigned a score, it is convenient to list them in column 1. Columns 2–5 pertain to only the failure times; $e(t_{(i)})$ is the cumulative value of $m_{(i)}/r_{(i)}$—Altshuler's (1970) estimate of the logarithm of the survivorship function multiplied by -1. For example, at $t_{(i)} = 18$, $e(t_{(i)}) = 0.100 + 0.125 = 0.225$; at $t_{(i)} = 19$, $e(t_{(i)}) = 0.225 + 0.333 = 0.558$. The last column, w_i, gives the score for every observation. For an uncensored observation $w_i = 1 - e(t_{(i)})$, for example, at $t_i = 18$, $w_i = 1 - 0.225 = 0.775$. Since $e(t_{(i)})$ is an estimate of a function of the survivorship function, which we assume to be constant between two consecutive failures, $e(t_i^+)$ is equal to $e(t_{(j)})$ for $t_{(j)} \leq t_i^+$. Thus w_i for censored observations t_i^+ equals $-e(t_{(j)})$ where $t_{(j)} \leq t_i^+$. For example, w_i for 16^+ is $-e(15)$, or -0.100, and that for 18^+ is $-e(18)$, or -0.225. Tied observations like the two 19's receive the same score: 0.442. The 10 scores w_i sum to zero, which can be used to check the computation.

The statistic $S = 0.900 + 0.775 + 0.442 + 0.442 + 0.192 = 2.751$. The variance of S computed by (5.8) is 1.210. Hence, the test statistic $L = 2.751/\sqrt{1.210} = 2.5$, and the p value is approximately 0.0064. Thus, the hypothesis is rejected at the 0.0064 level, the data showing that CMF treatment is superior.

The logrank statistic S can be shown to equal to the sum of the observed failures minus the conditional expected failures computed at each failure time, or simply the difference between the observed and expected failures in one of the groups. A similar version of the logrank test is a chi-square test, which compares the observed number of failures to the expected number of failures under the hypothesis. Let O_1 and O_2 be the observed numbers and E_1 and E_2 the expected numbers of death in the two treatment groups. The test statistic

$$X^2 = \frac{(O_1 - E_1)^2}{E_1} + \frac{(O_2 - E_2)^2}{E_2} \tag{5.10}$$

has approximately the chi-square distribution with one degree of freedom. A large X^2 value (e.g., $\geq \chi^2_{1,.05}$) would lead to the rejection of the null hypothesis in favor of the alternative that the two treatments are not equally effective at $(\alpha = 0.05)$.

To compute E_1 and E_2, we arrange all the uncensored observations in ascending order and compute the number of expected deaths at each uncensored time and sum them. The number of expected deaths at an uncensored time is obtained by multiplying the observed deaths at that time by the proportion of patients exposed to risk in that treatment group. Let d_t be the number of deaths at time t and n_{1t} and n_{2t} be the numbers of patients still exposed to risk of dying at time up to t in the two treatment groups. The expected deaths for groups 1 and 2 at time t are

$$e_{1t} = \frac{n_{1t}}{n_{1t} + n_{2t}} \times d_t \qquad e_{2t} = \frac{n_{2t}}{n_{1t} + n_{2t}} \times d_t \tag{5.11}$$

Then the total numbers of expected deaths in the two groups are

$$E_1 = \sum_{\text{all } t} e_{1t} \qquad E_2 = \sum_{\text{all } t} e_{2t}$$

In practice, we only need to compute the total number of expected deaths in one of the two groups, for example, E_1, since E_2 is the total number of deaths minus E_1. The following example illustrates the calculation procedure.

Example 5.4

Let us use the hypothetical data in Example 5.1 again. The remission times in months are:

CMF (group 1)   23, 16+, 18+, 20+, 24+

Control (group 2)   15, 18, 19, 19, 20 .

Consider the following null and alternative hypotheses:

$$H_0: S_1 = S_2 \quad \text{(two treatments equally effective)}$$
$$H_1: S_1 \neq S_2 \quad \text{(two treatments not equally effective)}$$

Table 5.4 gives the calculation of E_1. For example, at $t = 18$, four patients in group 1 and four in group 2 are still exposed to the risk of relapse, and there is one relapse. Thus, $d_t = 1$, $n_{1t} = n_{2t} = 4$, and $e_{1t} = 0.5$.

The total number of expected relapses is $E_1 = 3.75$. Since there are a total of six deaths ($O_1 = 1$, $O_2 = 5$) in the two groups, $E_2 = 6 - 3.75 = 2.25$. Using (5.10), we have

$$X^2 = \frac{(1 - 3.75)^2}{3.75} + \frac{(5 - 2.25)^2}{2.25} = 5.378$$

Using Table C-2, the p value corresponding to this X^2 value is less than 0.05 ($P \cong 0.02$). Therefore, we reach the same conclusion that there is a significant difference in remission duration between the CMF and control groups.

Table 5.4   Computation of E_1 of Logrank Test

Relapse time t	d_t	n_{1t}	n_{2t}	e_{1t}	e_{2t}
15	1	5	5	0.5	0.5
18	1	4	4	0.5	0.5
19	2	3	3	1.0	1.0
20	1	3	1	0.75	0.25
23	1	2	0	1.0	0
Total				3.75	2.25

5.1.4 Peto and Peto's Generalized Wilcoxon Test

Another generalization of Wilcoxon's two-sample rank sum test is described by Peto and Peto (1972). Similar to the logrank test, this test assigns a score to every observation. For an uncensored observation t, the score is $u_i = \hat{S}(t+) + \hat{S}(t-) - 1$, and for an observation censored at T, the score is $u_i = \hat{S}(T) - 1$, where $\hat{S}$ is the Kaplan–Meier estimate of the survival function. If we use the notation of Section 5.1.2, the score for an uncensored observation $t_{(i)}$ is $u_i = \hat{S}(t_{(i)}) + \hat{S}(t_{(i-1)}) - 1$ and $\hat{S}(t_{(0)}) = 1$ and that for a censored observation t_j^+ is $u_j = \hat{S}(t_{(i)}) - 1$, where $t_{(i)} \le t_j^+$. These generalized Wilcoxon scores sum to zero. The test procedure after the scores are assigned is the same as for the logrank test. The following example illustrates the computational procedures.

Example 5.5

Using the same data and hypotheses as in Example 5.1, the calculations of the scores u_i for Peto and Peto's generalized Wilcoxon test are given in Table 5.5

Using the scores of group 1, we obtain

$$S = -0.100 - 0.212 - 0.605 - 0.408 - 0.803 = -2.128$$

$$\text{Var}(S) = (5)(5)[(0.9)^2 + \cdots + (-0.803)^2]/(10 \times 9) = 0.765$$

Thus, $Z = -2.128/\sqrt{0.765} = -2.433 < -Z_{0.05} = -1.64$. We reject H_0 at the 0.05 level and reach the same conclusion as in the last three examples that the data show that CMB is more effective than no treatment.

5.1.5 Cox's *F*-Test

Cox's *F*-test (Cox 1964) is based on ordered scores from the exponential

Table 5.5 Computations of Peto and Peto's Generalized Wilcoxon Test

$t_{(i)}$	$\hat{S}(t)$	u_i
15	0.900	$1 + 0.900 - 1 = 0.900$
16+	—	$0.900 - 1 = -0.100$[a]
18	0.788	$0.900 + 0.788 - 1 = 0.688$
18+	—	$0.788 - 1 = -0.212$[a]
19	0.657	$0.788 + 0.657 - 1 = 0.445$
19	0.526	$0.526 + 0.657 - 1 = 0.183$
20	0.395	$0.395 + 0.526 - 1 = -0.079$
20+	—	$0.395 - 1 = -0.605$[a]
23	0.197	$0.197 + 0.395 - 1 = -04.08$[a]
24+	—	$0.197 - 1 = -0.803$[a]

[a]Group 1.

distribution. It is for singly censored or complete samples; it is not applicable to progressively censored data. The procedure is as follows:

1. Rank the observations in the combined sample.
2. Replace the ranks by the corresponding expected order statistics in sampling the unit exponential distribution $[f(t) = e^{-t}]$. Denote by t_{rn} the expected value of the rth observation in increasing order of magnitude,

$$t_{rn} = \frac{1}{n} + \cdots + \frac{1}{n-r+1} \qquad r = 1, \ldots, n \qquad (5.12)$$

where n is the total number of observations in the two samples. In particular,

$$t_{1n} = \frac{1}{n}$$

$$t_{2n} = \frac{1}{n} + \frac{1}{n-1} \qquad (5.13)$$

$$t_{nn} = \frac{1}{n} + \frac{1}{n-1} + \cdots + 1$$

For n not too large, they can easily be computed by using tables of reciprocals. When two or more observations are tied, the average of the scores is used.

3. For data without censored observations the entire set of n observations is replaced by the set of scores $\{t_{rn}\}$ so obtained. The sample mean scores denoted by $\bar{t}_1$ and $\bar{t}_2$ of the two samples with n_1, n_2 observations are then computed. The ratio $\bar{t}_1/\bar{t}_2$ has been shown to follow an F distribution with $(2n_1, 2n_2)$ degrees of freedom. Critical regions for testing $H_0: S_1 = S_2$ against $H_1(S_1 > S_2)$, $H_2(S_1 < S_2)$, and $H_3(S_1 \neq S_2)$ are, respectively, $\bar{t}_1/\bar{t}_2 > F_{2n_1,2n_2,\alpha}$, $\bar{t}_1/\bar{t}_2 < F_{2n_1,2n_2,1-\alpha}$, and $\bar{t}_1/\bar{t}_2 > F_{2n_1,2n_2,\alpha/2}$ or $\bar{t}_1/\bar{t}_2 < F_{2n_1,2n_2,1-\alpha/2}$.

4. The calculation of F is slightly different for singly censored data. Let r_1 and r_2 be the number of failures and $n_1 - r_1$ and $n_2 - r_2$ the number of censored observations in the two samples. Then there are $p = r_1 + r_2$ failures in the combined sample and $n - p$ censored observations. Cox (1964) suggests using scores $t_{1n}, \ldots, t_{pn}$ as before for the failures and $t_{(p+1)n}$ for all censored observations. The mean score, for example, for the first group is

$$\bar{t}_1 = \frac{r_1 \bar{t}_1' + (n_1 - r_1)t_{(p+1)n}}{r_1} \qquad (5.14)$$

where $\bar{t}_1'$ is the mean score of the failures. The mean score for the second group is calculated in a similar way. The F-statistic is $\bar{t}_1/\bar{t}_2$, which has an approximate F distribution with $(2r_1, 2r_2)$ degrees of freedom.

This test is for the hypothesis that the two samples are from populations with equal means. It can also determine if the second population mean is k times the first population mean, for a given k, by dividing the observations in the second sample by k before ranking and applying the test. The set of all values k not rejected in such a significance test forms a confidence interval. The following example illustrates the computation.

Example 5.6
In an experiment comparing two treatments for solid tumor, six mice are assigned to treatment A and six to treatment B. The experiment is terminated after 30 days. The following survival times in days are recorded.

$$\text{Treatment A} \quad 8, 8, 10, 12, 12, 13$$
$$\text{Treatment B} \quad 9, 12, 15, 20, 30+, 30+$$

That is, all of the mice receiving treatment A die within 13 days and two mice receiving treatment B are still alive at the end of the study. Do the data provide evidence that treatment B is more effective than treatment A? Our null and alternative hypotheses are $H_0: S_A = S_B$ and $H_1: S_A < S_B$.

To compute the test statistic, it is convenient to set up a table like Table 5.6. The first column lists all the observations in the two samples. The second column contains the ordered exponential scores t_{rn}. In this case, $n_1 = 6$, $n_2 = 6$, $n = 12$, $r_1 = 6$, and $r_2 = 4$. The scores are computed following (5.12) and (5.13). For example, t_{rn} for $t_i = 10$ is equal to $\frac{1}{12} + \frac{1}{11} + \frac{1}{10} + \frac{1}{9}$

Table 5.6   Computations of Cox's F-Test for Data in Example 5.6

t_i	t_{rn}	t_{rn} of Sample A	t_{rn} of Sample B
8	$\frac{1}{12} = 0.0831$ $\Big\}\,0.129$	0.129	—
8	$\frac{1}{12} + \frac{1}{11} = 0.174$	0.129	—
9	$\frac{1}{12} + \frac{1}{11} + \frac{1}{10} = 0.174 + 0.100 = 0.274$	—	0.274
10	$0.274 + \frac{1}{9} = 0.385$	0.385	—
12	$0.385 + \frac{1}{8} = 0.510$	0.661	—
12	$0.510 + \frac{1}{7} = 0.661$ $\Big\}\,0.661$	0.661	—
12	$0.653 + \frac{1}{6} = 0.820$	—	0.661
13	$0.820 + \frac{1}{5} = 1.020$	1.020	—
15	$1.020 + \frac{1}{4} = 1.270$	—	1.270
20	$1.270 + \frac{1}{3} = 1.603$	—	1.603
30+	$1.603 + \frac{1}{2} = 2.103$	—	2.103
30+	2.103	—	2.103
		2.985	8.014

or simply the previous t_{rn} plus $\frac{1}{9}$, that is $0.274 + \frac{1}{9} = 0.385$. The tied observations receive an average score: for example, for $t_i = 12$, $t_{rn} = \frac{1}{3}(0.510 + 0.653 + 0.820) = 0.661$. The last two columns of Table 5.6 give the scores for the two samples and the sums are entered at the bottom. Thus $\bar{t}_A = 2.985/6 = 0.498$ and $\bar{t}_B = 8.014/4 = 2.004$ according to (5.14) and

$$F = \frac{\bar{t}_A}{\bar{t}_B} = \frac{0.498}{2.004} = 0.249$$

with (12, 8) degrees of freedom. The critical region is $F < F_{12,8,0.95} = 1/F_{8,12,0.05} = 1/2.8486 = 0.351$ for $\alpha = 0.05$.[3] Hence, the data provide strong evidence that treatment B is superior to treatment A.

5.1.6 Comments on the Tests

The tests presented in Sections 5.1.1–5.1.5 are based on rank statistics obtained from scores assigned to each observation. The first four tests are applicable to data with progressive censoring. They can be further grouped into two categories: generalization of the Wilcoxon test (Gehan's and Peto and Peto's) and the non–Wilcoxon test (Cox–Mantel and the logrank test). In the logrank test, if the statistic S is the sum of w scores in group 2, it is the same as U of the Cox–Mantel test. This can be seen in Examples 5.2 ($U = 2.75$) and 5.3 ($S = 2.751$); the small discrepancy is due to rounding-off errors.

The only reason to choose one test over another in a given circumstance is if it will be more powerful, that is, more likely to reject a false hypothesis. When sample sizes are small ($n_1, n_2 \leq 50$), Gehan and Thomas (1969) show that Cox's F-test is more powerful then Gehan's generalized Wilcoxon test if samples are from exponential or Weibull distributions and if there are no censored observations or the observations are singly censored. Comparisons of Gehan's Wilcoxon test to several other tests are reported by Lee et al. (1975). They show that when samples are from exponential distributions, with or without censoring the Cox–Mantel and logrank tests are more powerful and more efficient than the generalized Wilcoxon tests of Gehan and Peto and Peto. There is little difference between the Cox–Mantel and the logrank tests and between the two generalized Wilcoxon tests. When the samples are taken from Weibull distributions with constant hazard ratio (i.e., the ratio of the two hazard functions does not vary with time), the results are essentially the same as in the exponential case. However, when the hazard ratio is nonconstant, the two generalizations of the Wilcoxon test have more power than the other tests. Thus, the logrank test is more powerful than the Wilcoxon tests in detecting departures when the two hazard functions are parallel (proportional hazards) or there is random but

---

[3] $F_{r_1,r_2,\alpha} = 1/F_{r_2,r_1,1-\alpha}$.

equal censoring and when there is no censoring in the samples (Crowley and Thomas 1975). The generalized Wilcoxon tests appear to be more powerful then the logrank test for detecting other types of differences, for example, when the hazard functions are not parallel and when there is no censoring and the logarithm of the survival time follows the normal distribution with equal variance but possibly different means.

The generalized Wilcoxon tests give more weight to early failures than later failures whereas the logrank test gives equal weight to all failures. Therefore, the generalized Wilcoxon tests are more likely to detect early differences in the two survival distributions whereas the logrank test is more sensitive to differences at the right tails. Prentice and Marek (1979) show that Gehan's Wilcoxon statistic is subject to a serious criticism when censoring rates are high. If heavy censoring exists, the test statistic is dominated by a small number of early failures and has low power.

There are situations in which neither the logrank nor the Wilcoxon tests are very effective. When the two distributions differ but their hazard functions or survivorship functions cross, neither the Wilcoxon nor the logrank test is very powerful and it will be sensible to consider other tests. For example, Tarone and Ware (1977) discuss general statistics of similar form (using scores) and Fleming and Harrington (1979) and Fleming et al. (1980) present a two-sample test based on a Smirnov-type statistic designed to measure the maximum distance between estimates of two distributions. The latter approach is shown to be more effective than the logrank or Wilcoxon tests when two survival distributions differ substantially for some range of t values but not necessarily elsewhere. These statistics have not been widely applied. Interested readers are referred to the original papers.

5.2 THE MANTEL AND HAENSZEL TEST

The Mantel–Haenszel (1959) test is particularly useful in comparing survival experience between two groups when adjustments for other prognostic factors are needed. The test has been used in many clinical and epidemiologic studies as a method of controlling the effects of confounding variables. For example, in comparing two treatments for malignant melanoma, it would be important to adjust the comparison for a possible confounding variable such as stage of the disease. In studying the association of smoking and heart disease, it would be important to control the effects of age. To use the Mantel–Haenszel test, the data are stratified by the confounding variable and cast into a sequence of 2×2 tables, one for each stratum.

Let s be the number of strata, n_{ji} be the number of individuals in group j, $j = 1, 2$, and stratum i, $i = 1, \ldots, s$, and d_{ji} the number of deaths (or failures) in group j and stratum i. For each of the s strata, the data can be represented by a 2×2 contingency table:

	Number of Deaths	Number of Survivors	Total
Group 1	d_{1i}	$n_{1i} - d_{1i}$	n_{1i}
Group 2	d_{2i}	$n_{2i} - d_{2i}$	n_{2i}
Total	D_i	S_i	T_i

The null hypothesis to be tested can be stated as

$$H_0: p_{11} = p_{12}$$

$$p_{21} = p_{22}$$

$$\vdots$$

$$p_{s1} = p_{s2}$$

where $p_{ij} = P$ (death|group j, stratum i). Thus, the test permits simultaneous comparison over all the s contingency tables of the difference in survival or death probabilities for the two groups.

The chi-square test statistic without continuity correction[4] is given by

$$\chi^2 = \frac{\left(\sum\limits_{i=1}^{s} d_{1i} - \sum\limits_{i=1}^{s} E(d_{1i}) \right)^2}{\sum\limits_{i=1}^{s} \text{Var}(d_{1i})} \tag{5.15}$$

where

$$E(d_{1i}) = \frac{n_{1i} D_i}{T_i} \tag{5.16}$$

$$\text{Var}(d_{1i}) = \frac{n_{1i} n_{2i} D_i S_i}{T_i^2 (T_i - 1)} \tag{5.17}$$

are the mean and variance, respectively, of the number of deaths in group i computed conditionally on the contingency table marginal totals. This statistic follows approximately the chi-square distribution with one degree of freedom. Thus, a computed chi-square value larger than the table chi-square value for the significance level chosen indicates a significant difference in survival between the two groups. The following two examples illustrate the use of the test.

---

[4]According to Grizzle (1967), the distribution of χ^2 without continuity correction is closer to the chi-square distribution than the χ^2 with continuity correction. His simulations show that the probability of Type I error (rejecting a true hypothesis) is better controlled without the continuity correction at $\alpha = 0.01, 0.05$.

Example 5.7

Five hundred and ninety-five persons participate in a case control study of the association of cholesterol and coronary heart disease (CHD). Among them, 300 persons are known to have CHD and 295 are free of CHD. To find out if elevated cholesterol is significantly associated with CHD, the investigator decides to control the effects of smoking. The study subjects are then divided into two strata: smokers and nonsmokers.

The following tables give the data:

Smokers

	With CHD	Without CHD	Total
Elevated cholesterol			
Yes	120	20	140
No	80	60	140
Total	200	80	280

Nonsmokers

	With CHD	Without CHD	Total
Elevated cholesterol			
Yes	30	60	90
No	70	155	255
Total	100	215	315

Using (5.16) and (5.17), we obtain

$$E(d_{11}) = \frac{140 \times 200}{280} = 100 \qquad E(d_{12}) = \frac{90 \times 100}{315} = 28.571$$

$$\text{Var}(d_{11}) = \frac{140 \times 140 \times 200 \times 80}{(280)^2(280 - 1)} = 14.337$$

$$\text{Var}(d_{12}) = \frac{90 \times 225 \times 100 \times 215}{(315)^2(315 - 1)} = 13.974$$

Using (5.15) and $d_{11} = 120$, $d_{12} = 30$, we have

$$\chi^2 = \frac{(150 - 128.571)^2}{14.337 + 13.974} = 16.220$$

which is significant at the 0.001 level. Thus, elevated cholesterol is significantly associated with CHD after adjusting for the effects of smoking.

Example 5.8

Table 5.7 gives survival data in life-table format of male cases with localized cancer of the rectum in Connecticut for the periods 1935–1944 and 1945–1954. We use Mantel and Haenszel's chi-square test to see if the survival distribution of patients diagnosed in 1935–1944 is the same as for patients diagnosed in 1945–1954. The null hypothesis is that the two survival distributions are the same. It is not necessary to set up 10 contingency tables for the 10 intervals. The chi-square value is easily calculated by constructing columns 7–12 directly from the life table. Using the sums in columns 1, 10, and 12, we obtain

$$\chi^2 = \frac{(330.0 - 246.50)^2}{132.491} = 52.624$$

which is significant at the 0.001 level. Thus, the data show a significant difference between the survival distributions of patients diagnosed in 1935–1944 and 1945–1954.

It should be noted that this chi-square test statistic, when applied to life tables, gives more weight to those deaths that occur in an early time interval rather than later. That is, if the two groups are subject to the same probability of surviving through the entire study period, (5.15)–(5.17) will give high mortality for the group in which early deaths occur. Mantel (1966) gives the following illustration.

Consider two groups of 100 individuals each. Both have 50 deaths. In group 1 all deaths occur in the first interval and in group 2 all deaths occur in the second interval. The contingency tables for the two intervals follow:

First Interval

	Deaths	Survivors	Total
Group 1	50	50	100
Group 2	0	100	100
Total	50	150	200

Second Interval

	Deaths	Survivors	Total
Group 1	0	50	50
Group 2	50	50	100
Total	50	100	150

From these two tables we have $E(d_{11}) = 100 \times \frac{50}{200} = 25$ and $E(d_{12}) = 50 \times \frac{50}{150} = 16.67$. The total expected deaths are $25 + 16.67 = 41.67$, so the 50

Table 5.7 Computational Procedure for Comparing Two Survival Distributions: Male Cases with Localized Cancer of Rectum in Connecticut

Interval	1935–1944 Deaths d_{1i}	Survivors $n_{1i}-d_{1i}$	Total n_{1i}	1945–1954 Deaths d_{2i}	Survivors $n_{2i}-d_{2i}$	Total n_{2i}	Combined Time Periods Deaths D_i	Survivors S_i	Total T_i	$E(d_{1i})$ $\dfrac{(3)\times(7)}{(9)}$	$n_{2i}S_i/T_i$ $\dfrac{(6)\times(8)}{(9)}$	$Var(d_{1i})$ $\dfrac{(10)\times(11)}{(9)-1.0}$
	(1)	(2)	(3)	(4)	(5)	(6)	(7)	(8)	(9)	(10)	(11)	(12)
1	167	220.0	387.0	185	559.0	744.0	352	779.0	1131.0	120.45	512.45	54.624
2	45	173.5	218.5	88	461.0	549.0	133	634.5	767.5	37.86	453.86	22.418
3	45	127.5	172.5	55	396.0	451.0	100	523.5	623.5	27.67	378.67	16.832
4	19	108.0	127.0	43	343.0	386.0	62	451.0	513.0	15.35	339.35	10.174
5	17	91.0	108.0	32	299.0	331.0	49	390.0	439.0	12.05	294.05	8.090
6	11	79.5	90.5	31	235.0	266.0	42	314.5	356.5	10.66	234.66	7.036
7	8	71.0	79.0	20	170.0	190.0	28	241.0	269.0	8.22	170.22	5.221
8	5	66.0	71.0	7	132.0	139.0	12	198.0	210.0	4.06	131.06	2.546
9	6	59.5	65.5	6	101.5	107.5	12	161.0	173.0	4.54	100.04	2.641
10	7	52.0	59.0	6	71.0	77.0	13	123.0	136.0	5.64	69.64	2.909
	330									246.50		132.491

deaths in group 1 is 20% larger than expected. Thus, a significant chi-square value may be obtained if early survival patterns differ significantly in the two groups.

5.3 COMPARISON OF K ($K > 2$) SAMPLES

In this section the two-sample problem is generalized to a situation in which the data consist of K ($K > 2$) samples, one sample from each of the K treatment populations. The problem is to decide whether the K independent samples can be regarded as coming from the same population, or in practical terms, to see if the survival data from patients receiving the K treatments provide enough evidence to conclude that the K treatments are equally effective. This problem has been considered by many statisticians, for example, Kruskal and Wallis (1952), Mantel and Haenszel (1959), Breslow (1970), and Peto and Peto (1972). In this section two nonparametric tests for the problem are presented. The first is Kruskal and Wallis's (1952) H-test for uncensored data. The second is a generalization of the H-test for censored data (Peto and Peto 1972). Both use ranks instead of the original observations and are simple to apply.

5.3.1 The Kruskal–Wallis Test

The Kruskal–Wallis H-test (Kruskal and Wallis 1952, Hollander and Wolfe 1973, and Marascuilo and McSweeney 1977), analogous to the F test in the usual analysis of variance, uses ranks rather than original observations; it is also called the Kruskal–Wallis one-way analysis of variance by ranks. It assumes that the variable (survival time) under study has an underlying continuous distribution.

Let N be the total number of independent observations in the K samples, n_j the number of observations in the jth sample, $j = 1, \ldots, K$, and t_{ij} the ith observation in the jth sample. The data can be arranged as in Table 5.8. The null hypothesis H_0 states that the K samples come from the same population (or clinically, the K treatments are equally effective).

In the computation of the Kruskal–Wallis H-test, we first rank all N

Table 5.8   General Data Format for One-Way Analysis of Variance

1	2	$\cdots$	K
t_{11}	t_{12}	$\cdots$	t_{1k}
$\vdots$	$\vdots$		$\vdots$
$t_{n_1 1}$	$t_{n_2 2}$	$\cdots$	$t_{n_k k}$

observations from smallest to largest. Let r_{ij} be the rank of t_{ij}. Compute, for $j = 1, K$,

$$R_j = \sum_{i=1}^{n_j} r_{ij}, \qquad \bar{R}_j = \frac{R_j}{n_j} \quad \text{and} \quad \bar{R} = \tfrac{1}{2}(N + 1) \qquad (5.18)$$

where R_j and $\bar{R}_j$ are, respectively, the sum of the ranks and the average rank of the jth treatment and $\bar{R}$ is the overall average rank. Then the Kruskal–Wallis H-statistic is

$$H = \frac{12}{N(N+1)} \sum_{j=1}^{K} n_j (\bar{R}_j - \bar{R})^2 \qquad (5.19)$$

$$= \frac{12}{N(N+1)} \sum_{j=1}^{K} \frac{R_j^2}{n_j} - 3(N+1) \qquad (5.20)$$

Under the null hypothesis, H has an asymptotic (n_j's approaching infinity or n_j's are large) chi-square distribution with $K - 1$ degrees of freedom. Thus, for large n_j's, the approximate test procedure at the α level is to reject H_0 if $H \geq \chi^2_{(k-1),\alpha}$. When $K = 3$ and the number of observations in each of the three samples is 5 or fewer, the chi-square approximation is not sufficiently close. For such cases, exact permutational distributions of H are available and percentage points χ_k are given in Table C-4 of Appendix C. The test procedure is to reject H_0 if $H \geq \chi_{k,\alpha}$, where $\chi_{k,\alpha}$ satisfies the equation $P(H \geq \chi_{k,\alpha} | H_0) = \alpha$.

When there are tied observations, each is assigned the average of the ranks. To correct for the effects of ties, H is computed by (5.20) and then divided by

$$1 - \frac{1}{N^3 - N} \sum_{j=1}^{g} T_j \qquad (5.21)$$

where g is the number of tied groups and $T_j = t_j^3 - t_j$ with t_j being the number of tied observations in a tied group. In counting g, an untied observation is considered as a tied group of size 1. Thus, a general expression of H corrected for ties is

$$H = \frac{\dfrac{12}{N(N+1)} \displaystyle\sum_{j=1}^{k} \dfrac{R_j^2}{n_j} - 3(N+1)}{1 - \dfrac{\displaystyle\sum_{j=1}^{g} T_j}{N^3 - N}} \qquad (5.22)$$

Note that when there are no ties, $g = N$, $t_j = 1$ for all j, and $T_j = 0$, and (5.22) reduces to (5.20). The following example illustrates the use of the test.

Table 5.9   Cholesterol Values of 12 Individuals on Three Different Diets

Diet 1	Diet 2	Diet 3
229	145	231
176	181	208
187	147	217
208	187	199

Example 5.9

In a study of the relationship between cholesterol level and diet, three diets are given randomly to 12 males. Table 5.9 shows the cholesterol levels of the 12 individuals after a given period of time. The purpose of the study is to decide if the three diets are equally effective in controlling cholesterol level.

    The null hypothesis H_0 states that there is no difference in cholesterol level of men having the three diets, and the alternative H_1 says that the cholesterol levels of men having the three different diets are different. To compute the H-statistic, we first rank the observations as in Table 5.10 and compute R_j. In this case $N = 12$, $n_1 = n_2 = n_3 = 4$, $g = 10$, and $T_j = 0$ except for $j = 5, 7$. Hence, $\Sigma_{j=1}^g T_j = 2(8 - 2) = 12$, and by (5.22),

$$H = \frac{\dfrac{12}{12(13)}\left(\dfrac{784}{4} + \dfrac{156.25}{4} + \dfrac{1406.25}{4}\right) - 3(13)}{1 - \dfrac{12}{1728 - 12}} = 6.168$$

Table 5.10   Computation of H for Data in Example 5.9

Ordered Observations	Ranks	Ranks of Diet 1	Ranks of Diet 2	Ranks of Diet 3
145	1	—	1	
147	2	—	2	
176	3	3		
181	4	—	4	
187	5.5	—	5.5	
187	5.5	5.5		
199	7	—	—	7
208	8.5	8.5		
208	8.5	—	—	8.5
217	10	—	—	10
229	11	11		
231	12	—	—	12
R_j		28	12.5	37.5

From Table C-4, we find that $P(H \geq 6.038 | H_0) = 0.037$ and $P(H \geq 6.269 | H_0) = 0.033$; we reject H_0 at the $\alpha \approx 0.035$ level. There is evidence of significant differences among the diets.

5.3.2 Multiple Comparisons Based on the Kruskal–Wallis Test

If the null hypothesis that the K samples are from the same distribution is rejected, we might ask which particular samples are from different distributions. In Example 5.9 we reject the null hypothesis that the three diets are similar. The investigator may be further interested in knowing which particular diets differ from one another. In this section we introduce some nonparametric methods for multiple comparison based on Kruskal–Wallis rank sums. An excellent treatment of multiple comparisons is given by Miller (1966).

To decide which treatments differ from one another, there are $\frac{1}{2} K(K - 1)$ decisions to make, one for each pair of treatments. The null hypothesis can be written as

$$H_0: \text{Samples } i \text{ and } j \text{ are from the same population}$$
$$\text{for } i = 1, \ldots, K - 1, \, j = i + 1, \ldots, K, i < j$$

Let the probability of at least one wrong decision when H_0 is true be controlled by α and the probability of making all correct decisions when H_0 is true be $1 - \alpha$. To make the $\frac{1}{2} K(K - 1)$ decisions, we introduce the following comparison procedures.

1. When sample sizes are equal, that is, $n_1 = n_2 = \cdots = n_K = n$, and n is small, we reject the hypothesis that the ith and jth samples, $i < j$, are from the same distribution if

$$|R_i - R_j| \geq y(\alpha, K, n) \tag{5.23}$$

where $y(\alpha, K, n)$ satisfies the equation

$$P(|R_i - R_j| \geq y(\alpha, K, n) | H_0, i < j) = 1 - \alpha \tag{5.24}$$

and $R_1, R_2, \ldots, R_k$ are given in (5.18). Some approximate values of y are given in Table C-5.

2. When sample sizes are equal to n and n is large, we introduce Miller's (1966) procedure, that is, to reject the hypothesis that the ith and jth samples, $i < j$, are from the same distribution if

$$|\bar{R}_i - \bar{R}_j| \geq q(\alpha, K)(\tfrac{1}{12} K(Kn + 1))^{1/2} \tag{5.25}$$

where $\bar{R}_1, \ldots, \bar{R}_k$ are given in (5.18) and $q(\alpha, K)$ is the upper α

percentile point of the range of K independent standard normal variables. Table C-6 gives the $q(\alpha, K)$ values for some K and α.

3. For cases of small unequal sample sizes $n_1, \ldots, n_K$, a conservative procedure is to reject the hypothesis that the ith and jth samples, $i < j$, are from the same distribution if

$$|\bar{R}_i - \bar{R}_j| \geq (x_{\alpha,k})^{1/2}(\tfrac{1}{12}N(N+1))^{1/2}(1/n_i + 1/n_j)^{1/2} \quad (5.26)$$

where N is the total number of observations. Values of $x_{\alpha,k}$ are given in Table C-4.

4. When $n_1, \ldots, n_k$ are large, Dunn (1964) suggests the following procedure. Reject the hypothesis that the ith and jth samples, $i < j$ are from the same distribution if

$$|\bar{R}_i - \bar{R}_j| > z_{\alpha/[K(K-1)]}(\tfrac{1}{12}N(N+1))^{1/2}(1/n_1 + 1/n_j)^{1/2} \quad (5.27)$$

where $z_{\alpha/[K(K-1)]}$ is the upper $100\alpha/[K(K-1)]$ percentage point of the standard normal distribution (see Table C-1).

Example 5.10

Let us use the data in Example 5.9. To examine which particular diets differ from one another, we apply (5.23). Since $K = 3$, there are three possible comparisons. The calculation is shown in Table 5.11. For $K = 3$, $n = 4$, and from Table C-5, $y(0.045, 3, 4) = 24$; hence for $(i, j) = (2, 3)$, (5.23) is satisfied. Thus, at $\alpha \cong 0.045$, we conclude that diets 2 and 3 are significantly different.

5.3.3 A Test for Censored Data

In Section 5.1 we introduced three nonparametric tests based on scores for comparing two samples with censored observations; Gehan's generalized Wilcoxon test (if Mantel's procedure is used), Peto and Peto's generalized Wilcoxon test, and the logrank test. The K-sample test discussed in this section can be considered an extension of these tests and the Kruskal–Wallis test.

The data format is the same as given in Table 5.8 except some of the t_{ij}'s may be censored observations.

Table 5.11   Multiple Comparisons for Data in Example 5.9

i	j	$|R_i - R_j|$	Decision
1	2	$|28 - 12.5| = 15.5$	Not significant
1	3	$|28 - 37.5| = 9.5$	Not significant
2	3	$|12.5 - 37.5| = 25.0$	Significant

Suppose we have a set of N scores $w_1, w_2, \ldots, w_N$ obtained according to the manner of scoring in one of the three tests mentioned above. The sum of the N scores is zero. Let S_j be the sum of the scores in the jth sample. The null hypothesis H_0 states that the K samples are from the same distribution. To test H_0, we calculate

$$X^2 = \frac{\displaystyle\sum_{j=1}^{K} \frac{S_j^2}{n_j}}{s^2} \tag{5.28}$$

where

$$s^2 = \frac{\displaystyle\sum_{i=1}^{N} w_i^2}{N-1} \tag{5.29}$$

Under the null hypothesis X^2 has approximately chi-square distribution with $K-1$ degrees of freedom (Peto and Peto 1972). Thus, we reject H_0 if X^2 exceeds the upper 100α percentage point of the chi-square distribution with $K-1$ degrees of freedom, that is, if $X^2 \geq \chi^2_{(K-1),\alpha}$.

Examples 5.11, using the scoring method of Mantel (1967) for Gehan's generalized Wilcoxon test, illustrates the K-sample test for censored data.

Example 5.11

Consider the initial remission times of leukemia patients (in days) induced by three treatments as given in Table 5.12. In this case, $K = 3$, $N = 66$, $n_1 = 25$, $n_2 = 19$, and $n_3 = 22$. A table similar to Table 5.1 may be set up to compute the score for every observation. The computation is left to the reader as an exercise. The sums of scores in the three samples are $S_1 = -273$, $S_2 = 170$, and $S_3 = 103$. The sum of squares of the scores $\sum w_i^2 = 89702$. Hence, from (5.28) and (5.29), $X^2 = 3.612$. From Table C-2, $\chi^2_{2,0.05} = 5.991$; thus we do not reject H_0. The data do not show significant differences among the three initial treatments.

Table 5.12   Initial Remission Times of Leukemia Patients

1	2	3
4, 5, 9, 10, 12, 13, 10, 23, 28, 28, 28, 29 31, 32, 37, 41, 41, 57, 62, 74, 100, 139, 20+, 258+, 269+	8, 10, 10, 12, 14, 20, 48, 70, 75, 99, 103, 162, 169, 195, 220, 161+, 199+, 217+, 245+	8, 10, 11, 23, 25, 25, 28, 28, 31, 31, 40, 48, 89, 124, 143, 12+, 159+, 190+, 196+, 197+, 205+, 219+

BIBLIOGRAPHICAL REMARKS

Gehan's test was first proposed in 1965. The Cox–Mantel test was first discussed by Cox in 1959, then by Mantel in 1966, and finally by Cox again in 1972. The scores for the logrank test were proposed by Peto and Peto in 1972 along with another generalization of the Wilcoxon test. In the same paper, they also discuss the K-sample test for censored data. The logrank test is also discussed by Peto et al. (1977). Cox's F-test was developed in 1964 and the Mantel–Haenszel chi-square test can be found in Mantel and Haenszel (1959) and Mantel (1966). The Kruskal–Wallis one-way analysis of variance can be found in most standard textbooks under nonparametric methods. Readers who are interested in the theoretical development or more properties of these tests should read the original papers cited above.

Applications of these tests are given in the original papers or can easily be found in medical and epidemiologic journals. The following are just a few examples: Freireich et al. (1963), Sullivan et al. (1975), Bonadonna et al. (1976), Hart et al. (1977), Kirk et al. (1980), Kashiwagi et al. (1985), and Oliver et al. (1988).

EXERCISES

The first five exercises are continuations of Exercise 4.1–4.4 and 4.6.

5.1 For the survival times given in Table 3.1, compare the survival distributions of the two treatment groups using:
   a. Gehan's generalized Wilcoxon test
   b. The Cox–Mantel test

5.2 For the remission data given in Table 3.1, compare the remission time distributions of the two treatment groups using:
   a. The logrank test
   b. The Peto and Peto generalized Wilcoxon test

5.3 For the data given in Table 3.4, compare the tumor-free time distributions of the three diet groups.

5.4 For the remission data of 42 leukemia patients given in Example 3.3, use the two generalized Wilcoxon test to see if 6-MP is more effective in prolonging remission time than a placebo.

5.5 For the first four skin tests given in Exercise Table 3.1, use the Cox–Mantel test and the logrank test to see if there is a significant difference in survival between patients with positive (≥ 5 mm for mumps, ≥ 10 mm for others) and negative (< 5 mm for mumps, < 10 mm for others) reactions.

5.6 Compute the test statistic W of Gehan's generalized Wilcoxon test by using (5.1) for the data in Example 5.1. Do you get the same result as in Example 5.1?

5.7 Consider the data in Example 5.11. Use Mantel's procedure for Gehan's generalized Wilcoxon test to compute a score for each observation and the sum of scores for each of the three treatment groups.

5.8 Using the data in Table 3.1, compare the age distributions of the two treatment groups using Cox's F-test.

5.9 In Asal et al. (1988a), the relationship between obesity and renal cell carcinoma is studied. Exercise Table 5.1 is reported.

Exercise Table 5.1

Percentage of Standard BMI[a]	Male		Female	
	Case	Control	Case	Control
<130	123	150	55	59
≥130	86	45	51	46

[a]Standard BMI: male, 22.1; female, 20.6. Percentage of standard BMI = (observed BMI/standard BMI) × 100.

Is elevated percentage of standard BMI associated with renal cell carcinoma after controlling the effects of sex?

5.10 Consider the survival data of males with angina pectoris in Table 4.6 and females with the same disease in Exercise Table 4.2. Is there a significant difference between the survival distributions of males and females?

5.11 In a study of noise level and efficiency, 18 studies were given a very simple test under three different noise levels. It is known that under normal conditions, they should be able to finish the test in 10 minutes. The students were randomly assigned to the three levels. Exercise Table 5.2 gives the time required to finish the test for the three levels. Are the three noise levels significantly different? If they are, determine which levels differ from one another.

Exercise Table 5.2

1	2	3
10.5	10.0	12.0
9.0	12.0	13.0
9.5	12.5	15.5
9.0	11.0	14.0
8.5	12.0	12.5
10.0	10.5	15.0

5.12 Exercise Table 5.3 gives the survival time in weeks of 30 brain tumor patients receiving four different treatments. Are the four treatments equally effective?

Exercise Table 5.3

1	2	3	4
4	1	3	5
5	4	7	15
9	9	14	20
12	12	20	31
20+	15	27	39
25	23	30	47
30+	30	32+	55+
		50+	67+

CHAPTER 6

Some Well-Known Survival Distributions and Their Applications

Usually, there are many physical causes that lead to the failure or death of an individual at a particular time. It is very difficult, if not impossible, to isolate these physical causes and mathematically account for all of them. Therefore, choosing a theoretical distribution to approximate survival data is as much an art as a scientific task. In this chapter, several theoretical distributions that have been widely used to describe survival time are discussed, their characteristics summarized, and their applications illustrated.

6.1 THE EXPONENTIAL DISTRIBUTION

The simplest and most important distribution in survival studies is the exponential distribution. In the late 1940s, researchers chose the exponential distribution to describe the life pattern of electronic systems. Davis (1952) gives a number of examples, including bank statement and ledger errors, payroll check errors, automatic calculating machine failure, and radar set component failure, in which the failure data are well described by the exponential distribution. Epstein and Sobel (1953) report why they select the exponential distribution over the popular normal distribution and show how to estimate the parameter when data are singly censored. Epstein (1958) also discusses in some detail the justification for the assumption of an exponential distribution.

The exponential distribution plays a role in lifetime studies analogous to that of the normal distribution in other areas of statistics. Applications in animal and human studies of chronic and infectious diseases can be found in Zelen (1966), Feigl and Zelen (1965), Zippin and Armitage (1966), Byar et al. (1974), DeWals and Bouckaert (1985), Shulz et al. (1986), Walle et al. (1987), and Eggermont (1988).

131

The exponential distribution is often referred to as a purely random failure pattern. It is famous for its unique "lack of memory," which requires that the age of the animals or individual not affect future survival. Although many survival data cannot be adequately described by the exponential distribution, an understanding of it facilitates the treatment of more general situations.

The exponential distribution is characterized by a constant hazard rate λ, its only parameter. A large λ indicates high risk and short survival while a small λ indicates low risk and long survival. Figure 6.1 depicts the survivorship function, the density function, and the hazard function of the exponential distribution with parameter λ. When $\lambda = 1$, the distribution is often referred to as the *unit exponential distribution*.

When the survival time T follows the exponential distribution with a parameter λ, the probability density function is defined as

$$f(t) = \begin{cases} \lambda e^{-\lambda t} & t \geq 0, \lambda > 0 \\ 0 & t < 0 \end{cases} \tag{6.1}$$

The cumulative distribution function is

$$F(t) = 1 - e^{-\lambda t} \qquad t \geq 0 \tag{6.2}$$

and the survivorship function is then

$$S(t) = e^{-\lambda t} \qquad t \geq 0 \tag{6.3}$$

So that, by (2.10), the hazard function is

$$h(t) = \lambda \qquad t \geq 0 \tag{6.4}$$

a constant, independent of t. Figure 6.1 gives the graphical presentation of the three functions.

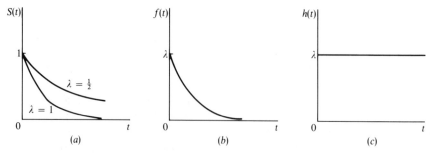

Figure 6.1 Exponential distribution: (*a*) survivorship function; (*b*) probability density function; (*c*) hazard function.

Because the exponential distribution is characterized by a constant hazard rate, independent of the age of the individual, there is no aging or wearing out and failure or death is a random event independent of time. When the natural logarithm of the survivorship function is taken, $\log_e S(t) = -\lambda t$, which is a linear function of t. Thus, it is easy to determine whether data come from an exponential distribution by plotting $\log_e \hat{S}(t)$ against t, where $\hat{S}(t)$ is an estimate of $S(t)$. A linear configuration indicates that the data follow an exponential distribution and the slope of the straight line is an estimate of the hazard rate λ.

The mean and variance of the exponential distribution with parameter λ are, respectively, $1/\lambda$ and $1/\lambda^2$. The median is $(1/\lambda)\log_e 2$. The coefficient of variation is 1.

A more general form of the exponential distribution is the *two-parameter exponential distribution* with probability density function

$$f(t) = \begin{cases} \lambda e^{-\lambda(t-G)} & t \geq G \\ 0 & t < G \end{cases} \tag{6.5}$$

Then

$$F(t) = \begin{cases} 1 - e^{-\lambda(t-G)} & t \geq G \\ 0 & t < G \end{cases} \tag{6.6}$$

$$S(t) = \begin{cases} e^{-\lambda(t-G)} & t \geq G \\ 0 & t < G \end{cases} \tag{6.7}$$

and

$$h(t) = \begin{cases} 0 & 0 \leq t < G \\ \lambda & t \geq G \end{cases} \tag{6.8}$$

The term G is a *guarantee time* within which no deaths or failures can occur, or a minimum survival time. If $G = 0$, (6.5)–(6.8) reduce to (6.1)–(6.4) for the one parameter exponential. The mean and the median of the two-parameter exponential distribution are, respectively, $G + 1/\lambda$ and $(\log_e 2 + \lambda G)/\lambda$.

Example 6.1

In a study of new anticancer drugs in the L1210 animal leukemia system, Zelen (1966) successfully used the exponential distribution as the model for survival time. The system consists of injecting a tumor inoculum into inbred mice. These tumor cells then proliferate and eventually kill the animal, but survival time may be prolonged by an active drug. Figure 6.2 shows the survival curve in a semilogarithmic scale of the untreated mice inoculated at different cell dilutions. Twenty-five mice were inoculated at each dilution. The reasonably linear configurations suggest that the survival distributions follow the exponential distribution quite well. The four straight lines fitted

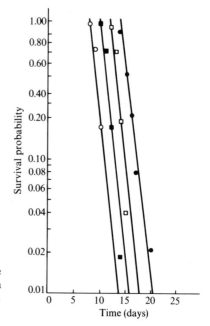

Figure 6.2 Survival curves of untreated mice inoculated with serial 10-fold dilutions of leukemia L1210: (○) 10^5 cells; (■) 10^3 cells; (●) 10^2 cells. [*Source*: Zelen (1966).]

to the points are almost parallel, indicating that the hazard rate was independent of the inoculum size. Table 6.1 gives the estimated values of the guarantee times G, hazard rates λ, and the mean survival times (in days) for the various dilutions. (Estimation procedures are discussed in Chapter 8.) Note that the estimated hazard rates λ are very close.

The probability that a mouse receiving 10^5 cells of inoculum will survive more than 20 days is, from (6.7),

$$S(20) = e^{-0.78(20-8)}0.0001$$

and the median survival time is 8.9 days.

Figure 6.3 gives the survival curves of mice treated with different doses of cyclophosphamide on day 3 after receiving a 10^5 tumor inoculum. Table 6.2 gives the estimates of G, λ, and the mean survival time. Mice receiving

Table 6.1   Estimates of G and λ and Mean Survival Time for Untreated Mice with Serial Leukemia Dilutions

Dilution	$\hat{G}$	$\hat{\lambda}$	Mean Survival Time
10^5	8.0	0.78	9.3
10^4	10.0	0.78	11.3
10^3	11.9	0.76	13.2
10^2	13.9	0.67	15.4

Source: Zelen (1966).

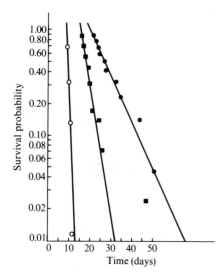

Figure 6.3 Survival curves of mice treated with cyclophosphamide on day 3 after inoculation of 10^5 tumor cells: ($\bigcirc$) control; ($\blacksquare$) 80 mg/kg; ($\bullet$) 160 mg/kg [*Source:* Zelen (1966).]

Table 6.2   Estimates of G, λ and Mean Survival Time for Treated Mice on Day 3

Dose (mg/kg)	$\hat{G}$	$\hat{\lambda}$	Mean Survival Time
Control	8.7	1.12	9.6
80	15.6	0.29	19.0
160	21.5	0.10	31.5

Source: Zelen (1966).

160 mg/kg of the drug show a remarkably improved survival pattern over the control group.

The probability that a mouse receiving 80 mg/kg of cyclophosphamide will survive more than 20 days is, from (6.7),

$$S(20) = e^{-0.29(20-15.6)} = 0.279$$

The median survival time is approximately 18 days.

6.2   THE WEIBULL DISTRIBUTION

The Weibull distribution is a generalization of the exponential distribution. However, unlike the exponential distribution, it does not assume a constant hazard rate and therefore has broader application. The distribution was proposed by Weibull (1939) and its applicability to various failure situations discussed again by Weibull (1951). It has been used, among other applications, to investigate the fatigue life of deep-groove ball bearings by Leiblein and Zelen (1956), to describe electron tube failures by Kao (1958), to analyze carcinogenesis experiments by Pike (1966), Peto et al. (1972), Peto and Lee (1973), and Williams (1978), to characterize early radiation response probabilities by Scott and Hahn (1980), and to model human disease-specific mortality by Juckett and Rosenberg (1990).

The Weibull distribution is characterized by two parameters, γ and λ. The value of γ determines the shape of the distribution curve and the value of λ determines its scaling. Consequently, γ and λ are called the shape and scale parameters, respectively. The relationship between the value of λ and survival time can be seen from Figure 6.4, which shows the hazard rate of the Weibull distribution with $\gamma = 0.5, 1, 2, 4$. When $\gamma = 1$, the hazard rate remains constant as time increases; this is the exponential case. The hazard rate increases when $\gamma > 1$ and decreases when $\gamma < 1$ as t increases. Thus, the Weibull distribution may be used to model the survival distribution of a population with increasing, decreasing, or constant risk. Examples of increasing and decreasing hazard rates are, respectively, patients with lung cancer and patients who undergo successful major surgery.

The probability density function and cumulative distribution functions are, respectively,

$$f(t) = \lambda \gamma (\lambda t)^{\gamma - 1} e^{-(\lambda t)^{\gamma}} \qquad t \geq 0,\ \gamma,\ \lambda > 0 \tag{6.9}$$

and

$$F(t) = 1 - e^{-(\lambda t)^{\gamma}} \tag{6.10}$$

The survivorship function is therefore

$$S(t) = e^{-(\lambda t)^{\gamma}} \tag{6.11}$$

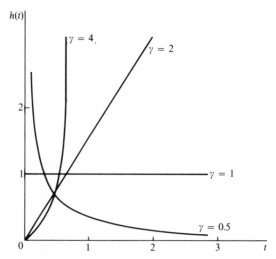

Figure 6.4 Hazard functions of Weibull distribution with $\lambda = 1$.

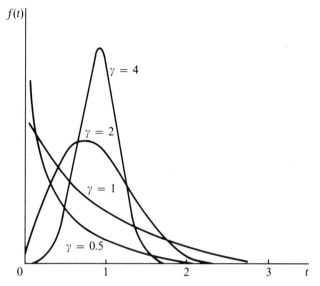

Figure 6.5 Density curves of Weibull distribution with $\lambda = 1$.

and the hazard function, the ratio of (6.9) to (6.11), is

$$h(t) = \lambda\gamma(\lambda t)^{\gamma - 1} \qquad (6.12)$$

Figure 6.5 gives the Weibull density function with scale parameter $\lambda = 1$ and several different values of the shape parameter γ.

For the survival curve, it is simple to plot the logarithm of $S(t)$,

$$\log_e S(t) = -(\lambda t)^\gamma \qquad (6.13)$$

Figure 6.6 gives $\log_e S(t)$ for $\lambda = 1$ and $\gamma = 1$, > 1, < 1. When $\gamma = 1$, $\log_e S(t)$ is a straight line with negative slope. When $\gamma < 1$, negative aging, $\log_e S(t)$ decreases very slowly from zero and then approaches a constant value. When $\gamma > 1$, positive aging, $\log_e S(t)$ decreases sharply from zero as t increases.

The mean of the Weibull distribution is

$$\mu = \frac{\Gamma(1 + 1/\gamma)}{\lambda} \qquad (6.14)$$

and the variance is

$$\sigma^2 = \frac{1}{\lambda^2}\left[\Gamma\left(1 + \frac{2}{\gamma}\right) - \Gamma^2\left(1 + \frac{1}{\gamma}\right)\right] \qquad (6.15)$$

where $\Gamma(\gamma)$ is the well-known gamma function defined as

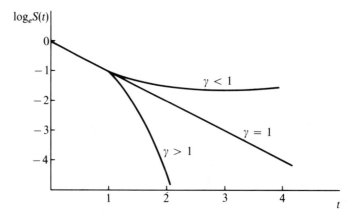

Figure 6.6   Curves of $\log_e S(t)$ of Weibull distribution with $\lambda = 1$.

$$\Gamma(\gamma) = \int_0^\infty x^{\gamma-1} e^{-x}\, dx \tag{6.16}$$

$$= (\gamma - 1)! \qquad \text{when } \gamma \text{ is a positive integer}$$

Values of $\Gamma(\gamma)$ can be found in Abramowitz and Stegun (1964). The coefficient of variation is then

$$CV = \left[\frac{\Gamma(1 + 2/\gamma)}{\Gamma^2(1 + 1/\gamma)} - 1 \right]^{1/2} \tag{6.17}$$

The Weibull distribution can also be generalized to take into account a guarantee time G during which no deaths or failures can occur. The three-parameter Weibull probability density function is

$$f(t) = \lambda^\gamma \gamma (t - G)^{\gamma-1} \exp[-\lambda^\gamma (t - G)^\gamma] \tag{6.18}$$

Consequently,

$$S(t) = \exp[-\lambda^\gamma (t - G)^\gamma] \tag{6.19}$$

and

$$h(t) = \lambda^\gamma \gamma (t - G)^{\gamma-1} \tag{6.20}$$

Example 6.2
Pike (1966) applied the Weibull distribution to a two-group experiment on vaginal cancer in rats exposed to the carcinogen DMBA. The two groups

were distinguished by a pretreatment regime. The times in days, after the start of the experiment, at which the carcinoma was diagnosed for the two groups of rats were as follows:

Group 1   143, 164, 188, 188, 190, 192, 206, 209, 213, 216,
           220, 227, 230, 234, 246, 265, 304, 216+, 244+

Group 2   142, 156, 173, 198, 205, 232, 232, 233, 233, 233, 233,
           239, 240, 261, 280, 280, 296, 296, 323, 204+, 344+

Data with a plus sign are from rats that did not satisfy the accepted criteria for diagnosis of carcinoma or were removed from the experiment for some other reason.

Assuming $G = 100$ and $\gamma = 3$, Pike obtained $\hat{\lambda}_1 = (4.51 \times 10^{-7})^{1/3}$ for group 1 and $\hat{\lambda}_2 = (2.38 \times 10^{-7})^{1/3}$ for group 2. Graphical and analytical estimation procedures are discussed in Chapters 7 and 8, respectively. Figure 6.7 plots the survival curves of the two groups. The step functions are nonparametric estimates similar to the Kaplan–Meier estimate. The smooth curves are obtained from the Weibull fits. Table 6.3 shows the calculation of the plotting points for group 2.

It is obvious that the Weibull distributions with $G = 100$, $\gamma = 3$, $\hat{\lambda}_1 = (4.51 \times 10^{-7})^{1/3}$, and $\hat{\lambda}_2 = (2.38 \times 10^{-7})^{1/3}$ fit the carcinoma-free time of the two groups of rats very well. Pike also considered three additional cases: with $G = 100$ assumed known; with $\gamma = 3$ assumed known; and with no parameters known. In all four cases, the Weibull distribution was shown to be an appropriate model of the carcinoma-free time of the rats exposed to carcinogen DMBA.

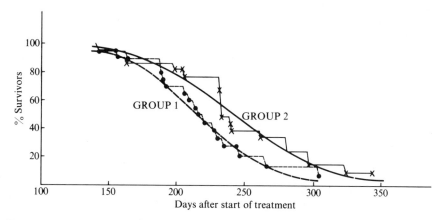

Figure 6.7 Survival curves of rats exposed to carcinogen DMBA [*Source*: Pike (1966). Reproduced with permission of the Biometrics Society.]

Table 6.3 Calculation of Survivorship Functions for Group 2 of Rats Exposed to DMBA

Time	r	$n-r+1$	$\dfrac{n-r}{n-r+1}$	Kaplan–Meier Estimates, $\hat{S}(t)$	Modified Kaplan–Meier Estimates[a], $\hat{S}(t)$ Plotted	$\hat{S}(t)$ Obtained from Weibull Fit
142	1	21	0.9524	0.9524	0.9546	0.9825
156	2	20	0.9500	0.9048	0.9091	0.9590
163	3	19	0.9474	0.8572	0.8637	0.9421
198	4	18	0.9444	0.8095	0.8182	0.7990
204+	—	17	1.0000	0.8095	0.8182	0.7647
205	6	16	0.9375	0.7589	0.7700	0.7588
232	7	15	0.9333	0.7083	0.7218	0.5778
232	8	14	0.9286	0.6577	0.6737	0.5778
232	9	13	0.9231	0.6071	0.6255	0.5706
233	10	12	0.9167	0.5565	0.5773	0.5706
233	11	11	0.9091	0.5059	0.5292	0.5706
233	12	10	0.9000	0.4553	0.4810	0.5706
239	13	9	0.8888	0.4047	0.4320	0.5271
240	14	8	0.8750	0.3541	0.3847	0.5198
261	15	7	0.8571	0.3035	0.3365	0.3697
280	16	6	0.8333	0.2529	0.2883	0.2489
280	17	5	0.8000	0.2023	0.2402	0.2489
296	18	4	0.7500	0.1517	0.1920	0.1660
296	19	3	0.6667	0.1011	0.1439	0.1660
323	20	2	0.5000	0.0506	0.0958	0.0710
344+	—	1	1.000	0.0506	0.0958	0.0313

[a]Instead of $(n-r)/(n-r+1)$. Pike uses $(n-r+1)/(n-r+2)$ in the Kaplan–Meier product-limit estimate to avoid $(n-r)/(n-r+1)=0$.

Source: Pike (1966). Reproduced with permission from the Biometric Society.

6.3 THE LOGNORMAL DISTRIBUTION

In its simplest form the lognormal distribution can be defined as the distribution of a variable whose logarithm follows the normal distribution. Its origin may be traced as far back as 1879 when McAlister (1879) described explicitly a theory of the distribution. Most of its aspects have since been under study. Gaddum (1945a,b) gave a review of its application in biology, followed by Boag's (1949) applications in cancer research. Its history, properties, estimation problems, as well as uses in economics have been discussed in detail by Aitchison and Brown (1957). Later, several other investigators such as Osgood (1958), Feinleib and MacMahon (1960), and Feinleib (1960) also observed that the distribution of survival time of several diseases such as Hodgkin's disease and chronic leukemia could be rather closely approximated by a lognormal distribution since they are markedly skewed to the right and the logarithms of survival times are approximately normally distributed. Recently, Horner (1987) showed that the distribution of age at onset of Alzheimer's disease followed the lognormal distribution.

Consider the survival time T such that $\log_e T$ is normally distributed with mean μ and variance σ^2. We then say that T is lognormally distributed and write T as $\Lambda(\mu, \sigma^2)$. It should be noted that μ and σ^2 are *not* the mean and variance of the lognormal distribution. Figure 6.8 gives the hazard function of the lognormal distribution with different values for the parameters. The hazard function increases initially to a maximum and then decreases (almost as soon as the median is passed) to zero as time approaches infinity (Watson and Wells 1961). Therefore, the lognormal distribution is suitable for

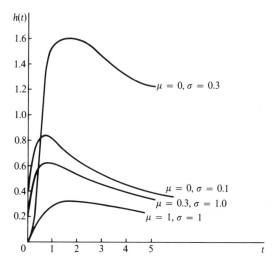

Figure 6.8 Hazard of the lognormal distribution with different parameters.

survival patterns with an initially increasing and then decreasing hazard rate. By a central limit theorem, it can be shown that the distribution of the product of n independent positive variates approaches a lognormal distribution under very general conditions, for example, the distribution of the size of an organism whose growth is subject to many small impulses, the effect of each of which is proportional to the momentary size of the organism.

The popularity of the lognormal distribution is in part due to the fact that the cumulative values of $y = \log_e t$ can be obtained from the tables of the standard normal distribution and the corresponding values of t are then found by taking antilogs. Thus, the percentiles of the lognormal distribution are easy to find.

The probability density function and survivorship function are, respectively,

$$f(t) = \frac{1}{t\sigma\sqrt{2\pi}} \exp\left[-\frac{1}{2\sigma^2}(\log_e t - \mu)^2\right] \qquad t>0, \sigma>0 \qquad (6.21)$$

and

$$S(t) = \frac{1}{\sigma\sqrt{2\pi}} \int_t^\infty \frac{1}{x} \exp\left[-\frac{1}{2\sigma^2}(\log_e x - \mu)^2\right] dx \qquad (6.22)$$

Let $a = \exp(-\mu)$. Then $-\mu = \log_e a$, (6.21) and (6.22) can be written as

$$f(t) = \frac{1}{t\sigma\sqrt{2\pi}} \exp\left[-\frac{1}{2\sigma^2}(\log_e ax)^2\right] \qquad (6.23)$$

and

$$S(t) = \frac{1}{\sigma\sqrt{2\pi}} \int_t^\infty \exp\left[-\frac{1}{2\sigma^2}(\log_e ax)^2\right] \frac{dx}{x} \qquad (6.24)$$

$$= 1 - G(\log_e ax/\sigma) \qquad (6.25)$$

where $G(y)$ is the cumulative distribution function of a standard normal variable

$$G(y) = \frac{1}{\sqrt{2\pi}} \int_0^y e^{-u^2/2} \, du \qquad (6.26)$$

The lognormal distribution is completely specified by the two parameters μ and σ^2. Time T cannot assume zero values since $\log_e T$ is not defined for $T = 0$. Figure 6.9 gives the lognormal frequency curves for $\mu = 0$, $\sigma^2 = 0.1$, 0.5, 2, from which an idea of the flexibility of the distribution may be obtained. It is obvious that the distribution is positively skewed and that the greater the value of σ^2, the greater the skewness. Figure 6.10 shows the

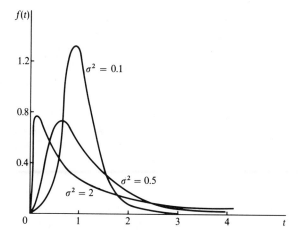

Figure 6.9 Lognormal density curves with $\mu = 0$.

frequency curves for $\sigma^2 = 0.5$, $\mu = 0, 0.5, 1.0$. It is also obvious that μ and σ^2 are, respectively, scale and shape parameters and *not* location and scale parameters as in the normal distribution. The hazard function, from (6.23) and (6.25), has the form

$$h(t) = \frac{\dfrac{1}{t\sigma\sqrt{2\pi}} \exp\left[-\dfrac{(\log_e at)^2}{2\sigma^2} \right]}{1 - G(\log_e at/\sigma)} \tag{6.27}$$

and is plotted in Figure 6.8.

The mean and variance of the two-parameter lognormal distribution are, respectively, $\exp(\mu + \frac{1}{2}\sigma^2)$ and $[\exp(\sigma^2) - 1]\exp(2\mu + \sigma^2)$. The coefficient of variation of the distribution is then $[\exp(\sigma^2) - 1]^{1/2}$. The median is e^μ and the mode is $\exp(\mu - \sigma^2)$.

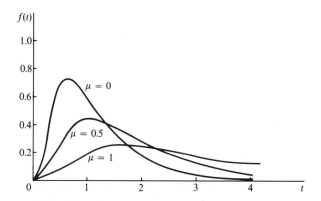

Figure 6.10 Lognormal density curves with $\sigma^2 = 0.5$.

The two-parameter lognormal distribution can also be generalized to a three-parameter distribution by replacing t with $t - G$ in (6.21). In other words, the survival time $\log_e(T - G)$ follows the normal distribution with mean μ and variance σ^2. In certain situations the value of G may be determined a priori and should not be regarded as an unknown parameter that requires estimation. If this is so, the variable $T - G$ may be considered in place of T and the distribution of $T - G$ has all the properties of the two-parameter lognormal distribution. However, the estimation procedures developed for the two-parameter case are not directly applicable to the distribution of $T - G$.

Example 6.3

In a study of chronic lymphocytic and myelocytic leukemia, Feinleib and MacMahon (1960) applied the lognormal distribution to analyze survival data of 649 white residents of Brooklyn diagnosed in the period 1943–1952. The analysis of several subgroups of patients follows. The survival time of each patient is computed from the date of diagnosis in months. Graphical and analytical methods are used to fit the lognormal distribution to the data. These methods will be discussed in Chapters 7 and 8.

Figure 6.11 plots the survival time of 234 male patients with chronic lymphocytic leukemia on lognormal probability graph paper, in which the horizontal axis for the survival time is in logarithmic scale and the vertical axis is in normal probability scale. When plotting $1 - S(t)$ on this graph paper, a straight line is obtained when the data follow a two-parameter lognormal distribution. An inspection of the graph shows that the observed distribution is concave. Gaddum (1945a,b) has pointed out that such a

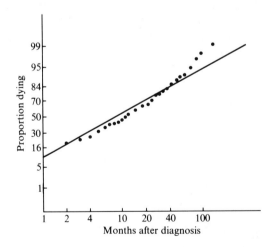

Figure 6.11 Lognormal probability plot of the survival time of 234 male patients with chronic lymphocytic leukemia. [*Source*: Feinleib and MacMahon (1960). Reproduced by permission of the publisher.]

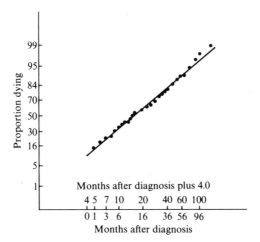

Figure 6.12. Lognormal probability plot of the survival time in months plus 4 of 234 male patients with chronic lymphocytic leukemia. [*Source:* Feinleib and MacMahon (1960). Reproduced by permission of the publisher.]

deviation can be corrected by subtracting an appropriate constant from the survival times. In other words, the three-parameter lognormal distribution can be used. Figure 6.12 give a similar plot in which the survival time of every patient plus 4 (or minus −4) is plotted. The configuration is linear, and hence empirically it seems valid to assume that the lognormal distribution is appropriate.

Similar graphs for male patients with chronic myelocytic leukemia and for female patients with chronic lymphocytic or myelocytic leukemia are given in Figures 6.13 and 6.14. Parameters of the lognormal distribution are estimated. Feinleib and MacMahon report that the agreement between the observed and calculated distributions is striking for each group except for females with chronic lymphocytic leukemia. The corresponding p values for the chi-square goodness-of-fit test are as follows:

	Chronic Myelocytic	**Chronic Lymphocytic**
Male	0.86	0.73
Female	0.57	0.016

Since a large p value indicates close agreement, it is concluded that the three-parameter lognormal distribution adequately describes the distribution of survival times for each subgroup except for females with chronic lymphocytic leukemia. The shape of the observed distribution for this latter group suggests that it might actually be composed of two dissimilar groups, each of whose survival times might fit a lognormal distribution.

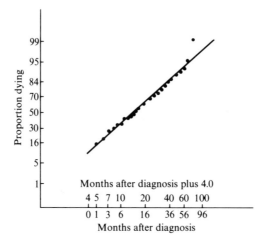

Figure 6.13 Lognormal probability plot of the survival time in months plus 4 of 162 Male patients with chronic myelocytic leukemia. [*Source:* Feinleib and MacMahon (1960). Reproduced by permission of the publisher.]

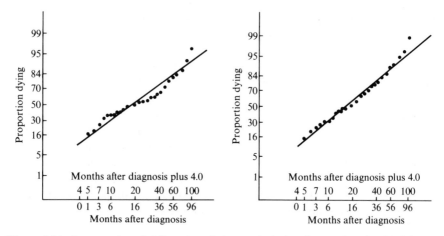

Figure 6.14 Lognormal probability plot of the survival time in months plus 4 of female patients with two types of leukemia. [*Source:* Feinleib and MacMahon (1960). Reproduced by permission of the publisher.]

6.4 THE GAMMA DISTRIBUTION

The gamma distribution, which includes the exponential and chi-square distribution, has been used by Brown and Flood (1947) to describe the life of glass tumblers circulating in a cafeteria and by Birnbaum and Saunders (1958) as a statistical model for life length of materials. Since then, this distribution has been used as a model for industrial reliability problems (e.g., Drenick 1960, Gupta 1960, and Gupta and Groll 1961), hepatograms in normal adults and in patients with cirrhosis and obstructive jaundice (Galli et al. 1983), speech amplitude (Niederjohn and Haworth 1986), and platelet survival (Bolin and Greene 1986).

Suppose that failure or death takes place in n stages or as soon as n subfailures have happened. At the end of the first stage, after time T_1, the first subfailure occurs; after that the second stage begins and the second subfailure occurs after time T_2, and so on. Total failure or death occurs at the end of the nth stage when the nth subfailure happens. The survival time T is then $T_1 + T_2 + \cdots + T_n$. The times $T_1, T_2, \ldots, T_n$ spent in each stage are assumed to be independently exponentially distributed with probability density function $\lambda \exp(-\lambda t_i)$, $i = 1, \ldots, n$. That is, the subfailures occur independently at a constant rate λ. The distribution of T is then called the Erlangian distribution. There is no need for the stages to have physical significance since we can always assume that death occurs in the n-stage process just described. This idea, introduced by A. K. Erlang in his study of congestion in telephone systems, has been widely used in queuing theory and life processes.

A natural generalization of the Erlangian distribution is to replace the parameter n restricted to the integers $1, 2, \ldots$ by a parameter γ taking any real positive value. We then obtain the *gamma distribution*.

The gamma distribution is characterized by two parameters, γ and λ. When $0 < \gamma < 1$, there is negative aging and the hazard rate decreases monotonically from infinity to λ as time increases from zero to infinity. When $\gamma > 1$, there is positive aging and the hazard rate increases monotonically from zero to λ as time increases from zero to infinity. When $\gamma = 1$, the hazard rate equals λ, a constant, as in the exponential case. Figure 6.15 illustrates the gamma hazard function for $\lambda = 1$ and $\gamma < 1$, $\gamma = 1, 2, 4$. Thus, the gamma distribution describes a different type of survival pattern where the hazard rate is decreasing or increasing to a constant value as time approaches infinity.

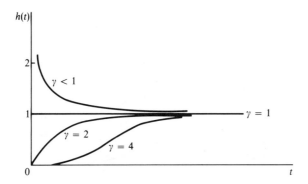

Figure 6.15   Gamma hazard functions with $\lambda = 1$.

The probability density function of a gamma distribution is

$$f(t) = \frac{\lambda}{\Gamma(\gamma)} (\lambda t)^{\gamma-1} e^{-\lambda t} \qquad t > 0, \gamma > 0, \lambda > 0 \qquad (6.28)$$

where $\Gamma(\gamma)$ is as defined in (6.16). Figures 6.16 and 6.17 show the gamma density function with various values of γ and λ. It is seen that varying γ changes the shape of the distribution while varying λ changes only the scaling. Consequently, γ and λ are shape and scale parameters, respectively. When $\gamma > 1$, there is a single peak at $t = (\gamma - 1)/\lambda$.

The cumulative distribution function $F(t)$ has a more complex form:

$$F(t) = \int_0^t \frac{\lambda}{\Gamma(\gamma)} (\lambda x)^{\gamma-1} e^{-\lambda x} \, dx \qquad (6.29)$$

$$= \frac{1}{\Gamma(\gamma)} \int_0^{\lambda t} u^{\gamma-1} e^{-u} \, du$$

$$= I(\lambda t, \gamma) \qquad (6.30)$$

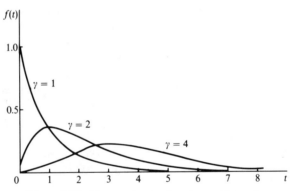

Figure 6.16   Gamma density functions with $\lambda = 1$.

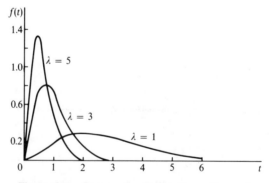

Figure 6.17   Gamma density functions with $\gamma = 3$.

where
$$I(s, \gamma) = \frac{1}{\Gamma(\gamma)} \int_0^s u^{\gamma-1} e^{-u}\, du \tag{6.31}$$

known as the *incomplete gamma function*, is tabulated in Pearson (1922 or 1957). For the Erlangian distribution, it can be shown that

$$F(t) = 1 - \sum_{k=0}^{n-1} \frac{e^{-\lambda t}(\lambda t)^k}{k!} \tag{6.32}$$

Thus, the survivorship function $1 - F(t)$ is

$$S(t) = \int_t^\infty \frac{\lambda}{\Gamma(\gamma)} (\lambda x)^{\gamma-1} e^{-\lambda x}\, dx \tag{6.33}$$

for the gamma distribution or

$$S(t) = e^{-t} \sum_{k=0}^{n-1} \frac{(\lambda t)^k}{k!} \tag{6.34}$$

for the Erlangian distribution.

Since the hazard function is the ratio of $f(t)$ to $S(t)$, it can be calculated from (6.28) and (6.34). When γ is an integer n,

$$h(t) = \frac{\lambda(\lambda t)^{n-1}}{(n-1)! \sum_{k=0}^{n-1} \frac{1}{k!} (\lambda t)^k} \tag{6.35}$$

When $\gamma = 1$, the distribution is exponential. When $\lambda = \frac{1}{2}$ and $\gamma = \frac{1}{2}\nu$, where ν is an integer, the distribution is chi square with ν degrees of freedom.

The mean and variance of the gamma distribution are, respectively, γ/λ and γ/λ^2, so that the coefficient of variation is $1/\sqrt{\gamma}$.

Many survival distributions can be represented, at least roughly, by suitable choice of the parameters λ and γ. In many cases, there is an advantage in using the Erlangian distribution, that is, in taking γ integer.

Example 6.4
Birnbaum and Saunders (1958) report an application of the gamma distribution to the lifetime of aluminum coupon. In their study, 17 sets of six strips were placed in a specially designed machine. Periodic loading was applied to the strips with a frequency of 18 cycles per second and a maximum stress of 21,000 psi. The 102 strips were run until all of them failed. One of the 102 strips tested had to be discarded for an extraneous reason, yielding 101 observations. The lifetime data are given in Table 6.4 in ascending order.

Table 6.4   Lifetimes of 101 Strips of Aluminum Coupon

370	1055	1270	1502	1763
706	1085	1290	1505	1768
716	1102	1293	1513	1781
746	1102	1300	1522	1782
785	1108	1310	1522	1792
797	1115	1313	1530	1820
844	1120	1315	1540	1868
855	1134	1330	1560	1881
858	1140	1355	1567	1890
886	1199	1390	1578	1893
886	1200	1416	1594	1895
930	1200	1419	1602	1910
960	1203	1420	1604	1923
988	1222	1420	1608	1940
990	1235	1450	1630	1945
1000	1238	1452	1642	2023
1010	1252	1475	1674	2100
1016	1258	1478	1730	2130
1018	1262	1481	1750	2215
1020	1269	1485	1750	2268
				2440

Source: Birnbaum and Saunders (1958).

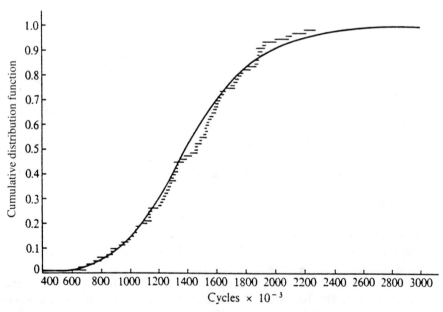

Figure 6.18   Graphical comparison of observed and fitted cumulative distribution functions of data in example 6.4. [*Source:* Birnbaum and Saunders (1958).]

From the data the two parameters of the gamma distribution were estimated (estimation methods are discussed in Chapter 8). They obtained $\hat{\gamma} = 11.8$ and $\hat{\lambda} = 1/(118.76 \times 10^3)$.

A graphical comparison of the observed and fitted cumulative distribution function is given in Figure 6.18, which shows very good agreement. A chi-square goodness-of-fit test (discussed in Chapter 7) yielded a χ^2 value of 4.49 with six degrees of freedom, corresponding to a significance level between 0.5 and 0.6. Thus, it was concluded that the gamma distribution was an adequate model for the life length of some materials.

6.5 OTHER SURVIVAL DISTRIBUTIONS

Many other distributions can be used as models of survival time, three of which we discuss briefly in this section: the *linear exponential*, the *Gompertz* (1825), and a distribution whose hazard rate is a step function.

The linear-exponential model and the Gompertz distribution are extensions of the exponential distribution. Both describe survival patterns that have a constant initial hazard rate. The hazard rate varies as a linear function of time or age in the linear-exponential model and as an exponential function of time or age in the Gompertz distribution.

Broadbent (1958), in demonstrating the use of the linear-exponential model, uses as an example the service of milk bottles that are filled in a dairy, circulated to customers, and returned empty to the dairy. The model was also used by Carbone et al. (1967) to describe the survival pattern of patients with plasmacytic myeloma. The hazard function of the linear-exponential distribution is

$$h(t) = \lambda + \gamma t \tag{6.36}$$

where λ and γ can be values such that $h(t)$ is nonnegative. The hazard rate increases from λ with time if $\gamma > 0$, decreases if $\gamma < 0$, and remains constant (an exponential case) if $\gamma = 0$, as depicted in Figure 6.19.

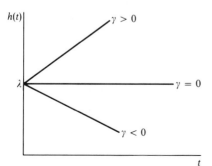

Figure 6.19 Hazard function of linear-exponential model.

The probability density function and the survivorship function are, respectively,

$$f(t) = (\lambda + \gamma t) \exp[-(\lambda t + \tfrac{1}{2}\gamma t^2)] \tag{6.37}$$

and

$$S(t) = \exp[-(\lambda t + \tfrac{1}{2}\gamma t^2)] \tag{6.38}$$

The mean of the linear-exponential distribution is $-(\lambda/\gamma) + (\gamma/2)^{-1/2}L(\lambda^2/2\gamma)$, where

$$L(x) = e^x \int_x^\infty y^{1/2} e^{-y} \, dy$$

is tabulated in Table 6.5. A special case of the linear-exponential distribution, the Rayleigh distribution, is obtained by replacing γ by $\tfrac{1}{2}\gamma$ (Kodlin, 1967). That is, the hazard function of the Rayleigh distribution is $h(t) = \lambda + \tfrac{1}{2}\gamma t$.

The Gompertz distribution is also characterized by two parameters, λ and γ. The hazard function

$$h(t) = \exp(\lambda + \gamma t) \tag{6.39}$$

is plotted in Figure 6.20. When $\gamma > 0$, there is positive aging starting from e^λ; when $\gamma < 0$, there is negative aging; and when $\gamma = 0$, $h(t)$ reduces to a

Table 6.5   Values of $L(x)$ and $G(x)$

x	$L(x)$	$G(x)$
0	0.886	∞
0.1	0.951	2.015
0.2	1.012	1.493
0.3	1.067	1.223
0.4	1.119	1.048
0.5	1.168	0.923
0.6	1.214	0.828
0.7	1.258	0.753
0.8	1.300	0.691
0.9	1.341	0.640
1	1.381	0.596
2	1.712	0.361
3	1.987	0.262

Source: Broadbent (1958).

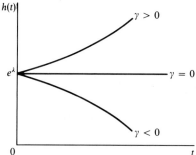

Figure 6.20   Gompertz hazard function.

constant, e^λ. The survivorship function of the Gompertz distribution is

$$S(t) = \exp\left[-\frac{e^\lambda}{\gamma} (e^{\gamma t} - 1) \right]$$

(6.40)

and the probability density function, from (6.39) and (2.14), is then

$$f(t) = \exp\left[(\lambda + \gamma t) - \frac{1}{\gamma} (e^{\lambda + \gamma t} - e^\lambda) \right]$$

(6.41)

The mean of the Gompertz distribution is $G(e^\lambda/\gamma)/e^\lambda$, where

$$G(x) = e^x \int_x^\infty y^{-1} e^{-y} \, dy$$

is tabulated in Table 6.5.

Lastly, we consider a distribution where the hazard rate is a step function:

$$h(t) = \begin{cases} a_1 & 0 \le t < t_1 \\ a_2 & t_1 \le t < t_2 \\ \vdots & \vdots \\ a_{k-1} & t_{k-2} \le t < t_{k-1} \\ a_k & t \ge t_{k-1} \end{cases}$$

(6.42)

where $t_1, t_2, \ldots, t_k$ are different time points. Figure 6.21 shows a typical hazard function of this nature for $k = 5$. Using (2.13) the survivorship function can be derived:

$$S(t) = \begin{cases} \exp(-a_1 t) & 0 \le t < t_1 \\ \exp[-a_1 t_1 - a_2(t - t_1)] & t_1 \le t < t_2 \\ \vdots & \vdots \\ \exp[-a_1 t_1 - a_2(t_2 - t_1) - \cdots \\ \quad - a_k(t - t_{k-1})] & t \ge t_{k-1} \end{cases}$$

(6.43)

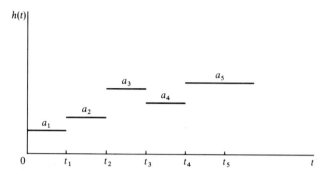

Figure 6.21   A step hazard function.

The probability density function $f(t)$ can then be obtained from (6.42) and (6.43) using (2.11):

$$f(t) = \begin{cases} a_1 \exp(-a_1 t) & 0 \le t < t_1 \\ a_2 \exp[-a_1 t_1 - a_2(t - t_1)] & t_1 \le t < t_2 \\ \quad \vdots & \quad \vdots \\ a_k \exp[-a_1 t_1 - a_2(t_2 - t_1) - \cdots \\ \qquad - a_k(t - t_{k-1})] & t \ge t_{k-1} \end{cases} \tag{6.44}$$

One application of this distribution is the life-table analysis discussed in Chapter 4. In a life-table analysis, time is divided into intervals and the hazard rate is assumed to be constant in each interval. However, the overall hazard rate is not necessarily constant.

The above seven distributions are, among others, reasonable models for survival time distribution. All have been designed by considering either a biological failure, a death process, or an aging property. They may or may not be appropriate for many practical situations, but the objective here is to illustrate the various possible techniques, assumptions, and arguments that can be used to choose the most appropriate model. If none of these distributions fits the data, investigators might have to derive an original model to suit the particular data, perhaps by using some of the ideas presented here.

BIBLIOGRAPHICAL REMARKS

In addition to the papers on the distributions cited in this chapter, Mann et al. (1974), Hahn and Shapiro (1967), Elandt-Johnson and Johnson (1980), Lawless (1982), Nelson (1982), Cox and Oakes (1984), and Gertsbakh (1989) also discuss statistical failure models including the exponential, Weibull, lognormal, and gamma distributions. Other references on survival distributions and their applications are Cox (1962), Weiss (1963), Johnson (1964), Hastings and Peacock (1974), Hokanson et al. (1977), Davis and Feldstein (1979), Lawless (1983), and Aitkin et al. (1983).

EXERCISES

6.1 Give a summary of the distributions discussed in this chapter, answering the following questions.

    a. What are the distributions that describe constant hazard rates? Give the range of the parameter values.

    b. What are the distributions that describe increasing hazard rates? If there are more than one, discuss the differences between them.

    c. What are the distributions that describe decreasing hazard rates? If there are more than one, discuss the differences between them.

6.2 Suppose that the survival distribution of a group of patients follows the exponential distribution with $G = 0$ (year), $\lambda = 0.65$. Plot the survivorship function and find:

    a. The mean survival time

    b. The median survival time

    c. The probability of surviving 1.5 years or more

6.3 Suppose that the survival distribution of a group of patients follows the exponential distribution with $G = 5$ (years) and $\lambda = 0.25$. Plot the survivorship function and find:

    a. The mean survival time

    b. The median survival time

    c. The probability of surviving six years or more

6.4 Consider the following two Weibull distributions as survival models:

    (i) $G = 0$, $\lambda = 1$, $\gamma = 0.5$

    (ii) $G = 0$, $\lambda = 0.5$, $\gamma = 2$

For each distribution, plot the survivorship function and the hazard function and find:

    a. The mean

    b. The variance

    c. The coefficient of variation

Which distribution gives the larger probability of surviving at least three units of time?

6.5 Suppose that the survival time follows the lognormal distribution with $\mu = 1$ and $\sigma = 0.5$. Find:

    a. The mean survival time

    b. The variance

    c. The coefficient of variation

    d. The median

    e. The mode

6.6 Suppose that pain relief time follows the gamma distribution with $\lambda = 1$, $\gamma = 0.5$. Find:

a. The mean

b. The variance

c. The coefficient of variation

6.7 Suppose that $\lambda = 1$, $\gamma = 2.0$. Plot the hazard function and find:

(a) The mean

(b) The probability of surviving longer than one unit of time

if the survival distribution is Gompertz and if it is linear exponential

6.8 Consider the survival times of hypernephroma patients given in Exercise Table 3.1. From the plot you obtained in Exercise 4.5, suggest a distribution that might fit the data.

CHAPTER 7

Graphical Methods for Survival Distribution Fitting and Goodness-of-Fit Tests

The use of probability models for survival experience has played an increasingly important role in biomedical sciences. Survival models will summarize the survival pattern, suggest further studies, and generate hypotheses. In this chapter we introduce two simple graphical methods for survival distribution fitting and four statistical tests for goodness-of-fit.

Section 7.1 discusses the advantages of the graphical techniques. In Section 7.2 we discuss probability plotting, including how to make probability plots as well as how to estimate parameters from them. In Section 7.3 we discuss the theory and applications of hazard plotting for censored data. Four statistical tests of good-of-fit are discussed in Section 7.4.

7.1 INTRODUCTION

Graphical methods have long been used for display and interpretation of data because they are simple and effective. Often used in place of or in conjunction with numerical analysis, a plot of data simultaneously serves a number of purposes that no numerical method can. The two graphical methods discussed in this chapter are *probability plotting* and *hazard plotting*. A number of authors have discussed the use of probability plots for informal evaluation of distributional assumptions and for estimation of parameters, for example, Chernoff and Lieberman (1954), Daniel (1959), Kimball (1960), Kao (1959), Wilk et al. (1962b), and King (1971). Probability plots require complete samples but the hazard plotting method (Nelson 1972) can handle censored observations.

If one chooses the appropriate distribution and makes a probability or hazard plot, the result will be a straight-line fit to the data. Parameters of the chosen distribution can be estimated from the plot without tedious

157

numerical calculations. Such estimates are adequate and useful for most purposes. However, prior information is often not sufficient to choose a suitable distribution, and the plot may not be a straight line.

A nonlinear plot can provide insight into the data. There are several possible interpretations. First, the wrong theoretical distribution might have been used. Second, the sample might be from a mixture of populations. If so, it is necessary to separate the data accordingly and make a separate plot for each population. If one or two points are way out of line, they might be the result of an error in collecting and recording the data or they might not be from the same population. Other reasons for peculier looking plots and interpretations of them are discussed by King (1971) and Hahn and Shapiro (1967).

Consider the normal probability plots in Figures 7.1(a) and (b). The plot in Figure 7.1(a) is convex, indicating that the data have a long tail to the left and could be from a distribution with a negatively skewed density function such as in Figure 7.2(a). On the contrary, the concave plot in Figure 7.1(b) indicates that the data have a long tail to the right and could be from a distribution with a positively skewed density function such as in Figure 7.2(b). From the discussion in Chapter 6, we may try to fit a lognormal or gamma distribution.

The advantages of the two graphical methods can be summarized as follows:

1. The plotting methods are fast and simple to use, in contrast with numerical methods, which may be computationally tedious and require considerable analytical sophistication. The additional accuracy of numerical methods is usually not great enough in practice to warrant the effort involved.

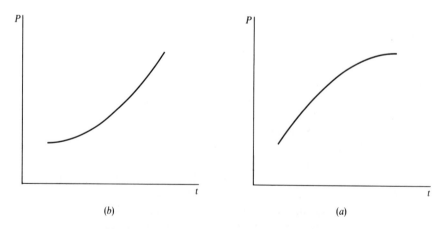

(b)   (a)

Figure 7.1 Two curved normal probability plots.

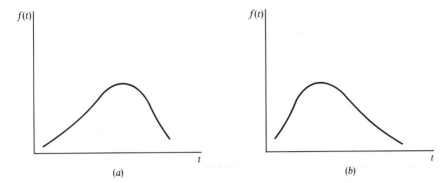

Figure 7.2 Two skewed density functions.

2. The plotting methods provide estimates of the parameters of the distribution by simple graphical means.
3. A probability or hazard plot allows one to assess whether a particular theoretical distribution provides an adequate fit to the data.
4. Peculiar appearance of a plot or points in a plot can provide insight into the data when the reasons for the peculiarities are determined.
5. A probability or hazard plot provides a visual representation of the data that is easy to grasp. This is useful not only for oneself but also in presenting data to others, since a plot allows one to assess conclusions drawn from the data by graphical or numerical means.

7.2 PROBABILITY PLOTTING

The basic ideas in probability plotting are illustrated by the following example.

Example 7.1
Consider the white blood cell counts (WBCs) of 23 pediatric leukemia patients given in Table 7.1, ranging from 8000 to 120,000. A *sample cumulative distribution* is constructed by ordering the data from smallest to largest, as shown in Table 7.1. A sample cumulative distribution curve can then be made by plotting each WBC value versus the percentage of the sample equal to or less than that value. That is, the ith ordered data value in a sample of n values is plotted against the percentage $100i/n$. For example, the 10th value of WBC, 50, is plotted against a percentage of $100 \times \frac{10}{23} = 43.5$. If WBC data for all the pediatric leukemia patients are available, one can make a similar *population cumulative distribution* function that contains information for the entire population. If a good statistical sample is obtained from the population, then the sample cumulative distribution function can

Table 7.1 Ordered WBC's and Cumulative Proportion for Example 7.1

Patient Number i	WBC ($\times 10^3$)	Cumulative Proportion $100i/n$	Cumulative Proportion $100(i - 0.5)/n$
1	8	4.3	2.2
2	8	8.7	6.5
3	10	13.0	10.9
4	15	17.4	15.2
5	20	21.7	19.6
6	30	26.1	23.9
7	50	30.4	28.3
8	50	34.8	32.6
9	50	39.1	37.0
10	50	43.5	41.3
11	50	47.8	45.7
12	60	52.2	50.0
13	60	56.5	54.3
14	75	60.9	58.7
15	75	65.2	63.0
16	80	69.6	67.4
17	80	73.9	71.7
18	90	78.3	76.1
19	90	82.6	80.4
20	90	87.0	84.8
21	100	91.3	89.1
22	110	95.7	93.5
23	120	100.0	97.8

provide an approximation of the population cumulative distribution function.

A plot of the cumulative distribution function for most large populations contains many closely spaced values and can be well approximated by a smooth curve drawn through the points. In contrast, a sample cumulative distribution function has a relatively small number of points and thus somewhat ragged appearance. To approximate the population cumulative distribution function, one draws a smooth curve through the data points, obtaining a best fit by eye. Such a curve for the WBC data is given in Figure 7.3. It is an estimate of the cumulative distribution function of the population and is used to obtain estimates and other information about the population.

An estimate of the population median (50th percentile) is obtained by entering the plot on the percentage scale at 50% going horizontally to the fitted line and then vertically down to the data scale to read the estimate of the median. For the WBC data, the estimate of the population median is

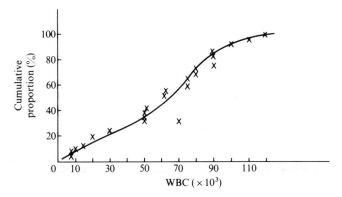

Figure 7.3 A sample cumulative distribution curve of the WBC data.

65,000. The median is a representative or nominal value for the population since half of the population values are above it and half below. An estimate of any other percentile can similarly be obtained by entering the plot at the appropriate point on the percentage scale going horizontally to the fitted line and then vertically down to the data scale where the estimate is read. For example, the estimate for the 25th percentile is 40,000.

One can obtain an estimate of the proportion of the population that has a WBC below a specific value in a similar way. For example, to find the proportion of the population with a WBC of 10,000 or less, you enter the plot on the horizontal axis at the given value, 10, go vertically up to the line fitted to the data, and then horizontally to the probability scale where the estimate of the population proportion is read, 8%. An estimate of the proportion above a given value is 100% minus the estimate of the proportion below the value. For example, the estimate of the proportion of the population with a WBC above 10,000 is $100 - 8 = 92\%$. An estimate of the proportion of a population between two given values is obtained by first getting an estimate of the proportion below each value and then taking the difference. For example, the estimate of the population proportion with WBC between 10,000 and 65,000 is $50 - 8 = 42\%$.

As mentioned above, a smooth curve can be fitted by eye to a sample cumulative distribution function to obtain an estimate of the population distribution function. Also, one can fit a theoretical cumulative distribution function to a sample cumulative distribution function and then use the theoretical curve as an estimate of the population cumulative distribution function. The distribution may be the normal, lognormal, exponential, Weibull, or gamma. Fitting such theoretical distributions to sample data can easily be carried out graphically by plotting a sample cumulative distribution function on probability scale. To make a probability plot, one usually uses a plotting position of $100(i - 0.5)/n$ or $100i/(n + 1)$ for the ith ordered value of the n observations in the sample. The $100(i - 0.5)/n$ plotting positions for the WBC data are given in Table 7.1. Probability plots can be made by using specially constructed probability papers[1] or by computer programming. The axis for the cumulative distribution is in probability scale and the axis for the observation is in linear or logarithmic scale.

The data and probability scales on a probability plot are so constructed that the theoretical cumulative distribution function will plot as a straight line. These theoretical distributions have one or more parameters that are determined to provide a fit to the sample data. This is carried out as follows. First, a theoretical distribution has to be selected. Second, the sample cumulative distribution function is plotted. If probability papers are used, this is done by simply plotting the observed values versus a sample cumulative function, such as $100(i - 0.5)/n$ or $100i/(n + 1)$, on the probability paper of the selected theoretical distribution. If the correct theoretical distribution is chosen, a straight line can then be drawn through the data points. The position of the straight line should be chosen to provide a fit to the bulk of the data and may ignore outliers in the tails of the plot or data points of doubtful validity. Figure 7.4 plots the WBC versus $100(i - 0.5)/n$ on normal probability paper. The plot is reasonably linear. This plot is called a *normal probability plot*.

The straight line fitted by eye in a probability plot can be used to estimate percentiles and proportions within given limits in the same manner as for the sample cumulative distribution curve. In addition, a probability plot provides estimates of the parameters of the theoretical distribution chosen. The mean (or median) WBC estimated from the normal probability plot in Figure 7.4 is 60,000, the 50th percentile. The 84th percentile is 91,000, which corresponds to the mean plus one standard deviation. Thus, the standard deviation is estimated as the difference between the 84th percentile and the mean, that is, 31,000.

We now discuss probability plots of the exponential, Weibull, lognormal, and gamma distributions.

1. *The Exponential Distribution.* The exponential cumulative distribution function is

$$F(t) = 1 - e^{-\lambda t} \qquad t > 0 \tag{7.1}$$

Probability paper for the exponential distribution is constructed on the basis of the relationship between a value t and $F(t)$,

$$t = \frac{1}{\lambda} \log_e \frac{1}{1 - F(t)} \tag{7.2}$$

obtained from (7.1). This relationship is linear between t and the function

---

[1] A large selection of probability, hazard, and other data analysis papers is offered in the catalog of Technical and Engineering Aids for Management (TEAM), Box 25, Tamworth, New Hampshire 03886.

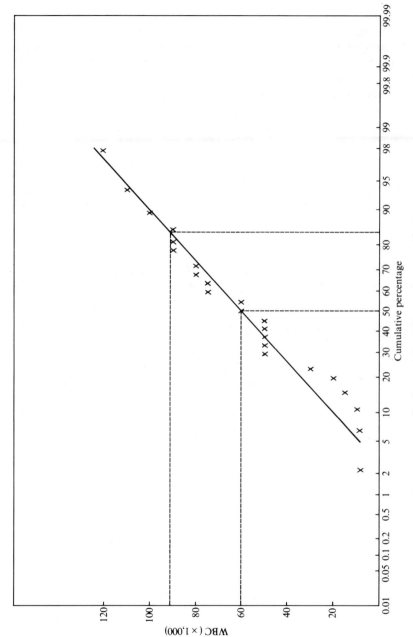

Figure 7.4 A normal probability plot of the WBC data in Example 7.1.

$\log_e\{1/[1 - F(t)]\}$. Thus, on the exponential probability plot, the data scale is linear and the probability scale has the value of $F(t)$ located at $\log_e\{1/[1 - F(t)]\}$. In other words, $F(t)$ is converted to $\log_e\{1/[1 - F(t)]\}$ by the probability paper. If probability papers are not available, this can easily be done by computer programming. For example, $F(t) = 0.20, 0.50$ are located on the probability scale at 0.22 [i.e., $\log_e(1/0.8)$] and 0.69 [i.e., $\log_e(1/0.5)$]. The probability axis is so constructed that if the distance from zero to 0.20 is 0.8 cm, the distance from zero to 0.50 is $(0.8/0.22) \times 0.69 = 2.5$ cm. An exponential probability plot is made by plotting the ith ordered observed survival time $t_{(i)}$ versus an estimate of $F(t_{(i)})$, for example, $100(i - 0.5)\%$, for $i = 1, \ldots, n$.

For the exponential distribution, the 63.2 percentile is equal to the mean $1/\lambda$. This fact can be used to estimate $1/\lambda$ or λ from the straight line fitted to the points. That is, the value corresponding to the 63.2 percentage point is an estimate of the mean and its reciprocal is an estimate of the hazard rate λ.

Example 7.2
Suppose that 21 patients with acute leukemia have the following remission times in months: 1, 1, 2, 2, 3, 4, 4, 5, 5, 6, 8, 8, 9, 10, 10, 12, 14, 16, 20, 24,

Table 7.2　Ordered Remission Times and Plotting Points for Example 7.2

t	i	$\dfrac{i - 0.5}{21} \times 100$
1	1	2.38
1	2	7.14
2	3	11.90
2	4	16.67
3	5	21.43
4	6	26.19
4	7	30.95
5	8	35.71
5	9	40.48
6	10	45.24
8	11	50.00
8	12	54.76
9	13	59.52
10	14	64.29
10	15	69.05
12	16	73.81
14	17	78.57
16	18	83.33
20	19	88.09
24	20	92.86
34	21	97.62

and 34. We would like to know if the remission time follows the exponential distribution. The ordered remission times and the plotting points $100(i - 0.5)/21$ are given in Table 7.2. The exponential probability plot is shown in Figure 7.5. A straight line is fitted to the points by eye, and the plot indicates that the exponential distribution fits the data very well. From Figure 7.5 the mean, estimated as the 63.2 percentile, is 9.4 months and thus the hazard rate is $\hat{\lambda} = 1/9.4 = 0.106$ per month.

2. *The Weibull Distribution*. The Weibull cumulative distribution function is

$$F(t) = 1 - \exp[-(\lambda t)^{\gamma}] \qquad t > 0, \gamma > 0, \lambda > 0 \qquad (7.3)$$

The probability paper for the Weibull distribution is based on the relationship

$$\log t = \log(1/\lambda) + (1/\gamma)\log\{\log_e[1/(1 - F)\} \qquad (7.4)$$

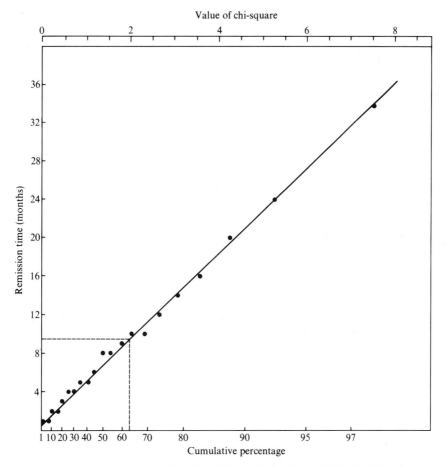

Figure 7.5 Exponential probability plot of the remission time of 21 leukemia patients.

between the value t and the cumulative distribution function F. This relationship, obtained from (7.3), is linear between the value $\log t$ and the function $\log\{\log_e[1/(1 - F)]\}$. Thus, on a Weibull probability plot, the data scale is logarithmic and the probability scale has the value F located at $\log\{\log_e[1/(1 - F)]\}$.

When Weibull probability paper is used, we plot the ith ordered survival time $t_{(i)}$ (which is converted to $\log t_{(i)}$ by the probability paper) versus an estimate of $F(t_{(i)})$, $100(i - 0.5)/n$, for $i = 1, \ldots, n$. The estimate of $F(t_{(i)})$ is converted to $\log\{\log_e[1/(1 - F)]\}$ by the probability paper.

The shape parameter γ is estimated graphically as the reciprocal of the slope of the straight line fitted to the data on the Weibull probability paper. The Weibull probability paper has an auxiliary scale that can be used to estimate γ. To obtain this estimate, we draw a line parallel to the fitted line from the origin at the top of the paper. The value given by the intersection of this line with the left-side auxiliary scale is the estimate. The 63.2 percentile is the value of $1/\lambda$ for the Weibull distribution. This fact can be used to estimate $1/\lambda$ and thus λ graphically from a Weibull probability plot. The following hypothetical example illustrates the use of the Weibull probability paper. The small number of observations used in the example is only for illustrative purposes. In practice, many more observations are needed in order to identify an appropriate theoretical model for the data.

Example 7.3
Six mice with brain tumors have survival times, in months of 3, 4, 5, 6, 8, and 10. The ordered survival times are plotted against $100(i - 0.5)/n$ for $i = 1, \ldots, 6$ on the Weibull probability paper in Figure 7.6. A straight line is fitted to the data point by eye. The 63.2 percentile is approximately 6.5 months, which is an estimate of $1/\lambda$; thus an estimate of λ is 0.154. A line parallel to the fitted line is drawn from the origin at the top. The intersection of this line with the auxiliary scale on the left-hand side gives a graphical estimate of γ, which is approximately 2.75. Having these estimates, the mean survival time can be estimated by (6.14), that is, $\mu = \Gamma(1.364)/0.154 = 0.8897/0.154 = 5.78$ months. [Note that the value of $\Gamma(x)$ can be found in Abramowitz and Stegun (1964).]

3. *The Lognormal Distribution.* If the survival time t follows a lognormal distribution with parameters μ and σ^2, then $\log t$ follows the normal distribution with means μ and variance σ^2. Consequently, $(\log t - \mu)/\sigma$ has the standard normal distribution. Thus, the lognormal cumulative distribution function can be written as

$$F(t) = G\left(\frac{\log t - \mu}{\sigma}\right) \qquad t > 0 \qquad (7.5)$$

where $G(\ )$ is the standard normal distribution function, μ is the mean of

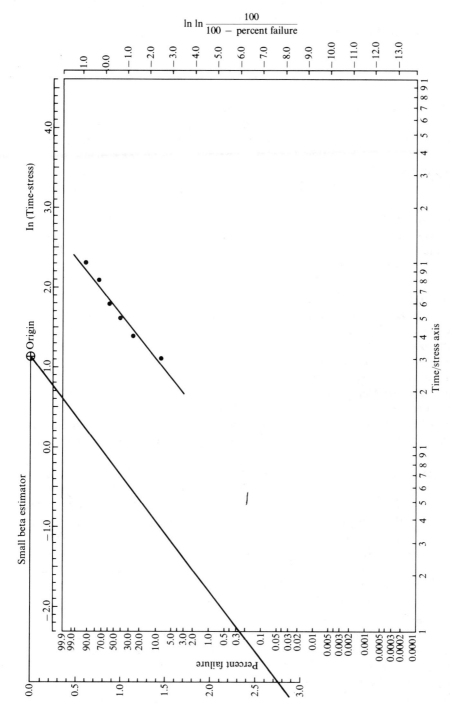

Figure 7.6 Weibull probability plot of survival time of six mice.

the logarithms of the data values, and $\sigma > 0$ is the standard deviation of the logarithms of the data values. Note that $\log t$ is base 10.

A probability plot for the lognormal distribution is based on the relationship

$$\log t = \mu + \sigma G^{-1}(F) \tag{7.6}$$

between a value of t and the cumulative distribution function F. The function $G^{-1}(F)$ is the inverse of the standard normal distribution function or its $100F$ percentile. This relationship, which is obtained from (7.5), is linear between the value $\log t$ and the function $G^{-1}(F)$. Thus, on the lognormal probability paper, the data scale is logarithmic and the probability scale has the value F located at $G^{-1}(F)$. The normal and lognormal probability papers have the same probability scale. The difference between the two is that normal papers have a linear data scale while lognormal papers have a logarithmic data scale. The lognormal probability paper is available in various cycle numbers: one-cycle has the logarithmic scale for t up to 10-fold, two-cycle for increases up to 100-fold, and so on.

Example 7.4

In a study of a new insecticide, 20 insects are exposed. Survival times in seconds are 3, 5, 6, 7, 8, 9, 10, 10, 12, 15, 15, 18, 19, 20, 22, 25, 28, 30, 40, and 60. Suppose that prior experience indicates that the survival time follows a lognormal distribution; that is, some insects might react to the insecticide very slowly and not die for a long time. The survival times versus their corresponding plotting positions $100(i - 0.5)/20$ are plotted on lognormal probability paper in Figure 7.7. The plot shows a reasonably straight line. The 50th percentile is approximately 14 seconds. Thus, an estimate of μ is log 14, or 1.15. The 84th percentile is approximately 30 seconds. Thus, an estimate of σ is log 30 − log 14, or 0.33.

Since $G^{-1}(0.5) = 0$, (7.6) reduces to $\log t = \mu$. Thus, the base 10 logarithm of the 50th percentile of a lognormal distribution is the value of the parameter μ. This fact can be used to estimate μ from a straight line fitted to data plotted on the lognormal probability paper. Similarly, since $G^{-1}(0.84) = 1$, (7.6) reduces to $\sigma = \log t - \mu$. Thus, the difference of the base 10 logarithms of the 84th and the 50th percentiles is the parameter σ. This fact can be used to estimate σ from the straight line fitted to data plotted on the lognormal probability paper. Note that μ and σ are the mean and standard deviation *not* of the distribution of the variable but, rather, of the logarithm of the variable.

4. *The Gamma Distribution.* The gamma distribution of (6.28) has a more complex cumulative distribution function (6.30). Gamma probability papers are available for $\gamma = 0.5$, 1, 1.5, 2, 3, 4, 5. The same probability papers can also be used to plot the chi-square distribution, a special case of the gamma distribution. In fact, this paper was first developed for the

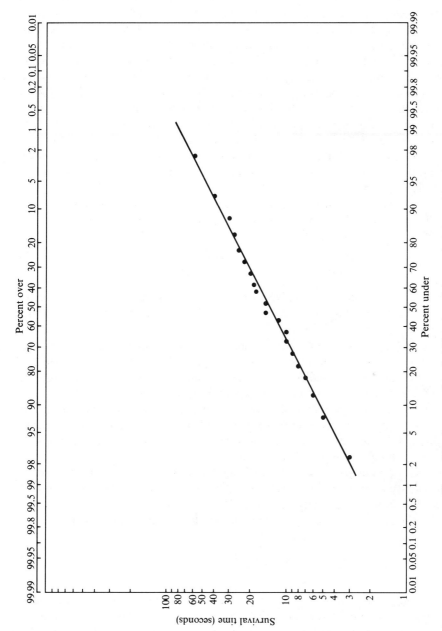

Figure 7.7 Lognormal probability plot of the survival time of 20 insects.

chi-square distribution and later extended to the more general gamma distribution. Because when $\lambda = \frac{1}{2}$ and $\gamma = \frac{1}{2}\nu$ the distribution is chi square with ν degrees of freedom, this paper is also applicable for plotting chi-square-distributed data with 1, 2, 3, 4, 6, 8, and 10 degrees of freedom. A special scale for the chi-square distribution is provided at the top of the paper. The vertical scale for the observations t is linear. The horizontal scale for the cumulative percentage differs with the value of γ. For $\gamma = 1$, the distribution is exponential or chi square with 2 degrees of freedom. As γ approaches 5, the gamma distribution is nearly equivalent to the lognormal distribution. For γ approaching 15, it approaches the normal distribution.

If the data follow a gamma distribution with the selected γ value, the probability plot should approximately be linear. Departure from linearity could be because the data not following the gamma distribution or the value of γ is not correct.

The parameter λ of a gamma distribution with known γ is estimated from a gamma probability plot as half the slope of the fitted line, or

$$\hat{\lambda} = \frac{x_2 - x_1}{2(t_2 - t_1)} \tag{7.7}$$

where t_1 and t_2 are two observed values on the vertical scale and x_1 and x_2 are the corresponding chi-square values on the special scale at the top of this paper.

The gamma probability plot is drawn in the same manner as an exponential, Weibull, or lognormal probability plot. However, because of the different probability scales for different values of the shape parameter, gamma paper should not be used to obtain arbitrary data fits. When there are good reasons for choosing the distribution, the value of the shape parameter has to be determined since it indicates the type of gamma paper to use. The likelihood method presented in Section 8.4 can be used to estimate the shape parameter.

Since gamma paper is not available for all values of γ, an alternative probability plotting method is proposed by Wilk et al. (1962b). This method can also be used for singly censored data.

Letting $x = \lambda t$ in (6.30), the cumulative distribution can be written as

$$F(x) = \frac{1}{\Gamma(\gamma)} \int_0^x u^{(\gamma - 1)} e^{-u} \, du \tag{7.8}$$

Suppose that $t_{(1)} \leq t_{(2)} \leq \cdots \leq t_{(n)}$ are n ordered survival times. Let $b_1, b_2, \ldots, b_n$ be n appropriately chosen percentage points of the gamma distribution corresponding to the ordered observations, for example, $b_i = (i - 0.5)/n$ for $i = 1, \ldots, n$. Then if $\hat{x}_i$ satisfies $F(\hat{x}_{(i)}) = b_i$, $i = 1, \ldots, n$, a plot of $(\hat{x}_{(i)}, t_{(i)})$, $i = 1, \ldots, n$, will tend to fall along a straight line passing through the origin with slope $1/\lambda$ if the observations are samples from a gamma distribution with density function given in (6.28).

The problem is to solve $F(\hat{x}_{(i)}) = b_i$ for $\hat{x}_{(i)}$ given the value of γ or an estimate. Wilk et al. (1962b) use an iterative procedure to obtain the quantiles $\hat{x}_{(i)}$. They provide a table (reproduced in Appendix C as Table C-7) of quantiles of the standardized gamma distribution (7.8) for various percentage values and a grid of γ values [0.1(0.1) 0.6(0.2) 5.0(0.5) 10.0 (1.0) 22.0]. For γ values and percentages other than those tabulated, linear (for γ) or bilinear (for a quantile value) interpolation is needed.

To prepare a gamma probability plot using this table, one must:

1. Order the survival time such that $t_{(1)} \leq t_{(2)} \leq \cdots \leq t_{(n)}$.
2. Compute $b_{(i)} = (i - 0.5)/n$ for $i = 1, \ldots, n$.
3. Determine the value of γ. Methods discussed in Section 8.4 may be used.
4. Find the corresponding quantiles $\hat{x}_{(i)}$ for each $b_{(i)}$.
5. Plot $t_{(i)}$ versus $\hat{x}_{(i)}$ on ordinary linear-by-linear graph paper.

Table 7.3 Ordered Arrival Times, Plotting Points, and Quantiles for Example 7.5

i	$t_{(i)}$	$100(i - 0.5)/27$	Quantile $(\hat{x}_{(i)})$
1	1	1.85	0.11
2	1	5.56	0.22
3	1	9.26	0.31
4	2	12.96	0.40
5	2	16.67	0.49
6	2	20.37	0.57
7	3	24.07	0.65
8	4	27.78	0.73
9	4	31.48	0.81
10	4	35.19	0.90
11	4	38.89	0.99
12	5	42.59	1.08
13	6	46.30	1.17
14	6	50.00	1.27
15	6	53.70	1.38
16	6	57.41	1.49
17	7	61.11	1.61
18	8	64.81	1.74
19	8	68.52	1.88
20	9	72.22	2.04
21	10	75.93	2.22
22	10	79.63	2.42
23	14	83.33	2.67
24	16	87.04	2.96
25	17	90.74	3.36
26	20	94.44	3.94
27	25	98.15	5.18

The computer program GAMPLOT given in Appendix B computes $b_{(i)}$ and $\hat{x}_{(i)}$ and plots $t_{(i)}$, versus $\hat{x}_{(i)}$, $i = 1, \ldots, n$, for a given value of γ.

The following hypothetical example illustrates the procedure of making a gamma probability plot.

Example 7.5

In a large hospital, the problem of clinical scheduling is investigated. The purpose is to find an optimal schedule such that the use of physicians' time is maximized and patients' waiting time is minimized. The following data are the arrival times in minutes of the first 27 patients at registration on a given weekday: 1, 1, 1, 2, 2, 2, 3, 4, 4, 4, 4, 5, 6, 6, 6, 6, 7, 8, 8, 9, 10, 10, 14, 16, 17, 20, and 25.

Using Wilk et al.'s method (described in Section 8.4), a maximum likelihood estimate of γ is found to be 1.59. Table 7.3 gives the ordered observations $100(i - 0.5)/n$ and the corresponding quantiles obtained from the computer program GAMPLOT.

Figure 7.8 shows a probability plot of the data on gamma probability

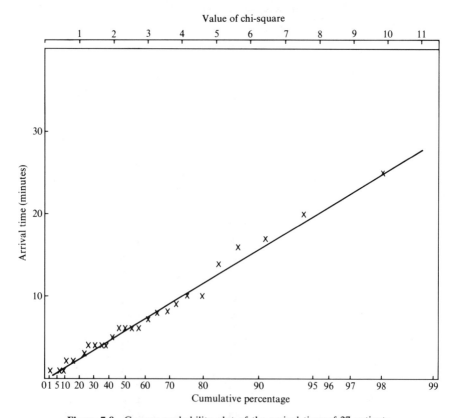

Figure 7.8 Gamma probability plot of the arrival time of 27 patients.

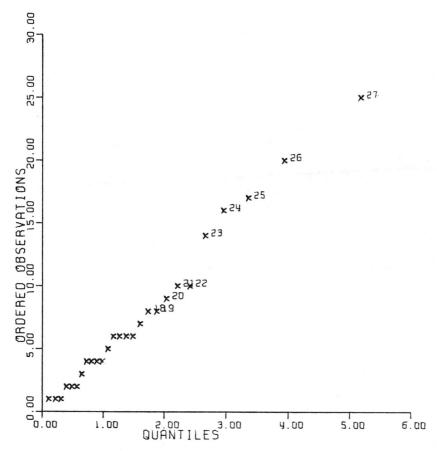

Figure 7.9 Gamma probability plot of the arrival time of 27 patients (method suggested by Wilk et al. 1962b).

paper with $\gamma = 1.5$. An estimate of λ obtained from two points on the fitted line by (7.7) is

$$\hat{\lambda} = \frac{8 - 4.02}{2(20 - 10)} = 0.199$$

Figure 7.9 gives the probability plot of the data following Wilk et al.'s method generated by GAMPLOT with $\gamma = 1.59$. The difference between these two plots is negligible. Both provide a reasonable fit to the data.

7.3 HAZARD PLOTTING

Hazard plotting (Nelson 1972) is analogous to probability plotting, the principal difference being that the observations are plotted against cumulative hazard values rather than cumulative probability values. Hazard plotting is designed to handle censored data, and the hazard rate can easily be determined from the plot. Standard probability statements can also be obtained from an auxiliary probability scale on the hazard plotting paper (or hazard paper). Similar to probability plotting, hazard plotting can also be programmed on a computer and estimates of parameters can be determined from the hazard plot with little computational effort.

Before discussing the theory underlying hazard plotting and hazard paper, we first describe how to make the plots by using hazard papers. Hazard papers are available for the exponential, lognormal, normal, and extreme-value distributions (the last two are not discussed here). Hazard papers have a bottom horizontal scale for cumulative percent hazard. The normal and exponential hazard papers have linear vertical scales for time. The Weibull and lognormal hazard papers have logarithmic vertical scales. The Weibull hazard paper is available in two cycle numbers: two-cycle for time up to hundredfold and three-cycle for time to thousandfold. The auxiliary probability scale is constructed according to the following relationship between the cumulative probability function $F(t)$ and the cumulative hazard function $H(t)$,

$$F(t) = 1 - e^{-H(t)} \qquad (7.9)$$

The probability scale is exactly the same as that on the corresponding probability paper and can be used to estimate the probability of failure or death in a similar way.

Hazard plotting includes the following six steps:

Step 1. Order the n observations in the sample from smallest to largest without regard to whether they are censored. In the list of ordered values, the censored data are each marked with a plus. If some exact and censored observations have the same value, they should be listed in random order.
Step 2. Number the ordered observations in reverse order with n assigned to the smallest data value, $n - 1$ to the second smallest, and so on. The numbers so obtained are called K values, or reverse-order numbers. For the exact observation K is the number of subjects still at risk at that time.
Step 3. Obtain the corresponding hazard value for each exact observation. Censored observations do not have a hazard value. The hazard value for an exact value is $100(1/K)$. This is the percentage of the K individuals who survived that length of time and then failed. It is an observed conditional failure probability for an exact observation.
Step 4. For each exact value, calculate the cumulative hazard value. This is the sum of the hazard values of the exact value and of all preceding exact values.
Step 5. Select a theoretical distribution to fit the data. Plot each exact value against its corresponding cumulative hazard value on the appropriate hazard

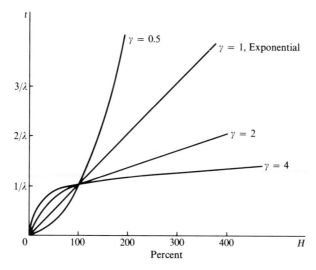

Figure 7.10 Cumulative hazard functions of the Weibull distribution with $\gamma = 0.5$, 1, 2, 4.

paper. The choice of distribution should be based on biomedical experience. If you do not know which hazard paper is most appropriate for a set of censored data, try different papers. First plot the sample cumulative hazard function on exponential hazard paper, which is just a square coordinate paper. Compare the plot with the theoretical cumulative hazard functions (see Figure 7.10) and choose the distribution with the cumulative hazard function most like the shape of the sample cumulative hazard function over the appropriate range of values.

Step 6. Examine the plot for linearity. If the plot is linear, fit a straight line to the data. Parameter estimates can then be obtained If nonlinearity or curvature occurs, investigate to determine the cause and/or try a hazard paper of a different distribution.

The following example illustrates the use of hazard papers.

Example 7.6

Consider the remission data of the 21 leukemia patients receiving 6-MP in Example 3.3. Following the first five steps, Table 7.4 and Figure 3.6 are constructed. In this case the remission data follow a reasonably straight line. Thus, the exponential distribution provides a reasonable fit.

The estimated probability of failure or death is obtained by entering the hazard plot on the vertical time axis at the time given, moving horizontally to the straight line that was fitted to the data, and then moving upward to the probability scale at the top of the paper where the probability is read as a percentage. In this example, the estimated probability of relapse before 16 weeks is obtained by entering the plot on the vertical scale at 16 weeks,

Table 7.4   Hazard Calculations

Patient Number	Remission Time	K	Hazard $(100/K)$	Cumulative Hazard
1	6	21	4.76	4.76
2	6+	20		
3	6	19	5.26	10.02
4	6	18	5.56	15.58
5	7	17	5.88	21.46
6	9+	16		
7	10	15	6.67	28.13
8	10+	14		
9	11+	13		
10	13	12	8.33	36.46
11	16	11	9.09	45.55
12	17+	10		
13	19+	9		
14	20+	8		
15	22	7	14.29	59.84
16	23	6	16.67	76.51
17	25+	5		
18	32+	4		
19	32+	3		
20	34+	2		
21	35+	1		

moving horizontally to the fitted line, and then moving upward to the percentage scale at the top where the probability is read as 38%. Estimates of distribution parameters are given by the slope and an intercept of the fitted line. The method for obtaining such estimates from a hazard plot is different for each distribution. We now discuss the method as well as the basic idea underlying hazard plotting for each distribution.

1. *The Exponential Distribution.* The exponential distribution in (6.1) has constant hazard function

$$h(t) = \lambda \qquad (7.10)$$

Thus, the cumulative hazard function is

$$H(t) = \lambda t \qquad (7.11)$$

From (7.11), the time to failure t can be written as a function of the cumulative hazard H,

$$t = \frac{1}{\lambda} H(t) \qquad (7.12)$$

Since time to failure is a linear function of the cumulative hazard, t plots as a straight-line function of H on square coordinate paper. Therefore, exponential hazard paper is merely square coordinate graph paper. The slope of the line is the mean survival time $1/\lambda$ of the distribution. More simply, $1/\lambda$ is the value of t when $H(t) = 1$. This fact is used to estimate $1/\lambda$ from data plotted on exponential hazard paper.

To find an estimate for the mean remission time of the leukemia patients in Example 7.6, we examine the slope of the fitted line since the time for which $H = 1$ (100%) is out of the range of the vertical scale. The slope of the fitted line is found to be 37 (weeks), which is an estimate of the mean remission time and thus an estimate of the hazard rate λ is 0.027.

2. *The Weibull Distribution.* The Weibull distribution given in (6.9) has the hazard function

$$h(t) = \lambda\gamma(\lambda t)^{\gamma-1} \qquad t > 0 \qquad (7.13)$$

The cumulative hazard function is

$$H(t) = (\lambda t)^{\gamma} \qquad t > 0 \qquad (7.14)$$

and is plotted in Figure 7.10 for four different values of γ: 0.5, 1, 2, and 4. From (7.14), the time to failure t can be written as a function of the cumulative hazard function,

$$t = \frac{1}{\lambda}[H(t)]^{1/\gamma} \qquad (7.15)$$

Taking the logarithm (base 10) of (7.15), we obtain

$$\log t = \log\frac{1}{\lambda} + \frac{1}{\gamma}\log H(t) \qquad (7.16)$$

Since $\log t$ is a linear function of $\log H(t)$, t plots as a straight-line function of $H(t)$ on log–log graph paper. Therefore, the Weibull hazard paper is merely log–log graph paper. The slope of the straight line is $1/\gamma$. This fact is used to estimate γ from data plotted on Weibull hazard paper. In addition, for $\log H(t) = 0$ or $H(t) = 1$, (7.16) reduces to $\log t = \log(1/\lambda)$, and thus the corresponding time t equals $1/\lambda$. This fact is used to estimate $1/\lambda$ and consequently λ.

To find an estimate of $1/\lambda$, enter the plot at the 100% hazard point on the horizontal cumulative scale, go up to the fitted line, and then go horizontally to the vertical time scale. The value read at the vertical time scale is the estimate. The estimate of γ can be found by drawing a line parallel to the fitted line and passing through the heavy dot in the upper left corner of the paper. The point where the line intersects the shape parameter scale at the top of the paper is an estimate of γ.

Example 7.7

Consider the following survival times in days of 14 patients: 20, 25, 30, 32+, 40, 45, 60, 80+, 90, 150, 150+, 200, 250+, 252. The data are plotted on Weibull hazard paper in Figure 7.11. The estimate of $1/\lambda$ is found by entering the plot at the 100% hazard point on the horizontal cumulative hazard scale at the bottom of the paper, going up to the fitted line, and then going horizontally to the vertical time scale where the value is approximately 170. The estimate of the shape parameter γ is found by drawing a line parallel to the fitted line that passes through the heavy dot in the upper left corner of the grid. The point where the line intercepts the shape parameter scale at the top of the paper is 1.25, the estimate of the shape parameter γ.

3. *The Lognormal Distribution.* The density function of a lognormal (base 10) distribution is

$$f(t) = \frac{1}{t\sigma \log_e 10\sqrt{2\pi}} \exp\left[-\frac{1}{2\sigma^2}(\log t - \mu)\right]^2$$

$$= \frac{0.4343}{t\sigma} g\left(\frac{\log t - \mu}{\sigma}\right) \qquad t > 0 \tag{7.17}$$

where $g(x)$ is the standard normal density function. The lognormal cumulative distribution function is

$$F(t) = G\left(\frac{\log t - \mu}{\sigma}\right) \qquad t > 0 \tag{7.18}$$

where $G(x)$ is the standard normal distribution function. Thus, by (2.10), the hazard function can be written as

$$h(t) = \frac{\dfrac{0.4343}{t\sigma} g\left(\dfrac{\log t - \mu}{\sigma}\right)}{1 - G\left(\dfrac{\log t - \mu}{\sigma}\right)} \tag{7.19}$$

The cumulative hazard function, plotted in Figure 7.12 for three values of σ, is

$$H(t) = -\log_e\left[1 - G\left(\frac{\log t - \mu}{\sigma}\right)\right] \tag{7.20}$$

Note that in (7.17)–(7.19) base 10 logarithms are used and in (7.20) base 10 and base e logarithms are used. From (7.20), the log of the survival time t as a function of the cumulative hazard H is

$$\log t = \mu + \sigma G^{-1}[1 - e^{-H(t)}] \tag{7.21}$$

where G^{-1} is the inverse of the standard normal distribution function G.

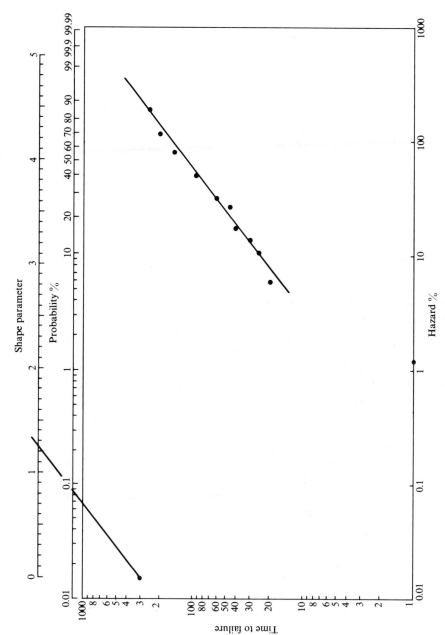

Figure 7.11 Hazard plot of data in Example 7.7.

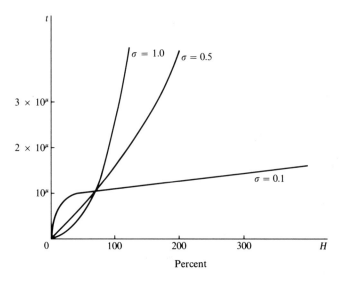

Figure 7.12   Cumulative hazard functions of the lognormal distribution with $\sigma = 0.1, 0.5, 1.0$.

Thus, $\log t$ is a linear function of $G^{-1}[1 - e^{-H(t)}]$. The vertical time scale of the lognormal hazard paper is logarithmic and the cumulative hazard value H is located on the horizontal axis at the position $G^{-1}[1 - e^{-H(t)}]$. The slope of the straight line is the value of the parameter σ. This fact is used to estimate σ from data plotted on the lognormal hazard paper. For $H(t) = \log_e 2$, $1 - e^{-H(t)} = 0.5$, and $G^{-1}(0.5) = 0$, (7.21) reduces to $\log t = \mu$. This fact is used to estimate μ.

To find an estimate of μ, enter the lognormal hazard plot at the 50% point on the horizontal probability scale at the top of the paper; go down to the fitted line and then horizontally to the vertical time scale. The logarithm (base 10) of the value on the vertical time scale is the estimate of μ. The estimate of the lognormal parameter can also be found in two steps. Take the logarithm of the 84th percentile. Then the difference between this value and the estimate of μ is an estimate of σ. This is because $G^{-1}(0.84) = 1$ and (7.21) becomes $\log t = \mu + \sigma$ or $\sigma = \log t - \mu$.

Example 7.8
Consider the following remission times in months of 18 cancer patients: 4, 5, 6, 7, 8, 9+, 12, 12+, 13, 15, 18, 20, 25, 26+, 28+, 35, 35+, 56. The data are plotted on lognormal hazard paper in Figure 7.13. To show how to obtain estimates of μ and σ, assume that the straight line was fitted to the data points. An estimate of μ is found by entering the plot at the 50% point on the horizontal probability scale, going down to the fitted line and then horizontally to the vertical time scale where the value 18 is read. The logarithm (base 10) of 18, or 1.26, is the estimate. The estimate of the

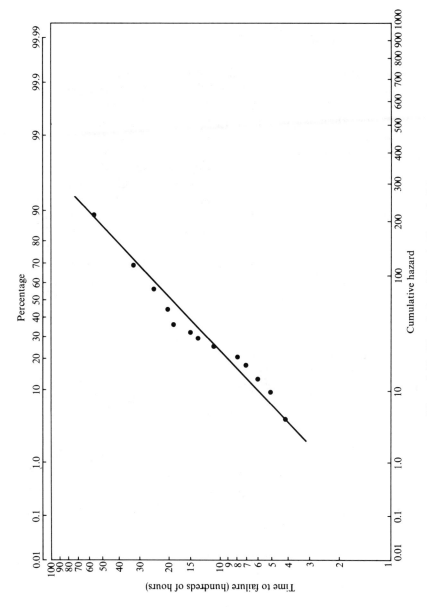

Figure 7.13 Hazard plot of data in Example 7.8.

lognormal parameter σ is found by taking the logarithm of the 84th percentile, which is $\log 48 = 1.68$. The difference between this value and the estimate of μ, that is, $1.68 - 1.26 = 0.42$, is the estimate of σ.

7.4   TESTS OF GOODNESS-OF-FIT

Probability plotting and hazard plotting are subjective methods in which the determination of whether the assumed distribution fits the data is based on visual examination rather than statistical test. Therefore, statistical tests are often used in conjunction with probability or hazard plotting.

A statistical test involves calculation of a test statistic from the data and the probability of obtaining the statistic if the correct distribution is chosen. If the probability of obtaining the calculated statistic is very low, we conclude that the assumed distribution does not provide an adequate fit to the data. This procedure allows us to reject an inadequate distribution but never allows us to prove that the distribution is correct. It gives us a probabilistic statement about the assumed distribution. The outcome of a statistical test of hypothesis depends on the amount of data available; the more data there are, the better are the chances of rejecting an inadequate distribution.

Let $F(t)$ be the underlying distribution of the data. The general null hypothesis is

$$H_0: F(t) = F_0(t)$$

where $F_0(t)$ is a specific distribution.

In this section, we discuss four statistical tests of goodness-of-fit, three for complete data (the WE-test for the exponential distribution, the W-test for the lognormal distribution, and a chi-square test), and one for censored data.

7.4.1   The WE-Test for the Exponential Distribution

Let $t_1, t_2, \ldots, t_n$ be the n observed times to failure. The test statistic of the WE-test for the null hypothesis that $F_0(t)$ is the one-parameter exponential distribution with unknown parameter λ is

$$WE_1 = \frac{\sum\limits_{i=1}^{n} (t_i - \bar{t})^2}{\left(\sum\limits_{i=1}^{n} t_i\right)^2} = \frac{\sum\limits_{i=1}^{n} t_i^2 - \left(\sum\limits_{i=1}^{n} t_i\right)^2 \Big/ n}{\left(\sum\limits_{i=1}^{n} t_i\right)^2} \tag{7.22}$$

where $\bar{t}$ is the sample mean (Shapiro and Wilk 1965, Hahn and Shapiro 1967).

Percentage points (95% and 90%) of the distribution of WE_1 have been tabulated for $n = 7, \ldots, 35$ shown in Table C-8 in Appendix C. The test is two-sided. Thus, if the computed value of WE_1 falls outside the 95% (or 90%) range, the probabilities are less than 0.05 (or 0.10) that the observed data are drawn from a one-parameter exponential distribution. Example 7.9 illustrates the test.

Example 7.9
Consider the remission data of the 21 leukemia patients receiving a placebo in Example 3.3. The 21 observed remission times in weeks are 1, 1, 2, 2, 3, 4, 4, 5, 5, 8, 8, 8, 8, 11, 11, 12, 12, 15, 17, 22, and 23. To determine whether the remission time of patients treated with a placebo follows the one-parameter exponential distribution, we apply the WE_1-test. From the data,

$$n = 21 \qquad \sum_{i=1}^{n} t_i = 182 \qquad \sum_{i=1}^{n} t_i^2 = 2414$$

By (7.22) we have

$$WE_1 = \frac{2414 - (182)^2/21}{(182)} = 0.025$$

From Table C-8 for $n = 21$, it is seen that $WE_1 = 0.025$ is within the 95% range. Thus the hypothesis that the data are from the exponential distribution is not rejected at the 0.05 level.

This conclusion is consistent with that obtained in Example 3.3 where the probability plotting technique is used.

For the two-parameter exponential distribution with unknown λ and G, the test statistic is

$$WE_2 = \frac{(\bar{t} - t_{(1)})^2}{\sum_{i=1}^{n} (t_i - \bar{t})^2} = \frac{\left(\dfrac{\sum_{i=1}^{n} t_i}{n} - t_{(1)}\right)^2}{\sum_{i=1}^{n} t_i^2 - \dfrac{\left(\sum_{i=1}^{n} t_i\right)^2}{n}} \tag{7.23}$$

where $t_{(1)}$ is the smallest observation.

Percentage points (95 and 90%) of the distribution of WE_2 have been tabulated for $n = 7, \ldots, 35$ in Table C-9. The test is two-sided. Thus, if the computed value of WE_2 falls outside the 95% (or 90%) range, the changes are less than 0.05 (or 0.10) that the observed data are drawn from a two-parameter exponential distribution. Example 7.10 illustrates the test.

Example 7.10

Suppose that 20 melanoma patients achieve remission after surgery. Their remission durations are 10, 8, 20, 28, 11, 14, 30, 20, 15, 24, 30, 12, 12, 28, 26, 24, 20, 28, 16, and 18 months. We want to determine whether the exponential distribution is an adequate model for the remission time. Since it is possible that the patient will remain in remission for some time, we choose the two-parameter exponential distribution and apply the WE_2-test. From the data,

$$n = 20 \qquad t_{(1)} = 8 \qquad \sum_{i=1}^{n} t_i = 394 \qquad \sum_{i=1}^{n} t_i^2 = 8754$$

By (7.23) we have

$$WE_2 = \frac{\left(\dfrac{394}{20} - 8\right)^2}{8754 - \dfrac{(394)^2}{20}} = 0.138$$

From Table C-9 for $n = 20$, it is seen that $WE_2 = 1.038$ is outside the 95% range. Thus, the hypothesis that the data come from an exponential distribution is rejected at the 0.05 level. Thus a two-parameter exponential distribution appears inadequate as a model for the remission time.

7.4.2   The W-Test for the Lognormal Distribution

The *W*-test of Shapiro and Wilk (1965) was originally proposed to test the assumption of a normal distribution. Because of the close relationship between the normal and lognormal distributions, the *W*-test may also be used to evaluate the assumption that F_0 is a lognormal distribution. In this case the test is applied to the logarithms (natural or common) of the observations.

Let $t_1, t_2, \ldots, t_n$ be the n observations and $x_i = \log t_i$ for $i = 1, \ldots, n$. Rearrange the x_i's in ascending order so that

$$x_{(1)} \le x_{(2)} \le \cdots \le x_{(n)}$$

If n is even, set $k = \frac{1}{2}n$; if n is odd, set $k = \frac{1}{2}(n - 1)$. Then let

$$b = a_n(x_{(n)} - x_{(1)}) + a_{n-1}(x_{(n-1)} - x_{(2)}) + \cdots + a_{(n-k+1)}(x_{(n-k+1)} - x_k)$$

$$= \sum_{i=1}^{n} a_{(n-i+1)}(x_{(n-i+1)} - x_i) \tag{7.24}$$

where the values of $a_{(n-i+1)}$ for $i = 1, \ldots, k$ are given in Table C-10 for $n = 3$–50. Then the *W*-test statistic is defined as

$$W = \frac{b^2}{\sum\limits_{i=1}^{n} (x_i - \bar{x})^2} = \frac{b^2}{\sum\limits_{i=1}^{n} x_i^2 - \frac{\left(\sum\limits_{i=1}^{n} x_i\right)^2}{n}} \tag{7.25}$$

Percentage points of the distribution of W are given in Table C-11. The table gives the minimum values of W that one would obtain with 1, 2, 5, 10, and 50% probability as a function of n if the data are from a lognormal distribution. The test is one-sided. That is, small values of W indicate that the lognormal distribution does not provide an adequate fit to the data. For example, if a computed W value is less than 0.842 for $n = 10$, the probability is less than 5% that the data could have come from a lognormal distribution.

An alternative is to find the probability of obtaining the computed value of W under the hypothesis of lognormal distribution. An approximate probability can be found by calculating the value of

$$Z = \gamma + \eta \log_e\left(\frac{W - \epsilon}{1 - W}\right) \tag{7.26}$$

using the value of γ, η, and ϵ given in Table C-12 for the appropriate sample size n. The value of Z follows the standard normal distribution under the hypothesis. The probability of obtaining a value less than or equal to the calculated Z value can then be obtained from Table C-1. This is the probability that the data are from a lognormal distribution. Example 7.11 illustrates the test.

Example 7.11

A sample of 10 individuals is involved in an experiment of testing the effectiveness of an analgesic. The pain relief times in minutes are 45, 64, 58, 92, 112, 52, 40, 76, 40, and 90. The W-test is applied to test the hypothesis that the data come from a lognormal distribution. Taking the logarithm (natural) of the observations and rearranging them in ascending order, we obtain $x_{(1)} = 3.69$, $x_{(2)} = 3.69$, $x_{(3)} = 3.81$, $x_{(4)} = 3.95$, $x_{(5)} = 4.06$, $x_{(6)} = 4.16$, $x_{(7)} = 4.33$, $x_{(8)} = 4.50$, $x_{(9)} = 4.52$, and $x_{(10)} = 4.72$.

Since $n = 10$, we set $k = 5$ in (7.24). From Table C-10, $a_{10} = 0.05739$, $a_9 = 0.3291$, $a_8 = 0.2141$, $a_7 = 0.1224$, and $a_6 = 0.0399$. From (7.24),

$$b = 0.5739(4.72 - 3.69) + 0.3291(4.52 - 3.69)$$

$$+ 0.2141(4.50 - 3.81) + 0.1224(4.33 - 3.95)$$

$$+ 0.0399(4.16 - 4.06)$$

$$= 1.0625$$

From the denominator of (7.25)

$$\sum_{i=1}^{n} x_i = 41.43 \qquad \sum_{i=1}^{n} x_i^2 = 172.85$$

Hence,

$$W = \frac{(1.0625)^2}{172.85 - (41.43)^2/10} = 0.936$$

This value of W exceeds the tabulated 5% value of $W = 0.842$ obtained from Table C-11.

To find the approximate probability of obtaining the computed value of $W = 0.881$, we find from Table C-12 that for $n = 10$, $\gamma = -3.262$, $\eta = 1.471$, and $\epsilon = 0.3660$. By (7.26),

$$Z = -3.262 + 1.471 \log_e \frac{0.936 - 0.366}{1 - 0.936}$$

$$= -0.045$$

From Table C-1, $P(Z \le -0.045) = 0.4821$. That is, the approximate probability of obtaining a value of W as low as 0.936 if the data come from a lognormal distribution is 0.4821. If a significance level of 0.05 or less is used, we conclude that, on the basis of the 10 observations, there is not enough evidence to reject the hypothesis that the data follow a lognormal distribution. Figure 7.14 gives a lognormal probability plot of the data. The linear configuration indicates that the lognormal distribution is appropriate.

The W-test is effective for testing lognormal against other nonlognormal alternatives, even for relatively small sample sizes. It has been shown that when $n = 20$ and the sample is from an exponential distribution, the power of the W-test is about 0.80.

7.4.3 The Chi-Square Goodness-of-Fit Test for Data without Censored Observations

The most widely used test of goodness-of-fit is the chi-square test introduced by Karl Pearson in 1900. It is a nonparametric statistical method and can be used to test any distributional assumption.

To use the chi-square test, the data have to be grouped into categories or "cells." The observed frequencies in the cells are then compared to the expected frequencies based on the hypothesized distribution. The question to be asked is whether the observed frequencies deviate significantly from the frequencies expected if the hypothesis were true. To answer the question, the following statistic called *chi square* is used as a measure of how far a sample distribution deviates form the hypothesized distribution:

$$\chi^2 = \sum_{i=1}^{k} \frac{(O_i - E_i)^2}{E_i} \qquad (7.27)$$

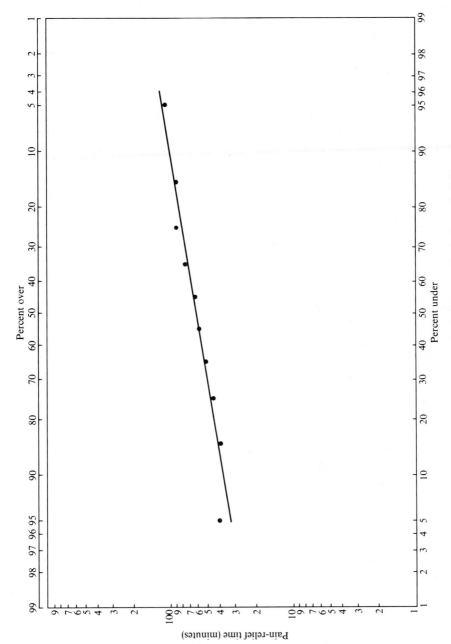

Figure 7.14 A lognormal probability plot of data in Example 7.11.

187

where k is the number of cells, O_i is the observed frequency in the ith cell, and E_i is the expected frequency in the ith cell if the hypothesized distribution is true.

It should be apparent, by examining (7.27), that large values of χ^2 result from large differences between the observed and expected frequencies. A calculated χ^2 value can be very small if the fit is good or very large if the fit is bad. To determine how large is too large to claim a good fit, we use the distribution of χ^2, which happens to be the chi-square distribution with $k - r - 1$ degrees of freedom, where r is the number of parameters needed to be estimated. Percentage points of the chi-square distribution are given in Table C-2. Thus, the procedure is to compute the χ^2 value in (7.27) from the data and compare it to the tabulated percentage points, $\chi^2_{k-r-1,\alpha}$, in Table C-2 for a selected significance level α. If the computed χ^2 value exceeds the tabulated value for the selected significance level and the appropriate degrees of freedom, we conclude that the data are not from the hypothesized distribution.

The expected frequency E_i in (7.27) is calculated by

$$E_i = np_i \qquad (7.28)$$

where p_i is the probability that an observation will fall into the ith cell if the hypothesized distribution is true. Thus the calculation of p_i differs with the distributions hypothesized and the numbers of cells k chosen.

There are two ways to determine the number of cells k. First, if the data are initially arranged in categories or cells, for example, data used in a life-table analysis, the number of cells and boundaries of the cells are determined. In this case, if the expected frequency in any cell is less than 5, the cell should be combined with an adjacent cell or cells so that the expected frequency in the combined cell is at least 5.

If the data are not grouped and consist of a large number of observations (say, 200 or more), Williams (1950) suggests computing the number of cells by the following formula:

$$k = 4\left[\frac{2(n-1)^2}{z^2}\right]^{1/5} \qquad (7.29)$$

where n is the number of observations in the sample, z is the 100α percentage point of the standard normal distribution (Table C-1), and α is the significance level chosen. For example, if $\alpha = 0.05$, the 5% point obtained from Table C-1 is 1.645. Equation (7.29) is then

$$k = 4\left[\frac{2(n-1)^2}{(1.645)^2}\right]^{1/5} = 4[0.74(n-1)^2]^{1/5} \qquad (7.30)$$

For moderate sample sizes the rule is to make k as large as possible subject to the constraint that the expected frequency in each cell is at least 5. A good initial trial is to set $k = \frac{1}{5}n$.

Having determined the number of cells k, the cell boundaries $t_1', t_2', \ldots, t_k'$ are determined from the cumulative distribution for the hypothesized distribution (using the estimated parameters) by solving

$$P(T \le t_1') = \frac{1}{k}$$

$$P(T \le t_2') = \frac{2}{k}$$

$$\vdots$$

$$P(T \le t_{k-1}') = \frac{k-1}{k} \qquad (7.31)$$

That is, the boundaries are set up in such a way that the probability of an observation falling into any one of the k cells is $1/k$, or

$$p_i = 1/k \qquad i = 1, \ldots, k \qquad (7.32)$$

Note that t_1' is the upper bound of the first cell and lower bound of the second cell, t_2' is the upper bound of the second cell and the lower bound of the third cell, and so on. The lower bound of the first cell is zero and the upper bound of the last cell is, theoretically, infinity

The procedure for the chi-square goodness-of-fit test may be summarized as follows:

1. Determine the number of cells k.
2. Estimate the parameters of the hypothesized distribution. Graphical methods discussed in this chapter or analytical methods discussed in Chapter 8 can be used.
3. Determine boundaries of the k cells using the estimated parameters of the hypothesized distribution and (7.31).
4. Determine the observed frequency in each of the k cells.
5. Determine the expected frequency in each of the k cells by (7.32) and (7.28).
6. Compute the χ^2 value by (7.27).
7. Compare the computed χ^2 value with the tabulated percentage points for a chi-square distribution with $k - r - 1$ degrees of freedom given in Table C-2, where r is the number of parameters estimated in step 2. If the computed χ^2 value exceeds the tabulated value for the appropriate degrees of freedom and significance level, we conclude that the data are not from the hypothesized distribution.

The following example illustrates the procedure.

Example 7.12

Consider the remission data of the 21 leukemia patients receiving placebo in Examples 3.3 and 7.9. We use the chi-square test to test the hypothesis that the data are from an exponential distribution. The procedure is as follows:

1. Since the sample size is small ($n = 21$), we first try $k = \frac{21}{5}$, or approximately 4.

2. Using the graphical method, we have obtained (Example 3.3), $\hat{\lambda} = 0.12$.

3. For the exponential distribution, $P(T \leq t) = F(t) = 1 - e^{-\lambda t}$. Hence, by (7.31)

$$P(T \leq t_1') = 1 - e^{-0.12 t_1'} = 0.25$$

$$\text{or} \quad t_1' = -\frac{1}{0.12} \log_e(1 - 0.25) = 2.4$$

$$P(T \leq t_2') = 1 - e^{-0.12 t_2'} = 0.50$$

$$\text{or} \quad t_2' = -\frac{1}{0.12} \log_e(1 - 0.5) = 5.8$$

$$P(T \leq t_3') = 1 - e^{-0.12 t_3'} = 0.75$$

$$\text{or} \quad t_3' = -\frac{1}{0.12} \log_e(1 - 0.75) = 11.6$$

4. The observed cell frequencies are given in Table 7.5.

5. Using (7.28) and (7.32), the expected cell frequencies are computed and given in Table 7.5.

6. By (7.27),

$$\chi^2 = \frac{(4 - 5.25)^2}{5.25} + \frac{(5 - 5.25)^2}{5.25} + \frac{(6 - 5.25)^2}{5.25} + \frac{(6 - 5.25)^2}{5.25}$$

$$= 0.524$$

Table 7.5   Observed and Expected Cell Frequencies Assuming Exponential Distribution for Remission Times

Cell (i)	Observed Frequency O_i	Expected Frequency E_i
≤ 2.39	4	5.25
2.4–5.79	5	5.25
5.8–11.59	6	5.25
≥ 11.60	6	5.25

7. Since only one parameter was estimated, $r = 1$, the degree of freedom of the chi-square distribution is $4 - 1 - 1 = 2$. Let $\alpha = 0.05$. The 5% point of the chi-square distribution with two degrees of freedom, from Table C-2, is $\chi^2_{2,0.05} = 5.99$. The computed χ^2 value is much smaller than the tabulated value. Thus, the given data do not provide enough evidence to reject the hypothesis that the data come from an exponential distribution.

7.4.4 A Goodness-of-Fit Test for Censored Data

The chi-square goodness-of-fit test is inappropriate when there are censored observations. Several goodness-of-fit procedures have been proposed for testing whether the data with randomly censored observations are from a specific distribution, for example, Koziol and Green (1976), Hyde (1977), and Hollander and Proschan (1979). In this section, the test developed by Hollander and Proschan is discussed.

Let $t_{(1)} < t_{(2)} < \cdots < t_{(n)}$ be a set of distinct ordered survival times and some of the $t_{(i)}$'s may be censored. If censored observations are tied with uncensored observations, treat the censored observations of the tie as being greater than the uncensored of the tie. Let $S(t)$ be the underlying survivorship function and $S_0(t)$ the survivorship function of the specified distribution. The null hypothesis is $H_0: S(t) = S_0(t)$.

Using the Kaplan–Meier product-limit method, $S(t)$ is estimated as

$$\hat{S}(t) = \prod_{\substack{t_{(i)} \leq t \\ t_{(i)} \text{uncensored}}} \frac{n-i}{n-i+1} \qquad t_{(i)} < t \leq t_{(i+1)}$$

Hollander and Proschan's test statistic for the null hypothesis that the data are from a distribution with survivorship function $S_0(t)$ is

$$C = \sum_{\substack{\text{all} \\ \text{uncensored} \\ \text{observations}}} S_0(t_{(i)})\hat{f}(t_{(i)}) \qquad (7.33)$$

where $\hat{f}(t_{(i)})$ is the jump of the Kaplan–Meier estimates at consecutive uncensored observations and at the largest observation, uncensored or not,

$$\hat{f}(t_{(i)}) = \frac{1}{n} \prod_{j=1}^{i-1} \left(\frac{n-j+1}{n-j} \right)^{1-\delta_{(j)}}$$

where $\delta_{(j)} = 1$ if $t_{(j)}$ is uncensored and $\delta_{(j)} = 0$ if $t_{(j)}$ is censored. Under the null hypothesis,

$$C^* = \sqrt{n}(C - \tfrac{1}{2})/\hat{\sigma} \qquad (7.34)$$

follows approximately the standard normal distribution, where $\hat{\sigma}$ is an estimate of the standard deviation of C and

$$\hat{\sigma}^2 = \tfrac{1}{16} \sum_{i=1}^{n} \frac{n}{n-i+1} [S_0^4(t_{(i-1)}) - S_0^4(t_{(i)})] \qquad (7.35)$$

To test $H_0: S = S_0$ versus $H_1: S > S_0$, we reject H_0 if $C^* < -Z_\alpha$; to test H_0 versus $H_1: S < S_0$, we reject H_0 if $C^* > Z_\alpha$; and to test H_0 versus $H_1: S \neq S_0$, we reject H_0 if $C^* > Z_{\alpha/2}$ or $C^* < -Z_{\alpha/2}$, where Z_α is the upper α percentile point of the standard normal distribution.

The procedure for the calculation of C^* can be summarized as follows.

1. Compute the Kaplan–Meier estimate $\hat{S}(t)$ for each uncensored observation.
2. Compute the jump of the Kaplan–Meier estimate at each uncensored observation, that is, $\hat{f}(t_{(i)})$, which is the difference of $\hat{S}(t)$ at two consecutive uncensored observations.
3. Compute $S_0(t_{(i)})$ for each observation.
4. Multiply $S_0(t_{(i)})$ by $\hat{f}(t_{(i)})$ and sum over all uncensored $t_{(i)}$'s to obtain C.
5. Compute $\hat{\sigma}^2$ according to (7.35) and consequently C^* according to (7.34).

The following example illustrates the test procedure.

Example 7.13
Consider the following survival times in weeks of 10 mice with a given tumor: 8, 5, 10+, 1, 3, 18, 22, 15, 25+, and 19. We wish to test that the survival time follows an exponential distribution with $\lambda = 0.06$. The null and alternative hypotheses are

$$H_0: S(t) = S_0(t) \qquad H_1: S(t) \neq S_0(t)$$

where $S_0(t) = e^{-0.06t}$.

Following the procedure outlined above, we first arrange the observations in ascending order and compute the Kaplan–Meier estimates as shown in column (d) of Table 7.6. The jumps are given in column (e). For example, the first jump is between $\hat{S}(0)$ and $\hat{S}(1)$ or $1 - 0.9 = 0.1$. Column (f) gives the survival function under the null hypothesis, for example, $S_0(3) = e^{-0.06 \times 3} = 0.835$. Following (7.33), column (g) gives the value of $C = 0.4808$. The last three columns are for the calculation of the estimated variance of C. Thus,

Table 7.6 Calculation of Test Statistic C^* for Data in Example 7.13

(a)	(b)	(c)	(d)	(e)	(f)	(g)	(h)	(i)	(j)
$t_{(i)}$	i	$\dfrac{n-i}{n-i+1}$	$\hat{S}(t)$	$\hat{f}(t_{(i)})$	$S_0(t_{(i)})$	(e) × (f)	$S_0^4(t_{(i)})$	$S_0^4(t_{(i-1)}) - S_0^4(t_{(i)})$	$\dfrac{n}{n-i+1} \times$ (i)
1	1	0.900	0.900	0.100	0.941	0.0941	0.7841	0.2159[a]	0.2159
3	2	0.889	0.800	0.100	0.835	0.0835	0.4861	0.2980	0.3311
5	3	0.875	0.700	0.100	0.741	0.0741	0.3015	0.1846	0.2308
8	4	0.857	0.600	0.100	0.619	0.0619	0.1468	0.1547	0.2210
10+	5	—	—	0	0.549	0	0.0908	0.0560	0.0933
15	6	0.800	0.480	0.120	0.407	0.0560	0.0274	0.0634	0.1268
18	7	0.750	0.360	0.120	0.340	0.0408	0.0134	0.0140	0.0350
19	8	0.667	0.240	0.120	0.320	0.0384	0.0105	0.0029	0.0097
22	9	0.500	0.120	0.120	0.267	0.0320	0.0051	0.0054	0.0270
25+	10	—	—	0	0.223	0	0.0025	0.0026	0.0260
						0.4808			1.3166

$^a S_0^4(t_{(0)}) = S_0^4(0) = 1.$

$$\hat{\sigma}^2 = \tfrac{1}{16}\,(1.3166) = 0.0823$$

$$\hat{\sigma} = 0.2869$$

and $C^* = \sqrt{10}\,(0.4808 - 0.5)/0.2869 = -0.2116$. For $\alpha = 0.05$, $Z_{\alpha/2} = 1.96$, C^* does not fall in the rejection region. From Table C-1, we obtain that the p value corresponding to $C^* = -0.2116$ is approximately 0.84. Therefore, we conclude that there is insufficient evidence to say that the data are not from an exponential distribution with $\lambda = 0.06$. Figure 7.15, which plots the Kaplan–Meier estimates and the hypothesized theoretical distribution $S_0(t)$, demonstrates a close agreement between the two.

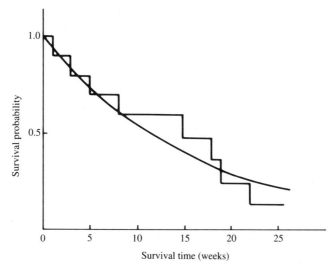

Figure 7.15 Kaplan–Meier estimator $\hat{S}(t)$ and the hypothesized survival function $S_0(t) = \exp(-0.06t)$.

7.5 COMPUTER PROGRAMS FOR THE GAMMA PROBABILITY PLOT

The computer program GAMPLOT, written in FORTRAN, computes the quantiles (or percentage points) of the gamma distribution and plots the quantiles against the ordered observations for a given value of the shape parameter γ. The value of γ is either known or estimated from the data by methods discussed in Chapter 8. The program consists of a main program and eight subroutines: GPLOT, GAMINV, SORTA, GAM, GTAILI, DETAZ, DAMLOG, and QNORM.

The main program, provided by the user, given the observations and the value of γ and calls the subroutine GPLOT. GPLOT calls the subroutine GAMINV to compute the quantiles and then plots the quantiles against the ordered observations. All of the other subroutines are necessary for the calculation of the quantiles. The functions of each subroutine are described in detail at the beginning of each subroutine listing given in Appendix B. A brief description is given below.

GPLOT   Plots the ordered observations against the standard gamma quantiles. This is the only subroutine called by the main program. The calling sequence is CALL GPLOT (GAMMA, N, T, XID, TITLE, K), where GAMMA is the value of the shape parameter γ, N is the number of observations, T(I) is the Ith observation, $I = 1, \ldots, N$ (note that the dimension of T is $N + 2$), XID(I) is the identification of the Ith observation, $I = 1, \ldots, N$, TITLE is the title of the project that will be shown on the gamma probability plot, and K is the number of the largest observations to be identified on the gamma probability plot.

GAMINV   Given a gamma distribution with shape parameter γ (assuming a scale parameter $\lambda = 1$) and a series of increasing fractions b_i, computes the corresponding quantiles x_i.

SORTA   Sorts the n observations t_i in array T in ascending order.

GAM   Computes the value of the gamma function $\Gamma(z)$ for a given z.

GTAILI   Finds the quantile corresponding to the probability of the gamma distribution with shape parameter γ and scale parameter $\lambda = 1$ when the probability is greater than 0.99.

DETAZ   Computes the incomplete gamma function $I(\gamma, z) = \int_0^1 t^{\gamma-1} e^{-zt}\, dt$.

DAMLOG   Computes the logarithm to the base e of $I(\gamma, z)$.

QNORM   Finds the quantiles of the standard normal distribution associated with a specified probability.

The program allows a maximum of 200 observations. The following example illustrates its use.

Example 7.14

Suppose that 27 patients have the following survival times in months: 1, 2, 10, 4, 7, 4, 1, 4, 8, 2, 6, 9, 3, 6, 1, 17, 8, 20, 6, 4, 2, 5, 16, 6, 10, 14, and 73. The gamma distribution is a possible model for the data and the shape parameter γ was estimated by the method discussed in Chapter 8. We obtained $\hat{\gamma} = 1.098$. GAMPLOT is used to make the gamma probability plot.

Table 7.7   Printed Output of GAMPLOT for Data in Example 7.14
27 POINTS FOR GAMMA PROBABILITY PLOT
GAMPLOT EXAMPLE
GAMMA = 1.098

I	Ident	Ordered Observations		Fractions	Quantiles
1	1	1.0000000E	00	1.8518515E − 02	2.7898546E − 02
2	15	1.0000000E	00	5.5555552E − 02	7.7685177E − 02
3	7	1.0000000E	00	9.2592537E − 02	1.2657529E − 01
4	2	2.0000000E	00	1.2962961E − 01	1.7596328E − 01
5	10	2.0000000E	00	1.6666663E − 01	2.2643673E − 01
6	21	2.0000000E	00	2.0370370E − 00	2.7839053E − 01
7	13	3.0000000E	00	2.4074072E − 01	3.3215201E − 01
8	8	4.0000000E	00	2.7777773E − 01	3.8802814E − 01
9	4	4.0000000E	00	3.1481481E − 01	4.4633317E − 01
10	6	4.0000000E	00	3.5185182E − 01	5.0742388E − 01
11	20	4.0000000E	00	3.8888884E − 01	5.7168007E − 01
12	22	5.0000000E	00	4.2592591E − 01	6.3955402E − 01
13	24	6.0000000E	00	4.6296293E − 01	7.1155214E − 01
14	14	6.0000000E	00	5.0000000E − 01	7.8833294E − 01
15	11	6.0000000E	00	5.3703701E − 01	8.7065077E − 01
16	19	6.0000000E	00	5.7407403E − 01	9.5944548E − 01
17	5	7.0000000E	00	6.1111110E − 01	1.0559502E 00
18	9	8.0000000E	00	6.4814812E − 01	1.1617231E 00
19	17	8.0000000E	00	6.8518513E − 01	1.2788658E 00
20	12	9.0000000E	00	7.2222221E − 01	1.4101868E 00
21	3	1.0000000E	01	7.5925922E − 01	1.5598679E 00
22	25	1.0000000E	01	7.9629624E − 01	1.7339878E 00
23	26	1.4000000E	01	8.3333331E − 01	1.9425011E 00
24	23	1.6000000E	01	8.7037033E − 01	2.2027340E 00
25	16	1.7000000E	01	9.0740740E − 01	2.5499382E 00
26	18	2.0000000E	01	9.4444442E − 01	3.0749979E 00
27	27	7.3000000E	01	9.8148143E − 01	4.19882887E 00

The main program required to obtain the gamma probability plot is:

```
C   PROGRAM GAMPLOT
    DIMENSION T (29), XID (27), TITLE (7), BUFFER (1024)
    DATA T/1.0,2.0,10.0,4.0,7.0,4.0,1.0,4.0,8.0,2.0,6.0,9.0,3.0,6.0,
1       1.0,17.0,8.0,20.0,6.0,4.0,2.0,5.0,16.0,6.0,10.0,14.0,73.0/
    DATA TITLE/4HGAMP,4HLOT,4HEXAM,4HPLE/
    DO 1 I = 1,27
1   XID (I) = I
    GAMMA = 1.098
    CALL PLOT (0.,0.,999)
    CALL GPLOT (GAMMA,27,T,XID,TITLE,5)
    CALL PLOTS (BUFFER,1024)
    STOP
    END
```

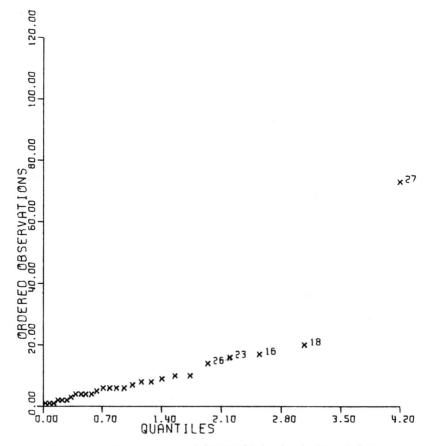

Figure 7.16 Program output of GAMPLOT for data in Example 7.14.

Note that subroutines PLOT and PLOTS are called to initiate and terminate the plotting procedure on the Calcomp plotter. These subroutines can be replaced to suit other types of plotters.

The printed output is given in Table 7.7 and the gamma probability plot is given in Figure 7.16.

BIBLIOGRAPHICAL REMARKS

Probability plotting has been widely used since Daniel's (1959) classical work on the use of half-normal plot. Kao (1959) uses a simple graphical method for estimating the parameters of a Weibull distribution for electron tube failures. Kimball (1960) throws more light on the problem of choosing plotting positions on probability papers. A quite complete and excellent treatment of probability plotting is given by King (1971). Although examples given are applications to industrial reliability, its simple interpretation of probability plots of many distributions, such as the uniform, lognormal, Weibull, and gamma, are applicable to biomedical research. A recent application is given by Horner (1987).

Hazard plotting was developed by Nelson (1972). More discussions of the technique and example are also given by Nelson (1982).

EXERCISES

7.1 Construct a probability scale for the Weibull probability paper.

7.2 Consider the following survival times of 16 patients in weeks: 4, 20, 22, 25, 38, 38, 40, 44, 56, 83, 89, 98, 110, 138, 145, and 27.

    a. Does the exponential distribution provide a reasonable fit to the survival data? Use the probability plotting technique.

    b. Estimate graphically the parameter λ of the exponential distribution and consequently the mean survival time.

    c. Use the WE_1-test to test the distributional hypothesis.

7.3 In order to computerize patients' records, a data clerk is hired to transcribe medical data from the patients' charts to computer coding forms. The number of correct entries between errors is listed in chronological order of occurrence over a period of five days as follows: 73, 12, 40, 65, 100, 15, 70, 40, 110, 64, 200, 6, 90, 102, 20, 102, 90, 34. The assumption is that the data clerk, during the five days, would not change her error rate appreciably.

    a. Use the technique of probability plotting to evaluate the above assumption. What is your conclusion?

    b. Use the WE_2-test to test the assumption.

7.4 Twenty-five rats were injected with a given tumor inoculum. Their times, in days, to the development of a tumor of a certain size are given below.

30	53	77	91	118
38	54	78	95	120
45	58	81	101	125
46	66	84	108	134
50	69	85	115	135

Which of the distributions discussed in this chapter provide a reasonable fit to the data? Estimate graphically the parameters of the distribution chosen.

7.5 In a clinical study, 28 patients with cancer of the head and neck did not respond to chemotherapy. Their survival times in weeks are given below.

1.7	8.3	14.0	22.7	6.0+	13.1+
5.1	9.6	15.9	33.0	7.4+	13.4+
5.3	11.3	16.7	3.7+	8.0+	16.1+
6.0	12.1	17.0	5.0+	8.3+	
8.3	12.3	21.0	5.9+	9.1+	

    a. Make a hazard plot for each of the following distributions: exponential, Weibull, and lognormal.

    b. Which distribution provides a reasonable fit to the data? Estimate graphically the parameters of the distribution chosen.

7.6 Thirty-one patients with advanced melanoma treated with combined chemotherapy, immunotherapy, and hormonal therapy have survival times as given below.

26.3+	16.1	24.0	4.3	31.3+
94.0	49.6	77.9	97.6+	17.6+
9.1	27.3	16.6+	7.3	16.3
34.6+	61.9+	3.4	75.6+	
9.4	46.6+	10.9	14.3	
25.7	22.4+	13.0	56.4	
88.7	7.1	64.4+	9.1	

a. Make a hazard plot for each of the following distributions: exponential, Weibull, and lognormal.

b. Which distribution provides a reasonable fit to the data? Estimate the parameters of the distribution chosen.

7.7   Consider the survival times of the hypernephroma patients in Exercise Table 3.1 (see Exercise 4.5). Make a hazard plot for the distribution you chose in Exercise 6.8. Did you make a good selection? If not, try two other distributions.

7.8   Consider the following survival times in weeks of 10 mice with injection of tumor cells: 5, 16, 18+, 20, 22+, 24+, 25, 30+, 35, 40+.

a. Do the data follow the exponential distribution with $\lambda = 0.02$? Plot the Kaplan–Meier estimator of $S(t)$ and the hypothesized distribution.

b. Make an exponential hazard plot and compare the graphical results with the test results obtained in (a).

7.9   Consider the following survival times in months of 25 patients with cancer of the prostate. Test the hypothesis that the survival time of prostate cancer patients follows the exponential distribution with $\lambda = 0.01$: 2, 19, 19, 25, 30, 35, 40, 45, 45, 48, 60, 62, 69, 89, 90, 110, 145, 160, 9+, 10+, 20+, 40+, 50+, 110+, 130+.

7.10 Make a gamma probability plot of the following data assuming $\gamma = 2$: 26, 15, 26, 10, 34, 36, 21, 22, 2, 3, 9, 21, 30, 100.

CHAPTER 8

Analytical Estimation Procedures for Survival Distributions

In this chapter we discuss some analytical, as opposed to graphical, procedures for estimating the most commonly used survival distributions. Section 8.1–8.4 are concerned with maximum likelihood estimates (MLEs) of the parameters of the exponential (one- and two-parameter), Weibull, lognormal, and gamma distributions. A regression method for estimating parameters of the exponential, Weibull, Gompertz, and linear-exponential distributions is presented in Section 8.5.

8.1 THE EXPONENTIAL DISTRIBUTION

8.1.1 The One-Parameter Exponential Distribution

The one-parameter exponential distribution has the following density, survivorship, and hazard functions:

$$f(t) = \lambda e^{-\lambda t} \qquad t \geq 0, \ \lambda > 0 \tag{8.1}$$

$$S(t) = e^{-\lambda t} \qquad t \geq 0 \tag{8.2}$$

$$h(t) = \lambda \qquad t \geq 0 \tag{8.3}$$

Obviously, the exponential distribution is characterized by one parameter, namely λ. The estimation of λ by maximum likelihood methods for data without censored observations will be given first followed by the case with censored observations.

A. Estimation of λ for Data without Censored Observations
Suppose that here are n individuals in the study and everyone is followed to death or failure. Let $t_1, t_2, \ldots, t_n$ be the exact survival times of the n individuals. The likelihood function is

200

$$L = \prod_{i=1}^{n} \lambda e^{-\lambda t_i}$$

and the MLE of λ is

$$\hat{\lambda} = \frac{n}{\sum_{i=1}^{n} t_i} \qquad (8.4)$$

Since the mean μ of the exponential distribution is $1/\lambda$ and MLEs are invariant under one-to-one transformations, the MLE of μ is

$$\hat{\mu} = \frac{1}{\hat{\lambda}} = \frac{\sum_{i=1}^{n} t_i}{n} = \bar{t} \qquad (8.5)$$

It can be shown that $2n\hat{\mu}/\mu$ has an exact chi-square distribution with $2n$ degrees of freedom (Epstein and Sobel 1953). Since $\lambda = 1/\mu$ and $\hat{\lambda} = 1/\hat{\mu}$, an exact $100(1 - \alpha)$ percent confidence interval for λ is

$$\frac{\hat{\lambda}\chi^2_{2n,1-\alpha/2}}{2n} < \lambda < \frac{\hat{\lambda}\chi^2_{2n,\alpha/2}}{2n} \qquad (8.6)$$

where $\chi^2_{2n,\alpha}$ is the 100α percentage point of the chi-square distribution with $2n$ degrees of freedom, that is, $P(\chi^2_{2n} > \chi^2_{2n,\alpha}) = \alpha$ (Table C-2). When n is large ($n \geq 25$, say), $\hat{\lambda}$ is approximately normally distributed with mean λ and variance λ^2/n. Thus, an approximate $100(1 - \alpha)$ percent confidence interval for λ is

$$\hat{\lambda} - \frac{\hat{\lambda}Z_{\alpha/2}}{\sqrt{n}} < \lambda < \hat{\lambda} + \frac{\hat{\lambda}Z_{\alpha/2}}{\sqrt{n}} \qquad (8.7)$$

where $Z_{\alpha/2}$ is the $100\alpha/2$ percentage point $[P(Z > Z_{\alpha/2}) = \alpha/2]$ of the standard normal distribution (Table C-1).

Since $2n\hat{\mu}/\mu$ has an exact chi-square distribution with $2n$ degrees of freedom, an exact $100(1 - \alpha)$ percent confidence interval for the mean survival time is

$$\frac{2n\hat{\mu}}{\chi^2_{2n,\alpha/2}} < \mu < \frac{2n\hat{\mu}}{\chi^2_{2n,1-\alpha/2}} \qquad (8.8)$$

The following example illustrates the procedures.

Example 8.1
Consider the remission data for the 21 patients with acute leukemia in Example 7.2. The remission durations in weeks are 1, 1, 2, 2, 3, 4, 4, 5, 5, 6,

8, 8, 9, 10, 10, 12, 14, 16, 20, 24, and 34. The probability plot of the data given in Figure 7.5 indicates that the exponential distribution is suitable for the remission times. Let us estimate the parameter λ by using the formulas given above and compare it with that obtained graphically.

According to (8.4), the MLE of the relapse rate, λ, is

$$\hat{\lambda} = \tfrac{21}{198} = 0.106 \text{ per week}$$

The mean remission time μ is then $\frac{198}{21} = 9.429$ weeks. These estimates are almost identical to those obtained graphically. Using the analytical procedures given above, confidence intervals for λ and μ can also be obtained, which was not possible by the graphical techniques.

A 95% confidence interval for the relapse rate λ, following (8.6), is approximately

$$\frac{(0.106)(24.433)}{42} < \lambda < \frac{(0.106)(59.342)}{42}$$

or (0.062, 0.150). A 95% confidence interval for the mean remission time, following (8.8), is

$$\frac{(42)(9.429)}{59.342} < \mu < \frac{(42)(9.429)}{24.433}$$

or (6.673, 16.208).

Once the parameter λ is estimated, other estimates can be obtained. For example, the probability of staying in remission for at least 20 weeks, estimated from (8.2), is $\hat{S}(20) = e^{-0.106(20)} = 0.120$. Any percentile of survival time t_p may be estimated by equating $S(t)$ to p and solving for t_p, that is $\hat{t}_p = -\log_e p / \hat{\lambda}$. For example, the median (50th percentile) survival time can be estimated by $\hat{t}_{0.5} = -\log_e 0.5 / \hat{\lambda} = 6.539$ weeks.

B. Estimation of λ for Data with Censored Observations

We first consider singly censored and then progressively censored data. Suppose that, without loss of generality, the study or experiment begins at time zero with a total of n subjects. Survival times are recorded and the data become available when the subjects die one after the other in such a way that the shortest survival time comes first, the second shortest second, and so on. Suppose that the investigator has decided to terminate the study after r of the n subjects have died and to sacrifice the remaining $n - r$ subjects at that time. Then the survival times for the n subjects are

$$t_{(1)} \le t_{(2)} \le \cdots \le t_{(r)} = t_{(1)}^{+} = \cdots = t_{(n-r)}^{+}$$

where a superscript plus indicates a sacrificed subject, and thus $t_{(i)}^{+}$ is a censored observation. In this case, n and r are fixed values and all of the $n - r$ censored observations are equal.

The likelihood function is

$$L = \frac{n!}{(n-r)!} \prod_{i=1}^{r} \lambda e^{-\lambda t_{(i)}} [e^{-\lambda t_{(r)}}]^{n-r}$$

and the MLE of λ is

$$\hat{\lambda} = \frac{r}{\sum_{i=1}^{r} t_{(i)} + \sum_{i=1}^{n-r} t_{(i)}^{+}} \tag{8.9}$$

The mean survival time $\mu = 1/\lambda$ can then be estimated by

$$\hat{\mu} = \frac{1}{\hat{\lambda}} = \frac{1}{r} \left[\sum_{i=1}^{r} t_{(i)} + \sum_{i=1}^{n-r} t_{(i)}^{+} \right] \tag{8.10}$$

It is shown by Halperin (1952) and $2r\lambda/\hat{\lambda}$ has a chi-square distribution with $2r$ degrees of freedom. The mean and variance of $\hat{\lambda}$ are $r\lambda/(r-1)$ and $\lambda^2/(r-1)$, respectively. The $100(1-\alpha)$ percent confidence interval for λ is

$$\frac{\hat{\lambda}\chi^2_{2r,1-\alpha/2}}{2r} < \lambda < \frac{\hat{\lambda}\chi^2_{2r,\alpha/2}}{2r} \tag{8.11}$$

When n is large, the distribution of $\hat{\lambda}$ is approximately normal with mean λ and variance $\lambda^2/(r-1)$. An approximate $100(1-\alpha)$ percent confidence interval for λ is then

$$\hat{\lambda} - \frac{\hat{\lambda}Z_{\alpha/2}}{\sqrt{r-1}} < \lambda < \hat{\lambda} + \frac{\hat{\lambda}Z_{\alpha/2}}{\sqrt{r-1}} \tag{8.12}$$

Epstein and Sobel (1953) show that $2r\hat{\mu}/\mu$ has a chi-square distribution with $2r$ degrees of freedom. Thus a $100(1-\alpha)$ percent confidence interval for μ (see also Epstein 1960b) is

$$\frac{2r\hat{\mu}}{\chi^2_{2r,\alpha/2}} < \mu < \frac{2r\hat{\mu}}{\chi^2_{2r,1-\alpha/2}} \tag{8.13}$$

They also develop test procedures for the hypothesis $H_0: \mu = \mu_0$ against the alternative $H_1: \mu < \mu_0$. One of the rules of action is to accept H_0 if $\hat{\mu} > c$ and reject H_0 if $\hat{\mu} < c$, where $c = (\mu_0\chi^2_{2r,\alpha})/2r$ and α is the significance level. Or if the estimated mean survival time calculated from (8.10) is greater than c, the hypothesis H_0 is rejected at the α level.

The following example illustrates the procedure.

Example 8.2

Suppose that in a laboratory experiment 10 mice are exposed to carcinogens. The experimenter decides to terminate the study after half of the mice are dead and to sacrifice the other half at that time. The survival times of the 5 expired mice are 4, 5, 8, 9, and 10 weeks. The survival data of the 10 mice are 4, 5, 8, 9, 10, 10+, 10+, 10+, 10+, and 10+. Assuming that the failure of these mice follows the exponential distribution, the survival rate λ and mean survival time μ are estimated, respectively, according to (8.9) and (8.10) by

$$\hat{\lambda} = \frac{5}{36 + 50} = 0.058 \text{ per week}$$

and $\hat{\mu} = 1/0.058 = 17.241$ weeks. A 95% confidence interval for λ by (8.11) is

$$\frac{(0.058)(3.247)}{(2)(5)} < \mu < \frac{(0.058)(20.483)}{(2)(5)}$$

or (0.019, 0.119). A 95% confidence interval for μ following (8.13) is

$$\frac{2(5)(17.241)}{20.483} < \mu < \frac{2(5)(17.241)}{3.247}$$

or (8.417, 53.098).

The probability of surviving a given time for the mice can be estimated from (8.2). For example, the probability that a mouse exposed to the same carcinogen will survive longer than 8 weeks is

$$\hat{S}(8) = e^{-0.058(8)} = 0.629$$

The probability of dying in 8 weeks is then $1 - 0.629 = 0.371$.

A slightly different situation may arise in laboratory experiments. Instead of terminating the study after the rth death, the experimenter may stop after a period of time T, which may be six months or a year. If we let the number of deaths between zero and T be r, the survival data may look as follows:

$$t_{(1)} \le t_{(2)} \le \cdots \le t_{(r)} \le t_{(1)}^{+} = \cdots = t_{(n-r)}^{+} = T$$

Mathematical derivations of the MLE of λ and μ are exactly the same and (8.9) can still be used.

The sampling distribution of $\hat{\mu}$ for singly censored data is also discussed by Bartholomew (1963).

Progressively censored data come more frequently from clinical studies where patients are entered at different times and the study lasts a predetermined period of time. Suppose that the study begins at time zero and terminates at time T and there are a total of n individuals entered. Let r be the number of individuals who die before or at time T and $n - r$ the number of individuals who are lost to follow-up during the study period or remain alive at time T. The data look as follows: $t_1, t_2, \ldots, t_r, t_1^+, t_2^+, \ldots, t_{n-r}^+$. Ordering the r uncensored observation according to their magnitude, we have

$$t_{(1)} \leq t_{(2)} \leq \cdots \leq t_{(r)}, t_1^+, t_2^+, \ldots, t_{n-r}^+$$

The likelihood function is

$$L = \prod_{i=1}^{r} \lambda e^{-\lambda t_{(i)}} \prod_{i=1}^{n-r} e^{-\lambda t_i^+}$$

and the MLE of the parameter λ is

$$\hat{\lambda} = \frac{r}{\displaystyle\sum_{i=1}^{r} t_{(i)} + \sum_{i=1}^{n-r} t_i^+} \tag{8.14}$$

Consequently,

$$\hat{\mu} = \frac{1}{\hat{\lambda}} = \frac{1}{r}\left[\sum_{i=1}^{r} t_{(i)} + \sum_{i=1}^{n-r} t_i^+\right] \tag{8.15}$$

is the MLE of the mean survival time. The sum of all of the observations, censored and uncensored, divided by the number of uncensored observations gives the MLE of the mean survival time. To overcome the mathematical difficulties arising when all of the observations are censored ($r = 0$), Bartholomew (1957) defines

$$\hat{\mu} = \sum_{i=1}^{n} t_i^+ \tag{8.16}$$

In practice, this estimate has little value.

Distributions of the estimators are discussed by Bartholomew (1957). The distribution of $\hat{\lambda}$ for large n is approximately normal with mean λ and variance:

$$\text{Var}(\hat{\lambda}) = \frac{\lambda^2}{\displaystyle\sum_{i=1}^{n} (1 - e^{-\lambda T_i})} \tag{8.17}$$

where T_i is the time that the ith individual is under observation. In other words, T_i is computed from the time the ith individual enters the study to the end of the study. If the observation times T_i are not known, the following quick estimate of $\text{Var}(\hat{\lambda})$ can be used:

$$\widehat{\text{Var}}(\hat{\lambda}) = \frac{\hat{\lambda}^2}{r} \qquad (8.18)$$

Thus an approximate $100(1 - \alpha)$ percent confidence interval for λ is

$$\hat{\lambda} - Z_{\alpha/2}\sqrt{\widehat{\text{Var}}(\hat{\lambda})} < \lambda < \hat{\lambda} + Z_{\alpha/2}\sqrt{\widehat{\text{Var}}(\hat{\lambda})} \qquad (8.19)$$

The distribution of $\hat{\mu}$ is approximately normal with mean μ and variance:

$$\text{Var}(\hat{\mu}) = \frac{\mu^2}{\sum\limits_{i=1}^{n}(1 - e^{-\lambda T_i})} \qquad (8.20)$$

Again, a quick estimate is

$$\widehat{\text{Var}}(\hat{\mu}) = \hat{\mu}^2/r \qquad (8.21)$$

An approximate $100(1 - \alpha)$ percent confidence interval for μ is then

$$\hat{\mu} - Z_{\alpha/2}\sqrt{\widehat{\text{Var}}(\hat{\mu})} < \mu < \hat{\mu} + Z_{\alpha/2}\sqrt{\widehat{\text{Var}}(\hat{\mu})} \qquad (8.22)$$

The exact distribution of $\hat{\mu}$ derived by Bartholomew (1963) is too cumbersome for general use and thus will not be included here.

Example 8.3
Consider the remission duration of the 21 leukemia patients receiving 6-MP in Example 3.3. The remission times in weeks were 6, 6, 6, 7, 10, 13, 16, 22, 23, 6+, 9+, 10+, 11+, 17+, 19+, 20+, 25+, 32+, 32+, 34+, and 35+. The hazard plot given in Figure 3.6 shows that the exponential distribution fits the data very well. Maximum likelihood estimates of the relapse rate and mean remission time can be obtained, respectively, from (8.14) and (8.15):

$$\hat{\lambda} = \frac{9}{109 + 250} = 0.025 \text{ per week} \qquad \hat{\mu} = \frac{1}{0.025} = 40 \text{ weeks}$$

The graphical estimate of λ obtained in Example 3.3 is 0.027, which is very close to the MLE. Thus, the remission duration of leukemia patients receiving 6-MP can be described by an exponential distribution with a constant relapse rate of 2.5% per week and a mean remission time of 40 weeks. The probability of staying in remission for a year or more is estimated by

$$\hat{S}(52) = e^{-0.025(52)} = 0.273$$

Using (8.18) and (8.21) for the variance of $\hat{\lambda}$ and $\hat{\mu}$, 95% confidence intervals for λ and μ are, respectively, (0.009, 0.041) and (13.867, 66.133).

8.1.2   The Two-Parameter Exponential Distribution

In the case where a two-parameter exponential distribution is more appropriate for the data (Zelen 1966), the density and survivorship functions are defined, respectively, as

$$f(t) = \begin{cases} \lambda e^{-\lambda(t-G)} & t \geq G \geq 0, \ \lambda \geq 0 \\ 0 & t < G \end{cases} \tag{8.23}$$

and

$$S(t) = \begin{cases} e^{-\lambda(t-G)} & t \geq G \geq 0, \ \lambda \geq 0 \\ 1 & t < G \end{cases} \tag{8.24}$$

where G is called the guarantee time, or the minimum survival time before which no deaths occur.

A. Estimation of λ and G for Data without Censored Observations

If $t_1, t_2, \ldots, t_n$ are the survival times of the n individuals, the MLE of λ is

$$\hat{\lambda} = \frac{n}{\displaystyle\sum_{i=1}^{n}(t_i - \hat{G})} \tag{8.25}$$

where $\hat{G}$ is an estimate of G that is the smallest observation in the data,

$$\hat{G} = \min(t_1, t_2, \ldots, t_n) \tag{8.26}$$

and the mean survival time is estimated by $\hat{\mu} = \hat{G} + 1/\hat{\lambda}$.

Example 8.4

Consider the survival times in months of 11 patients following initial pulmonary metastasis from ostenogenic sarcoma considered by Burdette and Gehan (1970). The data were 11, 13, 13, 13, 13, 13, 14, 14, 15, 15, and 17. Suppose that the two-parameter exponential distribution is selected. The guarantee time G is estimated by the smallest observation, that is, $\hat{G} = 11$, and the hazard rate λ estimated by (8.25) is

$$\hat{\lambda} = \frac{11}{(11-11) + (13-11) + \cdots + (17-11)} = 0.367$$

Thus, the exponential model tells us that the minimum survival time is 11 months and after that the chance of death per month is 0.367. Similarly, the probability of surviving a given amount of time can then be estimated from (8.24). For example, the estimated probability of surviving 18 months or longer is

$$\hat{S}(18) = e^{-0.367(18-11)} = 0.077$$

B. Estimation of λ and G for Data with Censored Observations

We first consider singly censored data. Suppose that an experiment begins with n animals and terminates as soon as the first r deaths occur. For this case, we introduce the estimation procedures derived by Epstein (1960a).

Let the first r survival times be $t_{(1)} \le t_{(1)} \le \cdots \le t_{(r)}$ and let T^* be the total survival observed between the first and the rth death:

$$T^* = (n-1)(t_{(2)} - t_{(1)}) + (n-2)(t_{(3)} - t_{(2)}) + \cdots + (n-r+1)(t_{(r)} - t_{(r-1)})$$

$$= -(n-1)t_{(1)} + t_{(2)} + t_{(3)} + \cdots + t_{(r-1)} + (n-r+1)t_{(r)}$$

$$= \sum_{i=1}^{r} t_{(i)} - nt_{(1)} + (n-r)t_{(r)} \tag{8.27}$$

The best estimates for G and μ in the sense that they are unbiased and have minimum variance are given by

$$\hat{G} = t_{(1)} - \hat{\mu}/n \tag{8.28}$$

and

$$\hat{\mu} = T^*/(r-1) \tag{8.29}$$

Then λ can be estimated by $\hat{\lambda} = 1/\hat{\mu}$.

Confidence intervals for the mean survival time μ are easy to obtain from the fact that $2(r-1)\hat{\mu}/\mu = 2T^*/\mu$ has a chi-square distribution with $2(r-1)$ degrees of freedom. Thus, for $r > 1$, the $100(1-\alpha)$ percent confidence interval for μ is

$$\frac{2(r-1)\hat{\mu}}{\chi^2_{2(r-1),\alpha/2}} < \mu < \frac{2(r-1)\hat{\mu}}{\chi^2_{2(r-1),1-\alpha/2}} \tag{8.30}$$

or

$$\frac{2T^*}{\chi^2_{2(r-1),\alpha/2}} < \mu < \frac{2T^*}{\chi^2_{2(r-1),1-\alpha/2}} \tag{8.31}$$

To find confidence intervals for G, we use the fact that $x_1 = 2n(t_{(1)} - G)/\mu$ and $x_2 = 2(r-1)\hat{\mu}/\mu$ are independent and have a chi-square distribution with 2 and $2(r-1)$ degrees of freedom, respectively. Thus the ratio

$$Y = \frac{x_1/2}{x_2/2(r-1)} = \frac{n(t_{(1)} - G)}{\hat{\mu}} = \frac{n(r-1)(t_{(1)} - G)}{T^*} \tag{8.32}$$

follows the F distribution with 2 and $2(r-1)$ degrees of freedom. Let $F_{2,2(r-1),\alpha}$ be the 100α percent point of the $F_{2,2(r-1)}$ distribution, that is,

$P(Y \geq F_{2,2(r-1),\alpha}) = \alpha$ (Table C-3 in Appendix C), and then a $100(1 - \alpha)$ percent confidence interval for G is

$$t_{(1)} - \frac{\hat{\mu}}{n} F_{2,2(r-1),\alpha} < G < t_{(1)} \tag{8.33}$$

or

$$t_{(1)} - \frac{T^*}{n(r - 1)} F_{2,2(r-1),\alpha} < G < t_{(1)} \tag{8.34}$$

Epstein and Sobel (1953) show that this interval is the shortest in the class of intervals being used. If for some particular values of r and α the value $F_{2,2(r-1)\alpha}$ is not tabulated in the F-table, Epstein (1960a) suggests using the following confidence intervals for G:

$$t_{(1)} - \frac{\hat{\mu}(r - 1)}{n} g_{1-\alpha} < G < t_{(1)} \tag{8.35}$$

or

$$t_{(1)} - \frac{T^*}{n} g_{1-\alpha} < G < t_{(1)} \tag{8.36}$$

where

$$g_{1-\alpha} = \left(\frac{1}{\alpha}\right)^{(1/r)-1} - 1 \tag{8.37}$$

is computable for any α and r. Example 8.5 illustrates the procedures.

Example 8.5

In a laboratory experiment 20 mice are injected with a tumor inoculum. These tumor cells multiply and eventually kill the animal. Suppose that the investigator decides to terminate the experiment after 10 deaths. The first occurs 30 days after the experiment starts. The total survival observed between the time when the first and tenth deaths occur is 600 animal days. Assuming that the survival distribution of these mice is exponential, the shortest 95% confidence interval for G can be obtained by (8.34). Since $F_{2,18,0.05} = 3.555$, the interval is

$$30 - \frac{600}{(20)(9)} (3.555) < G < 30$$

or (18.150, 30).

The mean survival time estimated by (8.29) is $\hat{\mu} = 66.667$ days and the 95% confidence interval for μ computed from (8.31) is

$$\frac{2(600)}{31.526} < \mu < \frac{2(600)}{8.231}$$

or (38.064, 145.790).

When data are progressively censored, Gehan (1970) derives an estimate for G and a modified MLE for the hazard rate λ. Suppose that r of the n individuals in the study die before the end of the study and $n - r$ individuals are alive at the time of the last follow-up or termination. The n survival times are denoted by

$$t_{(1)} \le t_{(2)} \le \cdots \le t_{(r)}, t_{(r+1)}^{+}, \ldots, t_{(n)}^{+}$$

An estimate of G is obtained by

$$\hat{G} = \max[t_{(1)} - 1/n\hat{\lambda}, 0] \tag{8.38}$$

and the variance of $\hat{G}$ is

$$\text{Var}(\hat{G}) = \frac{1}{(n\hat{\lambda})^2}\left(1 + \frac{1}{r-1}\right) \tag{8.39}$$

When n is large, G and $\text{Var}(\hat{G})$ can be estimated by

$$\hat{G} \cong t_{(1)} \tag{8.40}$$

and

$$\widehat{\text{Var}}(\hat{G}) \cong 1/(n\hat{\lambda})^2 \tag{8.41}$$

A modified MLE for λ is

$$\hat{\lambda} = \frac{r-1}{\displaystyle\sum_{i=1}^{r} t_{(i)} + \sum_{i=r+1}^{n} t_{(i)}^{+} - nt_{(1)}} \tag{8.42}$$

with variance

$$\text{Var}(\hat{\lambda}) = \lambda^2/(r-1) \tag{8.43}$$

Any percentile of survival time t_p may be estimated by equating $S(t)$ to p and solving for $\hat{t}_p$, that is, $\hat{t}_p = -(\log_e p)/\hat{\lambda} + \hat{G}$.

The following example illustrates the procedure.

Example 8.6
Suppose that 19 patients with brain tumor are followed in a clinical trial for a year. Their survival times in weeks are 3, 4, 6, 8, 8, 10, 12, 16, 17, 30, 33, 3+, 8+, 13+, 21+, 26+, 35+, 44+, and 45+. In this case $n = 19$, $r = 11$, $t_{(1)} = 3$, $\sum_{i=1}^{11} t_{(i)} = 147$, and $\sum_{i=12}^{19} t_{(i)}^{+} = 195$. The hazard rate λ per week and its variance may be estimated by (8.42) and (8.43) as

$$\hat{\lambda} = \frac{10}{147 + 195 - 19(3)} = 0.035$$

and

$$\widehat{\text{Var}}(\hat{\lambda}) = (0.035)^2/10 = 0.0001$$

The guarantee time G and its variance may then be estimated by (8.38) and (8.39):

$$\hat{G} = \max\left(3 - \frac{1}{19 \times 0.035}, 0\right) = 1.496$$

and

$$\widehat{\text{Var}}(\hat{G}) = \frac{1}{(19 \times 0.035)^2}(1 + \tfrac{1}{10}) = 2.487$$

Thus, after a guarantee time of approximately 1.5 weeks, the chance of death per week is 0.035. The estimated median survival time is

$$\hat{t}_{0.5} = -\frac{\log_e 0.5}{0.035} + 1.496 = 21.3 \text{ weeks}$$

The probability of surviving at least six months (or 26 weeks) is estimated by

$$\hat{S}(26) = e^{-0.35(26 - 1.496)} = 0.424$$

8.2 THE WEIBULL DISTRIBUTION

The Weibull distribution has the density function

$$f(t) = \gamma\lambda^\gamma t^{\gamma-1}\exp[-(\lambda t)^\gamma] \qquad t \geq 0, \gamma > 0, \lambda > 0 \qquad (8.44)$$

The MLE of the parameters λ and γ involves equations to be solved simultaneously. Numerical methods such as the Newton–Raphson iterative procedure (described in Appendix A) can easily be applied. We shall begin with the case where no censored observations are present.

Let $t_1, t_2, \ldots, t_n$ be the exact survival times of n individuals under investigation. If their survival times follow the Weibull distributions, the MLE of λ and γ in (8.44) can be obtained by solving the following two equations simultaneously:

$$n - \hat{\lambda}^{\hat{\gamma}} \sum_{i=1}^{n} t_i^{\hat{\gamma}} = 0 \qquad (8.45)$$

$$\frac{n}{\hat{\gamma}} + n \log_e \hat{\lambda} + \sum_{i=1}^{n} \log_e t_i - \hat{\lambda}^{\hat{\gamma}} \sum_{i=1}^{n} t_i^{\hat{\gamma}}(\log_e \hat{\lambda} + \log_e t_i) = 0 \qquad (8.46)$$

Next let us consider a typical laboratory experiment in which subjects are entered at the same time and the experiment is terminated after r of the n subjects have failed (or after a fixed period of time T). In both of these cases the data collected are singly censored. The ordered survival data are

$$t_{(1)} \leq t_{(2)} \leq \cdots \leq t_{(r)} = t_{(r+1)}^{+} = \cdots = t_{(n)}^{+}$$

If the time to failure follows the Weibull distribution with the density function given in (8.44), the MLE of λ and γ may be obtained by solving simultaneously the following two equations:

$$r - \hat{\lambda}^{\hat{\gamma}} \left[\sum_{i=1}^{r} t_i^{\hat{\gamma}} - (n - r)t_{(r)}^{\hat{\gamma}} \right] = 0 \tag{8.47}$$

$$\frac{r}{\hat{\gamma}} + r \log_e \hat{\lambda} + \sum_{i=1}^{r} \log_e t_i$$

$$- \hat{\lambda}^{\hat{\gamma}} \left[\sum_{i=1}^{r} t_i^{\hat{\gamma}} (\log_e \hat{\lambda} + \log_e t_i) + (n - r)t_{(r)}^{\gamma}(\log_e \hat{\lambda} + \log_e t_{(r)}) \right] = 0 \tag{8.48}$$

When data are progressively censored, we have

$$t_{(1)} \leq t_{(2)} \leq \cdots \leq t_{(r)}, t_{(r+1)}^{+}, \ldots, t_{(n)}^{+}$$

If the survival distribution is Weibull with density (8.44), the MLE of λ and γ may be obtained by solving simultaneously the following two equations:

$$r - \hat{\lambda}^{\hat{\gamma}} \left(\sum_{i=1}^{r} t_i^{\hat{\gamma}} + \sum_{i=r+1}^{n} t_i^{+\hat{\gamma}} \right) = 0 \tag{8.49}$$

$$\frac{r}{\hat{\gamma}} + r \log_e \hat{\lambda} + \sum_{i=1}^{r} \log_e t_i - \hat{\lambda}^{\hat{\gamma}} \sum_{i=1}^{r} t_i^{\hat{\gamma}}(\log_e \hat{\lambda} + \log_e t_i)$$

$$+ \hat{\lambda}^{\hat{\gamma}} \sum_{i=r+1}^{n} t_i^{+\hat{\gamma}}(\log \hat{\lambda} + \log_e t_i^{+}) = 0 \tag{8.50}$$

Other estimation procedures for the parameters of the Weibull distribution can be found in the vast literature on the subject. The following are a few examples: Cohen (1965), Harter and Moore (1965), Menon (1963), Mann (1968), Thoman et al. (1970), Billmann and Antle (1972), Bain (1972), and Engelhardt (1975).

8.3 THE LOGNORMAL DISTRIBUTION

If the survival time T follows the lognormal distribution with density function

$$f(t) = \frac{1}{t\sigma\sqrt{2\pi}} \exp\left[-\frac{1}{2\sigma^2} (\log_e t - \mu)^2 \right] \tag{8.51}$$

the mean and variance of T are, respectively, $\exp(\mu + \frac{1}{2}\sigma^2)$ and $[\exp(\sigma^2) - 1]\exp(2\mu + \sigma^2)$. Estimation of the two parameters μ and σ^2 has been investigated either by using (8.51) directly or by using the fact that $Y = \log_e T$ follows the normal distribution with mean μ and variance σ^2. In the following, we shall discuss the estimation of μ and σ^2 for samples with and without censored observations.

8.3.1 Estimation of μ and σ^2 for Data without Censored Observations

The estimation of μ and σ^2 for complete samples by maximum likelihood methods has been studied by many authors, for example, Cohen (1951) and Harter and Moore (1966). But the simplest way to obtain estimates of μ and σ^2 with optimum properties is by considering the distribution of $Y = \log_e T$. Let $t_1, t_2, \ldots, t_n$ be the survival times of n subjects. The MLE of μ is the sample mean of Y given by

$$\hat{\mu} = \frac{1}{n} \sum_{i=1}^{n} \log_e t_i \tag{8.52}$$

The MLE of σ^2 is

$$\hat{\sigma}^2 = \frac{1}{n}\left[\sum_{i=1}^{n} (\log_e t_i)^2 - \frac{\left(\sum_{i=1}^{n} \log_e t_i\right)^2}{n} \right] \tag{8.53}$$

The estimate $\hat{\mu}$ is also unbiased but $\hat{\sigma}^2$ is not. The best unbiased estimates of μ and σ^2 are $\hat{\mu}$ and the sample variance $s^2 = \hat{\sigma}^2[n/(n-1)]$. If n is moderately large, the difference between s^2 and $\hat{\sigma}^2$ is negligible.

One of the properties of the MLE is that if $\hat{\theta}$ is the MLE of θ, $g(\hat{\theta})$ is the MLE of $g(\theta)$ if $g(\theta)$ is a function with a single-valued inverse. Therefore, MLEs of the mean and variance of T are, respectively, $\exp(\hat{\mu} + \frac{1}{2}\hat{\sigma}^2)$ and $[\exp(\hat{\sigma}^2) - 1]\exp(2\hat{\mu} + \hat{\sigma}^2)$.

It is known that $\hat{\mu} = \bar{y}$ is normally distributed with mean μ and variance σ^2/n. Hence, if σ is known, a two-sided $100(1 - \alpha)$ percent confidence interval for μ is $\hat{\mu} \pm Z_{\alpha/2}\sigma/\sqrt{n}$. If σ is unknown, we can use Student's t distribution. A two-sided $100(1 - \alpha)$ percent confidence interval for μ is $\hat{\mu} \pm t_{\alpha/2,(n-1)}s/\sqrt{n-1}$, where $t_{\alpha/2,(n-1)}$ is the $100\alpha/2$ percentage point of Student's t distribution with $n - 1$ degrees of freedom (Table C-13 in Appendix C).

Confidence intervals for σ^2 can be obtained by using the fact that $n\hat{\sigma}/\sigma^2$ has a chi-square distribution with $n - 1$ degrees of freedom. A $100(1 - \alpha)$ percent confidence interval for σ^2 is

$$\frac{n\hat{\sigma}^2}{\chi^2_{(n-1),\alpha/2}} < \sigma^2 < \frac{n\hat{\sigma}^2}{\chi^2_{(n-1),1-\alpha/2}} \tag{8.54}$$

The following hypothetical example illustrates the procedure.

Example 8.7

Five melanoma (resected) patients receiving immunotherapy BCG are followed. The remission durations in weeks are, in order of magnitude, 8, 16, 23, 27, and 28. Suppose that the remission times follow a lognormal distribution. In this case parameters are estimated by (8.52) and (8.53) as follows:

t	$\log_e t$	$(\log_e t)^2$
8	2.079	4.322
16	2.773	7.690
23	3.135	9.828
27	3.296	10.864
28	3.332	11.102
	14.615	43.806

$$\hat{\mu} = \frac{14.615}{5} = 2.923$$

$$\hat{\sigma}^2 = \tfrac{1}{5}[43.806 - \tfrac{1}{5}(14.615)^2] = 0.217$$

$$s^2 = 5\hat{\sigma}^2/(5-1) = 0.271$$

The mean remission time is $\exp(2.923 + 0.217/2)$, or 20.728, weeks and the standard deviation of the remission time is $\{[\exp(0.217) - 1]\exp(5.846 + 0.217)\}^{1/2}$, or 10.204, weeks. A 95% confidence interval for μ is

$$2.923 - 2.776 \frac{0.521}{\sqrt{4}} < \mu < 2.923 + 2.776 \frac{0.521}{\sqrt{4}}$$

or $(2.200, 3.646)$. A 95% confidence interval for σ^2, following (8.54), is

$$\frac{5(0.217)}{11.1433} < \sigma^2 < \frac{5(0.217)}{0.4844}$$

or $(0.097, 2.240)$.

8.3.2   Estimation of μ and σ^2 for Data with Censored Observations

We first consider samples with singly censored observations. The data consist of r exact survival times $t_{(1)} \le t_{(2)} \le \cdots \le t_{(r)}$ and $n - r$ survival times that are at least $t_{(r)}$, denoted by $t_{(r)}^+$. Again we use the fact that $Y = \log_e T$ has normal distribution with mean μ and variance σ^2. Estimates of μ and σ^2 can be obtained from the transformed data $y_i = \log_e t_i$. Many authors have investigated the estimation of μ and σ^2, for example, Gupta (1952), Sarhan and Greenberg (1956, 1957, 1958, 1962), Saw (1959), and Cohen (1959, 1961). We shall discuss the methods of Sarhan and Greenberg and Cohen because of the available table that reduces computation time and efforts.

The best linear estimates of μ and σ proposed by Sarhan and Greenberg are linear combinations of the logarithms of the r exact survival times:

$$\hat{\mu} = \sum_{i=1}^{r} a_i \log_e t_{(i)} \tag{8.55}$$

and

$$\hat{\sigma} = \sum_{i=1}^{r} b_i \log_e t_{(i)} \tag{8.56}$$

where the coefficients a_i and b_i are calculated and tabulated by Sarhan and Greenberg for $n \leq 20$ and are partially reproduced in Table C-14 of Appendix C. The variance and covariance of $\hat{\mu}$ and $\hat{\sigma}$ are tabulated in Table C-15.

The following examples illustrates the procedure.

Example 8.8
Suppose that in a study of the efficacy of a new drug, 12 mice with tumors are given the drug. The experimenter decides to terminate the study after 9 mice have died. The survival times are, in weeks, 5, 8, 9, 10, 12, 15, 20, 21, 25, 25+, 25+, and 25+.

Assume that the times to death of these mice follow the lognormal distribution. In this case $n = 12$, $r = 9$, and $n - r = 3$. Using (8.55), (8.56), and Table C-14, $\hat{\mu}$ and $\hat{\sigma}$ can be calculated as

$$\hat{\mu} = 0.036 \log_e 5 + 0.0581 \log_e 8 + 0.0682 \log_e 9 + 0.0759 \log_e 10$$
$$+ 0.0827 \log_e 12 + 0.0888 \log_e 15 + 0.0948 \log_e 20$$
$$+ 0.1006 \log_e 21 + 0.3950 \log_e 25$$
$$= 2.811$$

$$\hat{\sigma} = -0.2545 \log_e 5 - 0.1487 \log_e 8 - 0.1007 \log_e 9 - 0.0633 \log_e 10$$
$$-0.0308 \log_e 12 - 0.0007 \log_e 15 + 0.0286 \log_e 20$$
$$+0.0582 \log_e 21 + 0.5119 \log_e 25$$
$$= 0.747$$

The variances of $\hat{\mu}$ and $\hat{\sigma}$ given in Table C-15 are, respectively, 0.0926 and 0.0723 and the covariance of $\hat{\mu}$ and $\hat{\sigma}$ is 0.0152.

Cohen's (1959, 1961) MLEs for the normal distribution can be used for $n > 20$. Let

$$\bar{y} = \frac{1}{r} \sum_{i=1}^{r} \log_e t_{(i)} \tag{8.57}$$

and

$$s^2 = \frac{1}{r}\left[\sum (\log_e t_{(i)})^2 - \frac{\left(\sum \log_e t_{(i)}\right)^2}{r}\right] \tag{8.58}$$

Then the MLEs of μ and σ^2 are

$$\hat{\mu} = \bar{y} - \hat{\lambda}(\bar{y} - \log_e t_{(r)}) \tag{8.59}$$

and

$$\sigma^2 = s^2 + \hat{\lambda}(\bar{y} - \log_e(t_{(r)})^2 \tag{8.60}$$

where the value of $\hat{\lambda}$ has been tabulated by Cohen (1961) as a function of a and b. The proportion of censored observations, b, is calculated as

$$b = (n - r)/n$$

and

$$a = [1 - Y(Y - c)]/(Y - c)^2$$

where $Y = [b/(1 - b)]f(c)/F(c)$, $f(c)$ and $F(c)$ being the density and distribution functions, respectively, of the standard normal distribution, evaluated at $c = (\log_e t_{(r)} - \mu)/\sigma$. Table 8.1 gives values of $\hat{\lambda}$ for $b = 0.01$ (0.01) 0.10 (0.05) 0.70 (0.10) 0.90 and for $a = 0.00$ (0.05) 1.00. For a censored sample, after computing $a = s^2/(\bar{y} - \log_e t_{(r)})^2$, and $b = (n - r)/n$, enter Table 8.1 with these values of a and b to obtain $\hat{\lambda}$. For values not tabulated, two-way linear interpolation can be used.

The asymptotic variances and covariance are the following:

$$\text{Var}(\hat{\mu}) \sim \frac{\sigma^2}{n} m_1$$

$$\text{Var}(\hat{\sigma}) \sim \frac{\sigma^2}{n} m_2 \tag{8.61}$$

$$\text{Cov}(\hat{\mu}, \hat{\sigma}) \sim \frac{\sigma^2}{n} m_3$$

where m_1, m_2, and m_3 are also tabulated by Cohen (1961). The table is reproduced in Table 8.2. For any censored sample, compute $\hat{c} = (\log_e t_{(r)} - \hat{\mu})/\hat{\sigma}$ and then enter the appropriate columns of Table 8.2 with $y = -\hat{c}$, and interpolate to obtain the required values of m_i, $i = 1, 2, 3$, if the experiment was terminated after a predetermined time. If the experiment was terminated after a given proportion of animals have died, enter Table 8.2 through the percentage censored column with percentage censored $= 100b$ and interpolate to obtain the required value of m_i.

Table 8.1 Estimated λ Values for $\hat{\mu}$ and σ^2

b \ a	.01	.02	.03	.04	.05	.06	.07	.08	.09	.10	.15	.20	.25	.30	.35	.40	.45	.50	.55	.60	.65	.70	.80	.90
.00	.010100	.020400	.030902	.041583	.052507	.063627	.074953	.086488	.09824	.11020	.17342	.24268	.31862	.4021	.4941	.5961	.7096	.8368	.9808	1.145	1.336	1.561	2.176	3.283
.05	.010551	.021294	.032225	.043350	.054670	.066189	.077909	.089834	.10197	.11431	.17935	.25033	.32793	.4130	.5066	.6101	.7252	.8540	.9994	1.166	1.358	1.585	2.203	3.314
.10	.010950	.022082	.033398	.044902	.056596	.068483	.080568	.092852	.10534	.11804	.18479	.25741	.33662	.4233	.5184	.6234	.7400	.8703	1.017	1.185	1.379	1.608	2.229	3.345
.15	.011310	.022798	.034466	.046318	.058356	.070586	.083009	.095629	.10845	.12148	.18985	.26405	.34480	.4330	.5296	.6361	.7542	.8860	1.035	1.204	1.400	1.630	2.255	3.376
.20	.011642	.023459	.035453	.047629	.059990	.072539	.085280	.098216	.11135	.12469	.19460	.27031	.35255	.4422	.5403	.6483	.7678	.9012	1.051	1.222	1.419	1.651	2.280	3.405
.25	.011952	.024076	.036377	.048858	.061522	.074372	.087413	.10065	.11408	.12772	.19910	.27626	.35993	.4510	.5506	.6600	.7810	.9158	1.067	1.240	1.439	1.672	2.305	3.435
.30	.012243	.024658	.037249	.050018	.062969	.076106	.089433	.10295	.11667	.13059	.20338	.28193	.36700	.4595	.5604	.6713	.7937	.9300	1.083	1.257	1.457	1.693	2.329	3.464
.35	.012520	.025211	.038077	.051120	.064345	.077756	.091355	.10515	.11914	.13333	.20747	.28737	.37379	.4676	.5699	.6821	.8060	.9437	1.098	1.274	1.476	1.713	2.353	3.492
.40	.012784	.025738	.038866	.052173	.065660	.079332	.093193	.10725	.12150	.13595	.21139	.29260	.38033	.4755	.5791	.6927	.8179	.9570	1.113	1.290	1.494	1.732	2.376	3.520
.45	.013036	.026243	.039624	.053182	.066921	.080845	.094958	.10926	.12377	.13847	.21517	.29765	.38665	.4831	.5880	.7029	.8295	.9700	1.127	1.306	1.511	1.751	2.399	3.547
.50	.013279	.026728	.040352	.054153	.068135	.082301	.096657	.11121	.12595	.14090	.21882	.30253	.39276	.4904	.5967	.7129	.8408	.9826	1.141	1.321	1.528	1.770	2.421	3.575
.55	.013513	.027196	.041054	.055089	.069306	.083708	.098298	.11308	.12806	.14325	.22235	.30725	.39870	.4976	.6051	.7225	.8517	.9950	1.155	1.337	1.545	1.788	2.443	3.601
.60	.013739	.027649	.041733	.055995	.070439	.085068	.099887	.11490	.13011	.14552	.22578	.31184	.40447	.5045	.6133	.7320	.8625	1.007	1.169	1.351	1.561	1.806	2.465	3.628
.65	.013958	.028087	.042391	.056874	.071538	.086388	.10143	.11666	.13209	.14773	.22910	.31630	.41008	.5114	.6213	.7412	.8729	1.019	1.182	1.366	1.577	1.824	2.486	3.654
.70	.014171	.028513	.043030	.057726	.072605	.087670	.10292	.11837	.13402	.14987	.23234	.32065	.41555	.5180	.6291	.7502	.8832	1.030	1.195	1.380	1.593	1.841	2.507	3.679
.75	.014378	.028927	.043652	.058556	.073643	.088917	.10438	.12004	.13590	.15196	.23550	.32489	.42090	.5245	.6367	.7590	.8932	1.042	1.207	1.394	1.608	1.858	2.528	3.705
.80	.014579	.029330	.044258	.059364	.074655	.090133	.10580	.12167	.13773	.15400	.23858	.32903	.42612	.5308	.6441	.7676	.9031	1.053	1.220	1.408	1.624	1.875	2.548	3.730
.85	.014775	.029723	.044848	.060153	.075642	.091319	.10719	.12325	.13952	.15599	.24158	.33307	.43122	.5370	.6515	.7761	.9127	1.064	1.232	1.422	1.639	1.892	2.568	3.754
.90	.014967	.030107	.045425	.060923	.076606	.092477	.10854	.12480	.14126	.15793	.24452	.33703	.43622	.5430	.6586	.7844	.9222	1.074	1.244	1.435	1.653	1.908	2.588	3.779
.95	.015154	.030483	.045989	.061676	.077549	.093611	.10987	.12632	.14297	.15983	.24740	.34091	.44112	.5490	.6656	.7925	.9314	1.085	1.255	1.448	1.668	1.924	2.607	3.803
1.00	.015338	.030850	.046540	.062413	.078471	.094720	.11116	.12780	.14465	.16170	.25022	.34471	.44592	.5548	.6724	.8005	.9406	1.095	1.267	1.461	1.682	1.940	2.626	3.827

For all values $0 \le a \le 1$, $\lambda = 0$.

Source: Cohen (1961).

Table 8.2   Estimated Values of m_1, m_2, and m_3 for Var($\hat{\mu}$), Var($\hat{\sigma}$), and Cov($\hat{\mu}$, $\hat{\sigma}$)

y	m_1	m_2	m_3	Percentage Censored
−4.0	1.00000	.500030	.000006	0.00
−3.5	1.00001	.500208	.000052	0.02
−3.0	1.00010	.501180	.000335	0.13
−2.5	1.00056	.505280	.001712	0.62
−2.4	1.00078	.506935	.002312	0.82
−2.3	1.00107	.509030	.003099	1.07
−2.2	1.00147	.511658	.004121	1.39
−2.1	1.00200	.514926	.005438	1.79
−2.0	1.00270	.518960	.007123	2.28
−1.9	1.00363	.523899	.009266	2.87
−1.8	1.00485	.529899	.011971	3.59
−1.7	1.00645	.537141	.015368	4.46
−1.6	1.00852	.545827	.019610	5.48
−1.5	1.01120	.556186	.024884	6.68
−1.4	1.01467	.568417	.031410	8.08
−1.3	1.01914	.582981	.039460	9.68
−1.2	1.02488	.600046	.049355	11.51
−1.1	1.03224	.620049	.061491	13.57
−1.0	1.04168	.643438	.076345	15.87
−0.9	1.05376	.670724	.094501	18.41
−0.8	1.06923	.702513	.116674	21.19
−0.7	1.08904	.739515	.143744	24.20
−0.6	1.11442	.782574	.176798	27.43
−0.5	1.14696	.832691	.217183	30.85
−0.4	1.18876	.891077	.266577	34.46
−0.3	1.24252	.959181	.327080	38.21
−0.2	1.31180	1.03877	.401326	42.07
−0.1	1.40127	1.13198	.492641	46.02
0.0	1.51709	1.24145	.605233	50.00
0.1	1.66743	1.37042	.744459	53.98
0.2	1.86310	1.52288	.917165	57.93
0.3	2.11857	1.70381	1.13214	61.79
0.4	2.45318	1.91942	1.40071	65.54
0.5	2.89293	2.17751	1.73757	69.15
0.6	3.47293	2.48793	2.16185	72.57
0.7	4.24075	2.86318	2.69858	75.80
0.8	5.2612	3.3192	3.3807	78.81
0.9	6.6229	3.8765	4.2517	81.59
1.0	8.4477	4.5614	5.3696	84.13
1.1	10.903	5.4082	6.8116	86.43
1.2	14.224	6.4616	8.6818	88.49
1.3	18.735	7.7804	11.121	90.32
1.4	24.892	9.4423	14.319	91.92
1.5	33.339	11.550	18.539	93.32

Table 8.2 (*Continued*)

y	m_1	m_2	m_3	Percentage Censored
1.6	44.986	14.243	24.139	94.52
1.7	61.132	17.706	31.616	95.54
1.8	83.638	22.193	41.664	96.41
1.9	115.19	28.046	55.252	97.13
2.0	159.66	35.740	73.750	97.72
2.1	222.74	45.930	99.100	98.21
2.2	312.73	59.526	134.08	98.61
2.3	441.92	77.810	182.68	98.93
2.4	628.58	102.59	250.68	99.18
2.5	899.99	136.44	346.53	99.38

Source: Cohen (1961).

To illustrate the use of Tables 8.1 and 8.2 for the computation of $\hat{\mu}$, $\hat{\sigma}^2$, Var($\hat{\mu}$), Var($\hat{\sigma}$), and Cov($\hat{\mu}, \hat{\sigma}$), consider Example 8.9, adapted from Cohen (1961).

When data are progressively censored, estimation of μ and σ is more complicated. Interested readers are referred to Gajjar and Khatri (1969) and Cohen (1963, 1976).

Example 8.9
Suppose that in a laboratory experiment 300 mice were followed until 119 died with $\bar{y} = 1304.832$ hours, $s^2 = 12128.250$, and $\log_e t_{(119)} = 1450.000$ hours. In this case $n = 300$ and $r = 119$. Accordingly,

$$\hat{a} = \frac{12128.250}{(1304.832 - 1450.000)^2} = 0.575 \qquad b = \frac{300 - 119}{300}$$

From Table 8.1 $\hat{\lambda}$ is approximately 1.36. Using (8.59) and (8.60), we obtain

$$\hat{\mu} = 1304.832 - 1.36(1304.832 - 1450.000)$$

$$= 1502.26 \text{ hours}$$

$$\hat{\sigma}^2 = 12128.250 + 1.36(1304.832 - 1450.000)^2$$

$$= 40788.55$$

and $\hat{\sigma} = 201.96$.

For the variance and covariance of $\hat{\mu}$ and $\hat{\sigma}$, we enter Table 8.2 with percentage censored $100b = 60.3$ and interpolate linearly to obtain $m_1 = 2.002$, $m_2 = 1.635$, and $m_3 = 1.051$. Substituting these values and $\hat{\sigma}^2 = 40788.55$ in to (8.61), we obtain

$$\text{Var}(\hat{\mu}) \cong \frac{40788.55(2.022)}{300} = 274.91$$

$$\text{Var}(\hat{\sigma}) \cong \frac{40788.55(1.635)}{300} = 222.30$$

$$\text{Cov}(\hat{\mu}, \hat{\sigma}) \cong \frac{40788.55(1.051)}{300} = 142.90$$

8.4 THE GAMMA DISTRIBUTION

The density function of the gamma distribution is

$$f(t) = \frac{\lambda}{\Gamma(\gamma)} (\lambda t)^{\gamma-1} \exp(-\lambda t) \qquad t \geq 0, \lambda, \gamma > 0 \qquad (8.62)$$

where

$$\Gamma(\gamma) = \begin{cases} \displaystyle\int_0^\infty x^{\gamma-1} e^{-x}\, dx \\ (\gamma - 1)! \qquad \text{if } \gamma \text{ is an integer} \end{cases}$$

In this section we discuss the MLE of λ and γ for data with no censored observations and data with singly censored observations. For progressively censored data, the equations to be solved for the MLE are very complicated and difficult and thus will not be discussed.

8.4.1 Estimation of λ and γ for Data without Censored Observations

Suppose that the n individuals under study are followed to death and their exact survival times $t_1, t_2, \ldots, t_n$ are known. The MLEs of λ and γ can be obtained by solving simultaneously

$$\frac{n\hat{\gamma}}{\hat{\lambda}} - \sum_{i=1}^{n} t_i = 0 \qquad (8.63)$$

and

$$n \log_e \hat{\lambda} - \frac{n\Gamma'(\hat{\gamma})}{\Gamma(\hat{\gamma})} + \sum_{i=1}^{n} \log_e t_i = 0 \qquad (8.64)$$

where $\Gamma'(\gamma)$ is the derivative of $\Gamma(\hat{\gamma})$,

$$\Gamma'(\gamma) = \int_0^\infty x^{\hat{\gamma}-1} \log_e(xe^{-x})\, dx \qquad (8.65)$$

From (8.63) we have

$$\hat{\lambda} = \frac{n\hat{\gamma}}{\displaystyle\sum_{i=1}^{n} t_i} \tag{8.66}$$

On eliminating λ, we substitute (8.66) into (8.64) and obtain

$$\frac{\Gamma'(\hat{\gamma})}{\Gamma(\gamma)} - \log_e \hat{\gamma} - \log_e \frac{\left(\displaystyle\prod_{i=1}^{n} t_i\right)^{1/n}}{\displaystyle\frac{\sum_{i=1}^{n} t_i}{n}} = 0 \tag{8.67}$$

to solve for $\hat{\gamma}$. This can be done by iterative procedures such as the Newton–Raphson method. Tables for the solution of (8.67) for $\hat{\gamma}$ as a function of R are given by Greenwood and Durand (1960), where R is the ratio of the geometric mean to the arithmetic mean of the n observations:

$$R = \frac{\left(\displaystyle\prod_{i=1}^{n} t_i\right)^{1/n}}{\displaystyle\frac{\sum_{i=1}^{n} t_i}{n}} \tag{8.68}$$

Wilk et al. (1962a) show that the relationship between $\hat{\gamma}$ and $1/(1 - R)$ is linear. A table of values of $\hat{\gamma}$ as a function of $1/(1 - R)$ given in their paper is reproduced in Appendix C (Table C-16). Thus if R and $1/(1 - R)$ are computed from the sample, a MLE of γ can be found from Table C-16. For values not tabulated, linear interpolation can be used. Having $\hat{\gamma}$ so determined, $\hat{\lambda}$ can be obtained from (8.66). Computer programs for solving (8.67) can be found in Roy et al. (1971).

In the method of moments (Fisher 1922), the estimators are obtained by simply equating the population mean and variance to the sample mean and variance. The moment estimators of γ and λ are

$$\lambda^* = \frac{\displaystyle\sum_{i=1}^{n} t_i}{\displaystyle\sum (t_i - \bar{t})^2} \tag{8.69}$$

and

$$\gamma^* = \frac{\left(\displaystyle\sum_{i=1}^{n} t_i\right)^2}{n \displaystyle\sum_{i=1}^{n} (t_i - \bar{t})^2} \tag{8.70}$$

Both types of estimators give biased estimates. The moment estimators are easy to calculate but are inefficient in the sense that their variances are larger than the variance of the MLE. To reduce the bias, Lilliefors (1971) suggests correction factors for these two types of estimators. The corrected MLEs of γ and λ are, respectively,

$$\hat{\gamma}_c = \frac{\hat{\gamma}}{1 + 3/n} \tag{8.71}$$

and

$$\hat{\lambda}_c = \frac{\hat{\gamma}_c}{t}\left(1 - \frac{1}{n\hat{\gamma}_c}\right) \tag{8.72}$$

The corrected moment estimators of γ and λ are

$$\gamma_c^* = \frac{\gamma^*}{1 + 2/n} - \frac{3}{n} \tag{8.73}$$

and

$$\lambda_c^* = \frac{\gamma_c^*}{t}\left(1 - \frac{1}{n\gamma_c^*}\right) \tag{8.74}$$

Lilliefors shows by the Monte Carlo method that the corrected MLE and the method-of-moments estimates are approximately unbiased. In addition, as long as $\gamma \geq 2$, the corrected moment estimators have no more bias than the corrected MLE and for $n = 10$ have considerably less bias. For $n = 10$, 20 and $\gamma \geq 2$, the variance is close to that of the MLE.

Example 8.10

Ten patients with melanoma achieve remission after surgery and therapy. They are followed to relapse. The durations of remission in months are recorded as follows: 5, 8, 10, 11, 15, 20, 21, 23, 30, and 40. Assuming that the distribution of remission duration is gamma, we first calculate the MLE of γ and λ according to Wilk et al. (1962a). To compute R, we obtain $\Sigma_{i=1}^{n} t_i = 183$ and $(\Pi_{i=1}^{n} t_i)^{1/10} = 15.43$. Therefore $R = 0.84$ and $1/(1 - R) = 6.25$. From Table C-16, $\hat{\gamma} = 2.89830$ for $1/(1 - R) = 6.0$ and $\hat{\gamma} = 3.14984$ for $1/(1 - R) = 6.5$. By linear interpolation, for $1/(1 - R) = 6.25$, $\bar{\gamma} = 3.02407$. From (8.66), $\hat{\lambda} = 0.16525$. The corrected MLEs obtained from (8.71) and (8.72) are $\hat{\gamma}_c = 2.326$ and $\hat{\lambda}_c = 0.122$.

The moment estimates of γ and λ are following (8.69) and (8.70) are $\lambda^* = 0.173$ and $\gamma^* = 3.171$. With the correction factors, $\lambda_c^* = 0.1225$ and $\gamma_c^* = 2.3425$, which are very close to the corrected MLE.

8.4.2 Estimation of λ and γ for Data with Censored Observations

When data are singly censored, the survival times can be ordered as

$$t_{(1)} \leq t_{(2)} \leq \cdots \leq t_{(r)} = t_{(r+1)}^+ = \cdots = t_{(n)}^+$$

where r individuals in the study have exact survival times recorded and $n - r$ others have their lives terminated after the rth death occurs. In this case, the maximum likelihood procedure becomes much more complicated.

Let $\eta = \lambda t_{(r)}$, $P = [\prod_{i=1}^{r} t_{(i)}]^{1/r}/t_{(r)}$, and $S = \sum_{i=1}^{r} t_{(i)}/rt_{(r)}$. The MLEs of η and γ can be obtained by solving simultaneously

$$\log_e P = \frac{n\Gamma'(\gamma)}{r\Gamma(\gamma)} - \frac{n}{r}\log_e \eta - \left(\frac{n}{r} - 1\right)\frac{J'(\gamma, \eta)}{J(\gamma, \eta)} \tag{8.75}$$

and

$$S = \frac{\gamma}{\eta} - \frac{1}{\eta}\left(\frac{n}{r} - 1\right)\frac{e^{-\eta}}{J(\gamma, \eta)} \tag{8.76}$$

where

$$J(\gamma, \eta) = \int_1^\infty t^{\gamma-1} e^{-\eta t} \, dt \tag{8.77}$$

and

$$J'(\gamma, \eta) = \frac{\partial}{\partial \gamma} J(\gamma, \eta) = \int_1^\infty t^{\gamma-1} \log_e t e^{-\eta t} \, dt \tag{8.78}$$

Wilk et al. (1962a) generate, for a grid of values of P and S and n/r, tables of values of $\hat{\gamma}$ and $\hat{\mu} = \hat{\gamma}/\hat{\eta}$ based on the solutions of (8.75) and (8.76). The tables are reproduced in Appendix C (Table C-17). Thus, to find $\hat{\gamma}$ and $\hat{\lambda}$, one needs to compute P and S first. For specific values of P, S, and n/r, $\hat{\gamma}$ and $\hat{\mu}$ may be found in Table C-17. Then $\hat{\lambda}$ can be obtained from $\hat{\lambda} = \hat{\gamma}/[\hat{\mu} t_{(r)}]$. Interpolations may be needed when any of the values of P, S, and n/r are not tabulated.

Computer programs for the MLE of the gamma parameters can be found in Roy et al. (1971).

Example 8.11 adapted from Wilk et al. (1962a) illustrates the procedure of calculating $\hat{\gamma}$, $\hat{\mu}$, and $\hat{\lambda}$ when Table C-17 is used.

This method can also be used in the case of a complete sample (no censored observations), that is, $r = n$. If it is obvious that some of the observations may be outliers (too large or too small), it is reasonable not to use them in the estimation of r. In this case, r is the number of observations used in the estimation procedure.

Example 8.11
Consider an experiment with $n = 34$ animals. The following data are the lifetimes t_i in weeks of 34 animals: 3, 4, 5, 6, 6, 7, 8, 8, 9, 9, 9, 10, 10, 11, 11, 11, 13, 13, 13, 13, 13, 17, 17, 19, 19, 25, 29, 33, 42, 42, 52, 52+, 52+, and 52+. The study is terminated when 31 animals have died and the other 3 are sacrificed. In our notation, $n = 34$ and $r = 31$.

1. Compute n/r, P, and S:

$$\frac{n}{r} = \frac{34}{31} = 1.10$$

$$S = \frac{\sum t_i}{rt_{(r)}} = \frac{487}{(31)(52)} = 0.30$$

To compute P, it is easier to first compute $\log P$:

$$\log P = \frac{1}{r} \sum \log t_{(i)} - \log t_{(r)}$$

$$= \tfrac{1}{31} \times 33.90207 - 1.716$$

$$= -0.622385$$

Hence $P = 0.24$.

2. Consider the entries for $n/r = 1.10$ and $P = 0.24$ in Table C-17:

$$S = 0.28: \quad \hat{\gamma} = 1.986 \quad \hat{\mu} = 0.365$$

$$S = 0.32: \quad \hat{\gamma} = 1.449 \quad \hat{\mu} = 0.410$$

Using linear interpolation, approximate estimates of γ and μ are $\hat{\gamma} = 1.72$ an $\hat{\mu} = 0.39$.

3. Finally, $\hat{\lambda} = 1.72/(0.39 \times 52) = 0.085$.

For a more accurate two-way interpolation, the reader is referred to Wilk et al. (1962a).

8.5   A REGRESSION METHOD FOR SURVIVAL DISTRIBUTION FITTING

Gehan and Siddiqui (1973) suggest a regression method for survival distribution fitting that requires data be arranged in intervals as in a life-table analysis. Much of the information required for a life-table analysis is needed. Four theoretical distributions are considered. The hazard functions are estimated by a nonparametric method for each interval and parameters of the distributions are then estimated by the method of weighted least squares. The best fit among the four distributions is selected by comparing the likelihood values of the observed data under the four distributions. The distribution that gives the largest likelihood value provides the best fit.

The four distributions considered are the exponential, Weibull, Gompertz, and linear exponential. The hazard functions of these distributions are the following:

1. *Exponential*: $h(t) = \lambda,\ \lambda > 0$
2. *Weibull*: $h(t) = \lambda'\gamma t^{\gamma - 1},\ \lambda',\ \gamma > 0,\ \lambda' = \lambda^{\gamma}$
3. *Gompertz*: $h(t) = \exp(\lambda + \gamma t),\ h(t) > 0$ $\qquad$ (8.79)
4. *Linear Exponential*: $h(t) = \lambda + \gamma t,\ h(t) > 0$

These four distributions share a common property. That is, the hazard function $h(t)$ or its logarithmic transform $\log_e h(t)$ is a linear function of t or $\log_e t$. The four models in (8.79) can be rewritten as

1. $h(t) = \lambda$
2. $\log_e h(t) = \log_e(\gamma\lambda') + (\gamma - 1)\log_e t$
3. $\log_e h(t) = \lambda + \gamma t$ $\qquad$ (8.80)
4. $h(t) = \lambda + \gamma t$

Let $y = h(t)$ or $\log h(t)$; the four models in (8.80) can be written in a general form:

$$y = a + bx \qquad (8.81)$$

For model 1 in (8.80),

$$y = h(t) \qquad a = \lambda \qquad b = 0$$

For model 2

$$y = \log_e h(t) \qquad a = \log_e(\gamma\lambda')$$
$$b = \gamma - 1 \qquad x = \log_e t$$

For model 3

$$y = \log_e h(t) \qquad a = \lambda \qquad b = \gamma \qquad x = t$$

For model 4

$$y = h(t) \qquad a = \lambda \qquad b = \gamma \qquad x = t$$

Thus, if x, y are known, the coefficients a and b in (8.81) can easily be estimated. Since the data are arranged in a life-table fashion, the hazard function can be estimated by (4.21) or (4.23), and t can be taken as the midpoint of the interval. Thus, for the ith interval, (8.81) is

$$y_i = a + bx_i$$

where y_i is the estimated hazard function or its logarithm and x_i is the midpoint of the interval or its logarithm. Having y_i and x_i for each interval, we are ready to estimate a and b.

Gehan and Siddiqui suggest using the weighted least-squares method. For a general discussion of this method, the reader is referred to Draper and Smith (1966) and Neter and Wasserman (1974).

Let s be the number of intervals. The weighted least-squares solution is defined so that the coefficients $\hat{a}$ and $\hat{b}$ are chosen to minimize the weighted sum of squares (WSS) of the differences between y_i and $\hat{y}_i = \hat{a} + \hat{b}x_i$:

$$\text{WSS} = \sum_{i=1}^{s} w_i(y_i - \hat{a} - \hat{b}x_i)^2 \tag{8.82}$$

where w_i is the weight. Gehan and Siddiqui consider three different weights, $w_i = 1$, $1/\hat{v}_i$, $n_i b_i$, where $\hat{v}_i$ is the estimated variance of h_i and b_i and n_i are, respectively, the width and number exposed to risk in the ith interval.

The weighted least-squares estimates for a and b are given by the formulas

$$\hat{b} = \frac{\displaystyle\sum_{i=1}^{s} w_i(x_i - \bar{x}')(y_i - \bar{y}')}{\displaystyle\sum_{i=1}^{s} w_i(x_i - \bar{x}')^2} = \frac{\displaystyle\sum_{i=1}^{s} w_i x_i y_i - \frac{\left(\displaystyle\sum_{i=1}^{s} w_i x_i\right)\left(\displaystyle\sum_{i=1}^{s} w_i y_i\right)}{\displaystyle\sum_{i=1}^{s} w_i}}{\displaystyle\sum_{i=1}^{s} w_i x_i^2 - \frac{\left(\displaystyle\sum_{i=1}^{s} w_i x_i\right)^2}{\displaystyle\sum_{i=1}^{s} w_i}} \tag{8.83}$$

and

$$\hat{a} = \bar{y}' - \hat{b}\bar{x}' \tag{8.84}$$

where

$$\bar{x}' = \frac{\displaystyle\sum_{i=1}^{s} w_i x_i}{\displaystyle\sum_{i=1}^{s} w_i}$$

and

$$\bar{y}' = \frac{\displaystyle\sum_{i=1}^{s} w_i y_i}{\displaystyle\sum_{i=1}^{s} w_i}$$

Having estimated the parameters of the four distributions, the selection of the best fit depends on the likelihood values of the observed data under the four distributions. Using the same notations as in the life-table analysis, the likelihood function is given by

$$L = \prod_{i=1}^{s-1} \left[1 - \frac{\hat{S}(t_{i+1})}{\hat{S}(t_i)} \right]^{d_i} \left[\frac{\hat{S}(t_{i+1})}{\hat{S}(t_i)} \right]^{n_i' - d_i}$$

Thus, the logarithm of the likelihood is

$$\log_e L = \sum_{i=1}^{s-1} d_i \log_e \left[1 - \frac{\hat{S}(t_{i-1})}{\hat{S}(t_i)} \right] + \sum_{i=1}^{s-1} (n_i' - d_i) \log_e \left[\frac{\hat{S}(t_{i+1})}{\hat{S}(t_i)} \right] \quad (8.85)$$

The survivorship functions for the four models are

1. $S(t) = \exp(-\lambda t)$
2. $S(t) = \exp[-(\lambda t)^\gamma]$
3. $S(t) = \exp\left\{ \frac{-\exp(\lambda)}{\gamma} [\exp(\gamma t) - 1] \right\}$ $\qquad\qquad$ (8.86)
4. $S(t) = \exp[-(\lambda t + \frac{1}{2}\gamma t^2)]$

Substituting the least-squares estimates of λ and γ in (8.86) and then in (8.85), we obtain estimated values of $\log_e L$ for each model.

Since models 2–4 reduce to model 1, the exponential distribution, the first question is whether the exponential distribution gives the best fit. To answer this question, consider twice the difference between the log-likelihood under the exponential distribution and the other distribution as approximately χ^2 distributed with one degree of freedom. That is, let $L(j)$ be the likelihood of the data under model j, $-2[\log_e L(j) - \log_e L(1)]$, for $j = 2, 3, 4$, follows approximately the chi square distribution with one degree of freedom. This test enables us to conclude whether the likelihood of the data under model 2, 3, or 4 is significantly greater than under model 1.

To choose among models 2, 3, and 4, we compare the log-likelihood values of the observed data under the various models. The model that gives the largest value could be chosen for further investigation.

Having chosen the best fitting model among the four, a goodness-of-fit test can be performed by considering twice the difference between the log-likelihood under the best fitting model and the sample data as approximately chi square distributed with $s - 1 - k$ degrees of freedom, where k is the number of parameters estimated in the model ($k = 1$ for model 1 and $k = 2$ for models 2, 3, and 4).

A computer program that performs all the calculations for all four models is available (Kennedy and Gehan 1971). The following example is from Gehan and Siddiqui (1973).

Table 8.3 Life-Table Analysis of 112 Patients with Plasma Cell Myeloma

Interval (months) ($t_i -$)	Midpoint t_{mi}	Number Entering Interval n_i'	Number Withdrawn in Interval l_i	Number Exposed to Risk n_i	Number Dying $n_i - r_i$	Proportion Dying $1 - p_i$	Proportion Surviving p_i	Hazard Function $\hat{h}_i$	Standard Error Hazard Function $\sqrt{\mathrm{Var}(\hat{h}_i)}$
0–	2.75	112	1	111.5	18	0.1614	0.8386	0.0319	0.0075
5.5–	8.0	93	1	92.5	16	0.1730	0.8270	0.0379	0.0094
10.5–	13.0	76	3	74.5	18	0.2416	0.7584	0.0550	0.0128
15.5–	18.0	55	0	55.0	10	0.1818	0.8182	0.0400	0.0126
20.5–	23.0	45	0	45.0	11	0.2444	0.7556	0.0557	0.0166
25.5–	28.0	34	1	33.5	8	0.2388	0.7612	0.0542	0.0190
30.5–	35.5	25	3	23.5	12	0.5106	0.4894	0.0686	0.0186
40.5–	45.5	10	4	8.0	3	0.3750	0.6250	0.0462	0.0259
50.5–	55.5	3	2	2.0	1	0.5000	0.5000	0.0667	0.0629
60.5–	—	0	0	0	0	—	—	—	—

Source: Gehan and Siddiqui (1973).

Example 8.12

Carbone et al. (1967) give survival data for 112 patients with plasma cell myeloma admitted to the National Cancer Institute between January 1953 and December 1964. The data and the estimated hazard functions using (4.23) are given in Table 8.3. Note that there is a generally increasing trend in the hazard function with time. Estimates of the parameters calculated for the four models using (8.83) and (8.84) are given in Table 8.4. Three methods of weighting were considered, namely, $w_i = 1$, $1/\hat{v}_i$, and $n_i b_i$.

To check if the data are exponentially distributed, we compute $-2[\log_e L(j) - \log_e L(1)]$ for $j = 2, 3, 4$, $w_i = 1/\hat{v}_i$, and $\log_e L(1) = -229.14$. The three chi-square values are 5.24, 4.48, and 4.74, respectively, for models 2, 3, and 4. The $\chi^2_{1,0.05}$ value is 3.84. Thus, these three values are all significant at the 0.05 level. It is concluded that the exponential distribution does not provide a good fit to the data.

The log-likelihood values obtained under models 2–4 are very close. The largest log-likelihood value is obtained from the Weibull distribution with weights $1/\hat{v}_i$. In this model, $\hat{\gamma} = 1.2520$ and $\hat{\lambda}' = 0.0196$; therefore, the hazard function is

$$h(t) = (0.0196)(1.2520)t^{(1.2520-1)} = 0.025t^{0.252}$$

Table 8.4   Estimated Parameters and Likelihood Values for Four Models

Model	Estimates of Parameters	Weight 1 $w_i = 1$	Weight 2 $w_i = 1/\hat{v}_i$	Weight 3 $w_i = n_i b_i$
Exponential, $h(t) = \lambda$	$\hat{\lambda}$	0.0507	0.0422	0.0454
	SE $(\hat{\lambda})$	0.0087	0.0044	0.0045
	$\log_e L$	−229.61	−229.14	−228.94
Weibull, $h(t) = \lambda'\gamma t^{\gamma-1}$	$\hat{\lambda}'$	0.0211	0.0196	0.0199
	$\hat{\gamma}$	1.2205	1.2520	1.2414
	SE $(\hat{\lambda}')$	0.0121	0.0077	0.0079
	SE $(\hat{\gamma})$	0.1773	0.1131	0.1150
	Cov $(\hat{\lambda}', \hat{\gamma})$	−0.0021	−0.0008	−0.0009
	$\log_e L$	−226.58	−226.52	−226.54
Gompertz, $h(t) = \exp(\lambda + \gamma t)$	$\hat{\lambda}$	−3.2786	−3.3524	−3.3855
	$\hat{\gamma}$	0.0105	0.0162	0.0167
	SE $(\hat{\lambda})$	0.2453	0.1741	0.1775
	SE $(\hat{\gamma})$	0.0129	0.0080	0.0082
	Cov $(\hat{\lambda}, \hat{\gamma})$	−0.0029	0.0011	−0.0012
	$\log_e L$	−227.19	−226.90	−226.91
Linear exponential, $h(t) = \lambda + \gamma t$	$\hat{\lambda}$	0.0378	0.0324	0.0334
	$\hat{\gamma}$	0.0005	0.0008	0.0008
	SE $(\hat{\lambda})$	0.0143	0.0066	0.0068
	SE $(\hat{\gamma})$	0.0008	0.0004	0.0004
	$\log_e L$	−227.09	−226.77	−226.73

Source: Gehan and Siddiqui (1973).

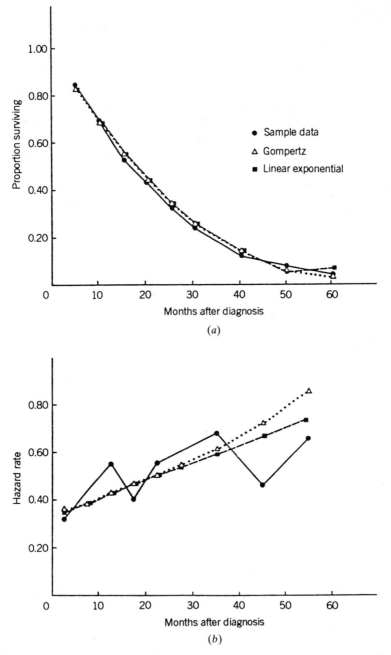

Figure 8.1   (*a*) Survivorship and (*b*) hazard functions for plasma cell myeloma patients.

which is zero at time zero and increases sharply to 0.0317 at 2.75 months. From medical knowledge of plasma cell myloma, it is unlikely that the hazard function behaves in this way. It is more likely that there is some hazard at time zero and the hazard function changes smoothly thereafter. Consequently, a Gompertz model with $w_i = 1/\hat{v}_i$ and the linear exponential model with $w_i = n_i b_i$ were chosen. The survivorship and hazard functions are plotted in Figures 8.1(a) and (b), respectively.

The log-likelihood for the sample data is -225.57. A test for goodness-of-fit for the Gompertz distribution gives $\chi^2 = -2[-226.90 - (225.57)] = 2.66$ with seven degrees of freedom and that for the linear-exponential model gives $\chi^2 = -2[-226.73 - (-225.37)] = 2.32$ with seven degrees of freedom. The probabilities of obtaining small χ^2 values such as these are between 0.90 and 0.95. Thus, both models provide an excellent fit to the data.

BIBLIOGRAPHICAL REMARKS

In addition to the papers cited in this chapter, Gross and Clark (1975) has a chapter on estimation and inference in the exponential distribution and a chapter on the estimation of parameters of three distributions including the Weibull and gamma. Mann et al. (1974), Lawless (1982), and Nelson (1982) also provide a chapter on the estimation procedures for survival distributions including the exponential, Weibull, gamma, and lognormal. Readers with a background in mathematical statistics and an interest in mathematical treatment of estimation procedures are referred to these books.

EXERCISES

8.1 Consider the survival times given in Exercise 7.2. Assuming they follow the one-parameter exponential distribution, obtain:
   a. The MLE of λ
   b. The MLE of μ
   c. The 95% confidence intervals for λ and μ. Compare your results in (a) and (b) with those obtained graphically in Exercise 7.2.

8.2 Assuming that the correct entries between errors in Exercise 7.3 follow the two-parameter exponential distribution, obtain:
   a. An estimate of G
   b. The MLE of λ
   c. The MLE of μ
   d. The probability of 100 correct entries between two errors

8.3 Consider the survival data in Exercise 7.5. Obtain the MLE of the parameter(s) and mean survival times, assuming:

a. A one-parameter exponential distribution

b. A two-parameter exponential distribution

8.4 In a study of deep venous thrombosis, the following blood clot lysis times in hours were recorded from 20 patients: 2, 3, 4, 5.5, 9, 13, 16.5, 17.5, 12.5, 7, 6, 17.5, 11.5, 6, 14, 25, 49, 37.5, 49, and 28.

a. Make a lognormal probability plot of the data and obtain graphical estimates of the parameters. Do the data follow a lognormal distribution?

b. Obtain MLEs of the parameters μ and σ^2. Compare your results with those obtained in (a).

c. Obtain 95% confidence intervals for μ and σ^2.

8.5 Consider the survival time of the 20 insects given in Example 7.4. The lognormal distribution provides a reasonable fit to the data as shown in Figure 7.7.

a. Obtain MLEs of the parameters μ and σ^2 and compare the results with those obtained in Example 7.4.

b. Obtain 95% confidence intervals for μ and σ^2.

8.6 Consider the following tumor-free times in days of 10 animals: 2, 3.5, 5, 7, 9, 10, 15, 20, 30, and 40.

a. Obtain a Weibull probability plot of the data and estimate γ and λ graphically.

b. Obtain MLEs of the parameters γ, and λ.

CHAPTER 9

Parametric Methods for Comparing Two Survival Distributions

In Chapter 5 we discussed several nonparametric tests for comparing two survival distributions. If the distributions follow a known model, parametric tests are more powerful than nonparametric, but their computation is more tedious.

In this chapter several parametric tests are presented for the comparison of two survival patterns that follow the exponential, Weibull, or gamma distribution. For lognormally distributed survival times, we can use the fact that the logarithmic transformation of the data, that is, $\log t_i$, follows the normal distribution. Standard tests based on normal distribution can then be applied.

9.1 COMPARISON OF TWO EXPONENTIAL DISTRIBUTIONS

Suppose that two survival distributions follow the exponential model with parameters λ_1 and λ_2, respectively. Two tests can compare the distributions when neither group has a guarantee time (i.e., the one-parameter exponential case): the likelihood ratio test and an F-test suggested by Cox (1953). These two tests can test the hypothesis that the two exponential distributions are equal whether or not the samples include censored observations. We shall present the tests for progressively censored data only since the singly censored and uncensored data are special cases.

9.1.1 The Likelihood Ratio Test

Suppose that there are n_1 and n_2 individuals in groups 1 and 2, respectively, $x_1, \ldots, x_{r_1}$ uncensored and $x_{r_1+1}^+, \ldots, x_{n_1}^+$ censored in group 1 and $y_1, \ldots, y_{r_2}$ uncensored and $y_{r_2+1}^+, \ldots, y_{n_2}^+$ censored in group 2. Thus, in group 1, there are r_1 uncensored and $n_1 - r_1$ censored observations. In group 2, there are r_2 uncensored and $n_2 - r_2$ censored observations. If it is known that the survival times of the two groups follow the exponential distribution with density function $f_i(t) = \lambda_i e^{-\lambda_i t}$, $i = 1, 2$, testing the equality of two exponential distributions is equivalent to testing the hypothesis $H_0: \lambda_1 = \lambda_2$. This is because the two exponential distributions are characterized by the two parameters λ_1 and λ_2. Thus, the null hypothesis is

$$H_0: \lambda_1 = \lambda_2 = \lambda$$

against the alternative hypothesis

$$H_1: \lambda_1 \neq \lambda_2$$

The test statistic for the likelihood ratio test is the ratio of two likelihood functions (or their logarithms) estimated from the sample data. Using conventional notations, let Λ be the likelihood ratio

$$\Lambda = \frac{L(\hat{\lambda}, \hat{\lambda})}{L(\hat{\lambda}_1, \hat{\lambda}_2)} \tag{9.1}$$

The denominator in (9.1) is the likelihood function for the two groups combined (the individual likelihood function is given in Section 8.1),

$$L(\hat{\lambda}_1, \hat{\lambda}_2) = \hat{\lambda}_1^{r_1} \hat{\lambda}_2^{r_2} \exp\left[-\hat{\lambda}_1\left(\sum_{i=1}^{r_1} x_i + \sum_{i=r_1+1}^{n_1} x_i^+ \right) - \hat{\lambda}_2\left(\sum_{i=1}^{r_2} y_i + \sum_{i=r_2+1}^{n_2} y_i^+ \right) \right] \tag{9.2}$$

where $\hat{\lambda}_1$ and $\hat{\lambda}_2$ are the MLE of λ_1 and λ_2 obtained, respectively, from groups 1 and 2. From Section 8.1

$$\hat{\lambda}_1 = \frac{r_1}{\displaystyle\sum_{i=1}^{r_1} x_i + \sum_{i=r_1+1}^{n_1} x_i^+} \qquad \hat{\lambda}_2 = \frac{r_2}{\displaystyle\sum_{i=1}^{r_2} y_i + \sum_{i=r_2+1}^{n_2} y_i^+} \tag{9.3}$$

The numerator in (9.1) is the likelihood function for the combined sample under the null hypothesis, that is, $\lambda_1 = \lambda_2 = \lambda$,

$$L(\hat{\lambda}, \hat{\lambda}) = \hat{\lambda}^{r_1+r_2} \exp\left[-\hat{\lambda}\left(\sum_{i=1}^{r_1} x_i + \sum_{i=r_1+1}^{n_1} x_i^+ + \sum_{i=1}^{r_2} y_i + \sum_{i=r_2+1}^{n_2} y_i^+ \right) \right] \tag{9.4}$$

where $\hat{\lambda}$ is the MLE of λ obtained from the combined sample,

$$\hat{\lambda} = \frac{r_1 + r_2}{\sum\limits_{i=1}^{r_1} x_i + \sum\limits_{i=r_1+1}^{n_1} x_i^+ + \sum\limits_{i=1}^{r_2} y_i + \sum\limits_{i=r_2+1}^{n_2} y_i^+} \tag{9.5}$$

It was shown that $-2\log_e\Lambda$ has an approximate chi-square distribution with one degree of freedom for samples of a least 25 $(n_1 + n_2 \geq 25)$ under the null hypothesis. Thus the hypothesis H_0 is rejected if $-2\log_e\Lambda$ exceeds the 100α percentage point $\chi^2_{1,\alpha}$, for the chi-square distribution with one degree of freedom $[P(-2\log_e\Lambda > \chi^2_{1,\alpha}) < \alpha]$.

The test procedure can be summarized as follows:

1. Compute $\hat{\lambda}_1$ and $\hat{\lambda}_2$ following (9.3).
2. Compute $L(\hat{\lambda}_1, \hat{\lambda}_2)$ in (9.2) using the given data and $\hat{\lambda}_1$ and $\hat{\lambda}_2$ obtained in step 1.
3. Compute $\hat{\lambda}$ following (9.5).
4. Compute $L(\hat{\lambda}, \hat{\lambda})$ in (9.4).
5. Compute Λ in (9.1) and $-2\log_e\Lambda$. If $-2\log_e\Lambda$ is greater than $\chi^2_{1,\alpha}$ for a selected α (Table C-2), reject H_0 and conclude that the two exponential survival distributions are not equal. Otherwise, the data do not provide enough evidence to reject the null hypothesis.

If there are no censored observations in the data, (9.1)–(9.5) are also applicable simply by letting $n_1 = r_1$, $n_2 = r_2$ and omitting the terms involving x_i^+ and y_i^+.

The likelihood ratio test is primarily for two-sided tests and is difficult to apply to a one-sided test. It is approximate and should be used with caution when the sample size is small. The power of the test, similar to other likelihood ratio tests, is not high. That is, if the likelihood ratio test is used regularly, one is more likely not to reject the null hypothesis when the two survival distributions are not equal.

Example 9.1
Consider the remission data of the two treatment groups given in Example 5.1. The remission times in months are as follows:

CMF   23, 16+, 18+, 20+, 24+
Control   15, 18, 19, 19, 20

Assume that the two distributions are exponential with parameters λ_1 and λ_2, respectively. Using the likelihood ratio test, we test the null hypothesis

$$H_0: \lambda_1 = \lambda_2 = \lambda \quad \text{(two treatments equally effective)}$$

against

$H_1: \lambda_1 \neq \lambda_2$ (two treatments not equally effective)

Following the above, we proceed as follows:

1. Compute $\hat{\lambda}_1$ and $\hat{\lambda}_2$ in (9.3). In this case,

$$n_1 = n_2 = 5 \qquad r_1 = 1 \qquad r_2 = 5$$

$$\sum_{i=1}^{r_1} x_i = 23 \qquad \sum_{i=r_1+1}^{n_1} x_i^+ = 78 \qquad \sum_{i=1}^{r_2} y_i = 91 \qquad \sum_{i=r_2+1}^{n_2} y_i^+ = 0$$

$$\hat{\lambda}_1 = \frac{1}{23+78} = \frac{1}{101} = 0.0099 \qquad \hat{\lambda}_2 = \frac{5}{91} = 0.0549$$

2. Compute $L(\hat{\lambda}_1, \hat{\lambda}_2)$ in (9.2):

$$L(\hat{\lambda}_1, \hat{\lambda}_2) = (0.0099)(0.0549)^5 \exp[-0.0099(101) - 0.0549(91)]$$
$$= 1.2290(10)^{-11}$$

3. Compute $\hat{\lambda}$ in (9.5):

$$\hat{\lambda} = \frac{1+5}{23+78+91} = \frac{6}{192} = 0.0313$$

4. Compute $L(\hat{\lambda}, \hat{\lambda})$ in (9.4):

$$L(\hat{\lambda}, \hat{\lambda}) = (0.0313)^6 \exp[-0.0313(192)]$$
$$= 2.3085(10)^{-12}$$

5. Compute Λ in (9.1):

$$\Lambda = \frac{2.3085(10)^{-12}}{1.2290(10)^{-11}} = \frac{0.23085}{1.2290} = 0.1878 \qquad -2\log_e \Lambda = 3.344$$

From Table C-2, we obtain $\chi^2_{1,0.05} = 3.84$. Thus we cannot reject H_0 at the 0.05 level. Recall that in Chapter 5 the null hypothesis was rejected at the 0.05 level by using the four nonparametric tests.

9.1.2 Cox's F-test for Exponential Distributions

When it can be assumed that the times to failure are exponentially distributed in both treatment groups, an F-test suggested by Cox (1953) can be used to test for treatment differences whether or not censored observations are present. Suppose we wish to test the hypothesis $H_0: \lambda_1 = \lambda_2$ against either the one-sided alternative $H_1: \lambda_1 < \lambda_2$ (or $H_2: \lambda_1 > \lambda_2$) or the two-sided alternative $H_3: \lambda_1 \neq \lambda_2$. An efficient test is to take $\bar{t}_1/\bar{t}_2$ as having an F distribution with $(2r_1, 2r_2)$ degrees of freedom, where

$$\bar{t}_1 = \frac{\sum\limits_{i=1}^{r_1} x_i + \sum\limits_{i=r_1+1}^{n_1} x_i^+}{r_1} \qquad (9.6)$$

$$\bar{t}_2 = \frac{\sum\limits_{i=1}^{r_2} y_i + \sum\limits_{i=r_2+1}^{n_2} y_i^+}{r_2} \qquad (9.7)$$

The test procedures are (1) for H_1, reject H_0 if $\bar{t}_1/\bar{t}_2 > F_{2r_1,2r_2,\alpha}$, (2) for H_2, reject H_0 if $\bar{t}_1/\bar{t}_2 < F_{2r_1,2r_2,1-\alpha}$, and (3) for H_3, reject H_0 if $\bar{t}_1/\bar{t}_2 > F_{2r_1,2r_2,\alpha/2}$ or $\bar{t}_1/\bar{t}_2 < F_{2r_1,2r_1,1-\alpha/2}$, where α is the significance level and $F_{2r_1,2r_2,\alpha}$ is the upper 100α percentage point of the F distribution with $(2r_1, 2r_2)$ degrees of freedom. Similarly, the hypothesis that $\lambda_1/\lambda_2 = k$ can be tested by referring $k\bar{t}_1/\bar{t}_2$ to the table of the F-distribution.

When there are no censored observations, that is, $n_1 = r_1$, $n_1 = r_2$, the second terms of the numerators in (9.6) and (9.7) are zero. Then the test statistic $\bar{t}_1/\bar{t}_2$ has an F distribution with $(2n_1, 2n_2)$ degrees of freedom.

Confidence intervals for the ratio λ_1/λ_2 can be obtained from the fact that $\lambda_1\bar{t}_1/\lambda_2\bar{t}_2$ has the F distribution with $(2r_1, 2r_2)$ degrees of freedom. It follows that a $100(1 - \alpha)$ percent confidence interval for the ratio of two hazard rates λ_1/λ_2 is

$$\frac{\bar{t}_2}{\bar{t}_1} F_{2r_1,2r_2,1-\alpha/2} < \frac{\lambda_1}{\lambda_2} < \frac{\bar{t}_2}{\bar{t}_1} F_{2r_1,2r_2,\alpha/2} \qquad (9.8)$$

Example 9.2

Thirty-six patients with glioblastoma multiforme were divided into two groups; the experimental group contained 21 patients who had surgery and chemotherapy and the control group contained 15 patients who had surgery only. The survival times in weeks are available about one year after the start of the study (Burdette and Gerhan 1970):

Experimental   1, 2, 2, 2, 6, 8, 9, 13, 16, 17, 29, 34, 2+, 9+, 13+, 22+,
25+, 36+, 43+, 45+

Control   0, 2, 5, 7, 12, 42, 46, 54, 7+, 11+, 19+, 22+, 30+, 35+,
39+

The hypotheses are

$H_0: \lambda_1 = \lambda_2$ (no difference in survival between experimental and control
groups)

$H_1: \lambda_1 < \lambda_2$ (difference in survival favoring experimental group)

In this case, $n_1 = 21$, $n_2 = 15$, $r_1 = 13$, $r_2 = 8$, $\Sigma\, x_i = 147$, $\Sigma\, x_i^+ = 195$, $\Sigma\, y_i = 168$, and $\Sigma\, y_i^+ = 163$. Hence

$$\bar{t}_1 = \frac{147 + 195}{13} = 26.308 \qquad \bar{t}_2 = \frac{168 + 163}{8} = 41.375$$

and $\bar{t}_1/\bar{t}_2 = 0.636$ with $(26, 16)$ degrees of freedom. For $\alpha = 0.05$, $F_{26,16,0.05}$ is approximately 2.23; hence the hypothesis H_0 is not rejected and the data do not provide enough evidence that the survival time is longer in the experimental group. A 95% confidence interval for the ratio λ_1/λ_2 is

$$\frac{41.375}{26.308}\,(0.419) < \frac{\lambda_1}{\lambda_2} < \frac{41.375}{26.308}\,(2.625)$$

or $(0.659, 4.128)$. The estimate of $\hat{\lambda}_1/\hat{\lambda}_2$ according to (9.3) is 1.58. Hence the data show that the death rate per week of the experimental group is close to that of the control group.

In Example 9.2, the guarantee time in both groups is zero. In the case where a group has a nonzero guarantee time, it can be subtracted from every observation in the group and the test then applied.

Monte Carlo studies (Gehan and Thomas 1969; Lee et al. 1975) show that when samples are from exponential distributions, with or without censoring, the F-test is the most powerful of the parametric and nonparametric tests discussed in this chapter and Chapter 5.

9.2 COMPARISON OF TWO WEIBULL DISTRIBUTIONS

If the survival time T has a Weibull distribution with shape parameter γ, then T^γ has an exponential distribution. Thus, when the observations in each sample are transformed such that each new observation is the γth power of the original observation, the likelihood ratio test and Cox's F-test described in Section 9.1 may be applied. Although Cox's F-test is still the most powerful, the transformation requires the knowledge of the values of γ_1 and γ_2. Hence, the use of the likelihood ratio test and Cox's F-test is limited. In the following, we introduce a two-sample test proposed by Thoman and Bain (1969) for samples without censoring.

Assume that independent random samples of equal size $(n_1 = n_2 = n)$ are obtained from Weibull distributions $f_1(t)$ and $f_2(t)$, where

$$f_i(t) = \lambda_i \gamma_i (\lambda_i t)^{\gamma_i - 1} \exp[-(\lambda_i t)^{\gamma_i}] \qquad i = 1, 2 \tag{9.9}$$

To test the equality of $f_1(t)$ and $f_2(t)$, it suffices to test $\gamma_1 = \gamma_2$ and $\lambda_1 = \lambda_2$. If the hypothesis $\gamma_1 = \gamma_2$ is rejected, we need not test the hypothesis $\lambda_1 = \lambda_2$. If the hypothesis $\gamma_1 = \gamma_2$ is not rejected, we do need to test $\lambda_1 = \lambda_2$.

In order to test $\gamma_1 = \gamma_2$, we use the property of the maximum likelihood estimator $\hat{\gamma}$ (Thoman et al. 1969, Thoman and Bain 1969). To test the null hypothesis H_0: $\gamma_1 = \gamma_2$ against H_1: $\gamma_1 > \gamma_2$, we use the fact that $(\hat{\gamma}_1/\gamma_1)/(\hat{\gamma}_2/\gamma_2) = \hat{\gamma}_1/\hat{\gamma}_2$ under H_0. The percentage points of $\hat{\gamma}_1/\hat{\gamma}_2$ are given in Table C-18 (reproduced from Thoman and Bain 1969). We compute the MLE of γ_1 and γ_2, that is, $\hat{\gamma}_1$ and $\hat{\gamma}_2$, and compare $\hat{\gamma}_1/\hat{\gamma}_2$ with percentage points for a given α in Table C-18. Reject H_0 at the α level if $\hat{\gamma}_1/\hat{\gamma}_2 > l_\alpha$. For example, if $n_1 = n_2 = n = 10$, a computed $\hat{\gamma}_1/\hat{\gamma}_2 > 1.897$ would lead to rejection of H_0 at a significance level of 0.05. For $\alpha \geq 0.50$, percentage points l_α can be found by using the relationship $l_\alpha = 1/l_{1-\alpha}$.

The above procedure can be generalized to test H_0: $\gamma_1 = k\gamma_2$ against H_1: $\gamma_1 = k'\gamma_2$. For the case when $k < k'$, the rejection region becomes $\hat{\gamma}_1/\hat{\gamma}_2 > kl_\alpha$, where α is the significance level.

If the hypothesis H_0: $\gamma_1 = \gamma_2$ is rejected, the two Weibull distributions are not the same. However, if the hypothesis is not rejected, then we need to test the equality of the two scale parameters λ_1 and λ_2. A test of H_0: $\lambda_1 = \lambda_2$ against H_1: $\lambda_1 < \lambda_2$ suggested by Thoman and Bain (1969) rejects H_0 if

$$G = \tfrac{1}{2}(\hat{\gamma}_1 + \hat{\gamma}_2)(\log_e \hat{\lambda}_2 - \log_e \hat{\lambda}_1) > z_\alpha \qquad (9.10)$$

where z_α is such that $P(G < z_\alpha/H_0) = 1 - \alpha$ and $\hat{\gamma}_1$, $\hat{\gamma}_2$, $\hat{\lambda}_1$, and $\hat{\lambda}_2$ are MLEs of γ_1, γ_2, λ_1, and λ_2, respectively. The percentage points z_α are given in Table C-19 of Appendix C (reproduced from Thoman and Bain 1969). For example, if the common sample size is 10, the hypothesis H_0: $\lambda_1 = \lambda_2$ is rejected if $G \geq 0.918$ at significance level 0.05.

A test of H_0: $\lambda_1 = \lambda_2$ against H_1: $\lambda_1 > \lambda_2$ can be constructed in a similar fashion. The critical points z_α needed can be obtained from Table C-19 by using the fact that $z_\alpha = -z_{1-\alpha}$.

Example 9.3 illustrates the test procedures. The data are adapted and modified from Harter and Moore (1965). Forty observations are generated from a Weibull distribution with $\lambda_1 = 0.01$ and $\gamma_1 = 2$ and another 40 from a Weibull distribution with $\lambda_2 = 0.01$ and $\gamma_2 = 3$. For illustrative purposes, we consider the two samples of size 40 as two treatment groups.

Example 9.3

Consider the survival times of the patients in the two treatment groups in Table 9.1. The null hypothesis is that the two populations have the same shape parameter, that is H_0: $\gamma_1 = \gamma_2$ against H_1: $\gamma_1 < \gamma_2$.

The MLEs are $\hat{\gamma}_1 = 1.945$, $\hat{\gamma}_2 = 2.715$, and hence $\hat{\gamma}_2/\hat{\gamma}_1 = 1.396$, which is significant at the 0.05 level ($l_{0.05} = 1.342$ for $n = 40$) but not significant at the 0.02 level ($l_{0.02} = 1.453$ for $n = 40$). If we choose $\alpha = 0.05$ and reject H_0, the decision is correct. An error of not rejecting H_0 would be committed if an α of 0.02 or 0.01 is chosen. This is because the two shape parameters are very close ($\gamma_1 = 2$, $\gamma_2 = 3$).

To illustrate the procedure of testing the equality of the scale parameters,

Table 9.1   Survival Times of Patients in Two Treatment Groups

Treatment 1	Treatment 2
5, 10, 17, 32, 32, 33, 34, 36, 43, 44, 44, 48, 48, 61, 64, 65, 65, 66, 67, 68, 82, 85, 90, 92, 92, 102, 103, 106, 107, 114, 114, 116, 117, 124, 139, 142, 143 151, 158, 195	20.9, 32.2, 33.2, 39.4, 40.0, 46.8, 57.3, 58.0, 59.7, 61.1, 61.4, 54.3, 66.0, 66.3, 67.4, 68.5, 69.9, 72.4, 73.0, 73.2, 88.7, 89.3, 91.6, 93.1, 94.2, 97.7, 101.6, 101.9, 107.6, 108.0, 109.7, 110.8, 114.1, 117.5, 119.2, 120.3, 133.0, 133.8, 163.3, 165.1

let us assume that the hypothesis $H_0: \gamma_1 = \gamma_2$ is not rejected. To test $H_0: \lambda_1 = \lambda_2$ against $H_1: \lambda_1 > \lambda_2$, we need the MLEs of λ_1 and λ_2. Harter and Moore obtain $\hat{\lambda}_1 = 0.010776$ and $\hat{\lambda}_2 = 0.010471$. From (9.10) we obtain

$$G = \tfrac{1}{2}(1.945 + 2.715)(4.559 - 4.530) = 0.068$$

From Table C-19, the critical region for $n = 40$ is $G > 0.404$. Hence we do not reject H_0. This decision is correct since $\lambda_1 = \lambda_2 = 0.01$. Note that the MLEs of γ_1, γ_2, λ_1, and λ_2 are very close to their real values.

9.3   COMPARISON OF TWO GAMMA DISTRIBUTIONS

Suppose that $x_1, \ldots, x_n$ and $y_1, \ldots, y_n$ are the survival times of patients receiving two different treatments and that they follow the gamma distribution with the density function given in (6.28). Let λ_1 and γ_1 be the parameters of the x population and λ_2 and γ_2 be those of the y population. The x_i's and y_i's are exact (uncensored) survival times. Under the assumption that γ_1 and γ_2 are known (usually assumed equal), a test of the null hypothesis $H_0: \lambda_1 = \lambda_2$ against $H_1: \lambda_1 \neq \lambda_2$ is available.

Let $\bar{x}$ and $\bar{y}$ be the sample mean survival times of the two groups. The test is based on the fact that $\bar{x}/\bar{y}$ has the F distribution with $2n\gamma_1$ and $2n\gamma_2$ degrees of freedom (Rao 1952). Thus the test procedure is to reject H_0 at the α level if $\bar{x}/\bar{y}$ exceeds $F_{2n\gamma_1, 2n\gamma_1, \alpha}$, the 100α percentage point of the F distribution with $(2n\gamma_1, 2n\gamma_2)$ degrees of freedom. Since the F-table gives percentage points for integer degrees of freedom only, interpolations (linear or bilinear) are necessary when either $2n\gamma_1$ or $2n\gamma_2$ is not an integer.

The following example illustrates the test procedure. The data are adapted and modified from Harter and Moore (1965). They simulated 40 survival times from the gamma distribution with parameters $\gamma_1 = \gamma_2 = \gamma = 2$, $\lambda = 0.01$. The 40 individuals are divided randomly into two groups for illustrative purposes.

Example 9.4
Consider the survival time of the two treatment groups in Table 9.2. The two populations follow the gamma distributions with a common shape

Table 9.2 Survival Times of 40 Patients Receiving Two Different Treatments

Treatment 1(x)	Treatment 2(y)
17, 28, 49, 98, 119	26, 34, 47, 59, 101,
133, 145, 146, 158, 160,	112, 114, 136, 154, 154,
174, 211, 220, 231, 252,	161, 186, 197, 226, 226,
256, 267, 322, 323, 327	243, 253, 269, 308, 465

parameter $\gamma = 2$. To test the hypothesis $H_0: \lambda_1 = \lambda_2$ against $H_1: \lambda_1 \neq \lambda_2$, we compute $\bar{x} = 181.80$, $\bar{y} = 173.55$, and $\bar{x}/\bar{y} = 1.048$. Under the null hypothesis, $\bar{x}/\bar{y}$ has the F distribution with $(80, 80)$ degrees of freedom. Use $\alpha = 0.05$, $F_{80,80,0.05} \cong 1.45$. Hence, we do not reject H_0 at the 0.05 level of significance. The result is what we would expect since the two samples are simulated from the same overall sample of 40 with $\lambda = 0.01$.

BIBLIOGRAPHICAL REMARKS

In addition to the papers cited in this chapter, readers are referred to Mann et al. (1974), Gross and Clark (1975), Lawless (1982), and Nelson (1982).

EXERCISES

9.1 Consider the remission data of the leukemia patients in Example 3.3. Assume that the remission times of the two treatment groups follows the exponential distribution. Test the hypothesis that the two treatments are equally effective using:

a. The likelihood ratio test

b. Cox's F-test

Obtain a 95% confidence interval for the ratio of the two hazard rates.

9.2 For the same data, test the hypothesis that $\lambda_2 = 5\lambda_1$.

9.3 Suppose that the survival time of two groups of lung cancer patients follows the Weibull distribution. A sample of 30 patients (15 from each group) were studied. Maximum likelihood estimates obtained from the two groups are, respectively, $\hat{\gamma}_1 = 3$, $\hat{\lambda}_1 = 1.2$ and $\hat{\gamma}_2 = 2$, $\hat{\lambda}_2 = 0.5$. Test the hypothesis that the two groups are from the same Weibull distribution.

9.4 Divide randomly the lifetimes of 100 strips (delete the last one) of aluminum coupon in Table 6.4 into two equal groups. This can be done by assigning the observations alternately to the two groups. Assume that the two groups follow a gamma distribution with shape parameter $\gamma = 12$. Test the hypothesis that the two scale parameters are equal.

9.5 Twelve brain tumor patients are randomized to receive radiation therapy or radiation therapy plus chemotherapy (BCNU) in a one-year clinical trial. The following survival times in weeks are recorded:

1. Radiation + BCNU: 24, 30, 42, 15+, 40+, 42+

2. Radiation: 10, 26, 28, 30, 41, 12+

Assuming the survival time follows the exponential distribution, use Cox's F-test for exponential distributions to test the null hypothesis $H_0: S_1 = S_2$ versus the alternative $H_1 = S_1 > S_2$.

9.6 Use one of the nonparametric tests discussed in Chapter 5 to test the equality of survival distributions of the experimental and control groups in Example 9.2. Compare you result with that obtained in Example 9.2.

CHAPTER 10

Identification of Prognostic Factors Related to Survival Time

Prognosis, the prediction of the future of an individual patient with respect to duration, course, and outcome of a disease, plays an important role in medical practice. Before a physician can make a prognosis and decide on the treatment, a medical history as well as pathologic, clinical, and laboratory data are often needed. Therefore, many medical charts contain a large number of patient characteristics (also called concomitant variables, independent variables, or covariates), and it is sometimes difficult to sort out which ones are most closely related to prognosis. The physician can usually decide which characteristics are irrelevant, but a statistical analysis is often needed to prepare a compact summary of the data that can reveal their relationship. In this chapter we discuss some statistical methods for the identification of prognostic factors (or characteristics or variables). Section 10.1 discusses briefly possible types of response and prognostic variables and things that can be done in a preliminary screening before a formal analysis. Sections 10.2 and 10.3 introduce, respectively, nonparametric and parametric methods for the identification of prognostic factors related to survival time.

10.1 PRELIMINARY EXAMINATION OF DATA

Information concerning possible prognostic factors can be obtained either from clinical studies designed mainly to identify them, sometimes called prognostic studies, or from ongoing clinical trials that compare treatments as a subsidiary aspect. The dependent variable (also called the response variable), or the outcome of prediction, may be dichotomous, polychotomous, or continuous. Examples of dichotomous dependent variables are response or nonresponse, life or death, and presence or absence of a given disease. Polychotomous dependent variables include different grades of symptoms (e.g., no evidence of disease, minor symptom, major symptom) and scores of psychiatric reactions (e.g., feeling well, tolerable, depressed, or very depressed). Continuous dependent variables may be length of survival from start of treatment or length of remission, both measured on a numerical scale by a continuous range of values. Of these dependent variables, response to a given treatment (yes or no), development of a specific disease (yes or no), length of remission, and length of survival are particularly common in practice. In this chapter, we focus our attention on continuous dependent variables such as survival time and remission duration. Dichotomous dependent variables will be discussed in Chapter 11.

A prognostic variable (or independent variable) or prognostic patient characteristic may either be numerical or nonnumerical. Numerical prognostic variables may be discrete, such as the number of previous strokes or number of lymph node metastases, or continuous, such as age or blood pressure. Continuous variables can be made discrete by grouping patients into subcategories (e.g., four age subgroups: <20, 20–39, 40–59, and ≥60). Nonnumerical prognostic variables may be unordered (e.g., race or diagnosis) or ordered (e.g., severity of disease may be primary, local, or metastatic). They can also be dichotomous (e.g., a liver either is or is not enlarged). Usually, the collection of prognostic variables includes some of each type.

Before a statistical calculation is done, the data have to be examined carefully for quality. If some of the variables are significantly correlated, then one of the correlated variables is likely to be as good a predictor as all of them. Correlation coefficients between variables can be computed to detect significantly correlated variables. In deleting any highly correlated variables, information from other studies has to be incorporated. If other studies show that a given variable has prognostic value, it should be retained.

In the next two sections we discuss statistical techniques, univariate and multivariate, that are useful in identifying prognostic factors. The multivariate techniques, which are all multiple-regression methods, involve a linear function of the independent variables or possible prognostic variables. The variables must be quantitative, with particular numerical values for each patient. This raises no problem when the prognostic variables are naturally quantitative (e.g., age) and can be used in the equation directly. However, if a particular prognostic variable is qualitative (e.g., a histological classification into one of three cell types A, B, or C), something needs to be done. This situation can be covered by the use of two dummy variables, for example, x_1, the first variable, taking the value 1 for cell type A and 0 otherwise; and x_2, the second variable, taking the value 1 for cell type B and 0 otherwise. Clearly, if there are only two categories (e.g., sex), only one dummy variable is needed: x_1 is 1 for a male, 0 for a female. Also, a better description of the data might be obtained by using transformed values of the prognostic variables (e.g., squares or logarithms) or by including products like $x_1 x_2$ (representing an interaction between x_1 and x_2). Transforming the dependent variable (e.g., taking the logarithm of a response time) can also improve the fit. It is often useful to do many regression analyses of the same data until an adequate description is obtained with as few variables as possible. Very often a variable of significant prognostic value in one study is unimportant in another. Therefore, confirmation in a later study is very important in identifying prognostic factors.

Another frequent problem in regression analysis is missing data. Three distinctions about missing data can be made: (1) dependent versus independent variables, (2) many versus few missing data, and (3) random versus nonrandom loss of data.

If the value of the dependent variable (e.g., survival time) is unknown, there is little to do but drop that individual from analysis and reduce the sample size. The problem of missing data is of different magnitude depending on how large a proportion of data, either for the dependent variable or for the independent variables, is missing. This problem is obviously less critical if 1% of data for one independent variable is missing than if 40% of data for several independent variables is missing. When a substantial proportion of subjects has missing data for a variable, we may simply opt to drop these individuals and perform the analysis on the remainder of the sample. It is difficult to specify "how large" and "how small," but no serious practical objection would be raised by dropping 10 or 15 cases out of several hundred. However, if missing data occur in a large proportion of individuals and the sample size is not comfortably large, a question of randomness may be raised. If individuals with missing data do not show significant differences in the dependent variable, the problem is not serious. If the data are not missing randomly, results obtained from dropping subjects will be misleading. Thus, dropping cases is not always an adequate solution to the missing data problem.

If the independent variable is measured on a nominal or categorical scale, an alternative method is to treat individuals in a group with missing information as another group. For quantitatively measured variables (e,g., age), the mean of the values available can be used for a missing value. This principle can also be applied to nominal data. It does not mean that the mean is a good estimate for the missing value, but it does provide convenience for analysis.

A more detailed discussion on missing data can be found in Cohen and Cohen (1975, Chapter 7).

10.2   NONPARAMETRIC METHODS

10.2.1   Univariate Analysis

A simple way to identify prognostic variables is to examine individually each variable's relationship to the length of survival or remission. To do this, the observed data are grouped according to various breakdowns. The survival experiences of patients in different groups can then be characterized by estimated survivorship or hazard functions. When there are many patients in each subgroup, a life-table analysis may be done separately for each subgroup. However, this could be very tedious with many prognostic variables. An easier procedure is to use the product-limit (PL) method of Kaplan and Meier (1958) discussed in Chapter 4 to estimate the survivorship function for each subgroup and each variable. Survival times of subgroups can be compared by the methods given in Chapter 5.

Example 10.1

Consider the survival experience of 1024 children with acute leukemia reported by George et al. (1973). The data are from seven clinical trials of the Southwest Oncology Group (formerly Southwest Cancer Chemotherapy Study Group). The 12 variables for each patient at the time of diagnosis considered to be of possible prognostic value were age; histologic type of leukemia; race; sex; initial peripheral blood leukocyte count (WBC); liver, spleen, and node enlargement; hemorrhagic tendencies; initial platelet count; hemoglobin; and percentage of blast cells. Summary survival data for the 12 variables are given in Table 10.1. For each variable, patients were grouped according to different breakdowns, the number of patients in each subgroup was counted, and survival percentiles in weeks were computed following the PL method. For example, the age groupings used in this study were <1 year, 1–2 years, 3–5 years, 6–10 years, and >10 years. The median survival times of these five groups were, respectively, 25, 57, 83, 66, and 46 weeks. An impressive 70% of children 3–5 years old survived for at least 1 year while only 16% of children under 1 survived for the same period. In addition, 40% of the children aged 3–5 lived for more than 2 years, while only 10% of the <1-year-old group were still alive at 2 years. Hence, the patients in the "middle" age group (roughly ages 2–9 years old) have the most favorable prognosis. Survival curves for these five age groups are given in Figure 10.1(a). The survivorship functions are estimated by the PL method.

   In addition to age, cell type and initial WBC were found to be of most prognostic importance. Figures 10.1(b) and (c) give the survival curves of patients in the three cell type groups and three WBC groups. The median survival time was 73 weeks for ALL (acute lymphocytic leukemia) patients and 70 weeks for AUL (acute unclassified leukemia) patients. The ALL and AUL patients had significantly longer survival than the AML (acute myelogenous leukemia) patients. The most natural groupings for WBC, suggested by survival patterns, were (in thousands per cubic millimeter of blood): <10, 10–99, and ≥100. Patients with an initial WBC of less than 10,000 had a median survival time of 94 weeks, those with WBC in the range of 10,000–99,000 had a median survival time of 58 weeks, and those with WBC of 100,000 and over had a median survival time of 30 weeks. All the differences were highly significant.

Table 10.1   Summary Survival Data for 12 Variables Related to Survival in Pediatric Acute Leukemia

Variable	Patients		Survival Percentiles (weeks)			Proportion Surviving		
	Number	Percent	25%	50%	75%	1 yr	2 yr	5 yr
Age (yr)								
<1	49	5	9	25	44	0.16	0.10	0.05
1–2	187	18	21	57	107	0.52	0.27	0.08
3–5	385	38	42	83	160	0.70	0.40	0.11
6–10	261	25	28	66	131	0.56	0.31	0.10
>10	140	14	19	46	84	0.45	0.18	0.07
Type (diagnosis)								
ALL	542	53	32	73	146	0.61	0.35	0.13
AUL	369	36	30	70	134	0.60	0.33	0.08
AML	107	11	9	25	60	0.31	0.12	0.06
Race								
Caucasian	865	86	30	71	139	0.61	0.33	0.12
Non-Caucasian	145	14	16	38	94	0.38	0.20	0.07
Sex								
Male	588	57	28	64	118	0.57	0.30	0.11
Female	436	43	28	68	146	0.57	0.34	0.10
WBC ($\times 10^3$)								
<10	341	35	43	94	174	0.71	0.43	0.12
10–99	520	54	27	58	112	0.55	0.27	0.09
≥100	109	11	16	30	50	0.24	0.13	0.02
Liver								
Enlarged	752	79	27	60	118	0.54	0.29	0.09
Not enlarged	204	21	32	81	150	0.66	0.36	0.12
Spleen								
Enlarged	667	70	28	57	109	0.53	0.26	0.08
Not enlarged	285	30	31	83	181	0.65	0.40	0.14
Nodes								
Enlarged	593	65	27	58	109	0.53	0.27	0.08
Not enlarged	327	35	32	81	181	0.65	0.38	0.15
Hemorrhage								
Yes	335	44	25	57	99	0.53	0.24	0.07
No	430	56	32	76	145	0.62	0.38	0.12
Platelet								
<50	505	54	25	57	105	0.52	0.25	0.07
50–199	316	34	31	70	139	0.60	0.35	0.12
≥200	107	12	39	95	161	0.70	0.46	0.12
Hemoglobin								
<4	88	9	24	54	99	0.51	0.22	0.08
4–6.9	329	34	28	67	118	0.59	0.30	0.07
7–9.9	359	37	28	68	146	0.57	0.36	0.12
≥10	191	20	27	62	114	0.59	0.27	0.11
Blast (%)								
<65	200	22	24	61	134	0.53	0.30	0.12
65–84	181	20	21	57	117	0.53	0.31	0.08
85–94	217	24	32	67	128	0.61	0.29	0.08
≥95	298	33	32	69	133	0.58	0.32	0.09

Source: George et al. (1973). Reprinted by permission of the editor.

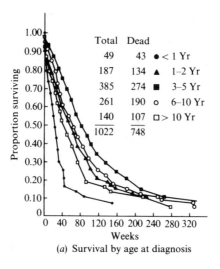

(a) Survival by age at diagnosis

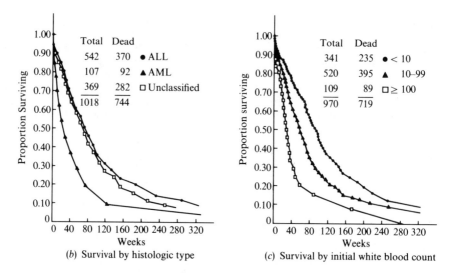

(b) Survival by histologic type   (c) Survival by initial white blood count

Figure 10.1   Survival curves of childhood leukemia patients.

In addition to examining each variable individually, the interaction of the three primary variables was checked by reporting survival in the various cell types by age by WBC subgroup. A total of 45 possible subgroups of patients were categorized. Median survival times are given in Table 10.2, which suggests a simple classification of patients with respect to prognosis at the time of diagnosis. Patients were arbitrarily classified as having a "good," "average," or "poor" prognosis on the basis of the median survival times (MSTs), which were $>1\frac{1}{2}$ years, $1-1\frac{1}{2}$ years, and <1 year, respectively.

Table 10.2 Median Survival Time (weeks) by Cell Type, Age, and WBC^a

WBC	<1 yr		1–2 yr		3–5 yr		6–10 yr		>10 yr		All Patients	
					ALL							
<10,000	47	(2), III	86	(32), I	109	(100), I	93	(49), I	68	(27), II	96	(210), I
10,000–99,999	31	(15), III	59	(57), II	77	(97), II	67	(65), II	52	(36), III	65	(270), II
≥100,000	18	(5), III	39	(12), III	132	(10), I	32	(17), III	30	(18), III	48	(62), III
	30	(22), III	65	(101), II	95	(207), I	72	(131), II	52	(81), III	73	(542), II
					AUL							
<10,000	15	(6), III	71	(23), II	98	(61), I	93	(47), I	139	(16), I	93	(153), I
10,000–99,999	26	(9), III	58	(39), II	64	(75), II	43	(43), III	48	(17), III	54	(183), II
≥100,000	28	(4), III	32	(6), III	31	(11), III	22	(8), III	2	(4), III	25	(33), III
	25	(19), III	60	(68), II	76	(147), II	65	(98), II	82	(37), I	70	(369), II
					AML							
≦10,000	3	(1), III	10	(4), III	25	(10), III	85	(11), I	8	(3), III	43	(29), III
10,000–99,999	9	(5), III	9	(9), III	46	(16), III	35	(16), III	25	(18), III	29	(64), III
≥100,000	12	(2), III	21	(3), III	16	(3), III	13	(5), III	4	(1), III	14	(14), III
	9	(8), III	12	(16), III	36	(29), III	49	(32), III	22	(22), III	25	(107), III

^{a}The number in parentheses is the number of patients in the given subgroup. (I) Represents a "good" prognosis (MST ≥ 1½ yr). (II) Represents an "average" prognosis (1 yr < MST < 1½ yr). (III) Represents a "poor" prognosis (MTS ≤ 1 yr).

Source: George et al. (1973). Reprinted by permission of the editor.

These time periods are arbitrary but can be considered as guides to the relative prognosis of patients in the various subgroups.

It is obvious that any grouping of patients using three or more variables would require a large number of patients to assign a reasonable number to each subgroup. Other limitations are that there is no indication of the relative importance of the variables, and weights are arbitrary. Thus the univariate approach can only provide a preliminary idea to which variable has prognostic importance. The simultaneous effects have to be investigated by multivariate techniques. In the following section we introduce Cox's regression model for survival data.

10.2.2 Multivariate Analysis: Cox's Proportional Hazards Model for Survival Data

The multiple-regression method is a conventional technique for investigating the relationship between survival time and possible prognostic variables. Let $x_1, x_2, \ldots, x_p$ be the p possible prognostic variables (covariates or explanatory variables). For the ith patient, observed values of the p variables are $x_{1i}, x_{2i}, \ldots, x_{pi}$. In the multiple-regression approach, the survival time of the ith patient, t_i, is the independent variable. We are interested in identifying a relationship of t_i or a function of t_i, say $w(t_i)$ and $(x_{1i}, x_{2i}, \ldots, x_{pi})$, that can be expressed in a regression function,

$$t_i = f_1(x_{1i}, x_{2i}, \ldots, x_{pi}) = f_1(\mathbf{x}_i)$$

or

$$w(t_i) = f_2(x_{1i}, x_{2i}, \ldots, x_{pi}) = f_2(\mathbf{x}_i)$$

Two examples of f_1 and f_2 are

$$f_1(\mathbf{x}_i) = \beta_1 x_{1i} + \beta_2 x_{2i} + \cdots + \beta_p x_{pi}$$
$$f_2(\mathbf{x}_i) = \exp(\beta_1 x_{1i} + \beta_2 x_{2i} + \cdots + \beta_p x_{pi})$$

Regression models proposed for survival distributions generally involve the assumption of proportional hazard functions (Lehmann 1953). A proportional hazards model possesses the property that different individuals have hazard functions that are proportional to one another, that is, $h(t \mid \mathbf{x}_1)/h(t \mid \mathbf{x}_2)$, the ratio of the hazard functions for two individuals with covariates $\mathbf{x}_1 = (x_{11}, x_{21}, \ldots, x_{p1})$ and $\mathbf{x}_2 = (x_{12}, x_{22}, \ldots, x_{p2})$, does not vary with time t. This implies that the hazard function given a set of covariates $\mathbf{x} = (x_1, x_2, \ldots, x_p)$ can be written as

$$h(t \mid \mathbf{x}) = h_0(t)g(\mathbf{x})$$

where $g(\mathbf{x})$ is a function of $\mathbf{x}$ and $h_0(t)$ can be considered as a baseline hazard function of an individual for whom $g(x) = 1$. One such model introduced by Cox (1972) is a general nonparametric model appropriate for the analysis of survival data with and without censoring. The model uses the hazard function as the dependent variables.

When survival times are continuously distributed and the possibility of ties can be ignored, the hazard is

$$h(t|\mathbf{x}) = h_0(t)\exp(\beta_1 x_1 + \beta_2 x_2 + \cdots + \beta_p x_p)$$

$$= h_0(t)\exp\left(\sum_{j=1}^{p} \beta_j x_j\right) \tag{10.1}$$

where $h_0(t)$ is the hazard function of the underlying survival distribution (arbitrary) when all the x variables are ignored, that is, all x's equal zero, and the β's are regression coefficients. In fact, $\exp(\sum_{j=1}^{p} \beta_j x_j)$ can be replaced by any known function of x_j's and β's. It is clear that Cox's model assumes that the hazard of the study group is proportional to that of the underlying survival distribution $h_0(t)$. It can be shown that (10.1) is equivalent to

$$S(t) = [S_0(t)]^{\exp\left(\sum_{j=1}^{p} \beta_j x_j\right)} \tag{10.2}$$

The use of (10.1) can be exemplified as follows.

1. *Two-Sample Problems.* Suppose that $p = 1$, that is, there is only one x variable, x_1, which is an indicator variable,

$$x_{1i} = \begin{cases} 0 & \text{if the } i\text{th individual is from sample 0} \\ 1 & \text{if the } i\text{th individual is from sample 1} \end{cases}$$

Then according to (10.1), the hazard functions of samples 0 and 1 are, respectively, $h_0(t)$ and $h_1(t) = h_0(t)\exp(\beta_1)$. The hazard function of sample 1 is equal to the hazard function of sample 0 multiplied by a constant $\exp(\beta_1)$, or the two hazard functions are proportional. In terms of the survivorship function,

$$S_1(t) = [S_0(t)]^c$$

where the constant $c = \exp(\beta_1)$ (Nadas 1970). The two-sample test developed from (10.1) is the Cox–Mantel test discussed in Chapter 5. It is now apparent that the test is based on the assumption of proportional hazard between the two samples.

2. *Two-Sample Problems with Covariates.* The x variables in (10.1) can either be indicator variables such as x_1 in the two-sample problem above or

concomitant variables. Having one or more x variables representing concomitant variables in (10.1) enables us to examine the relation between two samples adjusting for the presence of concomitant variables.

3. *Two-Sample Problems with Time-dependent Covariate.* In (10.1) one or more x variables can be functions of time. For example, suppose that, in addition to x_1 above, a time-dependent variable $x_2 = tx_1$ introduced. According to (10.1), the hazard in sample 1 is

$$h_1(t) = h_0(t)\exp(\beta_1 + \beta_2 t)$$
$$= ch_0(t)\exp(\beta_2 t)$$

and that in sample 0 remains $h_0(t)$.

4. *Regression Problems.* Dividing both sides of (10.1) by $h_0(t)$ and taking its logarithm, we obtain

$$\log_e \frac{h_i(t)}{h_0(t)} = \beta_1 x_{1i} + \beta_2 x_{2i} + \cdots + \beta_p x_{pi} = \sum_{j=1}^{p} \beta_j x_{ji} \qquad (10.3)$$

The left-hand side of (10.3) is a function of the hazard (or relative risk) for the ith patient, and the right-hand side is a linear combination of the concomitant variables $x_{1i}, \ldots, x_{pi}$ with coefficients $\beta_1, \ldots, \beta_p$, respectively. The x's can be indicator variables, covariates, and time-dependent covariates. If we let $y_i = \log_e[h_i(t)/h_0(t)]$, (10.2) is simply a standard multiple-regression equation with the concomitant variables as independent variables and a function of the hazard as the dependent variable.

In this section we focus our attention on the use of (10.1) in regression problems. Readers interested in its use in two-sample problems with covariates are referred to the papers cited in the Bibliographical Remarks at the end of the chapter.

Our main interest here is to identify important prognostic factors. In other words, we wish to identify from the p independent variables a subset of variables that relate significantly to the hazard, and consequently, the length of survival, of the patient. We are concerned with the regression coefficients. If β_i is zero, then the corresponding independent variable is not related to survival when adjustment is made for the other independent variables. Recall that in a standard multiple-regression problem a ranking of the variables in order of relative importance can be achieved by using a stepwise regression method. From the ranking and the significance test for each variable, we can select the most significant variables related to the dependent variable. Because of the close analogy between the standard multiple regression and (10.3), stepwise regression can also be applied to (10.3).

In addition to identifying prognostic factors, Cox's regression model can also define a prognostic index or ratio, namely, $\log_e[h_i(t)/h_0(t)]$, for each patient. This index can be used to compare two treatment groups as well as prognosis between patients. As mentioned earlier, $h_0(t)$ is the hazard function when all of the independent variables are ignored. If the independent variables are standardized about the mean and the model used is

$$\log_e \frac{h_i(t)}{h_0(t)} = \beta_1(x_1 - \bar{x}_1) + \cdots + \beta_p(x_p - \bar{x}_p) \tag{10.4}$$

where $\bar{x}_j$ is the average of the jth independent variable for all patients, then $h_0(t)$ is the hazard function when all variables are at their average values. The hazard index is the ratio of risk of failure for a patient with a given set of values $x_{1i}, \ldots, x_{pi}$ to that for an average patient who has an average value for every variable. Model (10.4) is more realistic and easier to interpret than model (10.3). This index or ratio can then be used to compare the relative risk for patients with different values of the independent variables.

To estimate the coefficients $\beta_1, \ldots, \beta_p$, Cox suggests a maximum likelihood procedure where the likelihood function is based on a conditional probability of failure. Suppose that $t_{(1)} < t_{(2)} < \cdots < t_{(k)}$ are the k exact failure times. Let $R(t_{(i)})$ be the risk set at time $t_{(i)}$. Here $R(t_{(i)})$ consists of all individuals whose survival times are at least $t_{(i)}$. For the particular failure at time $t_{(i)}$, conditionally on the risk set $R(t_{(i)})$, the probability that the failure is on the individual as observed is

$$\frac{\exp\left(\sum_{j=1}^{p} \beta_j x_{ji}\right)}{\sum_{l \in R(t_{(i)})} \exp\left(\sum_{j=1}^{p} \beta_j x_{jl}\right)} \tag{10.5}$$

Each failure contributes a factor and hence the required conditional log-likelihood is

$$LL(\beta) = \sum_{i=1}^{k} \sum_{j=1}^{p} \beta_j x_{ji} - \sum_{l=1}^{k} \log_e \left[\sum_{l \in R(t_{(i)})} \exp\left(\sum_{j=1}^{p} \beta_j x_{jl}\right) \right] \tag{10.6}$$

Maximum likelihood estimates of β's are obtained by solving simultaneously the p equations that are the derivatives of $LL(\beta)$ with respect to $\beta_1, \ldots, \beta_p$, respectively, equating to zero. The p equations are

$$U(\beta_1, \ldots, \beta_p) = \sum_{i=1}^{k} [x_{ui} - A_{ui}(\beta_1, \ldots, \beta_p)] = 0 \qquad u = 1, \ldots, p \tag{10.7}$$

where

$$A_{ui}(\beta_1, \ldots, \beta_p) = \frac{\sum\limits_{l \in R(t_{(i)})} x_{ul} \exp\left(\sum\limits_{j=1}^{p} \beta_j x_{jl}\right)}{\sum\limits_{l \in R(t_{(i)})} \exp\left(\sum\limits_{j=1}^{p} \beta_j x_{jl}\right)} \tag{10.8}$$

The p equations in (10.7) can be solved numerically by the Newton–Raphson method of iteration. In the Newton–Raphson method, the estimates of $\beta_1, \ldots, \beta_p$ are obtained by iterative use of (10.7) and the second derivative of (10.6):

$$I_{uv}(\beta_1, \ldots, \beta_p) = -\sum_{i=1}^{k} C_{(uvi)}(\beta_i, \ldots, \beta_p) \tag{10.9}$$

where

$$C_{(uvi)}(\beta_1, \ldots, \beta_p) = \frac{\sum\limits_{l \in R(t_{(i)})} x_{ul} x_{vl} \exp\left(\sum\limits_{j=1}^{p} \beta_j x_{jl}\right)}{\sum\limits_{l \in R(t_{(i)})} \exp\left(\sum\limits_{j=1}^{p} \beta_j x_{jl}\right)}$$
$$- A_{ui}(\beta_1, \ldots, \beta_p) A_{vi}(\beta_1, \ldots, \beta_p) \tag{10.10}$$

However, (10.1) assumes continuous survival time, which may not be practical since in practice survival times often involve ties. To cover this possibility, Cox generalizes (10.1) to discrete time by a logistic transformation,

$$\frac{h(t)\, dt}{1 - h(t)\, dt} = \frac{h_0(t)\, dt}{1 - h_0(t)\, dt} \exp\left(\sum_{j=1}^{p} \beta_j x_{ji}\right)$$

This model reduces to (10.1) in the continuous case.

Suppose that among the survival times $t_1, \ldots, t_n$, there are k distinct times. Let $t_{(1)} < \cdots < t_{(k)}$ be the k distinct failure times (uncensored observations). Let $m_{(i)}$ be the multiplicity of $t_{(i)}$, $m_{(i)} > 1$, if there is more than one observation with value $t_{(i)}$, $m_{(i)} = 1$, if there is only one observation with value $t_{(i)}$. Let $R(t_{(i)})$ denote the set of individuals at risk at time $t_{(i)}$. Here $R(t_{(i)})$ consists of those individuals whose survival times are at least $t_{(i)}$. Let $r_{(i)}$ be the number of such individuals. At time $t_{(i)}$, the probability that the individual fails as observed conditionally on the risk set $R(t_{(i)})$, from (10.1), is

$$\frac{\exp(\beta_1 z_{1i} + \beta_2 z_{2i} + \cdots + \beta_p z_{pi})}{\sum\limits_{l \in R(t_{(i)})} \exp(\beta_1 z_{1l} + \beta_2 z_{2l} + \cdots + \beta_p z_{pl})}$$

where z_{1i} is the sum of x_{1i}'s over the $m_{(i)}$ individuals failing at $t_{(i)}$, z_{2i} is the sum of x_{2i}'s over the $m_{(i)}$ individuals failing at $t_{(i)}$, and so on. The conditional log-likelihood function is then

$$LL(\beta) = \sum_{i=1}^{k} (\beta_1 z_{1i} + \cdots + \beta_p z_{pi})$$

$$- \sum_{i=1}^{k} \log \left(\sum_{l \in R(t_{(i)})} \exp(\beta_1 z_{1l} + \cdots + \beta_p z_{pl}) \right) \qquad (10.11)$$

The functions U and I in (10.7) and (10.9) become

$$U(\beta_1, \ldots, \beta_p) = \sum_{i=1}^{k} [z_{ui} - m_{(i)} A_{ui}] = 0 \qquad u = 1, \ldots, p \qquad (10.12)$$

and

$$I_{uv}(\beta_1, \ldots, \beta_p) = \frac{m_{(i)}[r_{(i)} - m_{(i)}]}{r_{(i)} - 1} C_{(cvi)}(\beta_1, \ldots, \beta_p) \qquad (10.13)$$

Standard errors of the estimates of β_i's can be estimated from (10.9) and (10.13). The $100(1 - \alpha)$ percent confidence interval for β_i is

$$\hat{\beta}_i \pm Z_{\alpha/2}(\text{estimated SE of } \hat{\beta}_i)$$

For a dichotomous variable, Cox's proportional hazards model can be used to estimate relative risk when adjustments are made for the other variables in the model. For example, if x_1 represents hypertension and is defined as

$$x_1 = \begin{cases} 1 & \text{if patient is hypertensive} \\ 0 & \text{otherwise} \end{cases}$$

then the hazard rate for hypertensive patients is $\exp(\hat{\beta}_1)$ times higher than for normotensive patients. That is, the related risk associated with hypertension is $\exp(\hat{\beta}_1)$. A $100(1 - \alpha)$ percent confidence interval for the relative risk can be obtained by using the confidence interval for β. Let (β_{1L}, β_{1U}) be the $100(1 - \alpha)$ percent confidence interval for β_1; a $100(1 - \alpha)$ percent confidence interval for the relative risk is $(\exp(\beta_{1L}), \exp(\beta_{1U}))$. This application of the proportional hazards model has been used extensively, particularly by epidemiologists.

In estimating $\beta_1, \ldots, \beta_p$, a stepwise procedure may be used to rank the variable. In a forward stepwise (or step-up) procedure, the independent variables are entered in the regression equation one at a time until the regression is satisfactory. The order of insertion is determined by using, for example, the maximum log-likelihood value, $LL(\hat{\beta})$, as a measure of the importance of variables not yet in the regression equation. Using the maximum log-likelihood value as a measure, it selects, as the first variable to enter the regression equation, the variable, say $x_{(1)}$, whose maximum log-likelihood is the largest. Let $LL(\hat{\beta}_i)$, $i = 1, \ldots, p$, be the maximum log-likelihood value obtained from fitting only the ith prognostic variable. Then $x_{(1)}$ is the first variable to enter the regression if

$$LL(\hat{\beta}_{(1)}) = \max_i [LL(\hat{\beta}_i)]$$

Now there are $p - 1$ prognostic variables not yet fitted. The maximum log-likelihood value $LL(\hat{\beta}_{(1)}, \hat{\beta}_{(i)})$ is computed for each of the $p - 1$ independent variables and the one that gives the largest log-likelihood value is the next variable to enter the regression equation. The procedure continues to fit one additional independent variable at a time until the regression is satisfactory. At every step a likelihood ratio test is performed to determine if the last variable entered adds significantly to the variables already selected.

At the first step, there is only one variable in the regression equation, that is,

$$\log_e \frac{h(t)}{h_0(t)} = \hat{\beta}_{(1)} x_{(1)}$$

where $x_{(1)}$, the most important single variable related to hazard, could be any one of $x_1, \ldots, x_p$. To test the significance of $x_{(1)}$, we test the hypothesis $H_0: \beta_{(1)} = 0$. For this we treat

$$\chi^2 = \frac{[U(\beta_{(1)})]^2}{I(\beta_{(1)})} \tag{10.14}$$

where U and I are given in (10.7) and (10.9), respectively, as chi-square distributed with one degree of freedom, or

$$Z = \frac{U(\beta_{(1)})}{\sqrt{I(\beta_{(1)})}} \tag{10.15}$$

as normally distributed with mean 0 and variance 1. A χ^2 (or Z) value larger than the 100α percentage point of the chi-square distribution with one degree of freedom (or the standard normal distribution) indicates that $x_{(1)}$ is significantly related to survival at the α level.

At subsequent steps, a likelihood ratio test is performed as follows: Let $LL(\beta_{(1)}, \ldots, \beta_{(k)})$ be the log-likelihood value in (10.6) at the kth step after k variables have been fitted. The significance of the kth variable is tested by considering

$$\chi^2 = -2[LL(\beta_{(1)}, \ldots, \beta_{(k-1)}) - LL(\beta_{(1)}, \ldots, \beta_{(k)})] \tag{10.16}$$

chi-square distributed with one degree of freedom. A χ^2 value exceeding the 100α percentage point of the chi-square distribution with one degree of freedom indicates that the kth variable entering the regression is significant at the α level.

In this procedure the first variable selected is the most important single variable in predicting survival time, the second variable entered is the second most important, and so on. The process thus provides a successive selection and ranking of the independent variables according to their relative importance.

The forward selection procedure is only one of the possible variable selection schemes that would result in a ranking of the independent variables. Others are the backward elimination (or stepdown) procedure and the stepwise procedure. In the backward procedure a regression on all p variables is done first and then variables are deleted from the regression equation one at a time until the regression is satisfactory. Advantages of the backward procedure have been discussed by Mantel (1970). The stepwise procedure allows a variable that has entered the regression to be removed at a later step if it is found to be no longer important.

The following two examples illustrate the use of Cox's regression model.

Example 10.2

Table 10.3 gives survival data from 30 patients with AML. Two possible prognostic factors are considered:

$$x_1 = \begin{cases} 1 & \text{if patient} \geq 50 \text{ years old} \\ 0 & \text{otherwise} \end{cases}$$

$$x_2 = \begin{cases} 1 & \text{if cellularity of marrow clot section is } 100\% \\ 0 & \text{otherwise} \end{cases}$$

Results of a Cox's regression analysis of this set of data is presented in Table 10.4. The positive signs of the regression coefficients indicate that the older patients (≥ 50 years) and patients with 100% cellularity of the marrow clot section have higher risk of dying. Furthermore, age is significantly related to survival after adjustment for cellularity. The coefficients of the binary covariates can be interpreted in terms of relative risk. The estimated risk of dying for patients at least 50 years of age is 2.75 times higher than that for patients less than 50. Patients with 100% cellularity have slightly higher risk (1.42) of dying than patients with less than 100% cellularity.

These estimates of regression coefficients and relative risks are point estimates. If the standard errors of the coefficients ($\hat{\beta}$) are not very small, these point estimates may not be accurate and interval estimates should be calculated. For example, the 95% confidence intervals for β_1 (age) and β_2 (cellularity) are $1.01 \pm 1.96 \, (0.46)$ or $(0.11, 1.91)$ and $0.35 \pm 1.96 \, (0.44)$ or $(-0.51, 1.21)$, respectively. Consequently, the 95% confidence intervals for the relative risks are $(e^{0.11}, e^{1.91})$ or $(1.12, 6.75)$ and $(e^{-0.51}, e^{1.21})$ or $(0.60,$

Table 10.3   Survival Times and Data of Two Possible Prognostic Factors of 30 AML Patients

Survival Time	x_1	x_2	Survival Time	x_1	x_2
18	0	0	8	1	0
9	0	1	2	1	1
28+	0	0	26+	1	0
31	0	1	10	1	1
39+	0	1	4	1	0
19+	0	1	3	1	0
45+	0	1	4	1	0
6	0	1	18	1	1
8	0	1	8	1	1
15	0	1	3	1	1
23	0	0	14	1	1
28+	0	0	3	1	0
7	0	1	13	1	1
12	1	0	13	1	1
9	1	0	35+	1	0

Table 10.4   Results of a Proportional Hazards Regression Analysis of Data in Table 10.3

Covariate	Regression Coefficient	Standard Error	P Value	exp(coefficient)
x_1 (age)	1.01	0.46	0.013	2.75
x_2 (cellularity)	0.35	0.44	0.212	1.42

3.35). The small number of patients (30) may have contributed to the wide confidence intervals. The lower bound of the confidence interval for age is only slightly above 1. This suggests that the importance of age should be interpreted carefully. In general, if the number of subjects is small and the standard errors of the estimates are large, the estimates may be unreliable.

When all of the covariates are considered simultaneously, the relative risk for a patient with $X_1 = x_1$ and $X_2 = x_2$ is calculated relative to patients with $x_1 = 0$ and $x_2 = 0$. Thus, the relative risk is $\exp(1.01 + 0.35) = 3.90$ for a patient who is over 50 years of age and whose cellularity is 100%, compared to patients who are less than 50 years and whose cellularity is less than 100%.

Example 10.3

In a study (Buzdar et al. 1978) to evaluate a combination of 5-fluorouracil, adriamycin, cyclophosphamide, and BCG (FAC–BCG) as adjuvant treatment in Stage II and III breast cancer patients with positive axillary nodes,

131 patients receiving FAC–BCG after surgery and radiation therapy were compared with 151 patients receiving surgery and radiation therapy only (control group).

Cox's regression model was used to identify prognostic factors and to evaluate the comparability of the two treatment groups. The model was fitted to the data from 151 patients to determine the variable related to length of remission. The possible prognostic variable considered were age (years), menopausal status (1, premenopausal; 0, other), size of primary tumor (2, <3 cm; 4, 3–5 cm; 7, >5 cm), state of disease (2, II; 3, III), location of surgery (1, M.D. Anderson Hospital; 0, other), number of involved nodes (2, <4; 7, 4–10; 12, >10), and race (1, Caucasian; 2, other). The model was fitted in forward stepwise fashion. Three variables, number of nodes involved, state of disease, and menopausal status, were significantly related to the disease-free time. Values of the maximum log-likelihood were used to determine the significance; that is, χ^2 in (10.16) was computed at successive steps and compared with the $\chi^2_{1,\alpha}$ value. The regression equation including these three variables only is

$$\log_e \frac{h_i(t)}{h_0(t)} = 0.1110(\text{number of nodes} - 6.16) + 0.8122(\text{stage} - 2.39)$$

$$+ 0.8720(\text{menopausal} - 0.26)$$

Table 10.5 gives the details of the fit. Relative risk was taken as $h_i(t)/h_0(t)$, the ratio of the risk of death per unit time for a patient with a given set of prognostic variables to the risk for a patient whose prognostic variables were at their average values. The relative risk for each variable was calculated by considering favorable or unfavorable values of that variable, assuming that other variables were at their average value. Note that the risk of relapse per unit time for a patient with 12 positive nodes is 3.04 (ratio of risk) times that for a patient with only two positive nodes. The risk of relapse per unit time for a Stage III patient was 3.25 times that of a Stage II patient.

Table 10.5 Patient Characteristics Related to Disease-free Time in Cox's Regression Model Fit to Control Patients

Prognostic Variable	Regression Coefficient	Significance Level (P)	Maximum Log likelihood	Relative Risk[a]		Ratio of Risks
				Favorable	Unfavorable	
Number of nodes	0.1110	<0.01	−257.407	0.63	1.91	3.04
Stage	0.8122	0.016	−254.533	0.73	1.64	3.25
Menopausal status	0.8720	<0.1	−250.576	0.80	1.91	2.39

[a]Favorable variables: number of nodes = 2, Stage II, postmenopausal. Unfavorable variables: number of nodes = 12, Stage III, premenopausal.

Source: Buzdar et al. (1978). Reprinted by permission of the editor.

Table 10.6   Hazard Ratio of Control and FAC–BCG Groups

Hazard Ratio	Control	FAC–BCG	P Value Control versus FAC–BCG
≥2	25 (17%)	29 (22%)	
1–1.9	47 (31%)	32 (24%)	
0.5–0.9	53 (35%)	49 (37%)	0.49
<0.5	26 (17%)	21 (16%)	

Source: Buzdar et al. (1978). Reprinted by permission of the editor.

The regression equation was calculated for the 151 control patients and the 131 FAC–BCG patients. The median hazard ratios [i.e., $h_i(t)/h_0(t)$] was 0.875 for both groups, and the distribution of hazard ratios given in Table 10.6 also indicates no real difference between groups. Hence the authors concluded that the distribution of number of positive nodes, stage of disease, and menopausal status was such that the overall risk of relapse in the two groups was comparable.

Among other things, the authors fitted Cox's regression model to the combined group of FAC–BCG and control patients, including as variables related to survival type of treatment (0, control; 1, FAC–BCG), menopausal status, size of primary tumor, and number of involved nodes. The regression equation with three significant variables obtained was as follows:

$$\log_e \frac{h_i(t)}{h_0(t)} = -1.8792(\text{treatment} - 0.47) + 0.9644(\text{menopausal status} - 0.33)$$

$$+ 0.1611(\text{size of primary tumor} - 4.04)$$

Table 10.7 gives the details of the fit. The most important variable in predicting survival time was the type of treatment (FAC–BCG favorable); other significantly important variables were menopausal status and size of primary tumor. The risk of death per unit time for a patient receiving no adjuvant treatment (control group) was 6.55 times that for a patient

Table 10.7   Patient Characteristics Related to Survival, Treatment Included

Prognostic Variable	Regression Coefficient	Significance Level (P)	Maximized Log-likelihood	Relative Risks[a] Favorable	Unfavorable	Ratio of Risks
Treatment	−1.8792	<0.01	−201.200	0.37	2.42	6.55
Menopausal Status	0.9644	0.01	−197.719	0.73	1.91	2.62
Size of primary Tumor	0.1611	0.05	−195.865	0.72	1.61	2.24

[a]Favorable variables: treatment—FAC–BCG, postmenopausal, size of primary tumor 2 cm. Unfavorable variables: no adjuvant treatment, premenopausal, size of primary tumor 7 cm.

Source: Buzdar et al. (1978). Reprinted by permission of the editor.

receiving the treatment, showing that FAC–BCG can prolong life consid-erably.

Based on the relationship between the hazard function and the survivor-ship function (Chapter 2), after estimates of the regression coefficients are obtained, the survivorship function $S(t)$ in (10.2) can be estimated. Details of the estimation procedure are not discussed here. Interested readers are referred to Kalbfleisch and Prentice (1980).

The proportional hazard model (10.1) assumes that the hazard rate of an individual with prognostic variables $\mathbf{x}$ is a constant multiple, $\exp(\Sigma \, \beta_j x_j)$, of the baseline hazard rate at all time. Though this assumption may be met in many situations, it is not reasonable in others. For example, in Example 10.2, information may not be complete for one to assume that the death rate for patients at least 50 years of age is a constant multiple of that for patients less that 50 years. To accommodate this nonproportional case, Cox's model can be generalized using the concept of stratification (Kalbfleisch and Prentice 1980). The data can be stratified by age (e.g., 1, ≥ 50 years; 2, other) and the model in (10.1) is modified as follows:

$$h_i(t \mid x) = h_{0i}(t)\exp\left(\sum \beta_j x_j \right) \qquad (10.17)$$

where $i = 1, 2$ for the two age strata. Notice that the underlying hazard function $h_{0i}(t)$ may be different for each stratum; however, the regression coefficients are the same for all strata. This is the stratified version of Cox's proportional hazards model for the case where patients in different strata (or levels) of a variable may not have proportional hazard rates, but within each stratum, the proportional hazards assumption still applies.

The regression coefficients can be estimated by using a marginal likeli-hood function with the Newton–Raphson method. Because the model is stratified by age and no specific relationship is assumed between the hazard rates for patients at least 50 years and others, tests of significance of the regression coefficients for the other variables are adjusted for age. Survivor-ship functions for the two age groups can be estimated [using (10.2)] at specific values of the other variables and plotted. Furthermore, to examine if the assumption of proportional hazards is met, the two survivorship functions estimated at given values of the covariates can be plotted on a $\log_e(-\log_e)$ scale. If the two curves are parallel over time, then the hazard rates are proportional and the inclusion of the variable in the model would be appropriate. The following example illustrates the use of the stratified model.

Example 10.4
Consider the data given in Example 10.2. Suppose that we are not sure if the risk of dying for patients at least 50 years of age is proportional to that for patients less than 50 years. Two regression equations can be defined from (10.17):

$$h_1(t \mid x_2) = h_{01}(t)\exp(\beta_2 x_2)$$

and

$$h_2(t \mid x_2) = h_{02}(t)\exp(\beta_2 x_2)$$

where $h_1(t \mid x_2)$ is the hazard function for patients under 50 years of age, $h_2(t \mid x_2)$ is the hazard function for patients at least 50 years as a function of cellularity, and $h_{01}(t)$ and $h_{02}(t)$ are the underlying hazard functions for the two groups. The results of the stratified analysis are $\hat{\beta}_2 = 0.22$, $SE(\hat{\beta}_2) = 0.44$, $p = 0.31$, and $\exp(\hat{\beta}_2) = 1.24$, which are close to those obtained earlier in the unstratified model. However, this may not always be the case.

The estimation procedure also allows an estimate of the survival function for the two age groups. Figure 10.2 gives the two estimated survival curves

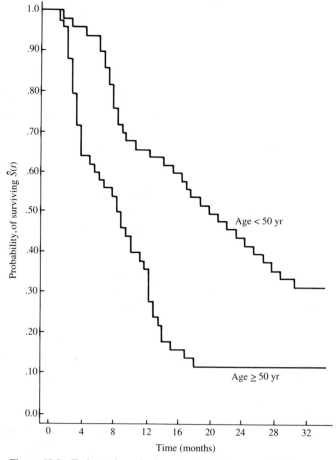

Figure 10.2   Estimated survival curves for patients stratified by age.

for patients with 100% cellularity. Similar survival curves can be obtained for patients with less than 100% cellularity.

In order to examine whether the proportional hazards model is appropriate for patients under 50 years and patients at least 50 years, the survival curves in Figure 10.2 can be plotted on a log–log scale (Figure 10.3). The two curves are roughly parallel. Therefore, it is appropriate to assume that the risks of dying for the two groups of patients are proportional over time.

Computer programs for Cox's model using the Breslow (1974) modification are available in BMDP (Dixon et al. 1988) and SAS (Harrell 1980a). The program in BMDP allows time-dependent variables and stratification including graphical presentation.

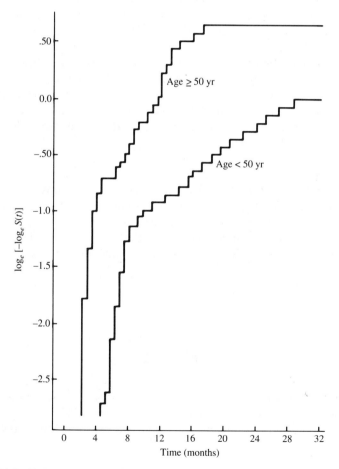

Figure 10.3 Estimated survival curves shown in Figure 10.2 plotted on a log–log scale.

10.3 PARAMETRIC REGRESSION METHODS

In recent years statisticians have become interested in developing mathematical models that include patient characteristics (covariates, or concomitant variables) for adjustments to predict and compare survival distributions. If no censoring is present, an obvious approach to adjusting for differences in concomitant variables is the least-squares linear regression and the conventional analysis of covariance. However, these methods require the assumption of a normal distribution. Although data can sometimes be transformed to achieve approximate normality, these methods are difficult to adapt for censored data. Therefore, regression models for distributions other than normal that are more easily adapted for censored data and more appropriate for survival distributions have been developed. In this section, three regression models for censored survival data are introduced for covariate adjustments and identifying prognostic factors. All three models assume that the underlying survival distribution is exponential.

Model I is the log-linear-exponential regression model of Glasser (1967). It was investigated by Prentice (1973) from the viewpoint of Fraser's (1968) structural inference and extended to the case of multiple convariates in a classical way by Breslow (1974). Model II is the linear-exponential model considered by Feigl and Zelen (1965) and extended to include censored data by Zippin and Armitage (1966). Model III is the linear-exponential model used by Byar et al. (1974).

Model I: Log-Linear Exponential

Suppose there are $n = n_1 + \cdots + n_k$ individuals in k groups. Let t_{ij} be the survival time and $x_{ij1}, x_{ij2}, \ldots, x_{ijp}$ the covariates of the jth individual in the ith group, where p is the number of covariates considered, $i = 1, \ldots, k$, and $j = 1, \ldots, n_i$. Model I assumes that the underlying distribution is exponential and the survivorship function for the jth individual in the ith group is

$$S_{ij}(t) = e^{-\lambda_{ij}t} \tag{10.18}$$

where

$$\lambda_{ij} = \exp\left(a_i + \sum_{l=1}^{p} b_l x_{ijl}\right) \tag{10.19}$$

The term $\exp(a_i)$ represents the "underlying hazard" of the ith group when covariates are ignored. To simplify the notation, we use the following indicator variables to distinguish censored observations from the uncensored:

$$\delta_{ij} = \begin{cases} 1 & \text{if } t_{ij} \text{ uncensored} \\ 0 & \text{if } t_{ij} \text{ censored} \end{cases}$$

The likelihood function for the data can then be written as

$$L(\lambda_{ij}) = \prod_{i=1}^{k} \prod_{j=1}^{n_i} (\lambda_{ij})^{\delta_{ij}} e^{-\lambda_{ij} t_{ij}}$$

Substituting (10.19) in the logarithm of the above function, we obtain the log-likelihood function of $\mathbf{a} = (a_1, \ldots, a_k)$ and $\mathbf{b} = (b_1, \ldots, b_p)$:

$$LL(\mathbf{a}, \mathbf{b}) = \sum_{i=1}^{k} \sum_{j=1}^{n_i} \left[\delta_{ij} \left(a_i + \sum_{l=1}^{p} b_l x_{ijl} \right) - t_{ij} \exp \left(a_i + \sum_{l=1}^{p} b_l x_{ijl} \right) \right]$$

$$= \sum_{i=1}^{k} \left[a_i r_i + \sum_{l=1}^{p} b_l s_{il} - e^{a_i} \sum_{j=1}^{n_i} t_{ij} \exp \left(\sum_{l=1}^{p} b_l x_{ijl} \right) \right]$$

$$(10.20)$$

where

$$s_{il} = \sum_{j=1}^{n_i} \delta_{ij} x_{ijl}$$

is the sum of the lth covariate corresponding to the uncensored survival times in the ith group and r_i is the number of uncensored times in that group.

Maximum likelihood estimates of a_i and b_l can be obtained by solving the following $k + p$ equations simultaneously:

$$r_i - e^{\hat{a}_i} \sum_{j=1}^{n_i} t_{ij} \exp \left(\sum_{l=1}^{p} \hat{b}_l x_{ijl} \right) = 0 \qquad i = 1, \ldots, k \qquad (10.21)$$

$$\sum_{i=1}^{k} \left[s_{il} - e^{\hat{a}_i} \sum_{j=1}^{n_i} t_{ij} x_{ijl} \exp \left(\sum_{l=1}^{p} \hat{b}_l x_{ijl} \right) \right] = 0 \qquad l = 1, \ldots, p \qquad (10.22)$$

This can be done by using the Newton–Raphson iterative procedure (see Appendix A). Let $\hat{\mathbf{a}}$ and $\hat{\mathbf{b}}$ be the MLEs of $\hat{\mathbf{a}}$ and $\hat{\mathbf{b}}$ so obtained. Substituting $\hat{\mathbf{a}}$ and $\hat{\mathbf{b}}$ in (10.20), $LL(\hat{\mathbf{a}}, \hat{\mathbf{b}})$ gives the overall maximum value of the likelihood. Substituting $\hat{\mathbf{b}} = \mathbf{0}$ (or $\hat{b}_l = 0$ for $l = 1, \ldots, p$) in (10.20), $LL(\hat{\mathbf{a}}, \mathbf{0})$ yields the unadjusted maximum likelihood value. The difference between $LL(\hat{\mathbf{a}}, \hat{\mathbf{b}})$ and $LL(\hat{\mathbf{a}}, \mathbf{0})$ may be used to test the overall significance of the covariates by considering

$$\chi^2 = -2[LL(\hat{\mathbf{a}}, \mathbf{0}) - LL(\hat{\mathbf{a}}, \hat{\mathbf{b}})] \qquad (10.23)$$

as chi-square distributed with p degrees of freedom. A χ^2 greater than the 100α percentage point of the chi-square distribution with p degrees of freedom indicates significant covariates. Thus fitting the model with subsets of the covariates $x_1, \ldots, x_p$ allows selection of significant covariates of prognostic variables. For example, to test the significance of x_2 after adjusting for x_1, compute

$$\chi^2 = -2[\mathrm{LL}(\hat{a}, \hat{b}_1) - \mathrm{LL}(\hat{a}, \hat{b}_1, \hat{b}_2)]$$

and consider it as chi square distributed with one degree of freedom. A significant χ^2 value indicates the importance of x_2. This can be done automatically by a stepwise procedure. In addition, if one or more of the covariates are for treatments, the equality of survival in specified treatment groups can be tested by comparing the resulting maximum log-likelihood values.

Having estimated the coefficients a_i and b_i, a survivorship function adjusted by covariates can then be estimated from (10.19) and (10.18).

The following example, adapted from Breslow (1974) illustrates how this model can identify important prognostic factors.

Example 10.5

Two hundred and sixty-eight children with newly diagnosed and previously untreated ALL were entered into a chemotherapy trial. After having successfully completed an induction course of chemotherapy designed to induce remission, the patients were randomized onto five maintenance regimens designed to maintain the remission as long as possible. Maintenance chemotherapy consisted of alternating eight-week cycles of 6-MP and methotrexate (MTX), to which actinomycin-D (A-D) or nitrogen mustard (NM) were added. The regimens are given in Table 10.8. Regimen 5 is the

Table 10.8   Summary Statistics for the Five Regimens

Regimen	Additive Therapy 6-MP Cycle	MTX Cycle	Number of Patients	Number in Remission	Geometric Mean[a] of WBC	Mean Age (yr)	Median Remission Duration
1	A-D	NM	46	20	9,000	4.61	510
2	A-D	A-D	52	18	12,308	5.25	409
3	NM	NM	64	18	15,014	5.70	307
4	NM	A-D	54	14	9,124	4.30	416
5	None	None	52	17	13,421	5.02	420
1, 2, 4	—	—	152	52	10,067	4.74	435
3, 5	—	—	116	35	14,280	5.40	340
ALL	—	—	268	87	11,711	5.02	412

[a] The geometric mean of $x_1, x_2, \ldots, x_n$ is defined as $(\Pi_{i=1}^{n} x_i)^{1/n}$. It gives a less biased measure of central tendency than the arithmetic mean when some observations are extremely large.

Source: Breslow (1974). Reproduced with permission of the Biometric Society.

control. Many investigators had a prior feeling that actinomycin-D was the active additive drug; therefore, pooled regimens 1, 2, and 4 (with actinomycin-D) were compared to regimens 3 and 5 (without actinomycin-D). Covariates considered were initial WBC and age at diagnosis. Analysis of variance showed that differences between the regimens with respect to these variables were not significant. Table 10.8 shows that the regimen with lowest (highest) WBC geometric mean has the longest (shortest) estimated remission duration. Figure 10.4 gives three remission curves by WBC; differences in duration were significant. It is well known that the initial WBC is an important prognostic factor for patients followed from diagnosis; however, it is interesting to know if this variable will continue to be important after the patient has successfully achieved remission.

To identify important prognostic variables, model (10.19) was used to analyze the effect of WBC and age at diagnosis. Previous studies (Pierce et al. 1969, George et al. 1973) showed that survival is longest for children in the middle age range (6–8 years), suggesting that both linear and quadratic terms in age be included. The WBC was transformed by taking the common logarithm. Thus, the number of covariates is $p = 3$. Let x_1, x_2, and x_3 denote log WBC, age, and age squared and b_1, b_2, and b_3 be the respective coefficients. Instead of using a stepwise fitting procedure, the model was fitted five times using different numbers of covariates. Table 10.9 gives the results. The regression coefficients were obtained by solving (10.21) and (10.22). Maximum log-likelihood values were calculated by substituting the regression coefficients obtained in (10.20). The χ^2 values were computed following (10.23), which show the effect of the covariates included. The first

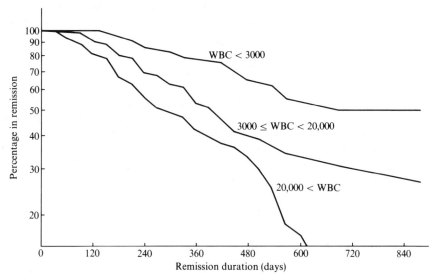

Figure 10.4 Remission curves of all patients by WBC at diagnosis. *Source*: Breslow (1974). Reproduced with permission of the Biometric Society.

Table 10.9 Regression Coefficients and Maximum Log-likelihood Values for Five Fits

Fit	Covariates Included	Maximum Log-likelihood	b_1	b_2	b_3	χ^2	df
			Regression Coefficient				
1	None	−1332.925					
2	x_1 (log WBC)	−1316.399	0.72			33.05	1
3	x_1, x_2 (age)	−1316.111	0.73	0.02		33.63	2
4	x_2, x_3 (age squared)	−1327.920		−0.24	0.018	10.01	2
5	x_1, x_2, x_3	−1314.065	0.67	−0.14	0.011	37.72	3

Source: Breslow (1974). Reproduced with permission of the Biometric Society.

fit did not include any covariates. The log-likelihood so obtained is the unadjusted value $LL(\hat{a}, 0)$ in (10.23). The second fit included only x_1 or log WBC, which yields a larger log-likelihood value than the first fit. Following (10.23), we obtain

$$\chi^2 = -2[LL(\hat{a}, 0) - LL(\hat{a}, \beta_1)]$$
$$= -2[-1332.925 + 1316.399]$$
$$= 33.05$$

with one degree of freedom. The highly significant ($p < 0.001$) χ^2 value indicates the importance of WBC. When age and age squared are included (fit 4) in the model, the χ^2 value, 10.01, is less than that of fit 2. This indicates that WBC is a better predictor than age as the only covariate. To test the significance of age effect after adjusting for WBC, we subtract the log-likelihood value of fit 2 from that of fit 5 and obtain

$$\chi^2 = -2(-1316.399 + 1314.065) = 4.668$$

with $3 - 1 = 2$ degrees of freedom. The significance of this χ^2 value is marginal ($p = 0.10$). Comparing the maximum log-likelihood value of fit 2 to that of fit 5, we find that log WBC accounts for the major portion of the total covariate effect. Thus, log WBC was identified as the most important prognostic variable. In addition, subtracting the maximum log-likelihood value of fit 5 from that of fit 3 yields

$$\chi^2 = -2(-1316.111 + 1314.065)$$
$$= 4.092$$

with one degree of freedom. This significant ($p < 0.05$) χ^2 value indicates that the age relationship is indeed a quadratic one with children 6–8 years old having the most favorable prognosis.

For a complete analysis of the data, the interested reader is referred to Breslow (1974).

Model II: The Linear Exponential Model of Feigl and Zelen

1. *Uncensored Data.* Let $t_1, \ldots, t_n$ be a sample of n survival times from patients whose individual survival distribution is exponential. Furthermore, let x_{ij}, $i = 1, \ldots, n$, $j = 1, \ldots, p$, be the observed value of the jth covariate such that the mean survival time of the ith patient is a linear function of the x_{ij}'s, that is,

$$\mu = 1/\lambda_i = b_0 + b_1 x_{i1} + \cdots + b_p x_{ip}$$

$$= \sum_{j=0}^{p} b_j x_{ij} \tag{10.24}$$

where $x_{i0} = 1$. The term b_0 represents the "underlying hazard" in the sense that $1/b_0$ is the hazard rate λ_i when covariates are ignored or all x_{ij}'s are zero. In an analogy to a regression problem, b_0 represents the intercept. Then the likelihood function of the n survival times can be written as

$$L(\mathbf{b}) = \prod_{i=1}^{n} f_i(t) = \prod_{i=1}^{n} \lambda_i e^{-\lambda_i t_i}$$

$$= \prod_{i=1}^{n} \left(\sum_{j=0}^{p} b_j x_{ij} \right)^{-1} \exp \left[-\sum_{i=1}^{n} t_i \left(\sum_{j=0}^{p} b_j x_{ij} \right)^{-1} \right] \tag{10.25}$$

having the log-likelihood function

$$LL(\mathbf{b}) = -\sum_{i=1}^{n} \log_e \left(\sum_{j=0}^{p} b_j x_{ij} \right) - \sum_{i=1}^{n} t_i \left(\sum_{j=0}^{p} b_j x_{ij} \right)^{-1} \tag{10.26}$$

The MLEs of $\mathbf{b} = (b_0, b_1, \ldots, b_p)$ can be found by solving simultaneously the $p + 1$ equations

$$-\sum_{i=1}^{n} \left(\sum_{j=0}^{p} \hat{b}_j x_{ij} \right)^{-1} + \sum_{i=1}^{n} t_i \left(\sum_{j=0}^{p} \hat{b}_j x_{ij} \right)^{-2} = 0$$

$$-\sum_{i=1}^{n} x_{ij} \left(\sum_{j=0}^{p} \hat{b}_j x_{ij} \right)^{-1} + \sum_{i=1}^{n} t_i x_{ij} \left(\sum_{j=0}^{p} \hat{b}_j x_{ij} \right)^{-2} = 0 \tag{10.27}$$

$$j = 1, \ldots, p$$

This can be done by an iterative procedure such as the Newton–Raphson method.

Having obtained estimates $\hat{b}_j$, $j = 0, 1, \ldots, p$, the log-likelihood function can then be used to test the significance of covariates. The procedure is exactly the same as for model I.

The survivorship function (for the ith patient) adjusted for the covariates can be obtained from

$$\hat{S}_i(t) = \exp(-\hat{\lambda}_i t) = \exp\left[-t\left(\sum_{j=0}^{p} \hat{b}_j x_{ij}\right)^{-1}\right] \tag{10.28}$$

A test of the adequacy of the model is provided by comparing the observed and expected number of patients dying in certain time intervals. Define $t_i(p)$ by

$$\hat{S}_i[t_i(p)] = \exp\left[-t_i(p)\left(\sum_{j=0}^{p} \hat{b}_j x_{ij}\right)^{-1}\right] = p \tag{10.29}$$

That is, the patient having the characteristics $x_i = (x_{i1}, \ldots, x_{ip})$ survives longer than $t_i(p)$ with probability p. We may choose p equal to the quartile probabilities; that is, $p = 0.25, 0.50, 0.75$. These in turn determine for the ith patient the four quartile intervals $[0, t_i(0.25)]$, $[t_i(0.25), t_i(0.50)]$, $[t_i(0.50), t_i(0.75)]$ and $[t_i(0.75), \infty]$. The probability is 0.25 for the ith patient to die in any one of these intervals. Such intervals can be determined for each patient. If the model gives a good fit, we would expect the survival times to be distributed equally in the four intervals. The observed and expected numbers of patients can be compared using a goodness-of-fit test. Obviously one is not restricted to the quartile intervals but can take arbitrary intervals depending on the data.

A computer program written by Morabito and Marubini (1976) is suitable for fitting this model.

Example 10.6

Consider the example of patients with AML given by Feigl and Zelen (1965). The patients were classified into two groups according to the presence or absence of a morphologic characteristic of white cells. Patients termed AG positive were identified by the presence of Auer rods and/or significant granulature of the leukemic cells in the bone marrow at diagnosis. These factors were absent for the AG-negative patients. Table 10.10 gives the survival time t_i in weeks from the date of diagnosis and the WBC at the time of diagnosis. Usually, the higher the WBC, the lower the probability of surviving a specified length of time. Here, the logarithm of the WBC is used as the possible prognostic variable. The MLEs for the two sets of data are summarized in Table 10.11. Computations of the log-likelihood values $LL(\hat{b}_0, \hat{b}_1)$ and $LL(\hat{b}_0, 0)$ and tests of significance of log WBC are left to the reader as an exercise.

A test of goodness-of-fit is given in Table 10.12, which gives the expected and observed numbers of patients in the quartile intervals. The chi-square values $\chi^2 = \Sigma (O - E)^2/E$ with two degrees of freedom are clearly not significant ($p = 0.85, 0.17$). Hence, the model provides a good fit to the

Table 10.10   Observed Survival Times and White Blood Counts

AG Positive $N = 17$		AG Negative $N = 16$	
White Blood Count	Survival Time (weeks)	White Blood Count	Survival Time (weeks)
2300	65	4400	56
750	156	3000	65
4300	100	4000	17
2600	134	1500	7
6000	16	9000	16
10,500	108	5300	22
10,000	121	10,000	3
17,000	4	19,000	4
5400	39	27,000	2
7000	143	28,000	3
9400	56	31,000	8
32,000	26	26,000	4
35,000	22	21,000	3
100,000	1	79,000	30
100,000	1	100,000	4
52,000	5	100,000	43
100,000	65		
Median 10,000 values	56	23,000	7.5

Source: Feigl and Zelen (1965). Reproduced with permission of the Biometric Society.

Table 10.11   Maximum Likelihood Estimates for the Model[a]

Patient Group	$\hat{b}_0$	$\hat{b}_1$	$\sqrt{\mathrm{Var}\,b_0}$	$\sqrt{\mathrm{Var}\,b_1}$	$\mathrm{Cov}(b_0, b_1)$
AG positive	240	-44	95.5	20.1	-1914
AG negative	30	-3	35.1	8.2	-284

[a]Variances and covariances are obtained by taking the second derivative of the log-likelihood function: $\mathrm{Var}\,\hat{b}_j = -E(\partial^2 \mathrm{LL}/\partial b_j^2)$, $j = 0, 1$, $\mathrm{Cov}(b_0, b_1) = E(\partial^2 \mathrm{LL}/\partial b_0 b_1)$, where $E(\ )$ denotes the expected value.

Source: Feigl and Zelen (1965). Reproduced with permission of the Biometric Society.

Table 10.12   Goodness-of-Fit Test of the Model

Group	Number of Patients	Quartile Interval				Total
		$[0, t(0.25)]$	$[t(0.25), t(0.50)]$	$[t(0.50), t(0.75)]$	$[t(0.75), \infty]$	
AG positive,						
$\chi^2 = 0.317\,(p = 0.85)$ Observed		5	4	3	5	17
Expected		4.25	4.25	4.25	4.25	17
AG negative,						
$\chi^2 = 3.50\,(p = 0.17)$ Observed		4	3	2	7	16
Expected		4	4	4	4	16

Source: Feigl and Zelen (1965). Reproduced with permission of the Biometric Society.

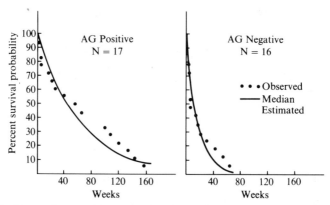

Figure 10.5 Observed and estimated survival distributions. *Source*: Feigl and Zelen (1965). Reproduced with permission of the Biometric Society.

data. Another check is on the empirical survival distribution with the estimated survival curve adjusted for WBC. Figure 10.5 gives the empirical survival distribution and the estimated survival curve associated with the median white blood count for each group. There is a close agreement between the two.

Figure 10.6 gives the estimate of the survivorship function $\hat{S}_i(t)$ for various WBCs for the two groups. Note that the survival probabilities for the AG-positive group depend strongly on WBC, but not the AG-negative group. Thus, WBC is a poor predictor for patients in the AG-negative group.

One way to compare the survival of the two groups adjusted for WBC is to make an approximate significance test for the difference of the two mean survival times. Let (b_{01}, b_{11}) and (b_{02}, b_{12}) be the parameters of the AG-positive and AG-negative groups, respectively. Then the hypothesis is

$$H_0: (b_{01} + b_{11}x) = (b_{02} + b_{12}x) \quad \text{or} \quad (b_{01} - b_{02}) + x(b_{11} - b_{12}) = 0$$

An approximate normal test is to reject H_0 if

$$|d/s_d| \geq Z_{\alpha/2} \tag{10.30}$$

where

$$d = (\hat{b}_{01} - \hat{b}_{02}) + x(\hat{b}_{11} - \hat{b}_{12})$$
$$s_d^2 = \mathrm{Var}(\hat{b}_{01}) + \mathrm{Var}(\hat{b}_{02}) + 2x[\mathrm{Cov}(\hat{b}_{01}, \hat{b}_{11}) + \mathrm{Cov}(\hat{b}_{02}, \hat{b}_{12})]$$
$$+ x^2[\mathrm{Var}(\hat{b}_{11}) + \mathrm{Var}(\hat{b}_{12})]$$

An approximate choice of a fixed value for x is the median value for both groups taken together. The median WBC for the two groups is 10,500.

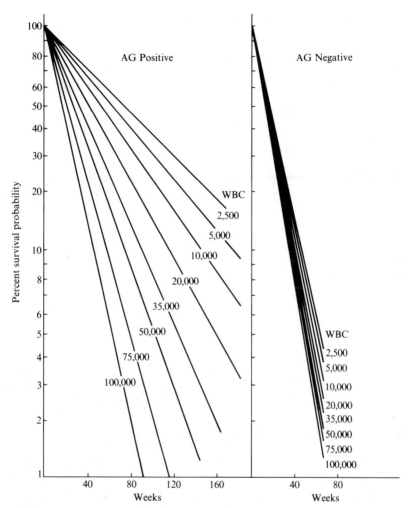

Figure 10.6 Estimated survival distributions for various WBCs. *Source*: Feigl and Zelen (1965). Reproduced with permission of the Biometric Society.

Substituting in (10.30) with $x = \log(10,500) = 4.02$, $d/s_d = 45.2/17.2 = 2.63$, which is significant at the 0.008 level. Hence, the survival experiences of the two groups adjusted for the WBC are significantly different.

2. *Censored Data.* Zippin and Armitage (1966) generalize the exponential model suggested by Feigl and Zelen to include censored observations. Suppose that n patients are entered on a study, r of these die, and $s = n - r$ are still alive at the end of the study. Let $t_1, \ldots, t_r$ be the exact survival times of the r deaths and $t_1^+, \ldots, t_s^+$ be the s censoring times. Then the likelihood function

$$L(\boldsymbol{\lambda}) = \prod_{i=1}^{r} \lambda_i e^{-\lambda_i t_i} \prod_{k=1}^{s} e^{-\lambda_k t_k^+}$$

under the linear-exponential model (10.24), becomes

$$L(\mathbf{b}) = \prod_{i=1}^{r} \left(\sum_{j=1}^{p} b_j x_{ij} \right)^{-1} \exp\left[-t_i \left(\sum_{j=1}^{p} b_j x_{ij} \right)^{-1} \right] \prod_{k=1}^{s} \exp\left[-t_k^+ \left(\sum_{j=1}^{p} b_j x_{kj} \right)^{-1} \right]$$
(10.31)

The log-likelihood is then

$$LL(\mathbf{b}) = -\sum_{i=1}^{r} \log_e \left(\sum_{j=1}^{p} b_j x_{ij} \right) - \sum_{i=1}^{r} t_i \left(\sum_{j=1}^{p} b_j x_{ij} \right)^{-1} - \sum_{k=1}^{s} t_k^+ \left(\sum_{j=1}^{p} b_j x_{kj} \right)^{-1}$$
(10.32)

The maximum likelihood estimators $\mathbf{b}$ may be obtained by solving simultaneously the $p + 1$ equations:

$$-\sum_{i=1}^{r} \left(\sum_{j=1}^{p} b_j x_{ij} \right)^{-1} + \sum_{i=1}^{r} t_i \left(\sum_{j=1}^{p} b_j x_{ij} \right)^{-2} + \sum_{k=1}^{s} t_k^+ \left(\sum_{j=1}^{p} b_j x_{ij} \right)^{-2} = 0$$

$$-\sum_{i=1}^{r} x_{ij} \left(\sum_{j=1}^{p} b_j x_{ij} \right)^{-1} + \sum_{i=1}^{r} t_i x_{ij} \left(\sum_{j=1}^{p} b_j x_{ij} \right)^{-2} + \sum_{k=1}^{s} t_k^+ x_{kj} \left(\sum_{j=1}^{p} b_j x_{kj} \right)^{-2} = 0$$

$$j = 1, \ldots, p \quad (10.33)$$

Again, this can be done by iterative procedures.

The exact survival times of the 17 AG-positive leukemia patients given by Feigl and Zelen are used by Zippin and Armitage to illustrate the use of the generalized model. Five of the 17 patients are selected at random to be "survivors" at the end of the study. Their exact survival times are used as censoring times. Table 10.13 shows the modified data. For these data, Morabito and Marubini (1976) obtain $\hat{b}_0 = 257.4$ and $\hat{b}_1 = -45.8$ by using an iterative procedure. (Zippin and Armitage obtain $\hat{b}_0 = 240$ and $\hat{b}_1 = -44$ following a response surface approach that is not discussed here.) The standard errors of $\hat{b}_0$ and $\hat{b}_1$ are 121.93 and 25.99, respectively, by Morabito and Marubini (104.2 and 23.5, respectively, by Zippin and Armitage).

Model III: The Linear Exponential Model of Byar et al.

Byar et al. (1974) developed another exponential model relating censored survival data to concomitant information for prostate cancer patients in which the individual hazard is linearly related to the possible prognostic variables.

Let $t_1, \ldots, t_n$ be the survival times of n individuals. Suppose that the ith individual under study has values of the p prognostic variables $x_{i0}, x_{i1}, \ldots, x_{ip}$. The first variable x_{i0} is a dummy variable and is set equal to 1 for all patients for notational symmetry (same as in model II), and for the other variables, $x_{ij} = 1$ if patient i has the characteristics j and $x_{ij} = 0$ otherwise, $j = 1, \ldots, p$. Supposing that the survival distribution is exponen-

Table 10.13 **White Blood Count and Survival Time of Leukemia Patients**

White Blood Count (thousands)	$\log_{10}$WBC (x)	Length of Survival t (weeks)
0.75	2.8751	156
2.3	3.3617	65
2.6	3.4150	13+
4.3	3.6335	100
5.4	3.7324	39
6.0	3.7782	16
7.0	3.8451	143
9.4	3.9731	34+
10.0	4.0000	121
10.5	4.0212	108
17.0	4.2304	4
32.0	4.5052	5+
35.0	4.5441	13+
52.0	4.7160	5
100.0	5.0000	1+
100.0	5.0000	1
100.0	5.0000	65

Data modified from Feigl and Zelen, 1965.

Source: Zippin and Armitage (1966). Reproduced with permission from the Biometric Society.

tial, Byar et al. suggests that the individual hazard λ_i is constant in time and linearly related to the possible prognostic variables, that is,

$$\lambda_i = b_0 x_{i0} + b_1 x_{i1} + \cdots + b_p x_{ip}$$

$$= \sum_{j=0}^{p} b_j x_{ij} \qquad (10.34)$$

Similar to the model of Feigl and Zelen, b_0 is the underlying hazard rate or force of mortality or the intercept.

Suppose that r of the n patients are dead and $s = n - r$ are still alive at the end of the study; then the likelihood function is

$$L = \prod_{i=1}^{r} \lambda_i e^{-\lambda_i t} \prod_{k=1}^{s} e^{-\lambda_k t_k^+}$$

$$= \prod_{i=1}^{r} \left(\sum_{j=0}^{p} b_j x_{ij} \right) \exp\left[-\left(\sum_{j=0}^{p} b_j x_{ij} \right) t_i \right] \prod_{k=1}^{s} \exp\left[-\left(\sum_{j=0}^{p} b_j x_{kj} \right) t_k^+ \right] \cdot$$

$$(10.35)$$

Taking the logarithms of (10.35), we obtain the log-likelihood function

$$LL(\mathbf{b}) = \sum_{i=1}^{r} \left[\log_e \left(\sum_{j=0}^{p} b_j x_{ij} \right) - \left(\sum_{j=0}^{p} b_j x_{ij} \right) t_i \right] - \sum_{k=1}^{s} \left(\sum_{j=0}^{p} b_j x_{kj} \right) t_k^+ \tag{10.36}$$

To obtain the MLEs of the b_j's, we need to solve simultaneously the following $p + 1$ equations:

$$\sum_{i=1}^{r} \left[\frac{x_{ij}}{\sum_{j=0}^{p} \hat{b}_j x_{ij}} - x_{ij} t_i \right] - \sum_{k=1}^{s} x_{kj} t_k^+ = 0 \qquad j = 0, 1, \ldots, p \tag{10.37}$$

These equations may be solved simultaneously by an iterative procedure. The following example illustrates the use of this model.

Example 10.7

Byar et al. (1974) used the model described above in a study of 1824 patients with advanced prostate cancer. The 11 variables along with the estimates of the regression coefficients b_j, $j = 0, 1, \ldots, 11$, and their standard deviations (SD) are given in Table 10.14. Each of the variables has a value of either 0 (without the characteristic) or 1 (with the characteristic). A test of significance of the variables can be performed in a similar way as in model I. However, some idea of the relative importance of the variables is illustrated by the magnitude of the regression coefficients since the

Table 10.14 Regression Coefficients and Their Standard Deviations

Number	Variable	$\hat{b}_j$[a]	Standard Deviation	$\hat{b}_{j/SD}$
0	Force of mortality	7.88	0.64	12.31
1	Pain due to cancer	−0.20	1.75	−0.11
2	Acid phosphatase[b] 1.1–2.0	4.15	1.40	2.96
3	Acid phosphatase[b] 2.1–5.0	6.94	1.93	3.60
4	Acid phosphatase[b] >5.0	11.04	2.09	5.28
5	Ureteral dilatation	4.68	2.31	2.03
6	Metastasis[c]	7.36	1.91	3.85
7	Partially bedridden	11.18	2.39	4.68
8	Totally bedridden	21.55	8.48	2.54
9	Weight <130 pounds	3.50	1.34	2.61
10	Hemoglobin <12 g/100 ml	5.44	1.49	3.65
11	Age ≥70 year	3.02	0.85	3.55

[a]Expressed in deaths per 1000 patient-months follow-up.
[b]Expressed in King Armstrong units.
[c]Osseous or soft-part metastasis.

Source: Byar et al. (1974).

covariates are dichotomous. For example, variable 8 (totally bedridden) appears to be the single most important variable. The prognosis for partially bedridden patients was better than that for totally bedridden patients $(\hat{b}_7 < \hat{b}_8)$. Also the importance of the acid phosphatase increases in the same order as the value of the underlying laboratory measurement. The standard deviations of the estimates were greatly affected by the number of patients in whom a variable was present. The underlying force of mortality, $\hat{b}_0$, had the smallest standard deviation since all patients contributed to its estimation, but variable 8 (totally bedridden) had a large standard deviation since very few patients were totally bedridden at the beginning of the study.

A predicted hazard rate can be calculated for each patient by simply adding together the regression coefficients corresponding to the characteristics present at diagnosis following (10.34). This is done for all the 1824 patients, and they are then ranked in order of increasing magnitude of the predicted hazard rate. The patients are then divided into five risk groups (Table 10.15). The first two (lowest risk) were natural groups in that the types of patients who belong in them are easily defined. Group 1 contains only patients with none of the variables present at diagnosis (271 patients) or with just variable 1, pain (10 patients). Risk group 2 is composed of four kinds of patients: 22 patients at least 70 years old with pain; 364 patients at least 70 years old without pain; three patients who weigh less than 130 pounds and have pain; and 43 patients whose only unfavorable characteristic is weight less than 130 pounds. Risk groups 3–5 are obtained by dividing the remaining patients into groups of roughly equal sizes. They contain patients with most of the characteristics present at diagnosis.

The observed actuarial survival rates for the five groups are plotted in Figure 10.7. The predicted survival curves (dotted lines) are obtained by considering each risk group as a mixture of simple exponential survival functions. In each group, each unique set of characteristics is weighted by the proportion of patients possessing them. Let m be the number of unique

Table 10.15 Predicted Hazard Rates for Five Risk Groups

Risk Group	Number of Patients	Number of Unique Sets of Characteristics Represented in Group	Predicted Hazard (deaths/1000 Patient-months) Range	Average
1	281	2	7.68–7.88	7.87
2	432	4	10.70–11.38	10.94
3	373	22	11.83–16.82	14.51
4	388	86	17.27–27.10	21.50
5	350	160	27.18–64.46	36.60
Totals	1824	274		

Source: Byar et al. (1974).

sets of characteristics in the group, n_k the number of patients possessing the kth set of characteristics, $k = 1, \ldots, m$, and $\Sigma_{k=1}^{m} n_k$, the total number of patients. If λ_k represents the hazard rate for the patients with a unique set of characteristics, the survivorship function is given by

$$S(t) = \sum_{k=1}^{m} \frac{n_k}{n} e^{-\lambda_k t} \qquad (10.38)$$

The predicted survivorship function following (10.34) is then

$$\hat{S}(t) = \sum_{k=1}^{m} \frac{n_k}{n} \exp\left[-t \sum_{j=0}^{p} \hat{b}_j x_{kj} \right] \qquad (10.39)$$

Figure 10.7 shows an excellent agreement between the observed and

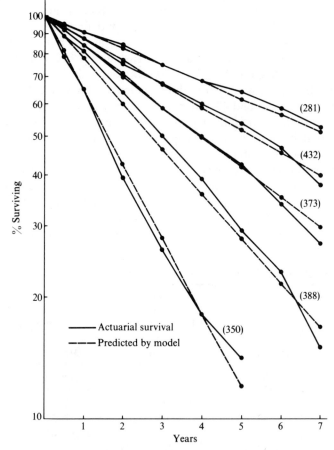

Figure 10.7 Actuarial and predicted survival curves for five risk groups. Numbers in parentheses are the number of patients in each group.

predicted survival, which confirms that the exponential model is appropriate for analyzing these data. The survival curves are approximately straight lines when plotted on a semilog scale, which suggests that there was near homogeneity of survival experience in each of the risk groups. The analysis allowed us to select, on the basis of pretreatment information alone, groups of patients with greatly differing survival experience as illustrated by the wide separation of the curves in Figure 10.7. For example, the observed five-year survival rates for the five groups were 65, 54, 42.5, 30, and 14%.

BIBLIOGRAPHICAL REMARKS

An excellent expository paper on statistical methods for the identification and use of prognostic factors has been written by Armitage and Gehan (1974). Many studies of prognostic factors have been published, including Rank et al. (1987), Bentzen et al. (1988), Droz et al. (1988), Farchi et al. (1987), Schwartzbaum et al. (1987), Ragland and Brand (1988), Lippman et al. (1988), Geller et al. (1989), Hertz-Picciotto et al. (1989), Kaplan et al. (1989), Claus et al. (1990), Schenker et al. (1990), Fisher et al. (1990), Mulders et al. (1990), and Aaby et al. (1990).

As mentioned earlier, Cox's regression model has stimulated the interest of many statisticians. A large number of papers on this model and related areas have been published since 1972. In addition to the articles cited earlier, the following are a few examples: Kalbfleisch (1974), Breslow (1975), Efron (1977), Kay (1977, 1984), Prentice and Gloeckler (1978), Liu and Crowley (1978), Prentice and Kalbfleisch (1979), Kalbfleisch and Prentice (1980), Tsiatis (1981), Anderson (1982), Wei (1984), Moreau et al. (1985), Gill and Schumacher (1987), Lagakos (1980), Arjas (1988), Segal and Bloch (1989), Ingram and Kleinman (1989), Gray (1990), and Liang et al. (1990).

EXERCISES

10.1   Consider the data given in Exercise Table 3.1. In addition to the five skin tests, age and sex may also have prognostic value. Examine the relationship between survival and each of the seven possible prognostic variables as in Example 10.1. For each variable, group the patients according to different cutoff points. Estimate and draw the survival function for each subgroup by the product-limit method and then use the methods discussed in Chapter 5 to compare survival distributions of the subgroups. Prepare a table similar to Table 10.1. Interpret your results. Is there a subgroup of any variable that shows significantly longer survival times? (For the skin test results, use the larger diameter of the two.)

10.2 Consider the seven variables in Exercise 10.1. Use Cox's model to identify the most significant variables. Compare your results with that obtained in Exericse 10.1.

10.3 Consider the data given in Exericse Table 3.3. Examine the relationship between (a) remission duration and (b) survival time and each of the nine possible prognostic variables; age, sex, family history of melanoma, and the six skin tests. Group the patients according to different cutoff points. Estimate and draw remission and survival curves for each subgroup. Compare remission and survival distributions of subgroups using the methods discussed in Chapter 5. Prepare tables similar to Table 10.1.

10.4 Use Cox's regression model to identify the significant variables in Exercise 10.3 for their relative importance to (a) remission duration and (b) survival time. Check the appropriateness of the proportional model using the significant variables identified.

10.5 Use the proportional hazards model to identify the most important factors that are related to survival time in the 157 diabetic patients given in Exericse Table 3.4. Check the appropriateness of the model.

10.6 (a) Construct a table similar to Table 3.8 using the data given in Table 3.6. (b) Use the proportional hazards model to identify the most important factors that are related to survival time.

10.7 Consider the data in Example 10.6. For each of the two groups (AG positive and AG negative), test the significance of log WBC by following the procedure given for model I.

CHAPTER 11

Identification of Risk Factors Related to Dichotomous Data

In biomedical research we are often interested in whether a certain health-related event will occur and the important factors that influence its occurrence. Examples of such events are the development of a given disease, response to a given treatment, and surviving longer than five years. In this chapter, we introduce statistical methods, univariate and multivariate, for identifying important factors related to such events.

To determine whether one is likely to develop a given disease, we need to know the important characteristics (or factors) related to its development. High- and low-risk groups can then be defined accordingly. Factors closely related to the development of a given disease are usually called *risk factors* or *risk variables* by the epidemiologist. We shall use this term in a broader sense to mean factors closely related to the occurrence of any health-related event.

For example, to find out if a woman will develop breast cancer because one of her relatives did, we need to know if a family history of breast cancer is an important risk factor. Therefore, we need to know the following:

1. Of such risk factors as age, race, family history of breast cancer, number of pregnancies, experience of breast feeding, and use of oral contraceptives, what are the most important?
2. Can we predict, on the basis of the important risk factors, whether the woman will develop breast cancer or whether she is more likely to develop breast cancer than another person?

In this chapter we discuss several methods for answering these questions. The general approach is to relate various patient characteristics (or independent variables) to the occurrence of the event (dependent variable) on the basis of data collected from those individuals to whom the event occurred (group 1) and those to whom the event did not occur (group 2). For example, in order to relate variables such as age, race, and number of pregnancies to the development of breast cancer, we need to collect information about these variables from a group of breast cancer patients as well as from a group of healthy normal women.

Often a large number of patient characteristics deserve consideration. These may be demographic variables, such as age; genetic, such as gene variant or phenotype; behavioral, such as smoking or drinking behavior and use of estrogen or progesterone medication; environmental, such as exposure to sun, air pollution, or occupational dust; or clinical, such as blood cell counts, weight, and blood pressure. The number of possible risk factors can be reduced through medical knowledge of the disease and careful examination of the possible risk factors individually. Section 11.1 presents two methods for individual variable examination. One is to compare the distribution of each possible risk variable of group 1 to that of group 2. The other method is the chi-square test for contingency tables. This test is particularly useful when the risk variables are categorical, for example, dichotomous or trichotomous. In this case, a $2 \times c$ contingency table can be set up and a chi-square test performed.

In Sections 11.2 and 11.3, we discuss two multivariate techniques for examining the possible risk variables simultaneously, namely, the linear discriminant function and Cox's (1970) linear regression model. Comparisons of these two methods are discussed at the end of Section 11.3.

11.1   UNIVARIATE ANALYSIS

11.1.1   Comparing the Distributions of Risk Variables for Two Groups

It is often convenient to name the binary or dichotomous observations "success" and "failure." Success may mean that the survival-related event occurred and failure that it failed to occur. Thus, a success may be a responding patient, a patient who survives more than five years after surgery, or an individual who develops a given disease. A failure may be a nonresponding patient, a patient who dies within five years after surgery, or an individual who does not develop a given disease. A preliminary examination of the data can compare the distribution of the risk variables in the success and failure groups. This method is especially appropriate if the risk variable is continuous. If, for example, the risk factor x is weight, and the dependent variable y is having cardiovascular disease, we may compare the weight distribution of the patients who have developed the disease to that of the disease free patients. If this disease group has significantly higher weights than the disease-free group, then we may consider weight an important risk factor. Commonly used statistical methods for comparing two distributions are the t-test for two independent samples if the assumption of normality holds and the Mann–Whitney U-test if the normality assumption is violated and a nonparametric test is preferred.

Table 11.1   Ages of 71 Leukemia Patients (yr)

Responders	20, 25, 26, 26, 27, 28, 28, 31, 33, 33, 36, 40, 40, 45, 45, 50, 50, 53, 56, 62, 71, 74, 75, 77, 18, 19, 22, 26, 27, 28, 28, 28, 34, 37, 47, 56, 19
Nonresponders	27, 33, 34, 37, 43, 45, 45, 47, 48, 51, 52, 53, 57, 59, 59, 60, 60, 61, 61, 61, 63, 65, 71, 73, 73, 74, 80, 21, 28, 36, 55, 59, 62, 83

Source: Hart et al. (1977). Data used by permission of the author.

Example 11.1

Consider the ages of 71 leukemia patients, 37 responders and 34 nonresponders (response is defined as complete response only), given in Table 11.1.

Figure 11.1 gives the estimated age distributions of the two groups. By using the Mann–Whitney U-test (or Gehan's generalized Wilcoxon test), we find that the difference in age between the responders and nonresponders is statistically significant ($p < 0.01$).

In consequence, the question may arise as to what age is critical. Can we say that patients under 50 may have a better chance of responding than patients over 50? To answer this question, one can dichotomize the age data and use the chi-square test discussed in the following section.

11.1.2   The Chi-Square Test and Odds Ratio

The chi-square test and odds ratio are most appropriate when the independent variable is categorical. If the independent variable is dichotomous, a 2×2 table can be used to represent the data. Any variables that are not dichotomous can be made so (with a loss of some information) by choosing a cutoff point, for example, age less than 50 years. The independent variables are then examined to find which ones (in some sense) provide the best risk associations with the dependent variable. To do that, we set up a 2×2 contingency table similar to Table 11.2 for each independent variable and look for a high degree of proportionality.

The first step is calculating the sample proportion of successes in the two risk groups a/C_1 and b/C_2. These are shown at the bottom of Table 11.2. Further analysis of the table is concerned with the precision of these proportions. A standard chi-square test can be used.

If the rates of success for the two groups E and $\bar{E}$ are exactly equal, the *expected* number of patients in the ijth cell (ith row and jth column) is

$$E_{ij} = N \times \frac{R_i}{N} \times \frac{C_j}{N} = \frac{R_i \times C_j}{N} \tag{11.1}$$

For example, in the top left cell, this expected number is

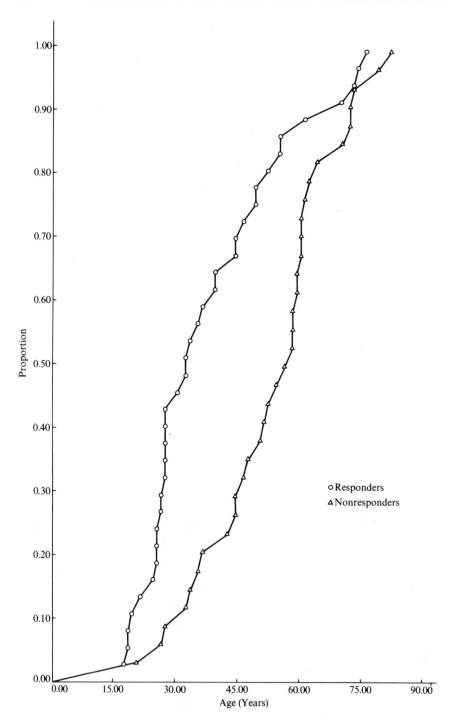

Figure 11.1   Age distribution of responders and nonresponders.

Table 11.2   General Setup of a 2 × 2 Contingency Table

	Risk Factor		
	Present (E)	Absent $(\bar{E})$	Total
Dependent variable			
Success	a	b	R_1
Failure	c	d	R_2
Total	C_1	C_2	N
Proportion of successes (success rate)	a/C_1	b/C_2	

$$E_{11} = \frac{R_1 \times C_1}{N}$$

since the overall success rate is R_1/N and there are C_1 individuals in the E group. Similar expected numbers can be obtained for each of the four cells. Let O_{ij} be the observed number of patients in the ijth cell. Then the discrepancies can be measured by the differences $(O_{ij} - E_{ij})$. In a rough sense, the greater the discrepancies, the more evidence we have against the null hypothesis that the success rates are the same in the two groups. The chi-square test is based on these discrepancies. Let

$$X^2 = \sum_{i=1}^{2} \sum_{j=1}^{2} \frac{(O_{ij} - E_{ij})^2}{E_{ij}} \tag{11.2}$$

under the hypothesis X^2 follows the chi-square distribution with one degree of freedom (df). The hypothesis of equal success rates for the E and $\bar{E}$ groups is rejected if $X^2 > \chi^2_{1,\alpha}$, where $\chi^2_{1,\alpha}$ is the 100α percentage point of the chi-square distribution with one degree of freedom. An alternative way to compute X^2 is

$$X^2 = \frac{(ad - bc)^2 N}{R_1 R_2 C_1 C_2} \tag{11.3}$$

In addition to the chi-square test, the odds ratio (Cornfield 1951) is a commonly used measure of association in 2 × 2 tables. The odds ratio (OR) is the ratio of two odds, namely the odds of success when the risk factor is present and the odds of success when the risk factor is absent. In terms of probabilities,

$$OR = \frac{P(\text{success} \mid E)/P(\text{failure} \mid E)}{P(\text{success} \mid \bar{E})/P(\text{failure} \mid \bar{E})} \tag{11.4}$$

Using the notation in Table 11.2, $P(\text{success} \mid E)$ and $P(\text{failure} \mid E)$ may be estimated by a/C_1 and c/C_1, respectively. Similarly, $P(\text{success} \mid \bar{E})$ and

$P(\text{failure} \mid \bar{E})$ may be estimated, respectively, by b/C_2 and d/C_2. Therefore, the numerator and denominator of (11.4) may be estimated, respectively, by

$$\frac{a/C_1}{c/C_1} = \frac{a}{c}$$

and

$$\frac{b/C_2}{d/C_2} = \frac{b}{d}$$

Consequently, the OR may be estimated by

$$\widehat{\text{OR}} = \frac{a/c}{b/d} = ad/bc \tag{11.5}$$

which is also referred to as the *cross-product ratio*.

Several methods are available for an interval estimate of OR, for example, Cornfield (1956) and Woolf (1955). Cornfield's method, which requires an iterative procedure, is considered more accurate but more complicated than Woolf's method. Woolf suggests using the logarithm of OR. The standard error of $\log_e \widehat{\text{OR}}$ may be estimated by

$$\widehat{\text{SE}}(\log_e \widehat{\text{OR}}) = \left(\frac{1}{a} + \frac{1}{b} + \frac{1}{c} + \frac{1}{d} \right)^{1/2} \tag{11.6}$$

Then a $100(1 - \alpha)$ percent confidence interval (CI) for $\log_e(\text{OR})$ is $\log_e \widehat{\text{OR}}$ $\pm Z_{\alpha/2} \widehat{\text{SE}}(\log_e \widehat{\text{OR}})$. The confidence interval for OR can be obtained by taking the antilog of the confidence limits for $\log_e(\text{OR})$. If $\log_e \text{OR}_U$ and $\log_e \text{OR}_L$ are the upper and lower confidence limits for $\log_e \text{OR}$, then $e^{\log_e \text{OR}_U}$ and $e^{\log_e \text{OR}_L}$ are the upper and lower confidence limits for OR.

Notice that, in (11.5), if b or c is zero, then $\widehat{\text{OR}}$ is undefined. If any one of the four cell frequencies is zero, the estimated standard error in (11.6) is also undefined. Should this occur, some statisticians (Haldane 1956, Fleiss 1979, 1981) suggest that 0.5 be added to each cell before using (10.5) and (10.6); however, others (e.g., Mantel 1977, Miettinen 1979) have concerns. It appears that the addition of 0.5 to each cell will solve the computational problem; however, if the cell frequencies are as small as zero, the addition of 0.5 will substantially affect the resulting estimate of OR and its standard error. The estimates so obtained must be interpreted with caution.

An odds ratio of 1 indicates that the odds of success is the same no matter whether the risk factor is present or not. An odds ratio greater than 1 means that the odds in favor of success is higher when the risk factor is present, and therefore there is a positive association between the risk factor and success. Similarly, an odds ratio less than 1 signifies a negative association between the risk factor and success. The interpretation should not be totally based on the point estimate. A confidence interval is always more meaningful, just like in any other estimation procedure.

The chi-square statistic in (11.2) may be used to test the null hypothesis that there is no association between the risk factor and success, or H_0: $OR = 1$. The following example illustrates the chi-square test and odds ratio.

Example 11.2

In the study of the response rate of 71 leukemia patients (Example 11.1), age is considered as one of the possible risk variables. The following 2×2 table is constructed.

	Age < 50	**Age ≥ 50**	**Total**
Response	27 (69%)	10 (31%)	37
Nonresponse	12	22	34
Total	39	32	71

The question is whether the response rates in the two age groups differ significantly or whether age is associated with response.

The X^2 value according to (11.3) is

$$X^2 = \frac{(594 - 120)^2(71)}{(37)(34)(39)(32)} = 10.16$$

with one degree of freedom. Reference to Table C-2 shows that the probability of getting an X^2 value of 10.16 if the two response rates are equal in the population is less than 0.01. Hence the difference between the two response rates is significant at the 1% level.

The estimated odds ratio, according to (11.5), is

$$OR = \frac{(27)(22)}{(10)(12)} = 4.95$$

The data show that the odds in favor of response is almost five times higher in patients under 50 years of age than in patients at least 50 years old. The difference is significantly different as indicated by the chi-square test above.

To obtain a confidence interval for OR, we first compute $\log_e \widehat{OR} = 1.60$. The estimated standard error of $\log_e \widehat{OR}$ following (11.6) is

$$\widehat{SE}(\log_e \widehat{OR}) = (\tfrac{1}{27} + \tfrac{1}{10} + \tfrac{1}{12} + \tfrac{1}{22})^{1/2} = 0.515$$

A 95% confidence interval for $\log_e OR$ is $1.60 \pm 1.96(0.515)$, or $(0.59, 2.61)$, and a 95% confidence interval for OR is $(e^{0.59}, e^{2.61})$, or $(1.80, 13.60)$. The wide interval may be due to the small cell frequencies. Note that the standard error of $\log_e \widehat{OR}$ is inversely related to the cell frequencies.

In this example, the cutoff point, 50, was chosen arbitrarily. It is often of interest to try more than one cutoff point if the number of observations in each cell is not too small.

There are cases where the independent variable has $k > 2$ classes. The chi-square test can be extended to $2 \times k$ tables. The expected frequencies are computed just like in (11.1) and the computation of X^2 [chi square distributed with $(2 - 1)(k - 1) = k - 1$ degrees of freedom] is the same as in (11.2) except that the sum is over all $2k$ cells. For details see Snedecor and Cochran (1967, Section 9.7). The odds ratio method can also be extended to handle polychotomous independent variables. It is done by selecting one of the classes as the reference class (the $\bar{E}$ group) and calculating the measure of association of each of the other classes relative to the reference class. The following example illustrates the procedures.

Example 11.3

Suppose that in the study of response rates of leukemia patients, another possible risk variable is the marrow absolute leukemic infiltrate, which is defined as the percentage of the total marrow that is either blast cells or promyelocytes. It is believed that patients should be classified into three classes: $\leq 45\%$, 46–90%, and $>90\%$. The 2×3 table is given below. Numbers in parentheses are expected frequencies. For example, $18.68 = (39)(34)/71$.

MARROW ABSOLUTE INFILTRATE

	≤45%	46–90%	>90%	Total
Response	4 (8.34)	20 (20.32)	13 (8.34)	37
Nonresponse	12 (7.66)	19 (18.68)	3 (7.66)	34
Total	16	39	16	71
Response rate, %	25	51	81	
$\widehat{OR}$	1	3.16	13.0	
95% Cl for OR		(0.86, 11.52)	(2.40, 70.46)	

The question is whether the difference in marrow absolute leukemic infiltrate is related to response. The value of X^2 is

$$X^2 = \frac{(4 - 8.34)^2}{8.34} + \frac{(12 - 7.66)^2}{7.66} + \cdots + \frac{(3 - 7.66)^2}{7.66} = 10.17$$

The number of degrees of freedom is $3 - 1 = 2$. With $X^2 = 10.17$ and two degrees of freedom, the probability that the three absolute infiltrate groups have the same response rate is less than 0.01. The data suggest that patients with a high percentage of marrow absolute infiltrate tend to have a high response rate. Marrow absolute infiltrate may be an important factor in predicting response.

The $\widehat{\text{OR}}$s given in the above table are calculated using the $\leq 45\%$ class as the reference class (or group). For example, for the $>90\%$ class, the odds ratio is $13 \times 12/4 \times 3 = 13$. The 95% confidence intervals for the ORs are obtained using (11.6). Though the odds ratio for the 46–90% group is larger than 1, the 95% confidence interval covers one. Therefore, the point estimate, 3.2, cannot be taken too seriously. It appears that the major difference is between the $>90\%$ and the $\leq 45\%$ group.

Computer programs for the chi-square test are available in many computer program packages, for example, the BMDP (Dixon et al. 1988), SPSS (SPSS 1988), and SAS (SAS Institute 1990).

Examination of each independent variable individually can only provide a preliminary idea of how important each variable is by itself. The relative importance of all the variables has to be examined simultaneously by some multivariate methods. In the following two sections we discuss two multivariate methods—the linear discriminant function and the linear logistic regression analysis. The linear discriminant function requires the assumption of a normal distribution for the independent variables while the logistic regression method is distribution free.

11.2   THE LINEAR DISCRIMINANT FUNCTION

The *linear discriminant function* first introduced by Fisher (1936) provides a rule of classifying an individual into one of several populations on the basis of a set of independent variables. Here we restrict ourselves to the case of two populations and the most fundamental concept of discriminant analysis. For advanced discussion of the subject, readers are referred to other books on multivariate analysis.

Suppose there are two groups of individuals, denoted by 1 (success) and 2 (failure), and on each individual, p independent variables $x_1, \ldots, x_p$ are measured. The populations of the p variables are assumed to be multivariate normal with a common covariance matrix but with different means for the two groups. A linear function of the p independent variables,

$$y = b_1 x_1 + b_2 x_2 + \cdots + b_p x_p \tag{11.7}$$

is called a *linear discriminant function* if it discriminates well between the two groups so that for any new individual known to come from one of the groups (the particular group being unknown), it could be used to assign the new individual to the most appropriate group.

Suppose that there are n_1 and n_2 individuals in groups 1 and 2, respectively. Let x_{ijk} denote the value of the kth independent variable of the jth individual in the ith group, $i = 1, 2$, $j = 1, \ldots, n_i$, $k = 1, \ldots, p$. The linear discriminant function in (11.7) for the jth individual in the ith group can be specified as

$$y_{ij} = b_1 x_{ij1} + b_2 x_{ij2} + \cdots + b_p x_{ijp} \qquad (11.8)$$

So, for each individual the values of the p independent variables can be transformed into an *index* or a *score*, y_{ij}. On the basis of y_{ij}, we classify the individual into group 1 or 2.

Now the problem is how to determine the coefficients $b_1, \ldots, b_p$ so that the linear discriminant function y_{ij} has the maximal discriminatory power. We shall present the method of obtaining the b's without going into the mathematical derivation. Interested readers are referred to Morrison (1967) and Kleinbaum, Kupper and Muller (1988).

Let $\bar{x}_{1k}$ and $\bar{x}_{2k}$ be the observed mean values of the kth independent variable in groups 1 and 2, respectively, that is,

$$\bar{x}_{ik} = \frac{1}{n_i} \sum_{j=1}^{n_i} x_{ijk} \qquad i = 1, 2, \; k = 1, \ldots, p$$

The pooled sample covariance matrix of the p variables is then

$$S = \begin{bmatrix} a_{11} & a_{12} & \cdots & a_{1p} \\ a_{21} & a_{22} & \cdots & a_{2p} \\ \vdots & \vdots & & \vdots \\ a_{p1} & a_{p2} & \cdots & a_{pp} \end{bmatrix} \qquad (11.9)$$

where

$$a_{kk} = \frac{\displaystyle\sum_{i=1}^{2} \sum_{j=1}^{n_i} (x_{ijk} - \bar{x}_{ik})^2}{n_1 + n_2 - 2}$$

is the variance of the kth variable and

$$a_{kk'} = \frac{\displaystyle\sum_{i=1}^{2} \sum_{j=1}^{n_i} (x_{ijk} - \bar{x}_{ik})(x_{ijk'} - \bar{x}_{ik'})}{n_1 + n_2 - 2}$$

is the covariance between variable k and k'. Compute the inverse of S, and let

$$S^{-1} = \begin{bmatrix} s_{11} & s_{12} & \cdots & s_{1p} \\ s_{21} & s_{22} & \cdots & s_{2p} \\ \vdots & \vdots & & \vdots \\ s_{p1} & s_{p2} & \cdots & s_{pp} \end{bmatrix} \qquad (11.10)$$

Further, let d_k be the difference of the two group means for variable k,

$$d_k = \bar{x}_{1k} - \bar{x}_{2k}$$

Then the coefficients $b_1, b_2, \ldots, b_p$ of the discriminant function are computed as follows:

$$b_1 = s_{11}d_1 + s_{12}d_2 + \cdots + s_{1p}d_p$$
$$b_2 = s_{21}d_1 + s_{22}d_2 + \cdots + s_{2p}d_p \qquad (11.11)$$
$$\vdots$$
$$b_p = s_{p1}d_1 + s_{p2}d_2 + \cdots + s_{pp}d_p$$

Thus, for a new individual, an index or a score that is the value of y in (11.7) can be computed on the basis of his values for $x_1, \ldots, x_p$ and $b_1, \ldots, b_p$. This value of y is used to classify the individual into one of the groups. In order to do it, a classification rule has to be established.

Let

$$\bar{y}_1 = b_1\bar{x}_{11} + b_2\bar{x}_{12} + \cdots + b_p\bar{x}_{1p}$$
$$\bar{y}_2 = b_1\bar{x}_{21} + b_2\bar{x}_{22} + \cdots + b_p\bar{x}_{2p} \qquad (11.12)$$

where $\bar{y}_1$ and $\bar{y}_2$ are the mean scores of groups 1 and 2, respectively. A symmetrical dividing line between the two groups is the average of the two mean scores:

$$y_0 = \tfrac{1}{2}(\bar{y}_1 + \bar{y}_2)$$
$$= \tfrac{1}{2}b_1(\bar{x}_{11} + \bar{x}_{21}) + \tfrac{1}{2}b_2(\bar{x}_{12} + \bar{x}_{22}) \qquad (11.13)$$
$$+ \cdots + \tfrac{1}{2}b_p(\bar{x}_{1p} + \bar{x}_{2p})$$

Then the classification rule is: If $\bar{y}_1 > \bar{y}_2$, assign a new individual to group 1 if $y > y_0$ and to group 2 if $y < y_0$.

Suppose it is known that an individual selected at random has a prior probability p_1 (say, 0.80) of being from group 1 and p_2 (say, 0.20) of being from group 2. It is possible to use this information in determining a classification rule under which the individual is more likely to be assigned to the group with the larger prior probability. For more details, the reader is referred Kleinbaum, Kupper, and Muller (1988).

After the individuals are classified into the two groups, a table such as Table 11.3 can be constructed to see how good the discriminant function is.

Table 11.3 Accuracy of Discriminant Function

	Assignment by Discriminant Function		
	Group 1	Group 2	Total
Actual group 1	n_{11}	n_{12}	n_1
Actual group 2	n_{21}	n_{22}	n_2

In this table n_{11} and n_{22} are the numbers of correct classifications and n_{12} and n_{21} are the numbers of misclassifications. The ratios n_{12}/n_1 and n_{21}/n_2 are estimates of the misclassification rates for groups 1 and 2, respectively.

The basic idea of the linear discriminant function is that if it is to discriminate well between the two groups, then we should expect the mean values $\bar{y}_1$ and $\bar{y}_2$ to be far apart relative to the variation of y within groups. That is, the b_i's are obtained so that the ratio

$$d^2 = \frac{(\bar{y}_1 - \bar{y}_2)^2}{\text{pooled within group variance of } y} \qquad (11.14)$$

is as large as possible. Therefore, a way to assess the effectiveness of the discrimination is to compute d^2. Its square root, d, is called the *generalized distance* between the two groups. If $d > 4$, the situation is like that in two univariate distributions whose means differ by more than four standard deviations. The overlap is therefore very small and the probability of misclassification is small.

Another measure of the "distance" between the two populations is the "Mahalanobis D^2," which is defined as

$$D^2 = \sum_{i=1}^{p} \sum_{j=1}^{p} d_i d_j s_{ij} \qquad (11.15)$$

Under the assumption that the two populations of the p variables follow a multivariate normal distribution with the same variances and covariances,

$$F = \frac{n_1 n_2 (n_1 + n_2 - p - 1) D^2}{(n_1 + n_2)(n_1 + n_2 - 2)p} \qquad (11.16)$$

follows approximately the F distribution with p and $n_1 + n_1 - p - 1$ degrees of freedom. This fact can be used to test whether there are significant differences between the two groups for all of the p variables considered simultaneously. A computed F value in (11.16) larger than $F_{p,(n_1+n_2-p-1),\alpha}$ indicates significant differences between the two groups.

The linear discriminant function possesses the property of *scale invariance*. That is, if all of the coefficients b_i, $i = 1, \ldots, p$, in (11.11) are multiplied by a constant factor, the use of the discriminant function is unaffected. Therefore, for simplicity, the S matrix in (11.9) can be replaced by the sum-of-products matrix:

$$S_p = \begin{bmatrix} a'_{11} & a'_{12} & \cdots & a'_{1p} \\ \vdots & \vdots & & \vdots \\ a'_{p1} & a'_{p2} & \cdots & a'_{pp} \end{bmatrix} \qquad (11.17)$$

where

$$a'_{kk} = \sum_{i=1}^{2} \sum_{j=1}^{n_i} (x_{ijk} - \bar{x}_{ik})^2$$

and

$$a'_{kk'} = \sum_{i=1}^{2} \sum_{j=1}^{n_i} (x_{ijk} - \bar{x}_{ik})(x_{ijk'} - \bar{x}_{ik}')$$

When S_p is used instead of S, the coefficients b_i obtained are equal to those given in (11.11) multiplied by $n_1 + n_2 - 2$.

Computer programs for the two-group discriminant analysis are widely available, for example the BMDP (Dixon et al. 1988), SPSS-X (SPSS 1988), and SAS (SAS Institute 1990) series. However, each gives a slightly different output.

The following example illustrates the computation procedures described above.

Example 11.4

Let us again use the data of the 71 leukemia patients. Thirty-seven of the patients responded to treatment and 34 did not. Suppose we wish to determine whether it is possible to discriminate between the responders and the nonresponders on the basis of two independent variables ($p = 2$); age (x_1) and labeling index percent (x_2). Table 11.4 gives the data.

The observed means of the two variables and differences in the two groups are given in Table 11.5. The sum-of-products matrix and its inverse are

$$S_p \cong \begin{bmatrix} 18160.72656 & -437.32202 \\ -437.32202 & 2336.29468 \end{bmatrix} \qquad S_p^{-1} \cong \begin{bmatrix} 0.000055 & 0.00001 \\ 0.00001 & 0.00043 \end{bmatrix}$$

The coefficients for the discriminant function obtained from (11.11) are $b_1 = 0.000055(-15.08) + 0.00001(3.98) = -0.00079$ and $b_2 = 0.00001(-15.08) + 0.00043(3.98) = 0.00156$. Thus, the linear discriminant function is

$$y = -0.00079x_1 + 0.00156x_2 \tag{11.18}$$

Based on (11.18), a score y is computed for every patient. Table 11.6 gives the scores in ascending order. The mean scores for the responders and nonresponders are $\bar{y}_1 = -0.01081$ and $\bar{y}_2 = -0.02896$. Thus $y_0 = -0.019885$. Since $\bar{y}_1 > \bar{y}_2$, using the symmetrical classification rule, we assign a patient to the responder group if $y > -0.019885$ and to the nonresponder group if $y < -0.019885$. The following table gives the counts of correct and incorrect

Table 11.4   Age, Labeling Index Percent and Response of 71 Patients with Leukemia

Age	Labeling Index Percent	Response[a]	Age	Labeling Index Percent	Response[a]
20	7	1	33	7	2
25	16	1	45	10	2
26	12	1	59	8	2
26	16	1	61	4	2
27	6	1	65	10	2
28	20	1	73	6	2
28	14	1	36	16	2
31	5	1	59	6	2
33	5	1	34	7	3
33	12	1	43	4	3
36	14	1	51	8	3
40	9	1	53	6	3
40	12	1	60	7	3
45	14	1	61	8	3
45	10	1	61	11	3
50	19	1	73	4	3
50	14	1	74	16	3
53	13	1	28	26	3
56	10	1	27	8	4
62	19	1	37	15	4
71	11	1	45	4	4
74	10	1	47	2	4
75	17	1	48	10	4
77	9	1	52	7	4
18	3	1	57	19	4
19	9	1	59	5	4
22	10	1	60	12	4
26	36	1	63	5	4
27	28	1	71	6	4
28	8	1	80	7	4
28	9	1	21	11	4
28	22	1	55	5	4
34	14	1	62	14	4
37	11	1	83	9	4
47	13	1			
56	15	1			
19	5	1			

[a]1, Responder; 2, 3, 4, nonresponder.
Source: Hart et al. (1977). Data used by permission of the author.

Table 11.5 Variable Means and Differences by Group

	Group 1 Responders	Group 2 Nonresponders	Difference
Age (1)	38.92	54.00	−15.08
Labeling			
index percent (2)	12.89	8.91	3.98

$$\bar{x}_{11} = 38.92 , \quad \bar{x}_{21} = 54.00 , \quad d_1 = -15.08$$
$$\bar{x}_{12} = 12.89 , \quad \bar{x}_{22} = 8.91 , \quad d_2 = 3.98$$

Table 11.6 Scores Obtained from Discriminant Function (11.18) for Leukemia Patients

Rank	First Group Values,[a] y_{1j}	Second Group Values,[b] y_{2j}	First Group Patient Number	Second Group Patient Number
1	0.03537		28	
2	0.02213		29	
3		0.01823		18
4	0.01201		32	
5	0.00890		6	
6	0.00506		2	
7	0.00427		4	
8		0.00045		31
9	−0.00043		7	
10	−0.00107		26	
11	−0.00189		27	
12	−0.00196		3	
13		−0.00366		7
14	−0.00497		1	
15	−0.00519		33	
16		−0.00601		20
17	−0.00677		11	
18	−0.00729		37	
19	−0.00751		10	
20	−0.00821		31	
21		−0.00897		19
22	−0.00961		25	
23	−0.00976		30	
24	−0.01010		16	
25	−0.01208		5	
26	−0.01223		34	
27	−0.01306		13	
28	−0.01391		14	
29		−0.01528		1
30		−0.01565		25
31		−0.01607		9

Table 11.6 (*Continued*)

Rank	First Group Values,[a] y_{1j}	Second Group Values,[b] y_{2j}	First Group Patient Number	Second Group Patient Number
32	−0.01681		8	
33	−0.01705		35	
34	−0.01772		12	
35	−0.01788		17	
36	−0.01839		9	
37	−0.01962		20	
38	−0.02013		15	
39		−0.02013		2
40	−0.02108		36	
41	−0.02181		18	
42		−0.02251		23
43		−0.02739		33
44		−0.02788		10
45		−0.02800		11
46	−0.02885		19	
47		−0.02892		27
48		−0.02946		21
49		−0.03035		24
50		−0.03126		15
51		−0.03270		12
52	−0.03304		23	
53		−0.03380		17
54		−0.03416		22
55		−0.03434		3
56		−0.03584		32
57		−0.03593		14
58		−0.03599		5
59		−0.03669		13
60		−0.03745		8
61		−0.03901		26
62	−0.03919		21	
63		−0.04215		4
64		−0.04218		28
65	−0.04313		22	
66		−0.04697		29
67	−0.04706		24	
68		−0.04856		6
69		−0.05167		16
70		−0.05182		34
71		−0.05255		30

[a]Responders.
[b]Nonresponders.

classification. The misclassification rates for the responders and nonresponders are, respectively, $\frac{8}{37} = 0.22$ and $\frac{8}{34} = 0.24$.

ASSIGNMENT BY DISCRIMINANT FUNCTION

Actual Group	Responder	Nonresponder	Total
Responder	29	8 (22%)	37
Nonresponder	8 (24%)	26	34

The Mahalanobis D^2 according to (11.15) is 1.25226, and consequently $F = 10.933$ for 2 and 68 ($37 + 34 - 2 - 1$) degrees of freedom, which is highly significant ($p < 0.01$). Thus, on the basis of the data, there are significant differences between the responders and nonresponders with respect to age and labeling index percent.

Figure 11.2 gives a scatter plot of age by labeling index percent (LI%). The diagonal line represents the cutoff points for which $y = y_0$. From (11.18), the equation for the line is

$$-0.019885 = -0.00079x_1 + 0.00156x_2$$

or

(11.19)

$$x_1 - 1.9747x_2 = 25.1709$$

The two dotted lines in Figure 11.2 represent the cutoff points when age or labeling index percent alone is considered. It is clear that discrimination by both age and labeling index percent is slightly better than by age alone

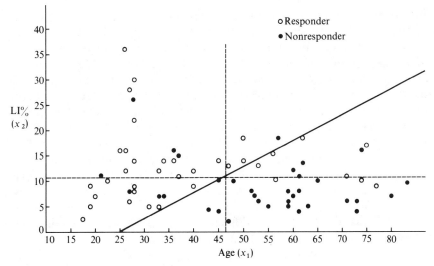

Figure 11.2 Age by labeling index percent for 71 leukemia patients.

$(y_0 = -0.3858, b_1 = -0.0083)$ and by labeling index percent alone $(y_0 = 0.01857, b_1 = 0.0017)$. This is confirmed by the higher misclassification rates given in the following table

ASSIGNMENT BY DISCRIMINANT FUNCTION

Actual Group	By Age Alone		By Labeling Index Percent Alone		Total
	Responders	Nonresponders	Responders	Nonresponders	
Responders	26	11 (30%)	22	15 (41%)	37
Nonresponders	10 (29%)	24	9 (26%)	25	34

The linear discriminant function that, based on the risk variables, classifies individuals into two groups can also provide information on the relative importance of the variables. The absolute magnitude of the b_i shows the contribution of the ith variable to the discriminant function. If $b_i > b_j$, discrimination by x_i is better than by x_j. However, the relative importance of the risk variables can better be evaluated by a stepwise discriminant analysis procedure that selects one variable at each step on the basis of the F value. The variable selected to be included in the discriminant function is the one that gives the most significant F value after adjusting for variables already in the equation. This stepwise procedure continues until no significant gain is obtained by adding new variable. Computer programs for stepwise discriminant analysis are available in BMDP, SPSS, and SAS packages.

11.3 THE LINEAR LOGISTIC REGRESSION METHOD

The linear discriminant analysis classifies an individual into one of two populations. Risk factors can be identified as independent variables that provide maximal discriminatory power. However, this technique is based on the assumption that the independent variables are normally distributed with equal variances. In most practical situations, some of the variables are qualitative or measured in nominal or ordinal scales and, often, the assumption of normality is violated. A practical method that does not require any distributional assumptions is Cox's (1970) *linear logistic regression method*. This method can be used to identify risk factors and predict the probability of success, for example, the probability of developing lung cancer or the probability of response, for individuals as a function of the risk factors. This probability can serve as an index of risk for a given disease or for not responding to a given treatment.

Suppose that there are n individuals to some of whom the health-related event occurred. They are called successes, and the others are failures. Let $y_i = 1$ if the ith individual is a success and $y_i = 0$ if the ith individual is a failure. Suppose that for each of the n individuals, p independent variables

$x_{i1}, x_{i2}, \ldots, x_{ip}$ are measured. These variables can be either qualitative, such as sex and race, or quantitative, such as blood pressure and white blood cell count. The problem is to relate the independent variables $x_{i1}, \ldots, x_{ip}$ to the dichotomous dependent variable y_i. One possible method is the ordinary linear regression technique. Assume that the y_i's are normally distributed with mean P_i and variance σ^2, and P_i, defined as the probability of success, or

$$P_i = P(y_i = 1 \mid x_{i1}, \ldots, x_{ip})$$

$$1 - P_i = P(y_i = 0 \mid x_{i1}, \ldots, x_{ip}) \qquad i = 1, \ldots, n \qquad (11.20)$$

is linearly dependent on x_{ij}'s The model may be written as

$$P_i = \sum_{j=1}^{p} b_j x_{ij} \qquad (11.21)$$

Then a least-squares technique is applied to estimate the coefficients b_j. Consequently, for a new individual, P_i can be estimated by substituting his x_{ij} values into (11.21). This method, treating the dichotomous dependent variable as if it is quantitative, has several limitations, including that the y_i's are not normally distributed and therefore no method of estimation that is linear in the y_i's will in general be fully efficient. Another limitation is that it is possible for the least-squares estimates obtained from (11.21) to lead to a fitted value that does not satisfy the condition $0 \le P_i \le 1$. In this case, modified least squares with the constraint that all P_i satisfy $0 \le P_i \le 1$ could be used; however, computations become very complicated.

Because of the above limitations, (11.21) is not an appropriate model for dichotomous dependent variables. The *linear logistic model* suggested by Cox is recommended.

11.3.1   The Logistic Regression Model and Estimation Procedures

In the linear logistic model, the dependence of the probability of success on independent variables is assumed to be

$$P_i = \frac{\exp\left(\sum_{j=0}^{p} b_j x_{ij}\right)}{1 + \exp\left(\sum_{j=0}^{p} b_j x_{ij}\right)} \qquad (11.22)$$

and

$$1 - P = \frac{1}{1 + \exp\left(\sum_{j=0}^{p} b_j x_{ij}\right)} \qquad (11.23)$$

where $x_{i0} = 1$ and b_j are unknown coefficients. The logarithm of the ratio of P_i and $1 - P_i$ is a simple linear function of the x_{ij}.

Let

$$\lambda_i = \log_e \frac{P_i}{1 - P_i} = \sum_{j=0}^{p} b_j x_{ij} \tag{11.24}$$

$\lambda_i = \log_e[P_i/(1 - P_i)]$ is called the logistic transform of P_i and (11.24) a linear logistic model. Another name for λ_i is *log odds*. Thus, the linear model relates the independent variables to the logistic transform of P_i or the log odds. The probability of success P_i can then be found from (11.24) or (11.22). In many ways (11.24) is the most useful analog for dichotomous response data of the ordinary regression model for normally distributed data.

To estimate the coefficients b_j's, Cox suggests the maximum likelihood method. Let $y_1, y_2, \ldots, y_n$ be the dichotomous observation on the n individuals. The likelihood function contains a factor (11.22) whenever $y_i = 1$ and (11.23) whenever $y_i = 0$. Thus, the likelihood is

$$L(b_0, b_1, \ldots, b_p) = \frac{\prod_{i=1}^{n} \exp\left(y_i \sum_{j=0}^{p} b_j x_{ij}\right)}{\prod_{i=1}^{n} \left[1 + \exp\left(\sum_{j=0}^{p} b_j x_{ij}\right)\right]}$$

$$= \frac{\exp\left(\sum_{j=0}^{p} b_j t_j\right)}{\prod_{i=1}^{n} \left[1 + \exp\left(\sum_{j=0}^{p} b_j x_{ij}\right)\right]} \tag{11.25}$$

where $t_j = \sum_{i=1}^{n} x_{ij} y_i$. The log-likelihood function is

$$LL(b_0, b_1, \ldots, b_p) = \sum_{j=0}^{p} b_j t_j - \sum_{i=1}^{n} \log_e\left[1 + \exp\left(\sum_{j=0}^{p} b_j x_{ij}\right)\right] \tag{11.26}$$

The maximum likelihood estimates of b_j's that maximize the log-likelihood function in (11.26) can be obtained by solving the following p equations simultaneously:

$$t_j - \sum_{i=1}^{n} \frac{x_{ij} \exp\left(\sum_{j=0}^{p} b_j x_{ij}\right)}{1 + \exp\left(\sum_{j=0}^{p} b_j x_{ij}\right)} = 0 \qquad j = 0, 1, \ldots, p \tag{11.27}$$

This can be done by an iterative procedure. If the Newton–Raphson method is used, we need the second derivative of LL, which is

$$I^*_{j_1 j_2} = -\sum_{i=1}^{n} \frac{x_{ij_1} x_{ij_2} \exp\left(\sum_{j=0}^{p} b_j x_{ij}\right)}{1 + \exp\left(\sum_{j=0}^{p} b_j x_{ij}\right)} \qquad j_1 = 0, \ldots, p \qquad j_2 = 0, \ldots, p$$

(11.28)

Let $I_{j_1 j_2} = (-1)I^*_{j_1 j_2}$. The inverse of the I matrix, I^{-1}, is the asymptotic covariance matrix of the b_j's.

The coefficients so obtained indicate the relationships between the variables and the log odds in favor of success. For a continuous variable, the corresponding coefficient gives the change in the log odds for an increase of one unit in the variable. For a categorical variable, the coefficient is equal to the log odds ratio (see Section 11.3.2).

An approximate $100(1 - \alpha)$ percent confidence interval for b_j can be obtained using (11.28),

$$\hat{b}_j \pm Z_{\alpha/2} \sqrt{I^{-1}_{jj}}$$

(11.29)

where $Z_{\alpha/2}$ is the $100(\alpha/2)$ percentile of the standard normal distribution.

To test the hypothesis that some of the b_j's are zero, a likelihood ratio test can be used. For example, to test $H_0: b_1 = 0$, the test statistic is

$$X^2 = -2[LL(\hat{b}_0, 0, \hat{b}_2, \ldots, \hat{b}_p) - LL(\hat{b}_0, \hat{b}_1, \hat{b}_2, \ldots, \hat{b}_p)]$$

(11.30)

where the first term is the maximized log-likelihood subject to the constraint $b_1 = 0$. If the hypothesis is true, X^2 is distributed asymptotically as chi-square with $p - 1$ degrees of freedom.

An alternative test for the significance of the coefficients in the Wald test (Rao 1973), which uses the following test statistic:

$$Z_w = \frac{\hat{b}_j}{\widehat{SE}(\hat{b}_j)}$$

(11.31)

Under the null hypothesis that $b_j = 0$, Z_w follows the standard normal distribution. A two-sided test would reject H_0 if $|Z_w| > Z_{\alpha/2}$. Though the Wald test is used by many, it is less powerful than the likelihood ratio test (Hauck and Donner 1977, Jennings 1986). In other words, the Wald test often misleads the user to conclude that the coefficient (consequently the corresponding risk factor) is not significant when it indeed is.

A number of computer programs for solving (11.24) are available, for example, Lee (1974), Morabito and Marubini (1976), GLIM (Numerical Algorithms Group 1987), SYSTAT (Wilkinson 1987), BMDP (Dixon et al. 1988), EGRET (Statistics and Epidemiology Research Corporation 1988), and SAS (SAS Institute 1990). Most of these computer programs have the stepwise regression capabilities. Using the stepwise procedure, the relative importance of the risk factors can be accessed. If the stepwise procedure is used, the likelihood ratio test can be performed to see if the additional variable is significant.

The independent variables x_{ij} in this model do not have to be the original variables. They can be any meaningful transforms of the original variables, for example, the logarithm of the original variable, $\log x_{ij}$, and the deviation of the variable from its mean, $x_{ij} - \bar{x}_j$.

From the estimated regression equation, a predicted probability of success can be computed by substituting the values of the risk factors in the equation. Using these predicted probabilities, a goodness-of-fit test can be performed to test the hypothesis that the model fits the data adequately. Several such tests are available (Lemeshow and Hosmer 1982), for example, the Pearson chi-square test, the Hosmer–Lemeshow (Hosmer and Lemeshow 1980) test, a test statistic suggested by Tsiatis (1980), and the score test of Brown (1982). In the following, we introduce the Hosmer–Lemeshow test.

Let p_i be the estimate of P_i obtained from the fitted logistic regression equation for the ith subject, $i = 1, \ldots, n$. The p_i's can be arranged in ascending order from the smallest to the largest. Those probabilities and the corresponding subjects are then divided into $g = 10$ groups with cutpoints equal to $k/10$, $k = 1, 2, \ldots, 10$. Thus, the first group contains all subjects whose estimated probabilities are less than or equal to 0.1, the second group contains all subjects whose estimated probabilities are less than or equal to 0.2, and so on. Let n_k be the number of subjects in the kth group. The estimated expected number of successes for the kth group is

$$E_k = = \sum_{j=1}^{n_k} p_j \qquad k = 1, 2, \ldots, g$$

The Hosmer–Lemeshow test statistic is defined as

$$C = \sum_{k=1}^{g} \frac{(O_k - E_k)^2}{n_k \bar{p}_k (1 - \bar{p}_k)} \tag{11.32}$$

where O_k is the observed number of successes in the kth group and $\bar{p}_k$ is the average estimated probability of the kth group, that is,

$$\bar{p}_k = \frac{1}{n_k} \sum_{j=1}^{n_k} p_j$$

Under the null hypothesis that the model is adequate, the distribution of C in (11.32) is well approximated by the chi-square distribution with $g - 2$ degrees of freedom. The test is basically a chi-square test of the discrepancy between the observed and predicted frequencies of success. Thus, a C value larger than the 100α percentage point of the chi-square distribution (or p value less than α) indicates that the model is inadequate.

Similar to other chi-square goodness-of-fit tests, the approximation depends on the estimated expected frequencies being reasonably large. If a large number (say far more than 20%) of the expected frequencies are less than 5, the approximation may not be appropriate and the p value must be interpreted carefully. If this is the case, adjacent groups may be combined to increase the estimated expected frequencies. However, Hosmer and Lemeshow warn that if less than six groups are used to calculate C, the test would be insensitive and would almost always indicate that the model is adequate. The computer program LR in BMDP gives several goodness-of-fit tests including the Hosmer and Lemeshow test.

The following example illustrates the procedure described above.

Example 11.5
In a study of 238 non-insulin-dependent diabetic patients, 10 covariates are considered possible risk factors for proteinuria (the response variable). The logistic regression method is used to identify the most important risk factors and to predict the probability of proteinuria on the basis of these risk factors. The 10 potential risk factors are age, sex (1, male; 2, female), smoking status (0, no; 1, yes), percentage of ideal body mass index, hypertension (0, no; 1, yes), use of insulin (0, no; 1, yes), glucose control (0, no; 1, yes), duration of diabetes mellitus (DM) in years, total cholesterol, and total triglyceride. Among the 238 patients, 69 have proteinuria ($y_i = 1$).

Using the stepwise procedure in BMDP (Dixon et al. 1988), it is estimated that, at step 1, $\hat{b}_0 = -0.896$ and $LL(\hat{b}_0)$ in (11.26) is -143.292. At step 2, duration of diabetes is selected to enter the regression because its maximum log-likelihood value is the largest among all the covariates. The MLEs of the two coefficients are $\hat{b}_0 = -1.467$ and $\hat{b}_1 = 0.055$, and $LL(\hat{b}_0, \hat{b}_1) = -139.429$. Since

$$X^2 = -2[LL(\hat{b}_0) - LL(\hat{b}_0, \hat{b}_1)] = 7.726$$

which is significant ($p = 0.005$), duration of DM is significantly related to the chance of proteinuria.

The Hosmer–Lemeshow test statistic for goodness-of-fit with only duration of DM in the model, $C = 9.814$ with eight degrees of freedom gives a p value of 0.278.

At step 3, sex is fitted to the equation because its addition yields the largest maximum log-likelihood value as compared to the other remaining covariates. The maximum log-likelihood value, $LL(\hat{b}_0, \hat{b}_1, \hat{b}_2) = -137.749$,

$\hat{b}_0 = 1.453$, $\hat{b}_1 = 0.060$, and $\hat{b}_2 = -0.279$. To test if sex is significantly related to proteinuria after duration of DM is considered, we perform the likelihood ratio test

$$X^2 = -2[LL(\hat{b}_0, \hat{b}_1) - LL(\hat{b}_0, \hat{b}_1, \hat{b}_2)] = 3.360$$

which is significant at $p = 0.067$. The stepwise procedure terminates after the third step because no other covariates are significant enough to enter the regression model, that is, none of the other covariate has a p value less than 0.15, which is set by the program. If any covariate already in the regression becomes insignificant after some other variables are in, the insignificant variable would be removed. The p values for entering and removing a variable can be determined by the user. The default values for entering and removing a variable are, respectively, 0.10 and 0.15. Thus, the procedure identifies duration of DM and sex as the two most important risk factors based on the given data. A question that may be raised at this point is whether one should include sex in the equation since its significance level is larger than the commonly used 0.05. The recommendation is to include it since it is close to 0.05 and since the p value should not be the only basis for determining whether a covariate should be included in the model. In addition, the Hosmer–Lemeshow goodness-of-fit test statistic $C = 5.036$, when sex is included, yields a p value of 0.754. Thus, the inclusion of this covariate improves considerably the adequacy of the model [for more discussion on model-building strategies, read Hosmer and Lemeshow (1989, Chapter 4)]. Thus, the final regression equation with the two significant risk factors is

$$\log_e \frac{P_i}{1 - P_i} = -1.453 + 0.060 \text{ (duration of DM)} - 0.279 \text{ (sex)}$$

Table 11.7 gives the details for the estimated coefficients.

The signs of the coefficients indicate that male patients and patients with longer duration of diabetes have a higher chance of proteinuria. Furthermore, for each increase of one year in duration of diabetes, the log odds increase by 0.060. Probabilities of proteinuria can be estimated following

Table 11.7 Estimated Coefficients for a Linear Logistic Regression Model Using Data from Diabetic Patients

Variable	Estimated Coefficient	Standard Error	Coefficient/SE	exp(coefficient)
Constant	−1.453	0.264	−5.504	0.234
Duration of DM	0.060	0.020	2.956	1.062
Sex	−0.279	0.152	−1.836	0.756

(11.22). For example, the probability of having proteinuria for a male patient who has had diabetes for 15 years is

$$P = \frac{e^{-0.832}}{1 + e^{-0.832}} = 0.303$$

where -0.832 is obtained by substituting the values of the two covariates in the fitted equation, that is, $-1.453 + 0.060(15) - 0.279(1) = -0.832$. Similarly, for a female patient who has the same duration of diabetes, the probability is 0.248.

In addition to individual variables, interaction terms can be included in the logistic regression model. If the association between an independent variable x_1 and the dependent variable y is not the same in different levels of another variable x_2, then there is interaction between x_1 and x_2. To check if there is interaction, one can include the product of x_1 and x_2 in the regression model and test the significance of this new variable. The following example illustrates the procedure.

Example 11.6

It is known that adriamycin is effective for treating certain types of cancer. It is also known that adramycin is highly toxic. Some patients develop congestive heart failure (CHF), but others who receive a similar dose of adriamycin do not. In an attempt to detect factors that would increase the risk of developing adriamycin cardiotoxicity, various patient characteristics of 53 cancer patients were studied. Seventeen of these patients developed CHF and 36 patients did not. After a careful investigation, it was found that the total dose (z_1) and percentage decrease in electrocardiographic QRS voltage (z_2) are most closely related to CHF. Table 11.8 shows the data and

Table 11.8 Total Dose and Percent Decrease in QRS of 53 Patients Receiving Adriamycin

Patient Number	CHF,[a] y	Total Dose, z_1	Percent Decrease in QRS, z_2
1	1	435	41
2	1	600	71
3	1	600	51
4	1	540	40
5	1	510	63
6	1	740	79
7	1	825	61
8	1	535	44
9	1	510	53
10	1	483	27
11	1	460	53
12	1	460	60

Table 11.8   (*Continued*)

Patient Number	CHF,[a] y	Total Dose, z_1	Percent Decrease in QRS, z_2
13	1	550	65
14	1	540	58
15	1	310	41
16	1	500	64
17	1	400	44
18	0	440	9
19	0	600	42
20	0	510	19
21	0	410	24
22	0	540	−24
23	0	575	39
24	0	564	35
25	0	450	10
26	0	570	6
27	0	480	6
28	0	585	21
29	0	420	14
30	0	470	1
31	0	540	33
32	0	585	33
33	0	600	4
34	0	570	2
35	0	570	5
36	0	510	12
37	0	470	−1
38	0	405	44
39	0	575	14
40	0	540	−10
41	0	500	−43
42	0	450	23
43	0	520	−1
44	0	495	29
45	0	585	40
46	0	450	30
47	0	450	23
48	0	500	12
49	0	540	−11
50	0	440	7
51	0	480	−22
52	0	550	20
53	0	500	19

[a]1, Yes; 0, no; $\bar{z}_1 = 517.679$, $\bar{z}_2 = 26.019$.
Source: Minow et al. (1977).

Table 11.9 Linear Logistic Regression Analysis Results of Data in Table 11.8

Variable	Estimated Coefficient	Standard Error	Coefficient/SE	Log-likelihood
Constant	−3.757	1.576	−2.384	−33.254
QRS	0.254	0.102	2.480	−10.185
TD	−0.024	0.021	−1.160	−9.225
TD × QRS	0.001	0.001	0.677	−8.803

some summary statistics. The following linear logistic regression model with transformed variables $x_1 = z_1 - \bar{z}_1$ and $x_2 = z_2 - \bar{z}_2$ is used:

$$\lambda = \log_e \frac{p}{1-p} = b_0 + b_1 x_1 + b_2 x_2 + b_3 x_1 x_2$$

The stepwise procedure selects percentage decrease in QRS as the most important variable, which is followed by the total dose (TD) and interaction (TD × QRS). The logistic regression analysis results are given in Table 11.9. The stepwise log-likelihood values given in the last column indicate that only QRS is significant since $2(-10.185 + 33.254) = 46.138$, which yields a p value less than 0.001. Neither the total dose nor the interaction is significant.

11.3.2 Odds Ratio and Coefficients of the Linear Logistic Regression Model

When the independent variables are dichotomous or polychotomous, the logistic regression coefficients can be linked with odds ratios. Consider the simplest case where there is one independent variable, x_1, which is either 0 or 1. The linear regression model in (11.22) and (11.23) becomes

$$P(y = 1 \mid x_1) = \frac{e^{b_0 + b_1 x_1}}{1 + e^{b_0 + b_1 x_1}}$$

$$P(y = 0 \mid x_1) = \frac{1}{1 + e^{b_0 + b_1 x_1}}$$

Values of the model when $x_1 = 0, 1$ are

$$P(y = 1 \mid x_1 = 0) = \frac{e^{b_0}}{1 + e^{b_0}}$$

$$P(y = 1 \mid x_1 = 1) = \frac{e^{b_0 + b_1}}{1 + e^{b_0 + b_1}}$$

$$P(y = 0 \mid x_1 = 0) = \frac{1}{1 + e^{b_0}}$$

$$P(y = 0 \mid x_1 = 1) = \frac{1}{1 + e^{b_0 + b_1}}$$

The odds ratio in (11.4) is

$$OR = \frac{P(y=1 \mid x_1 = 1)/P(y=0 \mid x_1 = 1)}{P(y=1 \mid x_1 = 0)/P(y=0 \mid x_1 = 0)}$$

$$= \frac{e^{b_0 + b_1}}{e^{b_0}} = e^{b_1} \tag{11.33}$$

and the log odds ratio is $\log_e(OR) = \log_e(e^{b_1}) = b_1$. Thus, the estimated logistic regression coefficient also provides an estimate of the odds ratio, that is, $\widehat{OR} = e^{b_1}$. If (b_{1L}, b_{1U}) is the confidence interval for b_1, the corresponding interval for OR is $(e^{b_{1L}}, e^{b_{1U}})$.

Example 11.7
Consider the age (x) and response (y) data from 71 leukemia patients in Table 11.10 (Examples 11.1 and 11.2). The logistic regression analysis results are given in Table 11.11. Notice that $\exp(b_1) = \exp(1.5994) = 4.95$, which is equal to the estimate of OR obtained in Example 11.2 using (11.5), and the standard error of $\hat{b}_1$ is the same as that of $\log_e \widehat{OR}$ except for a small rounding-off error. The confidence interval for OR can also be obtained from the logistic regression analysis results.

The relationship between the logistic regression coefficient and odds ratio can be extended to polychotomous variables by creating dummy variables (or design variables). The following example illustrates the procedure.

Example 11.8
Consider the data in Example 11.3. The variable marrow absolute infiltrate (MAI) has three levels. As in Example 11.3, the $\leq 45\%$ level is considered

Table 11.10   Age and Response Data of 71 Leukemia Patients

Response	$x < 50(1)$	$x \geq 50(0)$	Total
Yes (1)	27	10	37
No (0)	12	22	34
Total	39	32	71

Table 11.11   Results of Logistic Regression Analysis of Data in Table 11.10

Variable	Estimated Coefficient	Standard Error	Coefficient/SE	exp(coefficient)
Constant $(\hat{b}_0)$	-0.7885	0.3814	-2.067	0.45
Age $(\hat{b}_1)$	1.5994	0.5156	3.102	4.95

the reference group. In this case two design variables will be used and their values are assigned as follows:

$$D_1 = \begin{cases} 1 & \text{if MAI} = 46\text{--}90\% \\ 0 & \text{otherwise} \end{cases} \qquad D_2 = \begin{cases} 1 & \text{if MAI} > 90\% \\ 0 & \text{otherwise} \end{cases}$$

Thus, for MAI, the design variables and their values are

MAI (%)	D_1	D_2
≤ 45	0	0
46–90	1	0
>90	0	1

Using these design variables, the logistic regression analysis gives the results in Table 11.12.

The coefficient corresponding to D_1, 1.1499, is the log odds ratio between the 46–90% group and the ≤45% group. The odds ratio is $\exp(1.1499) = 3.16$, which is exactly equal to the estimate obtained in Example 11.3. Similarly, the coefficient corresponding to D_2 is the log odds ratio between the >90% group and the ≤45% group. The odds ratio obtained from the regression coefficient, 13.00, is the same as that obtained in Example 11.3.

The estimated standard error for D_1, 0.6603, is also the standard error of $\log_e(\widehat{OR})$. A 95% confidence interval for the coefficient is $1.1499 \pm 1.96(0.6603)$, or $(-0.1443, 2.4441)$, and consequently, a 95% confidence interval for OR is $(e^{-0.1443}, e^{2.4441})$, or $(0.86, 11.52)$, which is identical to that obtained in Example 11.3 using Woolf's estimate of SE($\log_e$OR).

For a continuous independent variable, the logistic regression coefficient gives the change in log odds for an increase of one unit in the variable. In general, for an increase of m units in the variable, the log odds ratio is equal to m times the logistic regression coefficient. The derivation is left to the reader as an exercise problem.

When more than one independent variable are included in the logistic regression model, each estimated coefficient can be interpreted as an estimate of the log odds ratio statistically adjusting for all the other variables. For example, in Example 11.5, the regression coefficient for the

Table 11.12 Results of Logistic Regression Analysis of Data in Example 11.3 and Two Design Variables

Variable	Estimated Coefficient	Standard Error	Coefficient/SE	exp(coefficient)
Constant	−1.0986	0.5774	−1.903	0.33
MAI (D_1)	1.1499	0.6603	1.742	3.16
MAI (D_2)	2.5649	0.8623	2.974	13.00

variable sex, -0.279, is an estimate of the log odds ratio for females versus males adjusting for duration of diabetes. Or the adjusted odds ratio for females versus males is estimated as $\exp(-0.279) = 0.76$; that is female diabetic patients have a lower risk of having proteinuria than male patients after adjusting for duration of diabetes. This interpretation, commonly used by epidemiologists, is appropriate if the linear relationship between the log odds and the independent variables holds.

11.3.3   Logistic Regression in Case–Control Studies

Since its early application in the Framingham heart study (Cornfield 1962, Truett et al. 1967), the logistic regression model has played an equally important role in epidemiologic studies and in clinical studies. In prospective studies, a simple random sample of subjects is taken and the values of the independent variables measured at a given time (usually called baseline measurements). The subjects are then followed for a given period of time and the outcome (dependent) variable is measured at the end of the follow-up. This sampling plan satisfies the requirements of the regression model. The purpose is to predict the binary outcome variable using the baseline measurements. In clinical trials, although the patients are randomized into different treatment groups, they are indeed followed for a given period of time. Essentially, clinical trails are prospective studies. The logistic regression method can easily be adapted by including treatment variables in the model. Similarly, the logistic regression model can be applied to stratified samples by adding to the model terms for strata.

However, in case–control studies, cases and controls are first identified according to the disease (outcome variable) of interest. The independent variables are then measured or collected for each case and each control. The purpose is to estimate the association between the dependent variables and the disease under study. The probability statements in (11.20) are no longer applicable. Instead, we are interested in $P(x_{i1}, \ldots, x_{ip} \mid y_i = 1)$ and $P(x_{i1}, \ldots, x_{ip} \mid y_i = 0)$. Fortunately, using Bayes theorem, these probabilities and consequently the likelihood function can be worked out so that the logistic regression model can be used for data from case–control studies in the same way as for data from prospective studies. The estimated regression coefficients, except the intercept (b_0), can be interpreted in the same manner as in prospective studies. That is, b_j is the estimated log odds ratio adjusted for the other variables. Theoretical derivations are given in Prentice (1976), Breslow and Powers (1978), Holford et al. (1978), Breslow and Day (1980), and Hosmer and Lemeshow (1989).

For matched case–control studies, the logistic regression is also applicable. Suppose that for each case, there are R (≥ 1) matched controls. Let x_{ijk} denote the observed value of the jth independent variable $(j = 1, \ldots, p)$ for the ith matched set $(i = 1, \ldots, n)$ and kth subject $(k = 1$ for case and $k > 1$ for controls, $k = 1, \ldots, R + 1)$. The n matched sets are considered samples from n strata on the basis of the matching variable. The contribution to the likelihood for each stratum is

$$\frac{\exp\left(\sum_{j=1}^{p} b_j x_{ij1}\right)}{\exp\left(\sum_{j=1}^{p} b_j x_{ij1}\right) + \sum_{k=2}^{R+1} \exp\left(\sum_{j=1}^{p} b_j x_{ijk}\right)} = \frac{1}{1 + \sum_{k=2}^{R+1} \exp\left[\sum_{j=1}^{p} b_j(x_{ijk} - x_{ij1})\right]}$$

(11.34)

When $R = 1$, that is, a one-to-one pair matching, the conditional likelihood obtained from (11.34) reduces to

$$L(b_1, \ldots, b_p) = \prod_{i=1}^{n} \frac{\exp\left[\sum_{j=1}^{p} b_j(x_{ij1} - x_{ij2})\right]}{1 + \exp\left[\sum_{j=1}^{p} b_j(x_{ij1} - x_{ij2})\right]}$$

(11.35)

(Holford et al. 1978, Breslow et al. 1978). Equation (11.35) has the same mathematical form as the likelihood function for ordinary logistic regression in (11.25). This fact permits the use of computer programs for ordinary logistic regression in one-to-one matched case–control studies. To do this, we use the following procedure:

1. Let n be the number of case–control pairs.
2. Use $x_{ij1} - x_{ij2}$, the difference between variables for the case and the control, as the independent variables in the model.
3. Let $y_i = 1$ (or 0) for all pairs since the disease status of all the pairs is the same, that is one case and one control.
4. Delete the intercept term b_0 from the model.

Most of the other features in ordinary logistic regression model fitting including the use of design (dummy) variables remain the same. However, the goodness-of-fit test of Hosmer and Lemeshow is not applicable to one-to-one matched designs. Readers who are interested in assessing the logistic regression model in matched case–control studies are referred to Pregibon (1984) and Moolgavkar et al. (1985).

The following example illustrates the basic procedure.

Example 11.9

To study the effect of obesity, family history of diabetes, and level of physical activity to non-insulin-dependent diabetes (NIDDM), 30 non-diabetic individuals are matched with 30 NIDDM patients by age and sex. Obesity is measured by body mass index (BMI), which is defined as weight in kilograms divided by height in meters squared. Family history of diabetes (FH) and levels of physical activity (PHY) are binary variables. Table 11.13 gives the partially fictitious data.

Following the procedure given above, the results of fitting the three variables using BMDP are given in Table 11.14.

Another computer program available for logistic regression analysis of matched case–control studies is written by Lubin (1981).

When $R > 1$, that is, there are two or more controls matched for each case, the conditional likelihood function can no longer be expressed in the

Table 11.13   Data of 30 Matched Case–Control Pairs

Pair	Case (Diabetic)			Control (nondiabetic)		
	BMI	FH[a]	PHY[b]	BMI	FH[a]	PHY[b]
1	22.1	1	1	26.7	0	1
2	31.3	0	0	24.4	0	1
3	33.8	1	0	29.4	0	0
4	33.7	1	1	26.0	0	0
5	23.1	1	1	24.2	1	0
6	26.8	1	0	29.7	0	0
7	32.3	1	0	30.2	0	1
8	31.4	1	0	23.4	0	1
9	37.6	1	0	42.4	0	0
10	32.4	1	0	25.8	0	0
11	29.1	0	1	39.8	0	1
12	28.6	0	1	31.6	0	0
13	35.9	0	0	21.8	1	1
14	30.4	0	0	24.2	0	1
15	39.8	0	0	27.8	1	1
16	43.3	1	0	37.5	1	1
17	32.5	0	0	27.9	1	1
18	28.7	0	1	25.3	1	0
19	30.3	0	0	31.3	0	1
20	32.5	1	0	34.5	1	1
21	32.5	1	0	25.4	0	1
22	21.6	1	1	27.0	1	1
23	24.4	0	1	31.1	0	0
24	46.7	1	0	27.3	0	1
25	28.6	1	1	24.0	0	0
26	29.7	0	0	33.5	0	0
27	29.6	0	1	20.7	0	0
28	22.8	0	0	29.2	1	1
29	34.8	1	0	30.0	0	1
30	37.3	1	0	26.5	0	0

[a]1, Yes; 0, no.
[b]1, physically active; 0, sedentary.

Table 11.14 Results of a Logistic Regression Analysis of Data in Table 11.13

Variable	Estimated Coefficient	Estimated Standard Error	Coefficient/SE	exp(coefficient)
BMI	0.090	0.065	1.381	1.094
FH	0.968	0.588	1.646	2.633
PHY	−0.563	0.541	−1.041	0.569

same form as in the ordinary logistic regression. Therefore, computer programs for ordinary logistic regression are not applicable. There are programs available for this case, for example, the EGRET Statistical Software (Statistics and Epidemiology Research Corporation 1988), PECAN (Lubin 1981), and STRAT (Breslow and Day 1980, Appendix V).

Press and Wilson (1978) compare the logistic regression method to the discriminant analysis and find that if the independent variables are normal with identical covariance matrices, discriminant analysis is preferred. Under nonnormality, the logistic regression method is preferred. In particular, if the independent variables are dichotomous, we cannot expect to predict accurately the probability of success with a discriminant function, even with a large amount of data. Their examples show that the logistic regression gives a higher correct classification rate.

BIBLIOGRAPHICAL REMARKS

The method of discriminant analysis can be found in most books on multivariate analysis, for example, Morrison (1967), Press (1972), and Kleinbaum, Kupper, and Muller (1988). Readers with less sophisticated mathematical background may find Kleinbaum, Kupper, and Muller easier to read than the other two books. Statisticians and readers who are interested in the mathematical development may prefer Morrison or Press. The linear logistic regression method is discussed extensively in two excellent books by Cox (1970) and by Hosmer and Lemeshow (1989). Cox's book provides the theoretical background while Hosmer and Lemeshow discuss the broad application of the method including modeling strategies and interpretation and presentation of analysis results. In addition to Press and Wilson (1978), comparisons of the two methods can be found in Anderson (1972), Efron (1975), Halperin et al. (1971), and McFadden (1976).

EXERCISES

11.1 Consider the study presented in Example 3.5 and the data for the 40 patients in Table 3.10.

a. Construct a summary table similar to Table 3.11.

b. Construct a table similar to Table 3.12.

c. Use the chi-square test to detect any differences in retinopathy rates among the subgroups obtained in part b.

d. On the basis of these 40 patients, identify the most important risk factors using Cox's linear logistic regression method.

11.2 Consider the data for the 33 hypernephroma patients given in Exercise Table 3.1. Let "response" be defined as stable, partial response, or complete response.

a. Compare each of the five skin test results of the responders with those of the nonresponders.

b. Use Cox's linear logistic regression method to identify the most important risk factors related to response.

   i. Consider the five skin tests only.

   ii. Consider age, sex, and the five skin tests.

11.3 Consider all of the nine risk variables (age, sex, family history of melanoma, and six skin tests) in Exercise 3.3 and Exercise Table 3.3. Identify the most important prognostic factors that are related to remission. Use both univariate and multivariate methods.

11.4 Consider the data of 58 hypernephroma patients given in Exercise Table 3.2. Apply the logistic regression method to response (defined as complete response, partial response, or stable disease). Include sex, age, nephrectomy treatment, lung metastasis, and bone metastasis as independent variables.

a. Identify the most significant independent variables.

b. Obtain estimates of odd ratios and confidence intervals when applicable.

11.5 Consider the case where there is one continuous independent variable X_1. Show that the log odds ratio for $X_1 = x_1 + m$ versus $X_1 = x_1$ is mb_1, where b_1 is the logistic regression coefficient.

CHAPTER 12

Planning and Design of Clinical Trials (I)

Clinical trials may be prophylactic or therapeutic. Prophylactic trials are conducted in preventive medicine. The purpose of a prophylactic trial is to assess the effectiveness of a preventive treatment. A therapeutic trial is designed to compare a new treatment with the best of the current treatments. In a typical therapeutic trial, patients with similar characteristics are divided into two groups; one is given the new treatment and the other, usually called the *control*, is given the current treatment. These patients are then observed over the same period of time to see which group does better. In a prophylactic trial, one group is given the prophylactic and the other is not. At the end of the study period, we investigate if the protected group has a lower incidence of the specific disease than the unprotected. Thus, basic principles for prophylactic and therapeutic trials are similar. In this chapter and the next, we focus our attention on the therapeutic clinical trial.

Before a clinical trial is conducted, the clinician must have in mind a primary objective, know how to prepare the plans, usually called a protocol. outlining the design and methods of conducting the clinical trial, and know how many patients (sample size) are needed and how to allocate them to different treatments. The clinician should have an understanding of what randomization is, the need for it, and how to use it as well as some knowledge of the statistical concepts used in the design and analysis of a clinical trial. In the following sections we attempt to provide such guidelines.

Deliberate experiments designed to assess the value of therapeutic procedures have been carried out for centuries. Section 12.1 gives a brief history of clinical trials. Different types of clinical trials are discussed in Sections 12.2 and 12.3. In Chapter 13, we shall discuss the use of prognostic factors in clinical trials, randomization, preparation of protocol, and controls in cancer clinical studies.

12.1 HISTORY OF CLINICAL TRIALS

From limited records, it is believed that typical ancient Egyptian treatments were a combination of ritual exhortation and mixtures of herbs and natural products. The Greeks believed that therapy was intended to assist the natural powers of healing by exercises and diet. In the Middle Ages when Arab merchants dominated the spice and drug trade, there were numerous pharmacologic experiments. Avicenna (980–1037) suggested, in his encyclopedic *Cannon*, that in the trial of a remedy, it should be used in its natural state upon uncomplicated disease, that two opposed cases should be observed, and that the time of action and of the reproducibility of the effects should be studied. He further suggested that the experiments must be done with a human body. These rules imply the very modern approach of clinical trials but there seems to be no record of their detailed application.

During the sixteenth century, although Leonardo da Vinci (1451–1519) had developed the theory and practice of modern scientific experiments, therapeutic experiments were performed only fortuitously. One such trial was forced upon Ambroise Paré (1510–1590) in 1537 after the capture of the castle of Villaine. He treated the wounded with boiling oil until he ran out of oil. He then applied in its place a digestive made of egg yolks, oil of rose, and turpentine: "I raised myself to visit them, when beyond my hope I found those to whom I had applied the digestive medicament feeling but little pain, their wounds neither swollen nor inflamed, and having slept through the night. The others were feverish with much pain and swelling about their wounds. Then I determined never again to burn thus so cruelly the poor wounded by arquebases" (Bull 1959, Packard 1925). During the seventeenth century there was a great growth of the application of scientific method in physics, chemistry, and biology. However, little development was made in therapeutic trials except that practical men did learn from comparative observations. A well-known story was the first expedition of four ships to India by the East India Company in 1600. Only one of the ships, that of General James Lancaster, provided lemon juice for its sailors. These sailors were almost free of scurvy while the others were very ill. Lemon juice was subsequently supplied to all their ships.

It was in the eighteenth century that therapeutic trials began to be developed. An important early trial concerned the use of smallpox inoculation. Maitland (1668–1748) and Lady Wartley-Montague persuaded King George I to permit a trial of six Newgate convicts in 1721. Their results were considered conclusive in favor of inoculation, and it became widely practiced. In 1747, a comparative trial of scurvy was conducted by a ship's surgeon, James Lind (1716–1794). He selected 12 patients with scurvy. The cases were as similar as possible. They lay together in one place and had the same diet. Two of them had two oranges and one lemon every day, two a quart of cider, two vinegar, two an elixir of vitriol, two a course of sea water, and the remaining two an electuary. After six days, the two who had oranges and lemons were well enough to take care of the sick. The cider seemed to be of some help; however, the other remedies were not effective at all. Forty-eight years later, the British Navy began to supply lemon juice to all its ships.

Many medical schools and hospitals were founded in the eighteenth century. These institutes then carried out many therapeutic experiments. For example, William Withering (1741–1799), First Physician at the Birmingham General Hospital, conducted a relatively large-scale study, involving 163 cases, of digitalis as a remedy for dropsy. In addition to the effectiveness of the drug, he also tried to establish the correct dosage. Other famous studies included those conducted by Jenner on vaccination and Pearson on syphilis treatment.

Statistical evaluation of clinical trials began in the nineteenth century. P.C.A. Louis (1787–1872) was one of the pioneers in the application of statistics in medicine. Animal experiments also began in this century. The rapid growth of bacteriology led to trials of immunization against rabies, therapeutic serum for diptheria, and inoculation for typhoid fever. The number of patients involved in these trials increased considerably, for example, a trial of diptheria included 300 patients in France and a trial of typhoid fever included more than 11,000 soldiers in India.

In the present century, the scientific study of therapeutic procedures has been greatly intensified by the two world wars, the increased number of facilities for their study in hospital, laboratory, and research organizations and the great advances in bacteriology, organic chemistry, pharmacology, statistics, computer science, and many other related disciplines. Numerous trials have been carried out for syphilis, diabetes, pneumonia, chronic cardiac disease, malaria, infectious hepatitis, cancer, AIDS, and so on. Large-scale prophylactic trials of tetanus antiserum, antityphoid and anticholera inoculations, and so on, have also been successful. One of the most well-known trials is that of penicillin by Abraham and associates. The many trials conducted by different investigators with and without control in infectious diseases and wounds with large and small groups of patients confirmed the effectiveness of the drug. Another well-known trial was set up by the British Medical Research Council under the leadership of Sir Austin Bradford Hill in 1946 to evaluate streptomycin in the treatment of pulmonary tuberculosis. This was the first large-scale randomized clinical trial, and Hill was largely responsible for introducing modern statistics into medicine. Since then, clinical trials have become a common practice in medical research. Well-known clinical trials include those for Salk polio vaccine, the University Group Diabetes Program, the Beta Block Heart Attack Trial, the Hypertension Detection and Follow-up Program, the National Surgical Adjuvant Breast Project, and many recent trials on AIDS. Today the planning, design, and analysis of clinical trials often involve multiple disciplines, for example, medicine, biochemistry, immunology, and biostatistics.

12.2 PHASE I AND II TRIALS

A clinical trial is a prospective study designed to determine the effectiveness of a treatment, a surgical procedure, or a therapeutic regimen administered to patients with a specific disease. Usually, the new drug or surgical procedure is first experimented on laboratory animals. If a tolerated dose can be determined and the drug or surgical procedure is considered to have potential efficacy, clinical trials on humans are warranted. We focus our attention on therapeutic clinical trials of drugs. The sequence of therapeutic clinical trials are given in Figure 12.1.

A phase I, or early trial, is an exploratory investigation of one of several surgical procedures or therapeutic regimens with different types of patients in search of one that can be used in later studies. Very few treatments are totally free of undesired side effects or without any hazard to the patients. For instance, almost all anticancer drugs designed to destroy cancer cells are also harmful to normal cells. Obviously, the possible dangers, for example, toxicity, of a treatment may need to be considered in relation to the dangers of the disease itself. A phase I trial is intended to determine a safe dose for further studies of therapeutic efficacy.

In general, the major objective of a phase I trial is to establish a maximum tolerated dose (MTD) for a new drug. The MTD will then be used in subsequent clinical trials to determine efficacy (phase II) and to compare with other drugs (phase III). In order to understand the new drug, qualitative and quantitative toxicity of the new drug are also investigated in phase I trials. Although efficacy is not the primary concern of phase I trials, the investigator would not totally ignore its importance. In addition, some of the basic pharmacology of the drug, including uptake, metabolism, and organ distribution of the drug, are often investigated.

The type of surgical operation or dose of drug to be used initially may be selected from antecedent experiments with laboratory animals. It was found by Freireich et al. (1966), for example, that, for cancer agents, MTD in

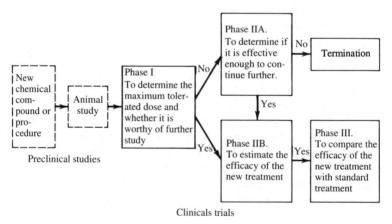

Figure 12.1 Sequence of clinical trial.

humans is comparable to that in five animal species, mouse, rat, hamster, dog, and monkey, when dosage was expressed per unit of surface area.

In a drug trial, the patient is started at a particular dosage level (usually low). For example, some phase I cancer trials use $\frac{1}{3}$(TDL), where TDL is the toxic dose low in most sensitive large animal species, denoted in milligrams per square meters. Another starting dose is $\frac{1}{3}$LD$_{10}$ (in mice), where LD$_{10}$ is the lethal dose for 10% of the experimental animals, expressed in milligrams per square meters. Many of the recent cancer trials have used $\frac{1}{10}$LD$_{10}$ (in mice) or $\frac{1}{3}$TDL (in dogs). The amount of drug per dose is then increased gradually, the trial duration increased, or the number of doses per unit time increased until a safe dosage level can be determined. For example, in the modified Fibonacci scheme (Carter et al. 1977, Hansen 1970), the starting dose is increased 100% initially, and subsequent dose levels are increased by 67, 50, 40, and 33%, consecutively. Eventually, a dosage regimen that can be tolerated by a majority of patients is chosen as the MTD.

No optimal design for phase I trials can be recommended. The knowledge and intuition of the investigator is very important as in the choice of patients and close observation. Very sick patients should not be selected for phase I studies since they have low tolerance to toxicity and are likely to show it at very low doses, thus giving possibly misleading results. However, this may not always be the case. In many cancer clinical trials, patients who are no longer amenable to other treatments are recruited for phase I studies. Patient's age should also be considered. Studies have shown that patients less than 18 years may tolerate higher doses of a drug than those older than 18 years. In cancer patients, younger (<50 years) patient's tumor may be more sensitive to chemotherapy than older ($\geq$50 years) patient's. Intervals between prior drug therapy and the administration of the phase I drug should be long enough to allow the toxic effect of the prior therapy to disappear. Other considerations for patient selection include concomitant medications and life expectancy long enough to complete the phase I study. The effect of concomitant drugs is often difficult to evaluate. Therefore, if possible, phase I patients should not be on medications other than the one being studied.

The schedule and route of administration are determined on the basis of the type of drug, mechanism of drug activity, patient convenience, and preclinical animal pharmacokinetic information. A measure of response or effectiveness has to be determined, for example, a decrease of tumor cells to one-half the starting volume. In addition, changes in other relevant characteristics of patients should be used in determining the effectiveness of treatment. In a two-stage procedure, if no effect is noted in phase I, a phase IIA trial may be undertaken. After a treatment is found to be effective in phase IIA, a phase IIB trial should be conducted. If the treatment is found to be ineffective again in phase IIB, it would be considered unworthy of further study.

The objective of a phase IIA or preliminary trial is to decide if a particular therapeutic regimen is effective enough to warrant further study.

The decision to be reached at the end of the preliminary trial is one of the two possibilities:

1. The treatment is unlikely to be effective in x percent of patients or more.
2. The treatment could be effective in x percent of patients or more.

One would like to reject an ineffective treatment as quickly as possible and investigate further those treatments with a higher likelihood of effectiveness. The preliminary trial is designed so that a minimum possible number of consecutive failures is observed before the study is terminated. Suppose that the relevant percentage of effectiveness is 30 ($x = 30$). Then the possible decisions for a preliminary trial are:

1. The treatment is unlikely to be effective in 30% of patients or more.
2. The treatment could be effective in 30% of patients or more.

The probability of consecutive treatment failure by number of patients, assuming that the true effectiveness of the treatment is 30%, is given below.

Consecutive Patients	Probability of Treatment Failure
1	0.7
2	$0.7 \times 0.7 = 0.49$
3	$0.7 \times 0.7 \times 0.7 = 0.343$
.	.
.	.
.	.
8	0.0576
9	0.0404

Thus if the treatment were at least 30% effective, there would be more than a 95% (1–0.0404) chance that one or more successes would be obtained in nine consecutive patients. If nine consecutive failures are observed, the treatment is unlikely to be effective in 30% of patients. Following this logic, Gehan (1961) gives the minimum number of patients necessary to decide whether a treatment is not of a given effectiveness or not worthy of further trial at a given level of rejection error (rejection error is defined as the chance of rejecting the treatment for further trial when it should have been accepted). Table 12.1 gives the sample size required for a

Table 12.1 Sample Size Required for Preliminary Trial

Rejection Error (%)	Therapeutic Effectiveness									
	5%	10%	15%	20%	25%	30%	35%	40%	45%	50%
5	59	29	19	14	11	9	7	6	6	5
10	45	22	15	11	9	7	6	5	4	4

Source: Gehan (1961).

preliminary trial of a new treatment for various levels of therapeutic effectiveness (percent) and two rejection error levels. For example, if one is interested in a treatment of 25% effectiveness and is willing to accept 5% rejection error, a sample of 11 patients is necessary. If all 11 fail to respond, the treatment is dropped. If one or more patients show response (treatment effectiveness), the treatment receives further trial.

The definition of therapeutic effectiveness requires careful assessment. In diseases for which previous procedures or therapies have been completely ineffective, the definition may pose no serious difficulties, and any objective benefit to the patient can be considered as a therapeutic effect. But if partial benefits are frequently observed, it is reasonable to require pronounced objective improvement in the patient's disease. The preliminary trial will eliminate regimens having little or no effectiveness.

When a regimen has passed a Phase II preliminary trial or has been sufficiently effective in a phase I trial, then a phase IIB or follow-up trial is recommended in order to provide a precise estimate of its effectiveness. An estimate of true effectiveness is the proportion of patients in the sample who are treated successfully. The number of patients required in the follow-up trial is given in Table 12.2. For example, if a sample of 11 patients had been taken in a preliminary trial (with a 10% rejection error) in search of a 20% effective treatment, and one treatment success was observed, then 60 additional patients would be required to guarantee an estimate of the true percentage effectiveness with a standard error of about 5%. If the required standard error was about 10% then 7 additional patients would be needed.

The sampling plan described above for the preliminary and follow-up trials is a two-stage procedure known as *double sampling* (Cox 1958). The initial sample is calculated to meet the probability of further trial. It requires at least a single treatment success before the second sample is taken. A complete phase II trial requires completion of both IIA and IIB if this sampling method is used, not just IIA. Although response rate is often used as the criterion for treatment success, it is not the only desirable statistic that can be obtained from phase II trials. Many phase II trials also estimate remission duration and survival time when applicable.

12.3 PHASE III TRIALS

The theme of phase III trials can be found in the well-known joke about the biostatistician who, when asked how his wife was, responded, "Compared to whom?" In medicine, the question might be, for example, how effective is BCG (an immunotherapy) as a cancer treatment? Compared to what? Placebo, chemotherapy, or other immunotherapy? At what dose, and for what type of cancer? Phase III clinical trials are designed to answer questions like these.

A phase III trial or comparative clinical trial is a planned experiment on human subjects involving two or more treatments in which the primary

Table 12.2 Number of Additional Patients Required in a Follow-up Trial to Estimate Therapeutic Effectiveness of Treatment with Specified Precision (Standard Error)

Rejection Error of Preliminary Trial	Required Standard Error	Number of Treatment Successes in Preliminary Trial	Therapeutic effectiveness					
			5%	10%	15%	20%	25%	30%
		Number of patients in preliminary trial	59	29	19	14	11	9
5%	5%	1	0	4	30	45	60	70
		2	0	17	45	63	78	87
		3	0	28	58	76	87	91
		4	0	38	67	83	89	91
		5	0	46	75	86	89	91
	10%	1	0	0	0	1	7	11
		2	0	0	0	6	12	15
		3	0	0	1	9	14	16
		4	0	0	3	11	14	16
		5	0	0	5	11	14	16
		Number of patients in preliminary trial	45	22	15	11	9	7
10%	5%	1	0	21	42	60	70	83
		2	0	35	60	78	87	93
		3	0	47	72	87	91	93
		4	4	57	81	89	91	93
		5	9	65	85	89	91	93
	10%	1	0	0	0	7	11	16
		2	0	0	4	12	15	18
		3	0	0	7	14	16	18
		4	0	0	9	14	16	18
		5	0	0	10	14	16	18

Source: Gehan (1961).

purpose is to determine the relative merits of the treatments. It is undertaken after the treatment has successfully passed a phase II trial. Comparative clinical trials have elicited great interest and discussion ever since they were introduced in the 1930s by A. B. Hill. Very often they are referred to as clinical trials. The objectives of the comparative clinical trial differ according to what one means by evaluating the relative effectiveness of the treatments. The primary objective may be the selection of the best treatment for future use in patients. It may be to estimate the effectiveness of each treatment with some degree of precision. Or, the objective can be twofold, to select the best treatment and to estimate the treatment effect. If the primary aim is to select the best treatment, then some type of sequential plan will be appropriate in which the decision to continue the study at any stage is determined by the results accumulated to that stage. As soon as it is clear which treatment is best, the study will be stopped, even though at that point it may be that only rather imprecise estimates of the effectiveness of each treatment can be made. If there is a combined aim of selecting the best treatment and learning something about each, a sufficient number of patients should be entered on each treatment so that effectiveness can be estimated with some precision. Additional patients might then be needed to satisfy the selection requirement. It is implicit that for such a trial it should be ethically justifiable to continue even after sufficient data have been collected to permit a decision about the best treatment. For each particular trial, the objectives need careful consideration, and clinicians should be aware of the types of study implied by a different choice of objectives. It usually is desirable to consult a statistician concerning determination of sample size and stopping rules for the study.

Phase III trials demand very careful planning. Loose plans and loose methods give loose results that may be misleading. In the following, some general guidelines are offered for planning and designing phase III trials.

12.3.1 Considerations in Planning

No complete, or completely satisfactory, list of considerations can yet be given. The following are a few important points.

1. *Time for Planning.* Several months to a year should be allowed for drawing up the protocol, or the document specifying the objective of the trial and the plan for carrying it out. Guidelines for preparing a protocol are given in Chapter 13. If the trial involves more than one institution, a longer planning time (a year or more) should be allowed.

2. *Number of Treatments Involved.* The number of treatments involved in a phase III trial is closely related to the number of patients per year that can be expected to enter a study. In order to guarantee enough patients in each treatment group, a clinical trial should involve a small number of treatments for comparison.

3. *Duration of the Trial.* The estimated duration of a trial includes the period for entry of patients and the follow-up period for the observation of response and survival. George and Desu (1974) derive the necessary duration of a clinical trial based on the assumptions that patients enter the trial according to a Poisson process and the survival time (or time to failure) is exponentially distributed. No rules are available for general cases. However, if a clinical trial extends over a long period of time, it is likely that other treatments will appear as candidates for comparison. In practice, clinical trials extending longer than five years must be thoroughly justified.

4. *Comparability of Patients.* The comparative clinical trial should be planned so that the only reasonable explanation for a difference between treatment groups is a result of the treatments. Patients must be comparable with respect to prognostic factors; otherwise, the results will very likely be misleading. A technique to achieve comparability of patients at the time of entry is formal randomization (see Section 13.2). However, randomization does not guarantee that patients in treatment groups will be comparable with respect to all prognostic factors. Although adjustments can often be made in the analysis, comparability has to be checked and assured before the analysis is done. In Section 13.3 we shall discuss the use of prognostic factors in a clinical trial to improve the precision of comparisons.

5. *Treatment Allocation Ratio.* Equal allocation of patients to the two treatments is a common practice. However, it may be advisable to allocate patients randomly in the ratio of 60 : 40 or 67 : 33 (2 : 1) when comparing a new treatment with an old or when one treatment is much more difficult or expensive to administer. The chance of detecting a real difference between the two treatments is not reduced much as long as the ratio is not more extreme then 2 : 1 (Peto et al. 1976).

6. *Use of a Concurrent or Historical Control Group.* A basic requirement of most clinical trials is the "controls", which is a group of patients corresponding in characteristics to the specially treated group but not given the treatment. An issue is whether a concurrent group should be used or whether historical controls suffice. The ethical question is whether it is proper to withhold from any patient a treatment that might give him benefit. Of course, the effectiveness of the treatment is not proven; if it were, there would be no need for a trial. On the other hand, the treatment had passed trials in phases I and II, so there must be some basis justifying a trial. The severity of the problem depends upon what is at stake. If, for example, the treatment in the trial is for fast relief of a common headache, then the morality of a rigidly controlled trial would not be seriously in doubt. However, it might be quite impossible to withhold, even temporarily, any treatment for a disease (e.g., cancer) in which life or death or serious aftereffects were at stake. On the other hand, it should be realized that a new treatment is certainly not always the best and by no means always free of danger. For example, certain antibiotics and hormones are not always innocuous. Thus, every comparative trial must be exhaustively weighted in

the ethical balance, each according to its own circumstances and its own problems.

The use of historical controls has been a debatable subject, especially in clinical trials of cancer. The use of historical controls allow all current patients in the trial to receive the new treatment and results compared with that of patients previously treated with the standard treatment. The major problem with historical control is that one is unable to ensure comparability between the groups of patients and methods of evaluation. This subject will be discussed in more detail in Section 13.4.

7. *Treatment Management.* It is important to manage patients on each treatment regimen in the same manner. Definitions of response and toxicity should be exactly the same for patients in each treatment group. The decision to remove patients from a trial should be applied the same way in each treatment group. If some investigators have a preference for one of the treatments so that patients are maintained on it longer or are classified as toxic only when toxicity is very severe, the results could be biased. On the other hand, patients who are on a treatment known to be more toxic may be more likely to report an adverse effect. One way to avoid physician's or patient's potential bias or different managements of treatment is to do a double-blind trial. In a double-blind trial, the treatments are prepared in identical forms so that neither the patient nor the physician conducting the trial is aware of which specific treatment is being given. However, double-blind trials can be difficult to organize and less effective than expected in reducing bias. For example, if one wants to compare a chemotherapy with no therapy for leukemia patients, if the therapy is in tablet form, it would be necessary to give tables to both sets of patients (dummy tables to the control group) and to take blood samples for each to measure the white blood cells. This could be troublesome to manage. Also, the physician is aware of the types of toxicity to be expected and can find out, by observing toxicity, which patients are receiving each treatment. In addition, it is difficult to explain a double-blind trial to a patient and his family. Therefore a double-blind study should be given careful consideration. In certain cases, they obviously cannot apply, such as for studying surgical procedures or radiation therapies.

8. *Ethical Considerations.* If the physician feels that one treatment is better than another for a particular patient, he or she cannot randomly assign a treatment. It is unethical not to treat a patient in a manner that the physician believes is best. Thus, if the physician is convinced that one of the treatments is better for the patient, the patient should not be entered in the study, and if at any time during the study there is a clear indication that one treatment is better, randomization of patients should be stopped.

To avoid termination of trials due to physician's bias before any statistical significance has been obtained, results of the trial should be kept from the participating physicians until a decision of whether or not to stop the trial has been reached by an advisory committee. This approach, placing the burden on the advisory committee, does not solve the ethical problem completely; however, it does provide a more objective judgment than that made only by the participating physician.

Ethical problems may arise in the beginning of a tial or they may also arise during the course of the trial. For example, when a patient's condition seems to deteriorate, it presents an ethical problem whether to withdraw the patient from study. It is important to take into serious consideration conditions under which a patient may be withdrawn from the study and to describe them in detail in the protocol before the trial begins.

Most institutions require approval from an institutional review board (IRB) for all research projects involving humans. Usually, the complete proposal along with a consent form must be submitted to the IRB at the time funding is sought. Prior to the start of the clinical trial, any changes made in the protocol must be submitted to the IRB for further approval. The responsibility of the IRB is to ensure that the clinical trial is worth conducting, in that benefits will outweigh risks and the potential subjects are satisfactorily informed. In general, the following information must be included in the consent form:

1. Purpose of the clinical trial.
2. Status of investigational drug, device, or procedure.
3. Description of the clinical trial including all of the procedures involved.
4. Potential benefits to the participants.
5. Possible risks to the participants.
6. Information concerning medical treatment and compensation in the event of injury or adverse effect.
7. Confidentiality and methods used to protect confidentiality.

In addition, the subject must be assured that participation is voluntary and the consent may be revoked and withdrawal is allowed at any time without penalty or loss of benefits. A very important note is that the consent form must be written using layman's terms without medical jargons, which is often neglected by the investigator.

9. *Others.* It is very important to prepare a good protocol, especially if the trial is cooperative. Every participating investigator or institution has to have a detailed protocol. It is also important that there be someone who has primary responsibility for keeping the trial running well. Very often, this falls upon a statistician rather than a clinician, mainly because of the time required to carry out this responsibility efficiently.

12.3.2 Designs for Comparative Trials

There are two types of experimental designs of comparative clinical trials: fixed-sample trials and sequential trials. In fixed-sample trials, the number of patients allocated to the two (or more) treatments is fixed before the study begins. In sequential trials, the decision whether to continue taking new patients is determined by the results accumulated to that time. In the following, we shall first discuss briefly some fixed-sample designs and then sequential designs.

1. *Simple Randomized Design.* In this simplest case patients are randomly assigned to the two (or more) treatments without considering their characteristics. Usually, the randomization is restricted to a fixed number of patients and each treatment group will be equal in size. The main advantage of this design is its simplicity and usefulness when important prognostic factors are unknown or the potential subjects are homogeneous with respect to patient characteristics. Its disadvantage is possible noncomparability among treatment groups and biased analysis. More details are discussed in Section 13.2

An example of this design is a double-blind study of comparing standard and low serum levels of lithium for maintenance therapy in patients with bipolar disorder (Gelenberg et al. 1989). Ninety four patients were randomized to two different doses of lithium for maintenance treatment: the standard dose, adjusted to produce a serum lithium concentration of 0.8–1.0 mmol/l, and a low dose, achieving a serum concentration of 0.4–0.6 mmol/l. The authors reported that the risk of relapse was 2.6 times higher (95% confidence interval, 1.3–5.2) in the "low-dose" group than in the "standard-dose" group. The conclusion was that doses producing serum lithium levels from 0.8 to 1.0 mmol/l were more effective than those resulting in lower serum lithium concentrations, although adverse side effects were more frequent in the higher dose group.

2. *Stratified Randomized Design.* If important prognostic factors are known and patients can be grouped into prognostic categories or "strata," comparability among treatment groups can be achieved. Within each stratum, patients are randomly assigned to the treatments. The simplest example is the pair comparison design that occurs when the strata are pairs of patients, one (chosen at random) given treatment A and the other treatment B (in the case of two treatments). The pairs are matched by one or more prognostic factors and must be treated alike in supplementary ways and, as far as possible, observed simultaneously. The main difficulty of the method is that, if there are two many categories, it is unlikely to have two patients available at the same time. As a result, this design is often limited to a chronic disease where a pool of patients is readily available and can be drawn upon at one specified time. The prognostic factors to be matched must be well thought out and defined in advance; for example, will any age between 50 and 60 years be classified into the same stratum?

Many clinical trials have used this design. For example, a study of the effect of propranolol for postoperative hypertension after surgical repair of coarctation of the aorta was conducted using a stratified randomized design (Leenen et al. 1987). Twenty-three patients (7 girls and 16 boys) aged 4–16 years undergoing coarctectomy participated in the study. The subjects were stratified by age: 4–10 and 10.1–16 years. Within each age stratum, the patients were randomly assigned to propranolol and placebo. Other examples of stratified randomized design are Crist et al. (1989), Eisenberger et al. (1989), Kaye et al. (1989), and Harrison et al. (1990).

3. *Crossover Design.* A crossover design (Jones and Kenward 1989) is a combination of the simple randomized design and the paired comparison design in which each patient serves as his own control. A common way of using this design is to give the sequence of treatments A followed by B to half the patients and sequence B followed by A to the other half. A patient is assigned to one of the two sequences by a random allocation, each treatment being given when the patient's disease is in a comparable state. Three comparisons of the two treatments are possible: between different patients in the first and second phases of the study and within the same patient. This design may be used when patients are in a relapse or remission state at successive times. For diseases that naturally decline or progress in severity rapidly with time, comparisons within patients have no advantage. For example, if the objective is to compare remission length, patients may not have two consecutive remissions or may have two remissions that differ in degree. There are other practical difficulties with the crossover design. If the objective is to compare treatments A and B within the same patient, some patients may not survive long enough to receive the second treatment in the sequence; alternatively, a patient may have an excellent response to the first treatment so that a long period of time would elapse before the second treatment would be given. Although comparing A to B in the same patient can confirm the results of the comparison of A to B in different patients [see Frei et al. (1961) and Freireich et al. (1963) for examples], one cannot learn the effect of each treatment on survival in this design.

The crossover design is good in measuring short-term reliefs of signs or symptoms. It does allow more precise comparison of the treatments. Sometimes, it is considered more ethical because each patient has a chance to receive both treatments. The possibility of within-patient comparison allows investigators to use smaller sample size because of the smaller variability. However, there are controversies (Brown 1980a,b). The major problem is the assumption of no carry-over effect. If the assumption is violated, the design is insensitive in detecting treatment differences. In considering a crossover design, the investigator must ask several questions. First, is the disease stable enough? Second, is the time at each stage long enough to allow the treatment to take effect? Third, is the time between the first and second treatments long enough to allow carry-over effects to disappear (Matthews 1987)? Fourth, is patient's cooperation and compliance sufficient to go through the two stages?

In order to ensure that the patient's condition is stable, a "run-in" period may be helpful before randomization is performed at the beginning of the trial. Similarly, allowing a "wash-out" period before the second treatment starts is, in general, recommended.

The crossover design is widely used by investigators. For example, a randomized, double-blind study was conducted to compare room air with 100% oxygen on performance of professional soccer players (Winter et al. 1989). Twelve professional soccer players performed two bouts of exhaustive exercise on a motor-driven treadmill separated by 5 minutes of a "recovery period." During the recovery period, the athletes breathed either room air or 100% oxygen, randomly assigned. The identity of the gas was unknown to both the athletes and investigators. Blood samples were drawn for lactate analysis before exercise, after exercise, but before breathing the unknown gas and after 4 minutes of inhalation of the unknown gas. The initial study was carried out during a morning session. At least 3–4 hours later on the same day, the complete procedure was repeated on each athlete with the opposite unidentified gas. The hours between the two sessions are utilized as a "washout" period. No significant differences were found between 100% oxygen and room air on recovery from exhaustive exercise or on subsequent exercise performance.

Other examples of the crossover design are Schade et al. (1987), Vega and Grundy (1989), and Malloy et al. (1991). The last study is an example of multiple crossover design with three treatment periods, each consisting of an active drug therapy [tetrahydroaminoacridine (THA)] and a placebo treatment, randomly assigned in a double-blind manner, to patients with Alzheimer's disease.

4. *Factorial Design*. In a $r \times s$ factorial design, one of the two treatments is administered at r levels, the other at s levels. The goal is to test various treatment effects as well as possible interaction effects. The simplest factorial design is a 2×2 design, in which two treatments are studied for their relationship to response and each is given at two levels, say at high dosage and low dosage or active drug and placebo. If the number of treatments and the number of levels are large, many patients would be required and the results might be difficult to interpret.

The Physicians' Health Study (Stampfer et al. 1985, Steeting Committee of the Physicians' Health Study Research Group 1989) used a randomized double-blind, placebo-controlled 2×2 factorial design to determine whether low-dose aspirin reduced mortality from cardiovascular disease and whether beta carotene decreases the incidence of cancer. A total of 22,071 physicians were first randomly assigned to active aspirin and aspirin placebo, and within each group the subjects were further randomized to receive either active carotene or carotene placebo. Thus, there were four equal-size groups of subjects, each of which received a different combination of drugs: (1) aspirin and carotene, (2) aspirin and carotene placebo, (3) aspirin placebo and active carotene, and (4) aspirin placebo and carotene placebo. Stampfer et al. point out that by using this design, results from each subject contribute to both questions of interest and the carotene hypothesis can be tested without materially affecting the aspirin study. In addition, only one-fourth of the study subjects receive only placebo, which may have made the study more acceptable to potential participants.

Other trials using the factorial design include a 3×3 factorial design reported by Hung et al. (1988) a $2 \times 2 \times 3$ design by Manninen (1988), a 2×4 design by Elaad and Ben-Shakhar (1989), and a 3×3 design by Davis et al. (1989).

If there is strong prior evidence that a particular treatment is more effective than all others, it is desirable to reach an early decision. The investigator can choose to use a sequential trial in which the observations are analyzed continuously and the decision whether or not to stop at any stage depends on the results so far obtained. Such a sequential design is frequently used in clinical trials since patients usually join a study serially, over a period of time, rather than all at the same time. Certain kinds of observations may lead to an early decision and thus close of the trial; others may lead to a long investigation. The classical sequential method was developed by Wald (1947). The application of sequential procedures to clinical trials was discussed by Armitage (1975). We shall briefly discuss the open sequential design and a closed sequential design.

1. *Open Sequential Design.* In this classical sequential procedure, patients are entered in the study in pairs; one is randomly assigned to receive treatment A and the other treatment B and preference of treatment is then recorded for each pair. In some cases, patients join the study sequentially and each patient receives the two treatments in adjacent periods of time, say weeks. The order of administration is random and the treatments are made indistinguishable to the patients. At the end of the second period each patient gives a preference for treatment A or B. This is known as an *untied pair*. In either case, the data are a series of preferences for A or B. If the two treatments give the same result in each pair of patients, this is known as a *tied pair*. The analysis is based entirely on the untied pair. If the two treatments are equally effective, one would expect to have equal number of preferences for A and B. The test is then based on the binomial distribution with n equal to the number of patients or number of pairs of patients and $p = \frac{1}{2}$. Table 12.3 gives an example of a typical open sequential design. Suppose two drugs A and B are given to a pair of patients. Column 1 gives the pair number, and column 2 the preference between the pair. The cumulative number of preferences for A (x) given in column 3 is to be compared with the critical values in column 4. The two-sided critical values are obtained from the binomial distribution at a 5% significance level. In this case, n is the number of pairs. The critical values (a, b) are obtained such that

$$P(x \leq a | n, p = \tfrac{1}{2}) \leq 0.025$$

and

$$P(x \geq b | n, p = \tfrac{1}{2}) \leq 0.025$$

Table 12.3 Open Sequential Design

Patient Pair Number	Preference	Cumulative Number of Preferences for A (x)	Critical Values
1	A	1	—
2	A	2	—
3	A	3	—
4	B	3	—
5	A	4	—
6	A	5	$(0, 6)$
7	A	6	$(0, 7)$
8	A	7	$(0, 8)$
9	B	7	$(1, 8)$
10	A	8	$(1, 9)$
11	A	9	$(1, 10)$
12	A	10	$(2, 10)$

The trial is terminated when $x \leq a$ or $x \geq b$ (critical region). No result is significant until $n = 12$ when $x = 10$ falls in the critical region. Thus, after the twelfth pair, the investigator can conclude that there is a significant difference between the treatments and terminate the trial.

Obviously, a large number of patients may be needed before a conclusion can be made and the trial terminated. There is no way to control the number of patients needed before the trial begins.

2. *Closed Sequential Design.* This is the design suggested by Armitage (1975). The main difference between this design and the open design is that it allows the investigator to terminate the trial concluding no preference between the two treatments.

Figure 12.2 is an example of a closed sequential design. The boundary points are obtained from tables of the binomial distribution or its normal

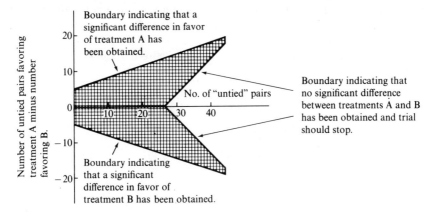

Figure 12.2 A closed sequential design.

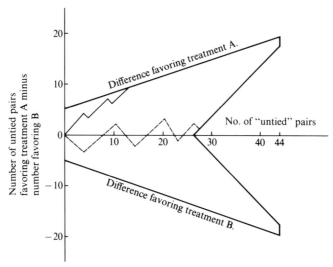

Figure 12.3 Two possible outcomes of a closed sequential design.

approximation when there are many untied pairs. The sequential results are represented by a zigzag line starting from the origin. For each A preference, the line is drawn one unit in a northeasterly direction to form a diagonal on the small square. For each B preference, the line is drawn one unit in a southeasterly direction to form a diagonal on the small square. The results continue to be entered until one of the boundary lines is cut. The solid line in Figure 12.3 is drawn on the basis of the data in Table 12.4, indicating a significant difference in favor of treatment A. However, a zigzag line like the broken line in Figure 12.3 indicates no appreciable difference between the two treatments and that the trial should stop.

Johnson and Pearce (1990) used this design in comparing two regimens to induce ovulation in patients with polycystic ovarian disease. The standard treatment was clomiphene, which may produce spontaneous abortion in 40–50% of pregnancies. The authors compared the rate of spontaneous

Table 12.4   Preference Expressed by Patients between Treatments A and B

Patient Number	Preference	Patient Number	Preference
1	A	9	A
2	No preference	10	A
3	A	11	B
4	A	12	A
5	A	13	A
6	B	14	A
7	A	15	A
8	A		

abortion after induction of ovulation with either clomiphene or pituitary suppression with buserelin followed by pure follicle stimulating hormone in 48 women with polycystic ovarian disease who had had recurrent primary spontaneous abortions. Pairs of patients were assigned to one or the other treatment by tossing a coin. A preference is defined as the case where both members of the pair became pregnant but one member miscarried. After the participation of 20 pairs, 11 preferences of buserelin were observed as compared to 2 preferences of clomiphene. No preference was observed in seven pairs. The boundary line indicating a preference for buserelin was cut and the result was significant at the 0.01 level.

An example of closed sequential design in which the analysis reaches the boundary indicating no treatment difference is given by Carlo et al. (1990).

In addition to using the binomial distribution to construct the boundary lines, other formulations can be used. Montaner et al. (1990) give an example in which a triangular test is used, which results in a triangular acceptance region. The test is based on the difference between the two treatments as measured by the log odds ratio when the outcome of interest is binary.

An investigator has to decide in advance whether sequential procedures are to be used. The advantages and disadvantages are as follows:

1. Advantages
   a. Sample sizes are on the average smaller than for the fixed-sample procedures.
   b. An early decision may be reached if there is a strong prior evidence that one of the treatments is more effective than the other.
   c. In most cases, sequential trials are easy to carry out.
2. Disadvantages
   a. Results from patients may not be available soon enough for decisions of whether to continue the trial. For example, if the response is survival time, months or years may elapse before an outcome can be determined.
   b. Although sample sizes are, on the average, smaller than in fixed-sample trials, there is no guarantee. In some situations, sequential trials also require a large number of patients.
   c. If an early decision is made from a small number of patients, the effectiveness of the treatments can only be estimated from the small group of patients.

12.3.3 Sample Size Determination in Fixed-Sample Designs

The essence of a successful clinical trial is a sufficient number of patients. Trials including only a small number of patients often give useless results unless one of the treatments shows an extremely higher effectiveness. A small p value (significance level) in a large trial usually indicates stronger evidence that the treatments are really different than the same p value in a small trial.

If the objectives of the comparative trial have been defined, the investigator must decide if the trial should involve a fixed number of patients or be conducted sequentially. The choice between these is to some extent related to the statistical philosophy adopted, for example, the Neyman–Pearson approach, the likelihood approach, or the Bayesian approach. The merits of the three viewpoints are still actively being investigated, and it is impossible to discuss them here in detail. These three approaches reach their conclusions by different logic; however, they do not lead to substantially different ways of conducting clinical trials. The likelihoodist considers only the likelihood of the observed set of data without regard to use of prior information. In the frequentist viewpoint, prior knowledge concerning the relative merits of the treatment is used informally both in the planning and analysis of clinical trials. In the Bayesian viewpoint, formal use is made of prior knowledge concerning the relative merits of the treatments. In the following, we discuss briefly the Neyman–Pearson approach for sample size determination.

Using the Neyman–Pearson approach to determine sample size, the clinical investigator must answer the following three questions:

1. How big a risk can be taken that the two treatments are incorrectly designated as significantly different? This is, in statistical terms, the level of significance α.
2. How big a risk can be taken that two treatments are incorrectly designated as not significantly different? This risk is referred to as the β error and $1 - \beta$ is defined as the power of a trial.
3. What is the smallest difference d between treatments (in terms of a quantitative difference such as a percentage of responders) that is important to detect?

Thus, in the Neyman–Pearson approach, a test of a null hypothesis is set up and carried out on the patient data. In a comparative trial of two treatments A and B, the null hypothesis might be that there is no real difference in the response rates of A and B, while the alternative hypothesis might be either that there is a difference or that the response rate is higher for A than for B. A two-sided alternative is usually used to decide whether the two treatments are different, and a one-sided alternative is used if prior information indicates that the new treatment is at least as good as the standard treatment.

A. Sample Size Calculations for Comparing Two Response Rates

The number of patients needed in an experimental and a control group for comparing two response rates based on α, β, and d have been investigated by many researchers. Cochran and Cox (1957, Table 2.1) calculate the required sample size for each treatment group by the arc sine transformation, which yields the approximate formula (Sillitto 1949)

$$n = \frac{(Z_\alpha + Z_\beta)^2}{2(\sin^{-1}\sqrt{P_1} - \sin^{-1}\sqrt{P_2})^2} \qquad (12.1)$$

where P_1 and P_2 are the response rates to treatments A and B, respectively, and Z_α and Z_β are the upper percentage points of the standard normal distribution corresponding to the significance level α and β. Later Gail and Gart (1973) used the exact test for the determination of sample sizes. Cochran and Cox's tables were then modified according to the exact test (Gehan and Schneiderman 1973). The difference between the two methods increases as n decreases. Tables 12.5 and 12.6 give the modified number of patients needed in each treatment group for a given probability of obtaining a significant result in a one-sided and two-sided test, respectively. For example, suppose 20% of patients are predicated to respond to a standard treatment, and the comparative trial is to determine whether a new treatment has a response rate of 50%. Then $P_1 = 0.20$, $P_2 = 0.50$, and $d = 0.30$. From Table 12.5, the number of patients needed in each treatment group for a one-sided test is 36 at a 5% significance level and 80% power and 76 at a 1% significance level and 95% power. The sample size required is roughly proportional to the desired power and selected significance level. Also, two-sided alternative hypotheses require larger sample sizes than one-sided alternatives.

Another formula for sample sizes in comparative trials can be found in Snedecor and Cochran (1967, pp. 111–114).

B. Sample Size Calculations for Comparing Two Survival Distributions

In many clinical trials, remission duration or survival time is used to compare two or more treatments. The sample size issue in this case has also been discussed by many, for example, George and Desu (1974), Wu et al. (1980), Lachin (1981), Schoenfeld and Richter (1982), Freedman (1982), Makuch and Simon (1982), Lachin and Foulkes (1986), and Moussa (1988). In the following, two sample size calculation methods are introduced for survival times involving censored observations.

The first method is based on the assumption that the patients are entered on the trial at a uniform rate until the trial ends in T years and the survival time follows an exponential distribution with parameter λ. Therefore, comparing the two treatments in terms of survival time is equivalent to comparing the two parameters of the exponential distributions. Let λ_1 and λ_2 be the parameters of the two exponential distributions. The total sample size needed for the two groups can be calculated by using the equation

Table 12.5 Number of Patients Needed in an Experimental and a Control Group for a Given Probability of Obtaining a Significant Result (One-Sided Test)

Smaller Proportion of Success (P_1)	d = Larger Minus Smaller Proportion of Success ($P_2 - P_1$)													
	0.05	0.10	0.15	0.20	0.25	0.30	0.35	0.40	0.45	0.50	0.55	0.60	0.65	0.70
0.05	330	105	55	40	33	24	20	17	13	12	10	9	9	8
	460	145	76	48	39	31	25	20	19	15	13	11	10	9
	850	270	140	89	63	37	41	34	21	25	22	18	16	14
0.10	540	155	76	47	37	30	23	19	16	13	11	11	9	8
	740	210	105	64	41	38	30	24	20	17	15	12	11	10
	1370	390	195	120	81	60	46	41	35	28	24	20	17	16
0.15	710	200	94	56	43	32	26	22	17	15	11	10	9	8
	990	270	130	77	52	43	34	26	23	19	16	12	11	10
	1820	500	240	145	96	69	52	41	37	30	24	22	18	16
0.20	860	230	110	63	42	36	27	23	17	15	12	10	9	8
	1190	320	150	88	58	46	36	29	23	18	16	12	11	10
	2190	590	280	160	105	76	57	44	39	30	27	22	18	16
0.25	980	260	120	69	45	37	31	23	17	15	12	10	9	
	1360	360	165	96	63	46	38	30	23	18	16	12	11	
	2510	660	300	175	115	81	60	46	40	33	27	22	17	

0.30	1080	280	130	73	47	37	31	23	17	15	11	10
	1500	390	175	100	65	46	38	30	23	18	16	12
	2760	720	330	185	120	85	61	47	39	32	24	20
0.35	1160	300	135	75	48	37	31	23	17	15	11	
	1600	410	185	105	67	46	38	30	23	18	15	
	2960	750	340	190	125	85	61	46	39	30	24	
0.40	1210	310	135	76	48	37	30	23	17	13		
	1670	420	190	105	67	46	38	30	23	17		
	3080	780	350	195	125	84	60	44	37	28		
0.45	1230	310	135	75	47	36	26	22	16			
	1710	430	190	105	65	44	36	26	20			
	3140	790	350	190	120	81	57	41	34			
0.50	1230	310	135	73	45	36	26	19				
	1710	420	185	100	63	41	35	24				
	3140	780	340	185	115	76	52	39				

Modified from Cochran and Cox (1957).

Upper figure: Test of significance at 0.05 for α, power equals 0.8 for $1 - \beta$.

Middle figure: Test of significance at 0.05 for α, power equals 0.09 for $1 - \beta$.

Lower figure: Test of significance at 0.01 for α, power equals 0.95 for $1 - \beta$.

Source: Gehan and Schneiderman (1973).

Table 12.6 Number of Patients Needed in an Experimental and a Control Group for a Given Probability of Obtaining a Significant Result (Two-Sided Test)

Smaller Proportion of Success (P_1)	d = Larger Minus Smaller Proportion of Success ($P_2 - P_1$)													
	0.05	0.10	0.15	0.20	0.25	0.30	0.35	0.40	0.45	0.50	0.55	0.60	0.65	0.70
0.05	420	130	69	44	36	31	23	20	17	14	13	11	10	8
	570	175	93	59	42	37	31	24	21	18	16	13	12	11
	960	300	155	100	72	54	42	38	33	27	24	20	18	16
0.10	680	195	96	59	41	35	29	23	19	17	13	12	11	8
	910	260	130	79	54	40	36	29	24	20	17	16	13	11
	1550	440	220	135	92	68	52	41	38	32	26	23	19	17
0.15	910	250	120	71	48	39	31	25	20	17	15	12	11	9
	1220	330	160	95	64	46	40	31	26	22	18	16	13	11
	2060	560	270	160	110	78	59	47	41	35	29	24	21	18
0.20	1090	290	135	80	53	42	33	26	22	18	16	12	11	9
	1460	390	185	105	71	51	43	33	28	23	18	16	13	11
	2470	660	310	180	120	86	64	50	44	36	27	24	21	17
0.25	1250	330	150	88	57	44	35	28	22	18	16	12	11	
	1680	440	200	115	77	56	45	36	29	23	18	16	12	
	2840	740	340	200	130	95	68	52	45	36	29	24	19	

0.30	1380	360	160	93	60	44	36	29	22	18	15	12
	1840	480	220	125	80	56	46	36	29	23	18	16
	3120	810	370	210	135	95	69	53	45	36	29	23
0.35	1470	380	170	96	61	44	36	28	22	17	13	
	1970	500	225	130	82	57	46	36	28	22	17	
	3340	850	380	215	140	96	69	52	44	35	26	
0.40	1530	390	175	97	61	44	35	26	20	17		
	2050	520	230	130	82	56	45	32	26	20		
	3480	880	390	220	140	95	68	50	41	32		
0.45	1560	390	175	96	60	42	33	25	19			
	2100	520	230	130	80	54	43	32	24			
	3550	890	390	215	135	92	64	47	38			
0.50	1560	390	170	93	57	40	31	23				
	2100	520	225	125	77	51	40	29				
	3550	880	380	210	130	86	59	45				

Modified from Cochran and Cox (1957).
Upper figure: Test of significance at 0.05 for α, power equals 0.8 for $1 - \beta$.
Middle figure: Test of significance at 0.05 for α, power equals 0.9 for $1 - \beta$.
Lower figure: Test of significance at 0.01 for α, power equals 0.95 for $1 - \beta$.
Source: Gehan and Schneiderman (1973).

$$N = \frac{(Z_\alpha + Z_\beta)^2 [\phi(\lambda_1) Q_1^{-1} + \phi(\lambda_2) Q_2^{-1}]}{(\lambda_1 - \lambda_2)^2} \tag{12.2}$$

where

$$\phi(\lambda) = \frac{\lambda^3 T}{\lambda T + e^{-\lambda T} - 1} \tag{12.3}$$

and Q_1 and Q_2 are the sample fractions for the two treatment groups (Lachin 1981). In other words, the sample sizes for groups 1 and 2 based on (12.2) are, respectively, $n_1 = NQ_1$ and $n_2 = NQ_2$. Thus, (12.2) allows unequal group sizes. For equal group sizes, $Q_1 = Q_2 = \frac{1}{2}$, (12.2) becomes

$$N = \frac{2(Z_\alpha + Z_\beta)^2 [\phi(\lambda_1) + \phi(\lambda_2)]}{(\lambda_1 - \lambda_2)^2} \tag{12.4}$$

and each treatment group needs to have $\frac{1}{2} N$ patients.

Example 12.1

Consider a clinical trial to compare a new drug with a standard drug. Suppose that the standard drug produces exponentially distributed survival times with hazard rate $\lambda_1 = 0.40$ or a median survival time of 1.7 year. The investigator estimates that the new drug will produce a hazard rate of $\lambda_2 = 0.25$ or a median survival time of 2.8 years. Assume that the trial will last 3 years and patients are to be entered during the entire 3 years. Let $\alpha = 0.05$ (one-sided) and $\beta = 0.10$ (or 90% power). Using (12.3), we have

$$\phi(\lambda_1) = \frac{(0.4)^3 (3)}{3(0.4) + e^{-(0.4)3} - 1} = 0.383$$

$$\phi(\lambda_2) = \frac{(0.25)^3 (3)}{3(0.25) + e^{-(0.25)3} - 1} = 0.070$$

For equal-sized groups, the required total sample size, according to (12.4), is

$$N = \frac{(1.64 + 1.28)^2 (0.383 + 0.070)}{2(0.4 - 0.25)^2} \cong 86$$

or 43 patients per group. If a two-sided test is perferred, $Z_\alpha = 1.96$, then $N = 106$, or 53 patients per group.

If, in a trial, the patients are recruited over an interval 0 to T_0 $(T_0 < T)$ and then followed to the end of the study period T, the desired sample size can still be obtained by using (12.2) but with $\phi(\lambda)$ defined as

$$\phi(\lambda) = \lambda^2 \left[1 - \frac{e^{-\lambda(T - T_0)} - e^{-\lambda T}}{\lambda T_0} \right]^{-1} \tag{12.5}$$

For the trial in Example 12.1, if the patients are entered in the first year of the 3-year study, $T_0 = 1$. Using (12.5) and (12.4), the total number of patients needed is 74, or 37 per group for a one-sided test.

If there is a real difference among the treatments, the more patients we enter into the trial, the more likely we are to detect it statistically. However, because of the large random differences between different groups of patients, clinical trials are not as sensitive as one would expect in detecting differences. The sensitivity of a comparative trial of survival time, for example, depends much more on the number of patients who die before the statistical analysis begins than on the total number of patients entered into the trial. An alternative method for sample size determination, proposed by Makuch and Simon (1982), gives the number of deaths (or failures) needed in each group. This method does not require any distributional assumption for the survival time.

Let K be the number of treatment groups, $K \geq 2$, and μ_j be the mean (or median) survival time for the jth treatment group. Assume that group 1 is the control group. Let $a_j = \mu_j/\mu_1$ and let a_k be the largest ratio of mean survival times among all possible $K(K-1)/2$ treatment group pairings; that is, the largest mean survival time ratio between treatment groups is between treatment group k and treatment group 1. In order to ensure a power of $1 - \beta$ for detecting a ratio of magnitude a_k at a significance level of α, the number of deaths (or failures) required in *each* group is

$$n_d = \frac{2\tau(K-1, \alpha, \beta)}{(\log_e a_k)^2}$$ (12.6)

where $\tau(K-1, \alpha, \beta)$ is a noncentrality parameter. (Discussion of the noncentrality parameter is beyond the level of this book and, therefore, not given here.) Some of its values are given in Table 12.7.

Example 12.2
Consider a cancer clinical trial to compare three treatments: placebo, radiation therapy, and radiation therapy plus chemotheraphy. Assume that the median survival times of the three treatment groups are estimated as 1.5, 2.0, and 3.0 years, respectively. In this case $a_2 = 2.0/1.5 = 1.33$,

Table 12.7 Values of $\tau(K-1, \alpha, \beta)$ Required for Sample Size Calculation

K	$\tau(K-1, 0.05, 0.05)$	$\tau(K-1, 0.05, 0.10)$	$\tau(K-1, 0.05, 0.20)$
2	12.995	10.507	7.849
3	15.443	12.654	9.635
4	17.170	14.171	10.903
5	18.572	15.405	11.935
6	19.780	16.469	12.828

Source: Makuch and Simon (1982).

Table 12.8   Number of Failures (n_d) Required to Detect a Significant Difference among K Treatment Groups

		a_k									
K	$1 - \beta$	1.2	1.3	1.4	1.5	1.6	1.7	1.8	1.9	2.0	2.5
3	0.90	762	368	224	154	115	90	74	62	53	31
	0.80	580	280	171	118	88	69	56	47	41	23
4	0.90	853	412	251	173	129	101	83	69	59	34
	0.80	656	317	193	133	99	78	64	53	46	26

Source: Makuch and Simon (1982).

$a_3 = 3.0/1.5 = 2.0$, and $a_1 = 1$. Among the three possible ratios of median survival times, a_3 is the largest. For $\alpha = 0.05$, $\beta = 0.10$ (or 90% power), from Table 12.7, $\tau(3 - 1, 0.05, 0.10) = 12.654$, and using (12.6),

$$n_d = \frac{2(12.654)}{(\log_e 2)^2} = 53$$

That is, 53 deaths (or failures) are required in each treatment group.

Table 12.8 gives the number of failures required to detect a significant difference among three and four groups at $\alpha = 0.05$, $1 - \beta = 0.9$, 0.8, and $a_k = 1.2$ (0.1) 2.0, and 2.5.

It must be pointed out that the Neyman–Pearson approach outlined above should only serve as a guide in planning a study. It should not bind the investigator.

12.3.4   Repeated Significance Testing or Group Sequential Design

In determining the best treatment, it is ethical to minimize the number of patients exposed to the inferior treatment and the duration of such exposure. If one treatment is found clearly superior to the other, the investigator has the ethical responsibility to terminate the trial and declare the finding so that more patients can be treated with the superior treatment. Naturally, the investigator may wish to analyze periodically the accumulating data during the clinical trial in order to detect significant difference and determine whether to terminate the trial early. Thus, a logical consequence is to design a clinical trial that would allow for planned interim data analysis. An important question is then how often the data should be analyzed and how repeated analyses would affect the significance level.

If a fixed sample design is used, the significance level is selected with the assumption that analysis would be performed only once at the end of the trial when all of the data have been collected. An α of 0.05 means that an observed difference has less than 5 chances out of 100 of having occurred by chance alone. If the investigator chooses to perform significance tests repeatedly during the trial as the data are accumulated, the chance of obtaining a significant result is much higher than 5 out of 100. In other words, if the accumulating data are tested again and again using a nominal significance level, eventually, a significant result will occur. Armitage et al. (1969) calculate the overall probability of achieving a significant result with a given nominal α after k repeated tests when there is no difference between the two treatments (Table 12.9). For example, a nominal α of 5% would be increased to 8.3% if two tests of significance are performed and it would be increased to 14.2% at the fifth test. Thus, the probability of erroneously declaring the difference between the two treatments increases as the number of analyses increases.

One way to overcome this problem is to use a nominal significance level α' small enough at each interim analysis so that the overall significance level, after repeated interim analyses, is still within the preset limit. Table 12.10 gives the nominal significance level necessary to achieve a given true level of significance. For example, if five tests are to be performed during the trial, in order to achieve a true significance level of 5%, the investigator would have to use a nominal significance level of 1.59% or less in any of the five tests.

However, it would be difficult to organize a trial without knowing how often the data should be analyzed. On the other hand, it would not be feasible to perform analysis after the entry of every pair of patients. A natural improvement suggested by Pocock (1977) is to apply the significance test at longer equally spaced intervals, say every $2m$ patients in the case of two treatments (m patients per treatment). Each $2m$ patients is considered a "group," and thus the approach is named *group sequential design*. In the following, we introduce Pocock's original group sequential design, which uses a constant nominal significance level α' (Table 12.10) for each test. This procedure illustrates the general concept of group sequential design.

Consider a clinical trial to compare two treatments A and B. Suppose that patients are entered sequentially and are randomized using a restricted (or block) method (described in Chapter 13) so that each consecutive group of $2m$ patients has m on each treatment. Assuming that the response

Table 12.9 Overall Probability of Achieving Significant Result with Given Nominal Significance Level after k Repeated Tests When There Is No Difference in Effects of Two Treatments

Nominal Significance Level (%)	Overall Probability (%)								
	$k = 1$	$k = 2$	$k = 3$	$k = 4$	$k = 5$	$k = 10$	$k = 25$	$k = 50$	$k = 200$
1	1	1.8	2.4	2.9	3.3	4.7	7.0	8.8	12.6
5	5	8.3	10.7	12.6	14.2	19.3	26.6	32.0	42.4
10	10	16.0	20.2	23.4	26.0	34.2	44.9	52.4	65.2

Source: Armitage et al. (1969).

Table 12.10 Value of Nominal Significance Level (σ') Necessary to Achieve Given True Level of Significance after k Repeated Tests

True Level of Significance (%)	Nominal Significance Level (%)												
	$k=1$	$k=2$	$k=3$	$k=4$	$k=5$	$k=6$	$k=7$	$k=8$	$k=9$	$k=10$	$k=15$	$k=20$	$k=100$
1	1	0.45	0.41	0.33	0.28	0.25	0.23	0.21	0.20	0.19	0.15	0.13	0.06
5	5	2.96	2.21	1.83	1.59	1.42	1.30	1.20	1.13	1.07	0.86	0.75	0.32
10	10	6.01	4.62	3.85	3.37	3.04	2.80	2.60	2.45	2.32	1.88	1.66	0.72

Source: McPherson and Armitage (1971).

variable is normal with known variance σ^2 and unknown means μ_A and μ_B, respectively, for treatments A and B. Let $\bar{x}_{Aj}$ and $\bar{x}_{Bj}$ be the observed mean responses to the two treatments in the jth *group* of m patients each. The average difference after k *groups*,

$$\bar{d}_k = \sum_{j=1}^{k} (\bar{x}_{Aj} - \bar{x}_{Bj})/k$$

is normally distributed with mean $\delta = \mu_A - \mu_B$ and variance $2\sigma^2/km$. Thus, the null hypothesis $H_0 : \mu_A = \mu_B$ $(H_1 : \mu_A \neq \mu_B)$ would be rejected if $|\bar{d}_k\sqrt{km}/\sqrt{2}\sigma| > Z_{\alpha'/2}$. In other words, the two-sided significance test applied after k groups has a significance level of

$$P_k = 2\left[1 - \Phi\left(\frac{\bar{d}_k\sqrt{km}}{\sqrt{2}\sigma}\right)\right]$$

where Φ is the distribution function of the standard normal distribution. If $P_k < \alpha'$, or $|\bar{d}_k\sqrt{km}/\sqrt{2}\sigma| > Z_{\alpha'/2}$, we reject the null hypothesis, claim evidence of treatment difference, and terminate the trial. Otherwise, the trial will continue.

The two decisions to be made at the design stage of a group sequential trial are (1) the maximum number of *groups* (or repeated significance test) K and (2) the number of patients needed in each *group*, $2m$. The maximum number of *groups* is required so that the decision of no evidence of treatment difference can be made if $P_k > \alpha'$ for all $k = 1, \ldots, K$. The selection of K may depend on the length of the trial, cost of interim analysis, and the response variable. However, there is little advantage in analyzing clinical trial data more than five times unless an extremely larger difference is expected between the two treatments (Pocock 1982). Pocock (1977) provides tables for obtaining suitable values of m and K given the overall significance level α, true difference δ, and statistical power $1 - \beta$. For any chosen value of K, $m = 2(\Delta\sigma/\delta)^2$, where $\Delta = \sqrt{m}\sigma/\sqrt{2}\delta$. Table 12.11 gives the value of Δ for various α, K, and $1 - \beta$. For example, if the overall significance level $\alpha = 0.05$, the required power is 0.90, and the investigator decides to conduct a maximum of four significance tests, from Table 12.11,

Table 12.11 Values of $\Delta = (\sqrt{\bar{m}}\delta)/(\sqrt{2}\sigma)$ for Normal Group Sequential Tests with Two Values of Overall Significance Level α Power $1 - \beta$, and Maximum Number of Groups K

	$\alpha = 0.5$				$\alpha = 0.01$			
K	$1 - \beta = 0.75$	$1 - \beta = 0.9$	$1 - \beta = 0.95$	$1 - \beta = 0.99$	$1 - \beta = 0.75$	$1 - \beta = 0.90$	$1 - \beta = 0.95$	$1 - \beta = 0.99$
1	2.634	3.242	3.605	4.286	3.250	3.858	4.221	4.920
2	1.967	2.404	2.664	3.152	2.405	2.839	3.099	3.584
3	1.647	2.007	2.221	2.622	2.006	2.362	2.575	2.973
4	1.449	1.763	1.949	2.297	1.760	2.070	2.255	2.600
5	1.311	1.592	1.759	2.071	1.588	1.866	2.032	2.341
6	1.207	1.464	1.617	1.903	1.460	1.714	1.866	2.149
7	1.125	1.364	1.506	1.770	1.359	1.595	1.735	1.998
8	1.058	1.282	1.415	1.662	1.277	1.498	1.630	1.875
9	1.002	1.214	1.339	1.573	1.209	1.417	1.541	1.773
10	0.955	1.156	1.275	1.497	1.150	1.348	1.466	1.686
11	0.914	1.105	1.219	1.431	1.100	1.289	1.401	1.611
12	0.878	1.061	1.170	1.373	1.056	1.237	1.344	1.502
15	0.791	0.959	1.053	1.235	0.950	1.112	1.209	1.389
20	0.691	0.835	0.919	1.077	0.829	0.970	1.053	1.209

Source: Pocock (1977).

we obtain $\Delta = 1.763$. Thus if the mean treatment difference to be detected (δ) is half of the standard deviation σ, that is, $\sigma/\delta = 2$, $m = 2(1.763 \times 2)^2 = 24.86$, or 50 patients per *group* (or 25 patients on each treatment).

A number of other multiple testing procedures have been suggested for comparing two treatments, for example, Peto et al. (1976), O'Brien and Fleming (1979), DeMets and Ware (1980), Pocock (1982), Fleming et al. (1984), and Whitehead and Stratton (1983). A summary of several of these group sequential designs are tabulated by Geller and Pocock (1987). For convenience, two of their tables, reduced, are reproduced in Tables 12.12 and 12.13. Table 12.12 gives the nominal significance level α' for two-sided group sequential designs with $\alpha = 0.05$ by Pocock (1977), O'Brien and Fleming (1979), Peto et al. (1976), and Pocock (1982). Table 12.13 provides the sample size needed per *group* ($2m$) in each of the four designs with power 0.75, 0.80, and 0.90 under the alternative hypothesis that $\mu_A - \mu_B = \delta$.

Using 72 studies from the Eastern Cooperative Oncology Group, Rosner and Tsiatis (1989) compare the fixed-sample design with four group sequential designs proposed by Pocock (1977), O'Brien and Fleming (1979), Fleming et al. (1984), and Whitehead and Stratton (1983) with respect to savings in time and information due to early termination and sample size requirements. Interim analyses were performed every six months using the logrank test at each analysis. The overall significance level and statistical

Table 12.12 Nominal Significance Levels α' for Two-Sided Group Sequential Designs with Overall Significance Level $\alpha = 0.05$

Number of groups	Analysis	Pocock (1977)	O'Brien and Fleming (1979)	Petol et al. (1976)	Pocock (1982), Optimal for $1 - \beta = 0.8$
2	1	0.029	0.005	0.001	0.025
	2	0.029	0.048	0.050	0.034
3	1	0.022	0.0005	0.001	0.014
	2	0.022	0.014	0.001	0.021
	3	0.022	0.045	0.050	0.030
4	1	0.018	0.0001	0.001	0.008
	2	0.018	0.004	0.001	0.017
	3	0.018	0.019	0.001	0.020
	4	0.018	0.043	0.049	0.029
5	1	0.016	0.00001	0.001	0.004
	2	0.016	0.0013	0.001	0.013
	3	0.016	0.008	0.001	0.017
	4	0.016	0.023	0.001	0.018
	5	0.016	0.041	0.049	0.028

Source: Geller and Pocock (1987).

Table 12.13 Sample Size per Group $2m^a$ for Group Sequential design

Design	Power	$K = 2$	$K = 3$	$K = 4$	$K = 5$
Pocock (1977)	0.75	15.48	10.85	8.40	6.87
	0.80	17.43	12.20	9.44	7.71
	0.90	23.12	16.11	12.43	10.14
O'Brien and	0.75	13.99	9.42	7.11	5.71
Fleming (1979)	0.80	15.82	10.65	8.04	6.45
	0.90	21.16	14.23	10.74	8.63
Peto et al	0.75	13.89	9.27	6.96	5.58
(1976)	0.80	15.71	10.49	7.88	6.31
	0.90	21.03	14.40	10.54	8.44
Pocock (1982)	0.75	15.05	10.22	7.70	6.17
	0.80	16.97	11.51	8.68	6.95
	0.90	22.56	15.29	11.51	9.23

[a]Multiply each entry by σ^2/δ^2.

Source: Geller and Pocock (1987).

power used were, respectively, 0.05 and 0.80. Decisions to reject or accept the null hypothesis were found to be consistent across the designs. The designs of O'Brien and Fleming (1979) and Whitehead and Stratton (1983) are competitive with respect to early decisions. The latter arrives at decisions sooner but it requires roughly 70% more patients than the fixed-sample design. The O'Brien and Fleming procedure requires only 3–5% more than the fixed-sample design. Another advantage of this design is that early analysis will terminate the trial only if the treatment difference is extremely large.

Interim analyses in cancer and other clinical trials are used commonly. Examples are Kemeny et al. (1984, 1988), Geller et al. (1984), Leyland-Jones et al. (1986), and Rowe-Jones et al. (1990). It should be kept in mind that the decision to stop a clinical trial early is not entirely statistical. Although statistical evidence plays a primary role, clinical knowledge—practical aspects of the treatments including adverse effects and costs, and future consequences must also be considered. Therefore, the decision of early termination should be made by a committee of experts, not by the statistician alone. Another important point is that interim test results should not be published or revealed publicly. There is always a tendency for the public to accept preliminary results as final. The investigators may face external pressure to stop the trial. Participants who are not on the "better" treatment may wish to stop the trial so that they can switch treatment. The trial could be terminated prematurely and the question still remains un-answered. Geller and Pocock (1987) give an example of how publicly revealing interim results can affect the future of a trial.

BIBLIOGRAPHICAL REMARKS

There are a large number of publications on the subject of clinical trials. It is impossible to cite all of them. An account of the historical development of clinical trials is given by Bull (1959). Classic works include Hill (1060a,b) and Witts (1964). Gehan and Schneiderman (1973), Burdette and Gehan (1970), Pocock (1983), Friedman et al. (1985), and Meinert (1986) give excellent general discussions on the subject. Another useful paper is Mainland (1960). For cancer clinical trials, the reader is referred to Staquet (1978) and Buyse et al. (1984).

An excellent paper written by a group of British and American statisticians on the subject of cooperative clinical trials is Peto et al. (1976). Part I of the paper discusses problems related to the design of such trials; Part II gives instructions for statistical analysis (life-table analysis and logrank test) and discusses the interpretation of trial results. This paper, written for the physician, addresses many practical problems arising in clinical trials.

In addition to Armitage (1975), Rümke (1963) gives a lucid introduction to the ideas and problems of sequential procedures in medicine. Group sequential procedures and interim analyses are also discussed by Tsiatis (1982), Slud and Wei (1982), Sellke and Siegmund (1983), Lan and DeMets (1983, 1989) DeMets and Lan (1984), Bristol (1988), and Enas et al. (1989).

EXERCISES

12.1 Define phase I and phase II clinical trials.

12.2 Suppose that in a phase II trial the investigator is interested in at least a 20% response for a new drug. He or she would allow only a 10% rejection error. How many patients are required?

12.3 What should be considered when planning a phase III trial?

12.4 What are the advantages and disadvantages of a double-blind trial? Give an example in your own practice in which a double-blind trial is appropriate and an example in which a double-blind trial is not suitable.

12.5 In determining sample size for a comparative trial, suppose that the investigator is willing to run a risk of $\alpha = 0.05$, and assume that the new treatment will produce at least a 20% higher response rate than the standard treatment. If the response rate to the standard treatment is known to be 15%, according to (12.1), how many patients are needed in each treatment group so that the chance of detecting the difference (or power) is 85%?

12.6 Assuming the survival times in two treatment groups follow the exponential distribution with median survival times 2.5 and 3.0 years, respectively, calculate the required sample size for a two-sided test with:

  a. $a = 0.05$, $\beta = 0.10$ and patients are recruited during the entire 5-year study period

  b. $\alpha = 0.10$, $\beta = 0.20$ and patients are recruited during the first 3 years of the 5-year study period

12.7 Consider a clinical trial to compare four treatments. The estimated mean remission times yielded by the four treatments are 3, 4, 4.5, and 6 years. Calculate the number of relapses required to detect a significant difference among the four treatments at $\alpha = 0.05$ and $\beta = 0.2$.

12.8 In designing a group sequential clinical trial, the investigator wishes to conduct 5 significance tests. The mean treatment difference to be detected is one-third of the standard deviation. How many patients are needed per *group* if

  a. $\alpha = 0.05$ and $1 - \beta = 0.90$?

  b. $\alpha = 0.01$ and $1 - \beta = 0.75$?

CHAPTER 13

Planning and Design of Clinical Trials (II)

In Chapter 12 we introduced the three phases of clinical trials. Among the three, a phase III or comparative trial is by far the most important since its results will affect future decisions concerning patients' lives or welfare. Several considerations in planning and designing comparative trials have been briefly discussed. In this chapter we focus our attention on several additional issues. Section 13.1 discusses the necessary elements of a protocol. Section 13.2 deals with the problem and methods of randomization. Section 13.3 is concerned with the use of prognostic factors. Finally, Section 13.4 discusses randomized versus historical controls in cancer clinical studies.

13.1 PREPARATION OF PROTOCOLS

Any scientific experiment requires a well-prepared plan. A protocol is a plan of scientific experiment or treatment according to Webster's New Collegiate Dictionary. There are no standard rules for the elements of a protocol, but in general, it should outline the purpose of the clinical study, details of experimental design, and the method of administering the treatment. Table 13.1 gives the major ingredients of a protocol for clinical trials.

A protocol should begin with its specific aims, a brief review of the history of the problem, and a rationale for doing the study. If other similar studies are being conducted or have been done, the results should be summarized along with the questions that remain unanswered.

Before undertaking a clinical trial, the investigator must have a clear objective in mind, and this objective must be well justified. A trivial question may not deserve a clinical trial. There should also be a statement of specific and well-defined objectives, which may be divided into primary and secondary. However, there should not be too many objectives, as such studies are too difficult to plan and carry out. The design of the study should

Table 13.1   Contents of a Protocol

1. Specific aims
2. Introduction, scientific background, literature review, and significance of the study
3. Objectives (primary and secondary) of the study
4. Patient population and inclusion and exclusion criteria
5. Experimental design of the study
6. Treatment administration programs
7. Clinical and laboratory procedures and data to be collected
8. Criteria for evaluating response, nonresponse, and toxicity
9. Trial monitoring and frequency of interim analysis
10. Procedures in the event of early significant results
11. Statistical considerations: sample size, interim and final analysis strategies
12. Informed consent
13. Data collection forms
14. References
15. Responsible investigator and telephone number

ensure that at least the primary objectives are attained. The major response variables must be identified according to the primary objectives. If there are multiple response variables, sample size estimation should be done for each variable and the largest estimate used. This could lead to a very large sample size and the trial could be very expensive. This is another good reason for a very careful consideration of the study objectives.

A protocol should define clearly the type of patients that will be entered in the study. This, of course, is related to the objectives of the study. For example, if the trial is to compare two treatments for stage III and IV melanoma, these stages must be precisely defined and only those patients who satisfy the definitions are eligible for the trial. Other questions the protocol must answer are: If the differential diagnosis of the disease is difficult, is a confirmed diagnosis of the disease required and what constitutes acceptable confirmation? If an incorrect diagnosis is discovered later, what is to be done? Are patients who have received previous treatment acceptable or are only untreated patients to be included? If the study involves more than one clinic or hospital, it is even more important to document the inclusion and exclusion criteria so that every participating investigator enters exactly the same kind of patients.

The investigator should keep a log book of all patients seen during the study period who have a diagnosis of the disease under study. If a patient is not entered into the study, the reason should be given. Information from these patients might be valuable in making statements about overall patient characteristics and about any differences between participants and nonparticipants. Since most data analyses are done by computer, it is convenient to have computerized data sheets to record patient information such as medical history and clinical and laboratory data. Every item on the form should be well defined and clear to the medical technicians who will fill out the forms, to the systems analyst who will build data files on the computer, and to the statisticians who will analyze the data.

The protocol must specify the experimental design. The investigator and the statistician must decide whether the trial is to be sequential or fixed sample, simple randomized or stratified randomized, open or blinded, and so forth. If patients are to be entered sequentially, a stopping rule should be given. If it is going to be fixed sample, sample size and treatment assignments have to be specified. Other questions that must be answered are: If a new treatment is to be compared with a standard treatment, will the controls be randomized or historical? If randomized controls are to be used, what type of randomization will be used? If historical controls are to be used, how will the selection procedure ensure that the historical patients are comparable to the current patients? Some estimate of the number of patients needed and duration of the trial should also be included in the protocol. Statistical considerations should be taken into account in determining sample size and the type of design. Statisticians can advise on these problems in the planning stage.

A comparative clinical trial is always a trial of an agent given according to a particular dosage schedule in a certain way to a certain type of patient; it is not a trial of the agent per se. A protocol should give a detailed description of the treatment program, that is, the regimen of treatment that each patient is to receive. This includes not only the total dose but also the method and administration schedule. If under certain circumstances, for example, severe adverse effect, the dosage is to be altered, the criteria for severity must be well defined and the altered dosage clearly specified. It is usually helpful to have schematic diagram outlining the entire treatment program.

In general, patients on a clinical trial are evaluated using clinical and laboratory procedures. Physical examination procedures as well as laboratory tests must be described in detail in the protocol. For example, if an electrocardiogram (ECG) is to be performed, the protocol should specify whether it will be a resting ECG or exercise ECG. Similarly, if blood will be drawn for glucose measurement, it should be specified whether the patients must fast for a given number of hours prior to the clinic visit. In addition, the schedule for clinical and laboratory examinations must be provided.

A protocol should contain criteria for evaluating treatment effectiveness. This includes response and adverse effects. The definition of response must be stated clearly. For example, *complete response* is usually defined as disappearance of all objective signs and symptoms of disease. All *objective signs and symptoms of disease* have to be listed explicitly. Other possible measures of response are time to recurrence, length of survival, time to development of metastasis, proportion of patients surviving a fixed time after treatment, or a certain percentage reduction in white blood cells. Some of these measures may take a long time to observe, for example, survival time for a chronic disease, and thus require a lengthy trial time. In this case, time to recurrence or increase of disease may be more feasible. A protocol should also provide procedures in the event of severe side-effects and toxicity. Rules for adjusting dosage or stopping treatment should be given. Information concerning the type of effects and degrees of toxicity should be recorded on the data forms.

To ensure that the protocol is being followed during the trial, all clinical trials should be monitored closely. Any deviations from the protocol should be corrected as early as possible. Adverse effects need to be monitored and reported so that the investigator can take prompt action in case of severe toxic reaction. In addition, whether interim analysis for treatment comparison is to be performed should be specified in the protocol. This is an issue involving ethical, practical, and statistical considerations. If the trial lasts a long duration, methods for retaining subjects' interest and for maintaining good compliance should also be discussed.

Statistical considerations include sample size determination and data analysis strategies. If interim analysis is to be performed, the protocol should describe when and how the data will be organized and how often it will be carried out. Decisions on the interpretation and actions to be taken should be well thought out and described in the protocol. The major objective for performing interim analysis is to avoid undue prolongation of a trial if the comparison between treatments is clear-cut so that more patients can be treated with a superior treatment. Therefore, it is critically important to determine which response variables are to be analyzed and what the stopping rules are. The stopping rules usually involve statistical significance level. However, it should not be the only rule.

In addition to the measure of response, required laboratory and clinical data should be specified in a protocol. Data collection forms may then be designed accordingly.

Other elements that should be included in a protocol are references, responsible investigator's name and telephone number, form of informed consent, and personnel who are to coordinate the execution of the protocol. The last item is especially important in a collaborative clinical trial. The coordinator should know the study thoroughly, be able to prevent trouble or catch it quickly when it happens, and always be ready to help when needed. The coordinator's duties often fall upon an applied statistician.

13.2 RANDOMIZATION

As expected, the results of a statistical analysis depend on the nature of the data. Results from a comparative trial depend on how patients were assigned to the treatments. It is easy to conduct a comparative trial in such a way that the results are useless. For example, suppose treatments A and B are assigned to two groups of children with leukemia. Patients receiving A are two to six years old and patients receiving B are either younger or older. It has been shown by several investigators that "middle" age children have a better prognosis than the others. So, if the trial is conducted in this way, the observed difference between groups A and B would be an estimate of treatment difference plus the difference due to age. If results of the trial show a significant difference between groups A and B, there is no way to find out whether the difference is due to treatment alone, age alone, or a combined effect of treatment and age. In this case, we have a bias that is not difficult to discover. In other trials where little is known about prognostic factors or type of variability existing, unexpected biases can lead to erroneous conclusions. Thus, in a comparative clinical trial, it is essential to have comparable groups of patients. *Randomization* is one way to achieve comparability.

From a statistical viewpoint, to reach valid conclusions about populations by inference from samples, statistical tests and procedures typically assume that the sample are obtained in a *random* fashion. In clinical trials the term randomization refers to the assignment of patients to treatments by a random process in such a way that each patient has an equal and independent chance of receiving any of the treatments under study. That is, not only must each member in the population have an equal chance of being selected, but the selection of any member of the population must in no way influence the selection of any other member. An essential feature of randomization is that it should be objective and impersonal.

Despite the above advantages, the role and appropriateness of randomization continue to be among the most controversial in clinical trials. Randomization is considered a requirement for a scientific experiment by the classically trained statistician but is seen frequently as an unethical act by the classically trained clinician. This controversial issue will be discussed more in Section 13.4. In this section we assume that a randomized controlled design is chosen and discuss several most commonly used procedures of randomization: simple, restricted, and stratified randomization. For simplicity, we will assume that there are two treatments under study, A and B. We will also assume that patients enter a study sequentially.

13.2.1 Simple Randomization

Simple randomization is the most elementary and simple to apply. There are many informal ways to assign treatments to patients randomly such as flipping a coin and shuffling numbered cards. However, the most common technique is to use a table of random numbers such as Table C-20. In this table of digits 0–9, each digit has an equal and independent chance of appearing in every entry, and each digit occurs with approximately equal frequency with no systematic pattern.

One acceptable scheme using a table of random numbers is to assign the even numbers to treatment A and the odd numbers of treatment B or numbers less than 5 to A and numbers greater than 4 to B. To illustrate the procedure, let us use the first row of Table C-20 to assign treatment to 20 patients. If the random number is even, assign treatment A; otherwise assign treatment B. The assignments are listed in Table 13.2. Of the 20 patients, 9 are assigned treatment A and 11 treatment B. In this case the ratio of A patients to B patients is very close to 1. However, simple randomization does not guarantee this, although the ratio approaches 1 as the number of patients increases. The main advantage of simple randomization is its simplicity to apply. Its main criticism is the possibility of unbalanced assignments.

13.2.2 Restricted Randomization (or Block Randomization)

One way to avoid unbalanced assignments is to use *restricted randomization* or *block randomization*. In this scheme, patients are grouped into several blocks of equal size according to their chronological entry time. Within each group of patients, the treatments are assigned so that there is a balanced allocation for each treatment. For example, if two treatments are to be assigned to 24 patients, it is reasonable to require balance after every four patients, assigning two to each treatment. We can divide the patients into six groups (or blocks) according to their entry times, the first four patients in group 1, the next four in group 2, and so forth. Then within each group we randomly assign two patients to each treatment. This will ensure that after

Table 13.2 Treatment Assignment by Simple Randomization

Chronological Patient Number	Random Number	Treatment Assignment	Chronological Patient Number	Random Number	Treatment Assignment
1	1	B	11	5	B
2	2	A	12	4	A
3	6	A	13	1	B
4	7	B	14	2	A
5	7	B	15	7	B
6	3	B	16	3	B
7	2	A	17	9	B
8	9	B	18	7	B
9	4	A	19	4	A
10	4	A	20	8	A

every forth patient assignment, there will be an equal number of patients on each treatment.

To implement restricted randomization, we consider all the six possible arrangements of assigning two treatments to four patients:

1. AABB
2. BBAA
3. ABAB
4. BABA
5. ABBA
6. BAAB

First, we assign randomly these six arrangements to the six groups of patients. This can be done by rolling a die. If the number has appeared previously, roll again until each side of the die appears only once. Suppose the results of the die rolling are 2, 5, 4, 3, 1, and 6; the first group of four patients will be assigned the second arrangement (BBAA), the second group of four patients will be assigned the fifth arrangement (ABBA), and so forth. The overall assignment is given in the first two columns of Table 13.3. The 24 patients are equally assigned to treatments A and B. The advantage of the restricted randomization is obvious for a study involving only one clinic or hospital. However, if there are several hospitals participating in the study and the restricted randomization is made from the headquarters, it is possible that although the treatment assignment over all hospitals would be balanced, within a hospital there might be a serious imbalance.

For example, suppose there are four hospitals participating. The hospitals are numbered 1, 2, 3, and 4 and are given in column 3 of Table 13.3. Then the treatment assignment in column 2 results in an imbalance within hospitals 2 and 4 as shown below.

Hospital	Number Assigned to A	Number Assigned to B	Total
1	3	4	7
2	0	6	6
3	4	2	6
4	5	0	5
Total	12	12	24

To ensure balance within a hospital, one could use the so-called *balanced restricted randomization* (or *balanced randomization*) in which treatment assignment balance within a hospital is ensured before the treatment is assigned. To implement it, we use a restricted randomization scheme with the help of an auxiliary table of random integers. These integers represent

Table 13.3   Treatment Assignment by Restricted Randomization

(1) Patient Number in order of Entrance Time	(2) (Tentative) Treatment Assigned	(3) Hospital	(4) n	(5) Final Assignment Hospital			
				1	2	3	4
1	B	1	1	B			
2	B	2	2		B		
3	A	1	2	A			
4	A	3	1			A	
5	A	4	3				A
6	B	2	3		B		
7	B	1	2	B			
8	A	1	2	A			
9	B	2	1		A		
10	A	3	1			B	
11	B	3	3			B	
12	A	4	1				B
13	A	4	3				A
14	B	2	1		A		
15	A	1	2	A			
16	B	1	3	B			
17	A	4	1				B
18	A	3	3			A	
19	B	2	2		B		
20	B	3	2			B	
21	B	1	1	B			
22	A	4	3				A
23	A	3	1			A	
24	B	2	1		A		

the difference in the number of treatments assigned to A and B that can be tolerated by the investigator. In practice, this auxiliary table may have only the integers 1 and 2 or 1, 2, and 3. The procedure is as follows: When a patient is entered, tentatively assign a treatment according to the restricted randomization scheme. Calculate the difference in number of patients assigned to A and B with the tentative assignment. Then choose a random integer from the auxiliary table. If the difference D in the treatment assignment is less than or equal to the random integer, the tentative assignment is to be used; otherwise the alternate treatment will be assigned.

Suppose an auxiliary random number table of integers 1, 2, and 3 is used for the example given above. Column 4 in Table 13.3 lists the random integers obtained from the auxiliary table. Column 5 gives the balanced restricted assignments of treatment for patients from the four hospitals. The first eight patients are assigned the same treatments as in the restricted randomization (column 2). For the ninth patient in hospital 2, the tentative assignment was B in column 2. However, since the previous two patients in hospital 2 were assigned treatment B, the difference in treatment assignment D is $2 - 0 = 2$, which is larger than the random integer 1 obtained from the auxiliary table. Therefore, the ninth patient is assigned the alternate treatment A. Similarly, for example, the fourteenth patient from hospital 2 is assigned treatment A since the difference would be $3 - 1 = 2 > 1$ if the tentative treatment B were assigned. The resulting treatment allocation is balanced as shown below.

Hospital	Number Assigned to A	Number Assigned to B
1	3	4
2	3	3
3	3	3
4	3	2
Total	12	12

When preparing a balanced randomization list, it is also reasonable to require a balance at every sixth patient. This can be done by numbering all 20 possible arrangements of assigning two treatments to six patients as follows:

00–04	AAABBB	50–54	BAAABB
05–09	AABABB	55–59	BAABAB
10–14	AABBAB	60–64	BAABBA
15–19	AABBBA	65–69	BBAAAB
20–24	ABAABB	70–74	BBAABA
25–29	ABABAB	75–79	BBABAA
30–34	ABABBA	80–84	BABAAB
35–39	ABBAAB	85–89	BABABA
40–44	ABBABA	90–94	BABBAA
45–49	ABBBAA	95–99	BBBAAA

Then select randomly a two-digit random number from the random-number table. This can be done by closing your eyes and pointing to the random-number table haphazardly and starting with the nearest number. Suppose the random numbers so obtained are 70, 12, 91, 69, ...; the first six patients will receive treatments BBAABA in sequence, the second six will receive treatments AABBAB, and so on.

This approach can be extended to three treatment cases (Peto et al. 1976). In order to obtain a random order of two A's, two B's, and two C's, we proceed as above but after selecting one of the 20 sequences, use the next two-digit random number to change one of the A's and one of the B's into C's. The following rules may be used.

Random Number	Change into C	Random Number	Change into C
01–11	First A, first B	56–66	Second A, third B
12–22	First A, second B	67–77	Third A, first B
23–33	First A, third B	78–88	Third A, second B
34–44	Second A, first B	89–99	Third A, third B
45–55	Second A, second B		

That is, if the random number is between 01 and 11, change the first A and the first B into C's; if the random number is between 12 and 22, change the first A and second B into C's; and so on. If the random number is 00, ignore it and use the next two-digit number. For example, suppose the first two-digit random number obtained is 21. The sequence selected would be ABAABB. Suppose the next random number is 57. We change the second A and third B into C's, and thus the treatment assignment is ABCABC.

If unbalanced allocation of two treatments is desired, the above procedure can easily be extended. Suppose a treatment allocation ratio of 2 : 1 or 1 : 2 is needed. After the above A, B, and C assignment, either change all the C's to A's (for a 2 : 1 ratio) or change all the C's to B's (for a 1 : 2 ratio).

13.2.3 Stratified Randomization

Prognostic factors that might influence response, when known, should be taken into account in the initial randomization. Not to do so may introduce biases into the data and thus lead to incorrect conclusions about the treatments under study. Discussions of the use and importance of prognostic factors will be given in Section 13.3.

Taking account of prognostic factors in the initial treatment assignment ensures that the distributions of patients with respect to the important prognostic factors are equal. For example, it is known that age is an important factor in childhood leukemia. In a study of two treatments for childhood leukemia, this fact should be seriously considered. Stratified randomization will ensure equal distributions of patients with regard to age in the two treatment groups. In other words, there should be an equal proportion of patients with good prognosis in each treatment group.

Suppose one decides to consider two strata, one consisting of patients with favorable prognoses and the other consisting of patients with poor prognoses. The simplest stratified randomization is to make up a separate restricted randomization schedule for each stratum. This can be done either in a single-clinic or multiclinic trial. In a multiclinic trial, adjustments are made within institutions as described earlier to prevent imbalances.

If there are several important prognostic factors involved and each has several subcategories, then the number of combinations and consequently the number of strata for randomization may be enormous. There may be very few patients in each stratum. Such a randomization is equivalent to one in which there is a single list for all patients or simple randomization. Sometimes, if there are several important prognostic factors, a summary prognostic index can be used as a single stratifying variable to achieve balance among the separate variables on which it is based. An example is the hazard index or hazard ratio $h_i(t)/h_0(t)$ or $\log_e[h_i(t)/h_0(t)]$ discussed in Section 10.2.2.

Zelen (1979) also proposed a new design for randomized trials. Patients are randomly assigned to two groups, A and B. Patients in group A receive the standard treatment; those in group B are asked if they will accept the experimental treatment; if they decline, they receive the standard treatment.

Statistical properties of various randomization methods in clinical trials have been discussed by many. For example, the December 1988 issue of the journal *Controlled Clinical Trials* is devoted to this topic. General recommendations are given regarding the use of several randomization methods. Among the recommendations offered, one is that the simple unrestricted complete randomization is desirable for large trials ($n > 200$ overall and within each subgroup) especially in unblinded trials. On the other hand, for small trials with $n < 100$ overall or within any subgroup, restricted randomization is better. The interested reader is referred to this issue of the journal.

13.3   THE USE OF PROGNOSTIC FACTORS IN CLINICAL TRIALS

The primary objective of most clinical trials is to obtain a precise comparison of treatments, usually a new treatment and a standard treatment. An important requirement for a precise comparison is that patients in the different treatment groups are comparable with respect to prognostic characteristics except the treatments being assigned. If a small number of prognostic variables have been identified, how can this information be used in a clinical trial so that treatment comparisons can be more precise? The information can be used to stratify patients at the stage of randomization (stratified randomization or prestratification) in order to achieve comparability in the treatment groups. It can also be used at the analysis stage (poststratification) to adjust for and minimize the effects of group differences. The use of prognostic factors in prestratification and poststratification has been discussed by many, for example, Armitage and Gehan (1974), Pocock and Simon (1975), Peto et al. (1976), Feinstein (1977), Simon (1979), Feinstein and Landis (1976), Armitage (1981), Meier (1981), Grizzle (1982), and McHugh and Matts (1983). In this section we briefly discuss the advantages and limitations of prestratification and poststratification. For simplicity, we assume that two treatments are to be compared.

As mentioned in Section 12.3.2, knowledge of prognostic variables could be used to group patients into prognostic strata. Patients within a stratum have similar values of the prognostic variable used to define the strata and patients among strata are substantially different. For example, if previous studies indicate that female patients tends to survive longer than male patients, patients could be divided into two strata, males and females. Within each strata, treatments A and B are assigned using simple randomization or block randomization. This would guarantee a comparable overall sex distribution in the two treatment groups and increase the precision of treatment comparisons. Another advantage is that in addition to an overall comparison of treatments A and B, comparisons of the two treatments could be carried out within stratum. For example, we could compare A and B within the male stratum and within the female stratum. If the results show that A is better than B in males and B is better than A in females (an extreme case), then there exists a strong interaction between treatment and sex (Armitage 1981). This information would be useful to physicians in treatment selection.

However, prestratification is not problem free. In order to use stratified randomization, values of the prognostic factors must be known exactly at the time of randomization. For variables such as age and sex, this requirements can easily be met with little risk of error. If any time-consuming laboratory test result is used to classify, the information may not be readily available at the time of randomization. Or if the variable used for stratification has no absolute definition or involves subjective judgment, for example, stress or degree of stenosis of a coronary artery, the chance of misclassification is high. It happens often that the physician does not have complete information about the patient or may have to make a personal judgment at the time of stratified randomization. The classification made at that time may be found erroneous later and the patient ineligible or being in the wrong stratum.

In large clinical trials that involve hundreds or thousands of subjects, the chance of achieving group comparability in prognostic factors is higher than in small trials. So, prestratification is more likely to be used in small studies. However, the advantage may be limited if the total number of strata is large. For example, age, sex, and race are used to stratify and there are three age levels and four race groups:

Age   40–49 years, 50–59 years, ≥60 years
Sex   Male, female
Race Caucasian, black, Hispanic, other

The total number of strata is $3 \times 2 \times 4 = 24$. For each of the 24 strata, we assign one-half the patients to A and one-half to B. If the total sample size is less than 100, each stratum will have no more than five patients. In this case, comparisons within stratum and the study of treatment–prognostic variable interaction would not be meaningful. Sometimes it may be difficult to find patients that fit in some of the strata.

In addition, some of the variables used to stratify may turn out to be insignificantly related to prognosis and other more important variables may be discovered. The effort of stratification at randomization may become totally unworthwhile. Therefore, the merit of stratified allocation is debatable and most statisticians agree that poststratification may be more desirable.

By poststratification we mean that the recognized prognostic variables are adjusted using statistical methods at the analysis stage. This method, often referred to as stratified analysis or covariate adjustment, is recommended for both randomized and nonrandomized studies. It not only minimizes the effect of imbalance between treatment groups and improves efficiency in comparison but also avoids erroneous conclusions due to confounding factors.

In order to minimize the effects of prognostic heterogeneity, the treatment groups can be divided into subgroups, or strata, the prognostic variable at the time of analysis. This is usually done if the prognostic variable is categorical, discrete, or convertable into intervals. For example, systolic blood pressure may be converted into two intervals (or strata): <160 mm Hg, ≥ 160 mm Hg. A comparison is then made within each stratum and then over all strata to obtain an overall comparison. If the outcome variable is also categorical (e.g., response or nonresponse), the stratified analysis can be performed using the Mantel–Haenszel chi-square test described in Chapter 5. The following example illustrates the importance of stratified analysis.

Example 13.1

Suppose that two treatments A and B are assigned randomly to 100 patients and the outcome variable of interest is response. At the end of the clinical trial, the following results are obtained:

Treatment	Response	Nonresponse	Total
A	30 (60%)	20	50
B	15 (30%)	35	50
	45	55	100

Sixty percent of the patients receiving treatment A and 30% of the patients receiving B responded. We might conclude that the two treatments are distinctly different. However, suppose that previous studies showed a better response rate in females and that sex might be a prognostic variable. When the patients are stratified by sex, we find the following results:

	Male			Female		
Treatment	Response	Nonresponse	Total	Response	Nonresponse	Total
A	2 (13%)	13	15	29 (80%)	7	35
B	4 (11%)	31	35	11 (70%)	4	15
Total	6	44	50	39	11	50

The Mantel–Haenszel chi-square test shows that there is no significant difference between the two treatments after adjusting for sex. Thus, the overall significant result obtained earlier is mainly due to the imbalance in sex distribution between the two treatment groups (70% females in group A and 30% females in group B). After the effect of sex is controlled, the difference between the two treatments becomes negligible. Similarly, stratified analysis can also be used to avoid false-negative results and to detect if the treatment difference exists only in a certain subgroup. Stratified analysis can be performed using variables that have not been reported as prognostic factors. Thus, it can be used to detect and identify important confounding variables as well as prognostic factors.

Although poststratification can improve comparison efficiency, when the sample size in a given stratum is small, the disproportion can cause average loss in efficiency. Meier (1981), in comparing the relative efficiency between a balanced and unbalanced design, concludes that the loss is not big. For sample sizes as large as 20 (10 in each group), the expected relative efficiency is close to 100%. Thus, for moderate sample sizes, poststratification is as efficient as prestratification.

In addition to poststratification, the linear logistic regression method discussed in Chapter 11 can be used to adjust for the effect of prognostic factors when the outcome variable is dichotomous and the prognostic variables are either categorical or continuous.

If the outcome variable is continuous, the analysis-of-covariance technique can be used to adjust for prognostic factors. This topic is not discussed in this book but can be found in most standard statistics textbooks, for example, Snedecor and Cochran (1967), Kleinbaum, Kupper, and Muller (1988), and Howell (1987). If time to a given event (e.g., survival time) is the outcome variable of interest and censored observations are involved, Cox's proportional hazards model discussed in Chapter 10 can be used to adjust for prognostic factors.

13.4   CONTROLS IN CANCER CLINICAL STUDIES

Cancer, one of the most feared, publicized, and politicized diseases in the United States, has entailed numerous clinical trials, especially in the past four decades. In cancer clinical trials, a new treatment, say A, frequently becomes available for patients with a certain type of cancer. If treatment B is the standard treatment generally accepted and used, though of very low effectiveness, how should one evaluate the new treatment A? Should a randomized controlled clinical trial be conducted in which half of the patients are treated with each treatment? Or should all the patients be given treatment A and a comparison made with patients previously treated with B? The latter is usually termed as unrandomized controlled or historical controlled clinical trial.

The two kinds of controls, randomized and historical, have drawn many discussions. Chalmers et al. (1972), Ingelfinger (1970), Hill (1971), Sacks et al. (1982), and Micciolo et al. (1985) have emphasized the importance and advantages of randomized controls. Gehan and Freireich (1974), Freireich and Gehan (1974, 1979), and Gehan (1982a, b) argue that there still exists a place for a careful choice of historical controls. Pocock (1976a,b) suggests a compromise in which the comparative evaluation of a new treatment uses both randomized and historical controls. In the following, we discuss briefly randomized controls and historical controls.

13.4.1 Randomized Controls

In the early 1950s, Daniels and Hill (1952) first employed randomization in their studies of streptomycin combined with *para*-aminosalicylic acid in the treatment of pulmonary tuberculosis in young adults. Since then, treatment randomization has been an important aspect of clinical trials. Methods of randomizations have been developed, four of which are discussed in Section 13.2. In general, the advantages of randomization (or randomized controls) are that it achieves comparability of patients among treatment groups, it avoids conscious or unconscious bias in the assignment of patients to treatment groups, and it provides a firm basis for the statistical evaluation of any apparent treatment effects.

Some disadvantages of randomization were mentioned earlier. Possible unbalanced assignment of treatments due to simple randomization can be eliminated by using restricted or block randomization. Some disadvantages of restricted randomization, in turn, can be avoided by stratified randomization. However, if importance prognostic factors are unknown or if there are too many prognostic factors, stratified randomization is not easy to apply. Although randomization does not ensure that patients in the different treatment groups are comparable for all important prognostic factors, the validity of statistical tests and significance levels based on randomization does not require this unachievable assumption. As mentioned earlier, the effects of important prognostic factors can be adjusted in the analysis in order to correct treatment comparisons for possible bias due to imbalances.

The most serious objections to randomized controls center around ethical responsibility, which requires a physician to administer the treatment that he or she believes is best for a patient. However, the relative merits of the treatments under study are yet to be determined. If it is known that one treatment is better than the other, then there is no need for comparative clinical trial. Therefore, researchers advocating randomized controlled clinical trials argue that a physician initiating a comparative trial makes the honest admission that the best treatment is unknown and therefore randomization is more ethical than a procedure in which the merits of a new therapy are determined from clinical impression and comparison with past experience.

Despite the disadvantages of randomized controlled clinical trials, much useful clinical information has been derived using them. They are complex, expensive, and time consuming but remain the most useful and acceptable methods for comparing treatments. Chalmers et al. (1970) even suggest that randomized controls be used in phase I and II studies (see also Chalmers 1975). They indicated that dosage can be randomized to give some insight into cumulative effects, and patients destined to receive multiple dosages of the new treatment could for ethical reasons be given a 50–50 chance of receiving conventional or placebo therapy. However, Gehan and Freireich (1974) disagree. They state that since comparing treatments is not an objective of phase I or II studies, randomized controls patients would only tend to obscure the real purpose of the studies and delay their completion. They propose that a selected rather than a randomized control group be used in certain circumstances.

13.4.2 Historical Controls

Historical controls include patients chosen from the literature, on a matching basis from a previous study, or from an immediately preceding trial in a sequence of studies. Gehan and Freireich (1974) state that knowledge of prognostic factors is the primary assumption in any selection of controls so that there is a firm basis for the selection of comparable groups and that the difference existing between the groups selected have little or no relation to the outcome of treatment. The investigator must hope that unknown prognostic factors are distributed equally to the two treatment groups.

A. *Literature Controls.* Suppose that treatments A and B are being compared in patients with a particular disease and that patients with the same disease who received one of the two treatments (say, A) have been reported in the literature. These patients can be selected as a historical control group to determine whether B is superior to A. In this case, all of the current patients are given treatment B. The comparison between A and B is valid provided that the literature control group and the current group are comparable in prognostic factors. Examples of literature controls can be found in Carbone et al. (1968), Frei et al. (1973), Reaman et al. (1987), and Donadio and Offord (1989).

Obviously, literature controls require careful checking for comparability and draw criticism when the patients either cannot be checked for comparability or are not comparable because the authors do not provide sufficient data. Today, many studies are conducted by large cooperative groups and computers are commonly used in analyzing data; it is possible to retain the raw data on magnetic tapes or diskettes so that other investigators may use the data. The detail work of checking for comparability could be very tedious.

B. *Matched Controls.* Matched controls are patients selected on a matching basis from a previous or concurrent study. Controls are matched with cases by using known prognostic factors. It is obvious that the use of this type of controls is feasible when important prognostic factors are known and there are not so many of them that pair matches are too difficult to find. Gehan and Freireich (1974) suggest selecting two control patients for each treated patient to test the selection process. These two control patients and the treated patient are as comparable as possible with respect to factors influencing prognosis. The randomness of the selection can be checked by comparing the two control groups of patients with respect to the endpoints in the analysis. If no differences are found, the validity of the selection process is confirmed. Examples of matched controls are given by Bodey et al. (1971), Hogan et al. (1987), and Ozer et al. (1989).

C. *Controls Selected from a Sequence of Studies.* Historical controls can also be selected from a sequence of studies, for example, Buzdar et al. (1978), Micciolo et al. (1985), Abdi et al. (1987), and Hersey et al. (1987). In many clinical research centers or cooperative groups for cancer research, a sequence of clinical trials is often conducted for a given disease. It is common to activate one trial for a given disease as soon as the previous one is terminated. Then the best treatment in the previous trial will usually be the standard control in the next trial. Again the patients selected must be comparable with respect to prognostic factors.

This type of historical control is reasonable, feasible, and probably the most acceptable. Since the controls have been treated in the recent past, criteria for diagnosis, types of patients, nature of the disease, means of treatment administration, supportive treatment, definition of response, and staff probably have not changed. Even if some changes have occurred, the nature of the changes would be known, and appropriate adjustments can be made. On the other hand, if a relatively long time has elapsed between studies or important changes have occurred in type of patients, supportive treatment, definition of response, or staff, controls from a previous study could lead to serious bias.

Important reasons for considering historical control groups are the following:

1. When patients are randomized to a control or treatment group, the comparability achieved with respect to prognostic factors between groups is an average. When historical controls are selected, all patients in the two groups are guaranteed comparable with respect to the characteristic influencing prognosis.

2. The use of historical controls requires fewer patients and therefore a shorter time and less money. Suppose two treatments A and B are to be compared according to the proportion of patients responding to each treatment (response rate $\hat{P}_A$ and $\hat{P}_B$). Suppose further that enough patients will receive each treatment that a statistical test of the

difference between treatments can be made at a given significance level α and power $1 - \beta$. From Table 12.5 or 12.6, the required patient number can then be found. Assume that n patients are needed in each treatment group. Alternatively, suppose B is the control treatment and its response rate is well known, say P. Then no patients need receive B in the trial. To make a statistical test of the difference between $\hat{P}_A$ and P at the same significance level and power assumed above, only $\frac{1}{2}n$ patients are needed on treatment A, that is, only *one-fourth* the total number of patients needed for the randomized comparative trial. This is evident by comparing relevant tables in Natrella (1963) and Table 12.6. Readers interested in the derivation are referred to Sillitto (1949).

3. A comparative trial should be undertaken only when the best available evidence from initial clinical studies suggests that the new treatment has a good chance of being equal to or better than the old treatment. Therefore, clinical investigators are not fulfilling their ethical responsibility if they plan a randomized comparative trial instead of administering the better treatment to consecutive patients. For example, there is no justification for adding to the anguish of a cancer patient by introducing the irrational concept of "flipping a coin." Evaluation of the treatment could proceed by comparison of results in concurrent patients with a historical control group. If the new treatment consists of the standard therapy plus an additional component and preliminary studies suggest that the new combination is at least as effective as the standard and possibly much more so, it is logical to consider historical controls.

The main objection to historical controls is that they provide limited protection against possible bias introduced by time changes in the nature of the patient population, diagnostic criteria, patient care, and exposure to pathologic agents. A subtle selection mechanism may require one to check a long list of prognostic factors to find a match. It is also very likely that prognostic data of historical controls are not available. Micciolo et al. (1985) give an example in which the results obtained by using historical control data in breast cancer after adjusting for differences in baseline characteristics given a biased conclusion in favor of the new treatment.

Pocock (1976a) gives the following conditions as requirements for an acceptable historical control group:

1. The historical control group must have a precisely defined standard treatment, that is, the same as the treatment for the randomized controls.
2. The group must have been part of a recent clinical study that contained the same requirements for patient eligibility.

3. The methods of treatment evaluation must be the same.
4. The distributions of important patient characteristics in the group should be comparable with those in the new trial.
5. The previous study must have been performed in the same organization with largely the same clinical investigators.
6. There must be no other indications leading one to expect different results between the randomized and historical controls.

13.4.3 The Combination of Randomized and Historical Controls

Pocock (1976a,b) suggests a compromise in which both randomized and historical controls are used. Let T, R, and H denote the groups of patients on the new treatment, the randomized control, and the historical control, respectively. Let N_t, N_r, and N_h be their respective sample sizes. The two questions discussed by Pocock using the Bayesian approach are:

1. How should one determine sample sizes N_t and N_r given the historical controls of size N_h?
2. How can one combine the data from the two sets of controls for comparison with data from the patients on the new treatment?

Suppose that the evaluation of treatment for a patient can be summarized by a single quantitative measurement X. For example, in a cancer clinical trial, X might be the patient's survival time, remission duration, or an ordered scale for the assessment of objective tumor response. Let x be the observed value of X for any patient in a particular treatment group. Suppose that the random variables associated with the three groups, X_t, X_r, and X_h, follow the normal distribution with unknown means μ_t, μ_r, and μ_h and variances σ_t^2, σ_r^2, and σ_h^2. The sample means of X_t, X_r, and X_h are $\bar{x}_t$, $\bar{x}_r$, and $\bar{x}_h$. The main objective of the comparative trial is to estimate the difference in effectiveness between the new treatment and the control treatment. In other words, the objective is to obtain an accurate estimate or $\mu_t - \mu_r$. In a randomized control trial without historical data, this is best estimated by $\bar{x}_t - \bar{x}_r$.

In the presence of historical data, $\bar{x}_h$ has to be incorporated. It should be noted that μ_h and μ_r are not necessarily equal. There exists a possible unknown bias $\delta = \mu_r - \mu_h$ in the historical control. We are unable to assess the exact value of δ. A possibility is to assume that δ is a normal random variable with mean zero and variance equal to some fixed quantity σ_δ^2. The choice of σ_δ is difficult in practice. On the side of caution a reasonably large value should be assigned. That is, it is better to place too little rather than too much confidence in historical data. Perhaps in practice one should consider several possible values of σ_δ in order to assess the effects on experimental design and analysis.

Sample Size Determination

Assume that the total sample size $N = N_t + N_r$ has been determined. Pocock suggests that the "optimal" value for N_r (which minimized the variance of $\mu_t - \mu_r$) given the values of N and N_h is

$$N_r = \frac{\sigma_r}{\sigma_r + \sigma_t}\left(N - \frac{\sigma_r \sigma_t}{\sigma_h^2/N_h + \sigma_\delta^2}\right) \tag{13.1}$$

where σ_r, σ_h, and σ_t are unknown and some estimates must be used. However, prior to the trial, we do not know the sample variance of σ_r^2 and σ_t^2. The three variances σ_r^2, σ_t^2, and σ_h^2 may all be set equal to the sample variance of the historical data s_h^2, where $s_h^2 = \Sigma\,(x - \bar{x}_h)^2/(N_h - 1)$. Thus (13.1) reduces to

$$N_r = \frac{1}{2}\left(N - \frac{N_h}{1 + N_h \sigma_\delta^2/\sigma_h^2}\right) \tag{13.2}$$

Having chosen N and N_r, N_r/N may be rounded off to some simple fraction (e.g., $\frac{1}{3}$ or $\frac{2}{5}$) so that small blocks can be formed (e.g., if $N_r/N \approx \frac{1}{3}$, then for every three patients it is ensured that one will receive the control treatment and two the new treatment). Let us use Pocock's example to illustrate the procedure.

Example 13.2

A trial is to be undertaken for advanced small cell lung cancer patients to compare the effectiveness of Cytoxan + CCNU with Cytoxan + CCNU + Procarbaxine. The main endpoint is tumor response. It is proposed that 200 patients be entered on the trial. In a preceding trial, Cytoxan alone was compared to Cytoxan + CCNU, and the latter treatment was found superior. In this earlier trial, 88 patients were treated with Cytoxan + CCNU. The tumor response rate was $\frac{37}{83} = 45\%$, with five nonevaluable cases. The question is how many of the 200 patients should be entered on the treatment Cytoxan + CCNU as randomized controls, for which there are already 83 patients in a historical group.

In this case, $N_h = 83$, $N = 200$, $\bar{x}_h = 0.45$, and from the binomial distribution, σ_h is estimated to be $\sqrt{0.45 \times (1 - 0.45)} = 49.7\%$. It is assumed that $\sigma_t = \sigma_r = \sigma_h = 49.7\%$. Suppose the standard deviation of the bias in the historical controls $\sigma_\delta = 3\%$; applying (13.2), we obtain

$$N_r = \frac{1}{2}\left[200 - \frac{83}{1 + 83(3/49.7)^2}\right] = 68.1$$

N_r is close to $\frac{1}{3}N = 66.6$ or $N_r/N \sim 3$, so that a convenient practical solution would be to randomize patients to Cytoxan + CCNU and Cytoxan + CCNU + Procarbaxine in a ratio of $1:2$.

In many clinical trials, more than one response variable may be used to evaluate treatment effect. In the trial described in Example 13.2 the survival time was also used for comparing the two treatments. In such cases, the sample size N_r for the randomized controls can be determined separately for each response variable and the eventual choice be a weighted mean of these values, with weights determined by the investigator according to the relative importance of the variables.

Combination of Controls in Analysis

When both randomized and historical controls are used, we have two estimators of μ_r, namely $\bar{x}_r$ and $\bar{x}_h$. The problem is then how to determine a combined estimator. Based on the assumptions that $\mu_r = \mu_h + \delta$ and that all of the variables are normal, the distribution of μ_r is normal with mean

$$\bar{x}_c = \frac{(\sigma_h^2/N_h + \sigma_\delta^2)\bar{x}_r + (\sigma_r^2/N_r)\bar{x}_h}{\sigma_r^2/N_r + \sigma_h^2/N_h + \sigma_\delta^2} \tag{13.3}$$

and variance

$$V_c = \left(\frac{N_r}{\sigma_r^2} + \frac{1}{\sigma_\delta^2 + \sigma_h^2/N_h}\right)^{-1} \tag{13.4}$$

Equation (13.3) can be written as

$$\bar{x}_c = \frac{\bar{x}_r + W\bar{x}_h}{1 + W} \tag{13.5}$$

where $W = (\sigma_r^2/N_r)/(\sigma_h^2/N_h + \delta_\delta^2)$. Thus, the combined estimate $\bar{x}_c$ for μ_r is a weighted sum of $\bar{x}_r$ and $\bar{x}_h$.

To test the hypothesis that the new treatment is as effective as the standard treatment, or $\mu_t = \mu_r$, we use the assumption that the distribution of μ_t is normal with mean $\bar{x}_t$ and variance σ_t^2/N_t. Hence the distribution of $\mu_t - \mu_r$ is normal with mean $\bar{x}_t - \bar{x}_r$ and variance $\sigma_t^2/N_t + V_c$.

In practice, the variances σ_t^2, σ_r^2, and σ_h^2 are unknown and either the sample variances s_t^2, s_r^2, and s_h^2 are substituted or the pooled variance

$$s_p^2 = \frac{(N_t - 1)s_t^2 + (N_r - 1)s_r^2 + (N_h - 1)s_h^2}{N_t + N_r + N_h - 3}$$

is used.

When sample observations follow the exponential distribution with mean μ, the maximum likelihood estimate $\hat{\mu}$ is the sum of all survival times, uncensored and censored, divided by the number of deaths, as given in (8.15). Let μ_r, μ_h, and μ_t be the true mean survival times n_r, n_h, and n_t the observed deaths, and $\bar{x}_r$, $\bar{x}_h$, and $\bar{x}_t$ the observed mean survival times of the R, H, and T groups, respectively. The historical bias $\delta = \log_e(\mu_r/\mu_h)$. To determine a combined estimator for μ_r and μ_h, $\bar{x}_c$ say, it can be shown that the distribution of $(\log_e \mu_r)$ is asymptotically normal with mean

$$\log_e \bar{x}_c = \frac{(1/n_h + \sigma_\delta^2)\log_e \bar{x}_r + (1/n_r)\log_e \bar{x}_h}{1/n_r + 1/n_h + \sigma_\delta^2} \tag{13.6}$$

and variance

$$\left(n_r + \frac{n_h}{1 + n_h \sigma_\delta^2}\right)^{-1} \tag{13.7}$$

For the new treatment the distribution of $\log_e \mu_t$ is normal with mean $\log_e \bar{x}_t$ and variance $1/n_t$. The difference between μ_r and μ_t can then be tested on the basis of $\log_e \mu_r - \log_e \mu_t$.

The following example, also from the Pocock, illustrates the procedure.

Example 13.3

This is a clinical trial to compare the survival experience of patients with advanced melanoma on two treatments, DTIC and TIC–Mustard. DTIC was also a treatment in a previous melanoma trial and this historical control group is considered acceptable. The survival experience for each treatment group is summarized in Table 13.4. Also, an exponential model seems reasonable according to survival plots of the data. The historical bias $\delta = \log_e(\mu_r/\mu_h)$ has been assigned a variance $\sigma_\delta^2 = 0.01$, which corresponds to μ_r/μ_h having standard deviation $\sqrt{e^{0.01}(e^{0.01} - 1)} = 0.1$. That is, the potential historical bias in the mean survival times may be of the order of 10%.

Applying (13.6) and (13.7), the distribution of log mean survival for the randomized controls (DTIC) is approximately normal with mean $\log_e \bar{x}_c = 3.28$ and variance 0.0112. Thus, the combined point estimate of mean survival time on DTIC is $\bar{x}_c = 26.5$ weeks. For the new treatment the log mean survival has an approximate normal distribution with mean 3.30 and variance 0.0204. Thus, $\log_e(\mu_r/\mu_t)$, the log of the ratio of true mean survival times for DTIC and TIC–Mustard, has an approximately normal distribution with mean -0.02 $(3.28 - 3.30)$ and standard deviation 0.18

Table 13.4 Summary Survival Data of Melanoma Patients

	Treatment	Number of Patients	Number of Deaths (n)	Estimated Mean Survival Time $\bar{x}$ (weeks)
Previous trial	DTIC (H)	57	44	23.2
Current trial	TIC–Mustard (T)	70	49	27.1
	DTIC (R)	80	59	28.4

Source: Pocock (1976a).

($\sqrt{0.0112 + 0.0204}$). This leads to a normal deviate for $-0.02/0.18 = -0.1$, which is not significant. Thus, there is no evidence of a survival difference between the two treatments.

Although Pocock's procedure allows the use of both randomized controls and historical controls, it is not free of problems. The determination of the bias, δ and σ_δ^2, in the historical data presents bias itself. To increase σ_δ^2, one places greater emphasis on the randomized controls and the historical data become of less value, and vice versa. In addition, the determination of σ_δ^2 is highly subjective.

Peto et al. (1976) suggest a much simpler allocation, that is, at least one-third of current patients should be randomized controls unless the new treatment is so superior to the standard that randomization is felt to be unethical. That is, one-third of the current patients are randomized controls and two-thirds are still available for comparison with whatever historical controls are chosen. For people who only believe randomized controls, this is as substantial a randomized trial as if the ordinary 1 : 1 randomization has been adopted. Other methods for combining historical and randomized controls include Tarone (1982) and Dempster et al. (1983).

BIBLIOGRAPHICAL REMARKS

In addition to the books cited in Chapter 12 and the medical and statistical journals that publish papers on practical and theoretical issues in clinical trials, a journal devoted to the subject, *Controlled Clinical Trials*, has been published since 1980. Readers who are interested in conducting clinical trials are referred to this journal for the most recent developments in this field.

EXERCISES

13.1 Use the random-number table in Appendix C (Table C-20) to assign two treatments A and B to 30 patients. Do you get equal numbers of patients in the two treatment groups by simple randomization? If not, use restricted randomization.

13.2 Suppose you are the coordinator of a study involving five hospitals. The purpose of the study is to compare the efficiency of treatments A and B. Set up a randomization scheme for 50 patients such that the treatment assignment over all hospitals and within a hospital will be balanced.

13.3 Use the random-number table (Table C-20) to assign three treatments A, B, and C to 45 patients in a way that the number of patients in each group is approximately equal.

13.4 Suppose a trial is to be conducted for advanced melanoma patients to compare chemotherapy only with chemotherapy plus immunotherapy. The main endpoint is response. In a previous trial, 60 patients were treated with the same chemotherapy alone. The response rate was 40%. For the present trial, it is determined to enter 120 patients. According to Pocock's formulas how many of these 120 patients should be entered on the randomized control for which there are already 60 patients in a historical control group? Let the standard deviation of the bias in the historical controls be (a) 1% and (b) 5%.

APPENDIX A

The Newton–Raphson Method

The Newton–Raphson method (Ralston and Wilf 1967, Carnahan et al. 1969) is a numerical iterative procedure that can be used to solve nonlinear equations. An iterative procedure is a technique of successive approximations, and each approximation is called an iteration. If the successive approximations approach the solution very closely, we say that the iterations converge. The maximum likelihood estimates of various parameters and coefficients discussed in Chapters 8, 10, and 11 can be obtained by using the Newton–Raphson method. In this section, we discuss and illustrate the use of this method, first considering a single nonlinear equation and then a set of nonlinear equations.

Let $f(x) = 0$ be the equation to be solved for x. The Newton–Raphson method requires an initial estimate of x, say $\hat{x}_0$, such that $f(\hat{x}_0)$ is close to zero, preferably, and then the first approximate iteration is given by

$$\hat{x}_1 = \hat{x}_0 - f(\hat{x}_0)/f'(\hat{x}_0) \qquad (A.1)$$

where $f'(\hat{x}_0)$ is the first derivative of $f(x)$ evaluated at $x = \hat{x}_0$. In general, the $(k + 1)$th iteration or approximation is given by

$$\hat{x}_{k+1} = \hat{x}_k - f(x_k)/f'(x_k) \qquad (A.2)$$

where $f'(\hat{x}_k)$ is the first derivative of $f(x)$ evaluated at $x = \hat{x}_k$. The iteration terminates at the kth iteration if $f(\hat{x}_k)$ is close enough to zero or the difference between $\hat{x}_k$ and $\hat{x}_{k-1}$ is negligible. The stopping rule is rather subjective. Acceptable rules are that $f(\hat{x}_k)$ or $d = \hat{x}_k - \hat{x}_{k-1}$ is in the neighborhood of 10^{-6} or 10^{-7}.

Example A.1
Consider the function

$$f(x) = x^3 - x + 2$$

374

We wish to find the value of x such that $f(x) = 0$ by the Newton–Raphson method. The first derivative of $f(x)$ is

$$f'(x) = 3x^2 - 1$$

Since $f(-1) = 2$ and $f(-2) = -4$, graphically (Figure A.1) we see that the curve cuts through the x axis $[f(x) = 0]$ between -1 and -2. This gives us a good hint of an initial value of x. Suppose that we begin with $\hat{x}_0 = -1$; $f(\hat{x}_0) = 2$ and $f'(\hat{x}_0) = 2$. Thus, the first iteration, following (A.1), gives

$$\hat{x}_1 = -1 - \tfrac{2}{2} = -2$$

and $f(\hat{x}_1) = -4$ and $f'(\hat{x}_1) = 11$. Following (A.2) we obtain the following:

Second iteration:

$$\hat{x}_2 = -2 + \tfrac{4}{11} = -1.6364$$
$$f(\hat{x}_2) = -0.7456 \qquad f'(\hat{x}_2) = 7.0334$$

Third iteration:

$$\hat{x}_3 = -1.6364 + \frac{0.7456}{7.0334} = -1.5304$$
$$f(\hat{x}_3) = -0.054 \qquad f'(\hat{x}_3) = 6.0264$$

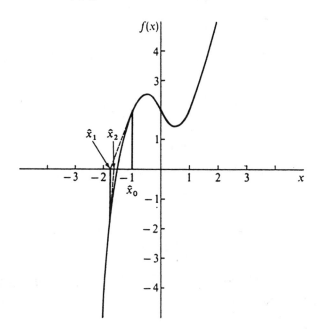

Fourth iteration:

$$\hat{x}_4 = -1.5304 + \frac{0.054}{6.0264} = -1.52144$$

$$f(\hat{x}_4) = -0.00036 \qquad f'(\hat{x}_4) = 5.9443$$

Fifth iteration:

$$\hat{x}_5 = -1.52144 + \frac{0.00036}{5.9443} = -1.52138$$

$$f(\hat{x}_5) = 0.0000017$$

At the fifth iteration, for $x = -1.52138$, $f(x)$ is very close to zero. If the stopping rule is that $f(x) \le 10^{-6}$, the iterative procedure would terminate after the fifth iteration and $x = -1.52138$ is the root of the equation $x^3 - x + 2 = 0$. Figure A.1 gives the graphical presentation of $f(x)$ and the iteration.

It should be noted that the Newton–Raphson method can only find the real roots of an equation. The equation $x^3 - x + 2 = 0$ has only one real root as shown in Figure A.1, the other two are complex roots.

The Newton–Raphson method can be extended to solve a system of equations with more than one unknown. Suppose that we wish to find values of $x_1, x_2, \ldots, x_p$ such that

$$f_1(x_1, \ldots, x_p) = 0$$
$$f_2(x_1, \ldots, x_p) = 0$$
$$\vdots$$
$$f_p(x_1, \ldots, x_p) = 0$$

Let a_{ij} be the partial derivative of f_i with respect to x_j, that is, $a_{ij} = \partial f_i / \partial x_j$. The matrix

$$J = \begin{bmatrix} a_{11} & \cdots & a_{1p} \\ a_{21} & \cdots & a_{2p} \\ \vdots & & \vdots \\ a_{p1} & \cdots & a_{pp} \end{bmatrix}$$

is called the *Jacobian matrix*. Let the inverse of J, denoted by J^{-1}, be

$$J^{-1} = \begin{bmatrix} b_{11} & \cdots & b_{1p} \\ b_{21} & \cdots & b_{2p} \\ \vdots & & \vdots \\ b_{p1} & \cdots & b_{pp} \end{bmatrix}$$

Let $x_1^k, x_2^k, \ldots, x_p^k$ be the approximate root at the kth iteration and let $f_1^k, \ldots, f_p^k$ be the corresponding values of the functions $f_1, \ldots, f_p$, that is,

$$f_1^k = f_1(x_1^k, \ldots, x_p^k)$$
$$\vdots$$
$$f_p^k = f_p(x_1^k, \ldots, x_p^k)$$

and b_{ij}^k be the ijth element of J^{-1} evaluated at $x_1^k, \ldots, x_p^k$. Then the next approximation is given by

$$x_1^{k+1} = x_1^k - (b_{11}^k f_1^k + b_{12}^k f_2^k + \cdots + b_{1p}^k f_p^k)$$
$$x_2^{k+1} = x_2^k - (b_{21}^k f_1^k + b_{22}^k f_2^k + \cdots + b_{2p}^k f_p^k)$$
$$\vdots$$
$$x_p^{k+1} = x_p^k - (b_{p1}^k f_1^k + b_{p2}^k f_2^k + \cdots + b_{pp}^k f_p^k)$$
$$\text{(A.3)}$$

The iterative procedure begins with a preselected initial approximate x_1^0, $x_2^0, \ldots, x_p^0$, proceeds following (A.3), and terminates either when $f_1, f_2, \ldots, f_p$ are close enough to zero or when differences in the x values at two consecutive iterations are negligible.

Example A.2

Suppose we wish to find the value of x_1 and x_2 such that

$$x_1^2 + x_1 x_2 - 2x_1 - 1 = 0 \qquad x_1^3 - x_1 + x_2 - 2 = 0$$

In this case, $p = 2$:

$$f_1 = x_1^2 + x_1 x_2 - 2x_1 - 1 \qquad f_2 = x_1^3 - x_1 + x_2 - 2$$

Since $\partial f_1 / \partial x_1 = 2x_1 + x_2 - 2$, $\partial f_1 / \partial x_2 = x_1$, $\partial f_2 / \partial x_1 = 3x_1^2 - 1$, and $\partial f_2 / \partial x_2 = 1$, the Jacobian matrix is

$$J = \begin{bmatrix} 2x_1 + x_2 - 2 & x_1 \\ 3x_1^2 - 1 & 1 \end{bmatrix} \qquad \text{(A.4)}$$

Let the initial estimates be $x_1^0 = 0$, $x_2^0 = 1$, $f_1^0 = -1$, and $f_2^0 = -1$:

$$J = \begin{bmatrix} -1 & 0 \\ -1 & 1 \end{bmatrix} \qquad J^{-1} = \begin{bmatrix} -1 & 0 \\ -1 & 1 \end{bmatrix}$$

Iteration 1: Following (A.3), we obtain

$$x_1^1 = 0 - [(-1)(-1) + 0(-1)] = -1 \qquad x_2^1 = 1 - [(-1)(-1) + 1(-1)] = 1$$

With these values, $f_1^1 = 1$, $f_2^1 = -1$, and

$$J = \begin{bmatrix} -3 & -1 \\ 2 & 1 \end{bmatrix} \qquad J^{-1} = \begin{bmatrix} -1 & -1 \\ 2 & 3 \end{bmatrix}$$

Iteration 2: From (A.3) we obtain

$$x_1^2 = -1 - [(-1)(1) + (-1)(-1)] = -1 \qquad x_2^2 = 1 - [(2)(1) + (3)(-1)] = 2$$

With these values $f_1^2 = 0$ and $f_2^2 = 0$. Therefore the iteration procedure terminates and the solution of the two simultaneous equations is $x_1 = -1$, $x_2 = 2$.

The number of iterations required depends much on the initial values chosen. In Example A.2, if we use $x_1^0 = 0$, $x_2^0 = 0$, it requires about 11 iterations to find the solution. Interested readers may try it as an exercise.

Computer Program GAMPLOT

```
C       PROGRAM GAMPLOT
        DIMENSION T(29),XID(27),TITLE(7),BUFFER(1024)
        DATA TITLE/4HEXER,4HCISE,4H 71.,4H0     /
        DATA T/26.,15.,26.,10.,34.,36.,21.,22.,2.,3.,9.,21.,30.,100./
        DO 1 I=1,14
   1    XID(I)=I
        GAMMA=2.
        CALL PLOT(0.,0.,999)
        CALL GPLOT(GAMMA,14,T,XID,TITLE,5)
        CALL PLOTS(BUFFER,1024)
        STOP
        END


        SUBROUTINE GPLOT(ETA,N,X,XID,TITLE,NID)
C
C       SUBROUTINE TO PLOT ARRAY X AGAINST STANDARD GAMMA QUANTILES
C
C       ETA = VALUE OF THE SHAPE PARAMETER FOR THE PLOT
C         N = NUMBER OF POINTS IN ARRAY X
C         X = THE ARRAY OF POINTS TO BE PLOTTED
C       XID = IDENTIFICATION OF PONITS IN X
C     TITLE = TITLE FOR THE PLOT
C        NT = NO. OF LETTERS IN TITLE
C       NID = NO. OF POINTS TO BE IDENTIFIED (STARTING FROM LARGEST)
C
        DIMENSION X(1)  ,PR(200),Q(202),XID(1)  ,TITLE(1) ,FMTV(5),FMTH(3)
       1,FMT(6),ID(200)
        DATA FMTH/4HQUAN,4HTILE,4HS    /
        DATA FMTV/4HORDE,4HRED ,4HOBSE,4HRVAT,4HIONS/
        DATA FMT/4HGAMM,4HA PR,4HOBAB,4HILIT,4HY PL,4HOT   /
        XN=N
        DO 1 I=1,N
        XI=I
   1    PR(I)=(XI-0.5)/XN
        CALL GAMINV(PR,Q,N,ETA)
        CALL SORTA(X,N,XID)
        WRITE(6,2) N,(TITLE(I),I=1,7),ETA
   2    FORMAT(1H1,I4,' POINTS FOR GAMMA PROBABILITY PLOT'/2X, 7A4/5X,'GAM
       1MA =',F7.3)
        DO 3 I=1,N
   3    ID(I)=XID(I)
        WRITE(6,4) (I,ID(I),X(I),PR(I),Q(I),I=1,N)
   4    FORMAT(1H0,6X,'I',5X,'IDENT',6X,'ORDERED OBSERVATIONS',7X,'FRACTIO
       1NS',14X,'QUANTILES'//(5X,I3,6X,I3,1P3E23.7))
C
C       PLOT
C
```

```
      CALL PLOT(0.,0.,-3)
      CALL SCALE(Q,6.,N,1)
      CALL SCALE(X,6.,N,1)
      CALL AXIS(0.,0.,FMTH,-12,6.,0.,Q(N+1),Q(N+2))
      CALL AXIS(0.,0.,FMTV,20,6.,90.,X(N+1),X(N+2))
      CALL LINE(Q,X,N,1,-1,4)
      CALL PLOT(0.3,8.0,3)
      CALL SYMBOL(0.3,8.0,.14,TITLE,0.,28)
      CALL PLOT(0.3,7.5,3)
      CALL SYMBOL(0.3,7.5,.14,FMT,0.,24)
C
C     LABEL LARGEST NID POINTS
C
      IF (NID) 100,100,70
  70  N1=N-NID+1
      DO 75 IPRT=N1,N
      XX=(Q(IPRT)-Q(N+1))/Q(N+2)+0.1
      YY=(X(IPRT)-X(N+1))/X(N+2)
      CALL NUMBER(XX,YY,0.1,XID(IPRT),0.,-1)
  75  CONTINUE
 100  CALL PLOT(10.,0.,-3)
      CALL SORTA(XID,N,X)
      RETURN
      END



      SUBROUTINE GAMINV(B,Y,N,ETA)
C
C         PURPOSE:
C
C             GIVEN A GAMMA DIST WITH SHAPE PARAMETER ETA, (ASSUMING A
C             SCALE PARAMETER, LAMBDA.EQ.1) AND A SERIES OF INCREASING
C             FRACTIONS IN THE B ARRAY, GAMINV COMPUTES THE
C             CORRESPONDING QUANTILES, AND STORES THEM IN THE Y ARRAY.
C             IF LAMBDA .NE. 1, THEN MAKE THE FOLLOWING STATEMENTS:
C             (WHERE THE ACTUAL PROBABILITES ARE IN THE ARRAY P)
C                     DO 10 I=1,N
C                  10 PL(I) = P(I)*LAMBDA
C                     CALL GAMINV(PL,Y,N,ETA)
C             NOTE - YOU MAY USE THIS ROUTINE WITH N=1,
C                     AND B AND Y SIMPLE VARIABLES RATHER THAN ARRAYS.
C
C         ARGUMENT DESCRIPTION:
C
C             B   = INPUT, ARRAY OF FRACTIONS - IN ASCENDING ORDER
C             Y   = OUTPUT, ARRAY OF QUANTILES COMPUTED BY GAMINV
C             N   = INPUT, THE NUMBER OF FRACTIONS
C             ETA = INPUT, VALUE OF THE SHAPE PARAMETER OF THE GAMMA
C                     DIST.
C
C         EXTERNAL I/O:
C
C             ERROR MESSAGE - FRACTIONS IN THE P ARRAY ARE NOT
C                     IN ASCENDING ORDER (IF THIS OCCURS
C                     A STOP STATEMENT IS EXECUTED; SO
C                     A JOB CANNOT CONTINUE IF THIS ERROR
C                     OCCURS.
C
C         OTHER PROGRAMS CALLED: DAMLOG,DETAZ,GAM,GAMLOG,GTAILI
C
      DIMENSION B(1),Y(1)
  20  N1=1
      N2=N
      IF(B(1).GT.0.) GO TO 25
      Y(1)=0.0
      N2=N-1
```

```
 25    IF(B(N).LT.1.0) GO TO 30
       Y(N)=1.0E38
       N1=2
 30    IF(N2-N1) 800,60,40
 40    DO 50 I=2,N
       IF(B(I).LT.B(I-1)) GO TO 890
 50    CONTINUE
 60    ASSIGN 110 TO IB
       EX=1.0/ETA
       NN=N+1
       IF(ETA.GT.20.0) GO TO 70
       IGAM=1
       GAMMA=GAM(ETA)
       GO TO 75
 70    IGAM=2
       GLOG=GAMLOG(ETA)
 75    IGO=IGAM
       DO 200 I=N1,N2
       II=NN-I
       IF(ETA.GT.2.0) IF(B(II)-.99) 61,61,404
 61    GO TO (105,106),IGO
 404   PR = B(II)
       CALL GTAILI(PR,ETA,QUANT)
       Y(II)=QUANT
       GO TO 200
 105   R=GAMMA*B(II)
       YZERO=(R*ETA)**EX
       GO TO 108
 106   RLOG=GLOG+ALOG(B(II))
       YZERO=EXP(EX*(ALOG(ETA)+RLOG))
 108   GO TO IB, (110,150)
 110   IF(YZERO) 900,112,113
 112   YDELT=1.0E-38
       GO TO 114
 113   YDELT=YZERO
       YZERO=0.0
 114   YSTAR=YDELT
 115   YLOG=ETA*ALOG(YSTAR)
       GO TO (116,119),IGO
 116   IF(YLOG.LE.87.0) IF(YSTAR-87.0) 1117,1117,118
 118   IGO=2
       RLOG=ALOG(R)
 119   FLOG=YLOG+DAMLOG(ETA,YSTAR)
       IF(FLOG-RLOG) 120,120,125
 1117  F=EXP(YLOG)*DETAZ(ETA,YSTAR)
       IF(F.GT.R) GO TO 125
 120   YZERO=YSTAR
       YSTAR=YSTAR+YDELT
       GO TO 115
 125   ASSIGN 150 TO IB
 150   IF((YSTAR-YZERO)/YSTAR.LE.1.0E-5) GO TO 175
       YMID=(YZERO+YSTAR)/2.0
       GO TO (156,158),IGO
 156   FT=YMID**ETA*DETAZ(ETA,YMID)
       IF(FT-R) 160,160,165
 158   YLOG=ETA*ALOG(YMID)
       FTLOG=YLOG+DAMLOG(ETA,YMID)
       IF(FTLOG.GT.RLOG) GO TO 165
 160   YZERO=YMID
       GO TO 150
 165   YSTAR=YMID
       GO TO 150
 175   Y(II)=YSTAR
       IF(IGAM-IGO) 180,200,900
 180   IF(YLOG.GT.87.0) GO TO 200
       IGO=IGAM
 200   CONTINUE
 800   RETURN
```

```
  890      PRINT 500
  500      FORMAT(37H0FRACTIONS ARE NOT IN ASCENDING ORDER)
  900      STOP
           END


           FUNCTION DAMLOG(ETA,Z)
C
C              PURPOSE:
C
C                  DAMLOG COMPUTES LOG TO THE BASE E OF D(ETA,Z), WHERE:
C                      D(ETA,Z)= INTEGRAL FROM 0 TO 1 OF
C                      T**(ETA-1) E**(-Z*T) DT
C                  (D(ETA,Z) IS THE INCOMPLETE GAMMA FUNCTION)
C
C              ARGUMENT: DESCRIPTION:
C
C                  ETA  = FIRST PARAMETER OF THE FUNCTION
C                  Z    = SECOND PARAMETER OF THE FUNCTION
C
           KQ=1
           TQ=0.0
           ETAM=ETA-1.0
           ZMETA=Z-ETA
           DTERML=-Z-ALOG(ETA)
   25      TQ=TQ+EXP(DTERML)
           IF(FLOAT(KQ).GT.ZMETA) GO TO 75
   50      KQ=KQ+1
           DTERML=DTERML+ALOG(Z/(ETAM+FLOAT(KQ)))
           GO TO 25
   75      FACTL=DTERML+ALOG(Z/(FLOAT(KQ)-ZMETA))
           FACT=EXP(FACTL)
           IF(FACT.GT.TQ*1.0E-05) GO TO 50
           DAMLOG=ALOG(TQ)
           RETURN
           END


           SUBROUTINE GTAILI(PR,ETA,QUANT)
C
C              PURPOSE:
C
C                  GTAILI FINDS THE QUANTILE, QUANT, CORRESPONDING TO THE
C                  PROBABILITY, PROB. OF THE GAMMA DISTRIBUTION WITH SHAPE
C                  PARAMETER ETA AND SCALE PARAMETER LAMBDA = 1.
C                  IT SHOULD BE USED IN THE CASES WHERE PROB .GT. .99. IT
C                  IS BASED ON A PROCEDURE BY RUBIN AND ZIDEK
C              ARGUMENT DESCRIPTION:
C
C                  PR = INPUT, PROBABILITY, GT..99
C                  ETA  = INPUT, SHAPE PARAMETER OF THE GAMMA DIST
C                  QUANT = OUTPUT, QUANTILE CORRESPONDING TO PR
C
C              OTHER PROGRAMS CALLED: QNORMS
C
           ZP=QNORMS(PR)
           ZP2=ZP*ZP
           ZP3=ZP2*ZP
           ZP4=ZP3*ZP
           ZP5=ZP4*ZP
           ZP6=ZP5*ZP
           C2=-(ZP3+3.0*ZP)/108.
           C3=(ZP4+4.*ZP2+8.0)/405.
           C4=-(ZP5+4.*ZP3+15.*ZP)/2592.
           C5=(ZP6-15.*ZP4-18.*ZP2+48.)/76545.
           V=ETA-1./3.
           R=ZP+C2/V+C3/(V**1.5)+C4/(V*V)+C5/(V**2.5)
           QUANT=(V**.33333333+R/(3.*V**.16666666))**3
           RETURN
           END
```

```
      FUNCTION DETAZ(ETA,Z)
C
C          PURPOSE:
C
C                DETAZ COMPUTES D(ETA,Z) WHERE:
C                    D(ETA,Z) = INTEGRAL FROM 0 TO 1 OF
C                    T**(ETA-1) E**(-Z*T) DT
C                (D(ETA,Z) IS THE INCOMPLETE GAMMA FUNCTION)
C
C          ARGUMENT: DESCRIPTION:
C
C                ETA  = FIRST PARAMETER OF THE FUNCTION
C                Z    = SECOND PARAMETER OF THE FUNCTION
C
      KQ=1
      TQ=0.0
      ETAM=ETA-1.0
      ZMETA=Z-ETA
      DTERM=EXP(-Z)/ETA
   25 TQ=TQ+DTERM
      IF(FLOAT(KQ).GT.ZMETA) GO TO 75
   50 KQ=KQ+1
      DTERM=DTERM*(Z/(ETAM+FLOAT(KQ)))
      GO TO 25
   75 IF(DTERM*(Z/(FLOAT(KQ)-ZMETA))-TQ*1.0E-05) 100,100,50
  100 DETAZ=TQ
      RETURN
      END


      FUNCTION GAM(X1)
C
C          PURPOSE:
C
C                GAM COMPUTES GAMMA(X1), GAMLOG COMPUTES LOG TO THE BASE
C                E OF GAMMA(X1), USING HASTINGS RATIONAL FUNCTION
C                APPROXIMATION FOR X1 .LE. 30.  FOR X1 .GT. 30 AN
C                IMPROVED VERSION OF STIRLINGS APPROXIMATION IS USED FOR
C                GAMMA(X1) =(X1-1) FACTORIAL.
C
C          ALTERNATE ENTRY POINT:
C
C                GAMLOG(X1) - SEE PURPOSE
C
C          ARGUMENT DESCRIPTION
C
C                X1 = ARGUMENT FOR THE GAMMA FUNCTION
C
C          EXTERNAL I/O:
C                ERROR MESSAGE - X1 IS TOO LARGE TO COMPUTE GAM(X1),
C                LOG(GAM(X1)) HAS BEEN RETURNED
C
C          OTHER PROGRAMS CALLED: NONE
C
      DIMENSIONA(8)
      DATAA/-0.577191652,0.988205891,-.897056937,0.918206857,-0.75670407
     18,0.482199394,-0.193527818,0.035868343/
C      XLSQ2P = ALOG ( SQRT (2 * PI ) )
      DATA XLSQ2P/.91893854/
C
      IL=1
      GOTO50
      ENTRY GAMLOG (X1)
      IL=2
      IF(X1 .LE. 30.) GOTO50
   56 ETAM=X1-1.
```

```
C       GAM USED HERE IS REALLY GAMLOG
        GAM=XLSQ2P+(ETAM+.5)*ALOG(ETAM)-ETAM+1./(12.*ETAM)+1./(288.*ETAM*ET
     1TAM)
        RETURN
55      WRITE(6,12)ETA
12      FORMAT(8H0GAMMA =F6.3,67H IS TOO LARGE TO COMPUTE GAM(GAMMA).  LOG
     1(GAM(GAMMA)) WAS RETURNED.   )
        GOTO56
50      ETA=X1
        IF(ETA .GT. 33.)GOTO55
        FETA=AMOD(ETA,1.)
        IF(FETA .GT. 0.) GOTO 41
        G1FETA=1.
        GOTO 42
41      G1FETA=1.+FETA*(A(1)+FETA*(A(2)+FETA*(A(3)+FETA*(A(4)+FETA*(A(5)+F
     1ETA*(A(6)+FETA*(A(7)+FETA*A(8))))))))
42      IF(ETA .LT. 1.)GOTO21
        IF(ETA .GT. 2.)GOTO22
        GAM=G1FETA
        GOTO30
21      GAM=G1FETA/FETA
        GOTO30
22      TEST=ETA-.5
        PROD=1.
        TERM=1.+FETA
35      PROD=PROD*TERM
        TERM=TERM+1.
        IF(TERM .LT. TEST)GOTO35
        GAM=PROD*G1FETA
30      IF(IL .EQ. 2)GAM=ALOG(GAM)
        RETURN
        END



        SUBROUTINE SORTA(X,N,XIDEN)
C
C       THIS ROUTINE SORTS THE FIRST N NUMBERS IN ARRAY X IN ASCENDING
C       ORDER, XIDEN WHICH IS SORTED ALONG WITH X IS THE IDENT. OF X
C
        DIMENSION X(200),XIDEN(200)
        DO 1 I=1,N
        J=N-I+1
        JJ=J-1
        IF (JJ .LE.1) GO TO 1
        DO 2 K=1,JJ
        IF (X(K) .LE. X(J)) GO TO 2
        X1=X(J)
        XID=XIDEN(J)
        X(J)=X(K)
        XIDEN(J)=XIDEN(K)
        X(K)=X1
        XIDEN(K)=XID
2       CONTINUE
1       CONTINUE
        RETURN
        END



        FUNCTION QNORMS(PR)
C
C           PURPOSE:
C
C                 QNORMS IS A FUNCTION WHICH FINDS THE QUANTILE OF THE
C                 STANDARD NORMAL DISTRIBUTION ASSOCIATED WITH A SPECIFIC
```

```
C                    PROBABILITY, PR, USING HASTINGS APPROXIMATION.
C
C             ARGUMENT DESCRIPTION:
C
C                 PR = THE PROBABILITY OF WHICH THE CORRESPONDING
C                          QUANTILE WILL BE FOUND
C
C             OTHER PROGRAMS CALLED: NONE
C
      IF(PR .NE. 0.5) GO TO 10
      QNORMS =0.0
      RETURN
   10 P = PR
      IF (P .LT. 0.5) GO TO 100
      P=1.-P
  100 ETA=SQRT(-2.*ALOG(P))
      TERM=((.010328*ETA+.802853)*ETA+2.515517)/(((.001308*ETA+.189269)*
     1ETA+1.432788)*ETA+1.0)
      QNORMS = TERM - ETA
      IF(PR .GT. 0.5) QNORMS = -QNORMS
      RETURN
      END
```

APPENDIX C

Statistical Tables

TABLE C-1   NORMAL CURVE AREAS

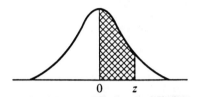

z	.00	.01	.02	.03	.04	.05	.06	.07	.08	.09
0.0	.0000	.0040	.0080	.0120	.0160	.0199	.0239	.0279	.0319	.0359
0.1	.0398	.0438	.0478	.0517	.0557	.0596	.0636	.0675	.0714	.0753
0.2	.0793	.0832	.0871	.0910	.0948	.0987	.1026	.1064	.1103	.1141
0.3	.1179	.1217	.1255	.1293	.1331	.1368	.1406	.1443	.1480	.1517
0.4	.1554	.1591	.1628	.1664	.1700	.1736	.1772	.1808	.1844	.1879
0.5	.1915	.1950	.1985	.2019	.2054	.2088	.2123	.2157	.2190	.2224
0.6	.2257	.2291	.2324	.2357	.2389	.2422	.2454	.2486	.2517	.2549
0.7	.2580	.2611	.2642	.2673	.2704	.2734	.2764	.2794	.2823	.2852
0.8	.2881	.2910	.2939	.2967	.2995	.3023	.3051	.3078	.3106	.3133
0.9	.3159	.3186	.3212	.3238	.3264	.3289	.3315	.3340	.3365	.3389
1.0	.3413	.3438	.3461	.3485	.3508	.3531	.3554	.3577	.3599	.3621
1.1	.3643	.3665	.3686	.3708	.3729	.3749	.3770	.3790	.3810	.3830
1.2	.3849	.3869	.3888	.3907	.3925	.3944	.3962	.3980	.3997	.4015
1.3	.4032	.4049	.4066	.4082	.4099	.4115	.4131	.4147	.4162	.4177
1.4	.4192	.4207	.4222	.4236	.4251	.4265	.4279	.4292	.4306	.4319
1.5	.4332	.4345	.4357	.4370	.4382	.4394	.4406	.4418	.4429	.4441
1.6	.4452	.4463	.4474	.4484	.4495	.4505	.4515	.4525	.4535.	.4545
1.7	.4554	.4564	.4573	.4582	.4591	.4599	.4608	.4616	.4625	.4633
1.8	.4641	.4649	.4656	.4664	.4671	.4678	.4686	.4693	.4699	.4706
1.9	.4713	.4719	.4726	.4732	.4738	.4744	.4750	.4756	.4761	.4767
2.0	.4772	.4778	.4783	.4788	.4793	.4798	.4803	.4808	.4812	.4817
2.1	.4821	.4826	.4830	.4834	.4838	.4842	.4846	.4850	.4854	.4857
2.2	.4861	.4864	.4868	.4871	.4875	.4878	.4881	.4884	.4887	.4890
2.3	.4893	.4896	.4898	.4901	.4904	.4906	.4909	.4911	.4913	.4916
2.4	.4918	.4920	.4922	.4925	.4927	.4929	.4931	.4932	.4934	.4936
2.5	.4938	.4940	.4941	.4943	.4945	.4946	.4948	.4949	.4951	.4952
2.6	.4953	.4955	.4956	.4957	.4959	.4960	.4961	.4962	.4963	.4964
2.7	.4965	.4966	.4967	.4968	.4969	.4970	.4971	.4972	.4973	.4974
2.8	.4974	.4975	.4976	.4977	.4977	.4978	.4979	.4979	.4980	.4981
2.9	.4981	.4982	.4982	.4983	.4984	.4984	.4985	.4985	.4986	.4986
3.0	.4987	.4987	.4987	.4988	.4988	.4989	.4989	.4989	.4990	.4990

Abridged from Table 1 of "Statistical Tables and Formulas" by A. Hald, John Wiley & Sons, 1952. Reproduced by permission of John Wiley & Sons.

TABLE C-2
PERCENTAGE POINTS OF THE χ^2-DISTRIBUTION

ν d.f.	99.5%	97.5%	5%	2.5%	1%	0.5%
1	392704×10^{-10}	982069×10^{-9}	3.84146	5.02389	6.63490	7.87944
2	0.0100251	0.0506356	5.99147	7.37776	9.21034	10.5966
3	0.0717212	0.215795	7.81473	9.34840	11.3449	12.8381
4	0.206990	0.484419	9.48773	11.1433	13.2767	14.8602
5	0.411740	0.831211	11.0705	12.8325	15.0863	16.7496
6	0.675727	1.237347	12.5916	14.4494	16.8119	18.5476
7	0.989265	1.68987	14.0671	16.0128	18.4753	20.2777
8	1.344419	2.17973	15.5073	17.5346	20.0902	21.9550
9	1.734926	2.70039	16.9190	19.0228	21.6660	23.5893
10	2.15585	3.24697	18.3070	20.4831	23.2093	25.1882
11	2.60321	3.81575	19.6751	21.9200	24.7250	26.7569
12	3.07382	4.40379	21.0261	23.3367	26.2170	28.2995
13	3.56503	5.00874	22.3621	24.7356	27.6883	29.8194
14	4.07468	5.62872	23.6848	26.1190	29.1413	31.3193
15	4.60094	6.26214	24.9958	27.4884	30.5779	32.8013
16	5.14224	6.90766	26.2962	28.8454	31.9999	34.2672
17	5.69724	7.56418	27.5871	30.1910	33.4087	35.7185
18	6.26481	8.23075	28.8693	31.5264	34.8053	37.1564
19	6.84398	8.90655	30.1435	32.8523	36.1908	38.5822
20	7.43386	9.59083	31.4104	34.1696	37.5662	39.9968
21	8.03366	10.28293	32.6705	35.4789	38.9321	41.4010
22	8.64272	10.9823	33.9244	36.7807	40.2894	42.7956
23	9.26042	11.6885	35.1725	38.0757	41.6384	44.1813
24	9.88623	12.4011	36.4151	39.3641	42.9798	45.5585
25	10.5197	13.1197	37.6525	40.6465	44.3141	46.9278
26	11.1603	13.8439	38.8852	41.9232	45.6417	48.2899
27	11.8076	14.5733	40.1133	43.1944	46.9630	49.6449
28	12.4613	15.3079	41.3372	44.4607	48.2782	50.9933
29	13.1211	16.0471	42.5569	45.7222	49.5879	52.3356
30	13.7867	16.7908	43.7729	46.9792	50.8922	53.6720
40	20.7065	24.4331	55.7585	59.3417	63.6907	66.7659
50	27.9907	32.3574	67.5048	71.4202	76.1539	79.4900
60	35.5346	40.4817	79.0819	83.2976	88.3794	91.9517
70	43.2752	48.7576	90.5312	95.0231	100.425	104.215
80	51.1720	57.1532	101.879	106.629	112.329	116.321
90	59.1963	65.6466	113.145	118.136	124.116	128.299
100	67.3276	74.2219	124.342	129.561	135.807	140.169

From "Tables of the Percentage Points of the χ^2-Distribution" by Catherine M. Thompson, *Biometrika*, Vol. 32(1941), pp. 188–189. Reproduced by permission of the editor of *Biometrika*.

TABLE C-3

5% POINTS OF THE *F*-DISTRIBUTION

Degrees of Freedom for Numerator

v_2 \ v_1	1	2	3	4	5	6	7	8	9
1	161.45	199.50	215.71	224.58	230.16	233.99	236.77	238.88	240.54
2	18.513	19.000	19.161	19.247	19.296	19.330	19.353	19.371	19.385
3	10.128	9.5521	9.2766	9.1172	9.0135	8.9406	8.8868	8.8452	8.8123
4	7.7086	6.9443	6.5914	6.3883	6.2560	6.1631	6.0942	6.0410	5.9988
5	6.6079	5.7861	5.4095	5.1922	5.0503	4.9503	4.8759	4.8183	4.7725
6	5.9874	5.1433	4.7571	4.5337	4.3874	4.2839	4.2066	4.1468	4.0990
7	5.5914	4.7374	4.3468	4.1203	3.9715	3.8660	3.7870	3.7257	3.6767
8	5.3177	4.4590	4.0662	3.8378	3.6875	3.5806	3.5005	3.4381	3.3881
9	5.1174	4.2565	3.8626	3.6331	3.4817	3.3738	3.2927	3.2296	3.1789
10	4.9646	4.1028	3.7083	3.4780	3.3258	3.2172	3.1355	3.0717	3.0204
11	4.8443	3.9823	3.5874	3.3567	3.2039	3.0946	3.0123	2.9480	2.8962
12	4.7172	3.8853	3.4903	3.2592	3.1059	2.9961	2.9134	2.8486	2.7964
13	4.6672	3.8056	3.4105	3.1791	3.0254	2.9153	2.8321	2.7669	2.7144
14	4.6001	3.7389	3.3439	3.1122	2.9582	2.8477	2.7642	2.6987	2.6458
15	4.5431	3.6823	3.2874	3.0556	2.9013	2.7905	2.7066	2.6408	2.5876
16	4.4940	3.6337	3.2389	3.0069	2.8524	2.7413	2.6572	2.5911	2.5377
17	4.4513	3.5915	3.1968	2.9647	2.8100	2.6987	2.6143	2.5480	2.4943
18	4.4139	3.5546	3.1599	2.9277	2.7729	2.6613	2.5767	2.5102	2.4563
19	4.3808	3.5219	3.1274	2.8951	2.7401	2.6283	2.5435	2.4768	2.4227
20	4.3513	3.4928	3.0984	2.8661	2.7109	2.5990	2.5140	2.4471	2.3928
21	4.3248	3.4668	3.0725	2.8401	2.6848	2.5727	2.4876	2.4205	2.3661
22	4.3009	3.4434	3.0491	2.8167	2.6613	2.5491	2.4638	2.3965	2.3419
23	4.2793	3.4221	3.0280	2.7955	2.6400	2.5277	2.4422	2.3748	2.3201
24	4.2597	3.4028	3.0088	2.7763	2.6207	2.5082	2.4226	2.3551	2.3002
25	4.2417	3.3852	2.9912	2.7587	2.6030	2.4904	2.4047	2.3371	2.2821
26	4.2252	3.3690	2.9751	2.7426	2.5868	2.4741	2.3883	2.3205	2.2655
27	4.2100	3.3541	2.9604	2.7278	2.5719	2.4591	2.3732	2.3053	2.2501
28	4.1960	3.3404	2.9467	2.7141	2.5581	2.4453	2.3593	2.2913	2.2360
29	4.1830	3.3277	2.9340	2.7014	2.5454	2.4324	2.3463	2.2782	2.2229
30	4.1709	3.3158	2.9223	2.6896	2.5336	2.4205	2.3343	2.2662	2.2107
40	4.0848	3.2317	2.8387	2.6060	2.4495	2.3359	2.2490	2.1802	2.1240
60	4.0012	3.1504	2.7581	2.5252	2.3683	2.2540	2.1665	2.0970	2.0401
120	3.9201	3.0718	2.6802	2.4472	2.2900	2.1750	2.0867	2.0164	1.9588
∞	3.8415	2.9957	2.6049	2.3719	2.2141	2.0996	2.0096	1.9384	1.8799

Degrees of Freedom for Denominator

TABLE C-3 (*continued*)

Degrees of Freedom for Numerator

ν_2 \ ν_1	10	12	15	20	24	30	40	60	120	∞
1	241.88	243.91	245.95	248.01	249.05	250.09	251.14	252.20	253.25	254.32
2	19.396	19.413	19.429	19.446	19.454	19.462	19.471	19.479	19.487	19.496
3	8.7855	8.7446	8.7029	8.6602	8.6385	8.6166	8.5944	8.5720	8.5494	8.5265
4	5.9644	5.9117	5.8578	5.8025	5.7744	5.7459	5.7170	5.6878	5.6581	5.6281
5	4.7351	4.6777	4.6188	4.5581	4.5272	4.4957	4.4638	4.4314	4.3984	4.3650
6	4.0600	3.9999	3.9381	3.8742	3.8415	3.8082	3.7743	3.7398	3.7047	3.6688
7	3.6365	3.5747	3.5108	3.4445	3.4105	3.3758	3.3404	3.3043	3.2674	3.2298
8	3.3472	3.2840	3.2184	3.1503	3.1152	3.0794	3.0428	3.0053	2.9669	2.9276
9	3.1373	3.0729	3.0061	2.9365	2.9005	2.8637	2.8259	2.7872	2.7475	2.7067
10	2.9782	2.9130	2.8450	2.7740	2.7372	2.6996	2.6609	2.6211	2.5801	2.5379
11	2.8536	2.7876	2.7186	2.6464	2.6090	2.5705	2.5309	2.4901	2.4480	2.4045
12	2.7534	2.6866	2.6169	2.5436	2.5055	2.4663	2.4259	2.3842	2.3410	2.2962
13	2.6710	2.6037	2.5331	2.4589	2.4202	2.3803	2.3392	2.2966	2.2524	2.2064
14	2.6021	2.5342	2.4630	2.3879	2.3487	2.3082	2.2664	2.2230	2.1778	2.1307
15	2.5437	2.4753	2.4035	2.3275	2.2878	2.2468	2.2043	2.1601	2.1141	2.0658
16	2.4935	2.4247	2.3522	2.2756	2.2354	2.1938	2.1507	2.1058	2.0589	2.0096
17	2.4499	2.3807	2.3077	2.2304	2.1898	2.1477	2.1040	2.0584	2.0107	1.9604
18	2.4117	2.3421	2.2686	2.1906	2.1497	2.1071	2.0629	2.0166	1.9681	1.9168
19	2.3779	2.3080	2.2341	2.1555	2.1141	2.0712	2.0264	1.9796	1.9302	1.8780
20	2.3479	2.2776	2.2033	2.1242	2.0825	2.0391	1.9938	1.9464	1.8963	1.8432
21	2.3210	2.2504	2.1757	2.0960	2.0540	2.0102	1.9645	1.9165	1.8657	1.8117
22	2.2967	2.2258	2.1508	2.0707	2.0283	1.9842	1.9380	1.8895	1.8380	1.7831
23	2.2747	2.2036	2.1282	2.0476	2.0050	1.9605	1.9139	1.8649	1.8128	1.7570
24	2.2547	2.1834	2.1077	2.0267	1.9838	1.9390	1.8920	1.8424	1.7897	1.7331
25	2.2365	2.1649	2.0889	2.0075	1.9643	1.9192	1.8718	1.8217	1.7684	1.7110
26	2.2197	2.1479	2.0716	1.9898	1.9464	1.9010	1.8533	1.8027	1.7488	1.6906
27	2.2043	2.1323	2.0558	1.9736	1.9299	1.8842	1.8361	1.7851	1.7307	1.6717
28	2.1900	2.1179	2.0411	1.9586	1.9147	1.8687	1.8203	1.7689	1.7138	1.6541
29	2.1768	2.1045	2.0275	1.9446	1.9005	1.8543	1.8055	1.7537	1.6981	1.6377
30	2.1646	2.0921	2.0148	1.9317	1.8874	1.8409	1.7918	1.7396	1.6835	1.6223
40	2.0772	2.0035	1.9245	1.8389	1.7929	1.7444	1.6928	1.6373	1.5766	1.5089
60	1.9926	1.9174	1.8364	1.7480	1.7001	1.6491	1.5943	1.5343	1.4673	1.3893
120	1.9105	1.8337	1.7505	1.6587	1.6084	1.5543	1.4952	1.4290	1.3519	1.2539
∞	1.8307	1.7522	1.6664	1.5705	1.5173	1.4591	1.3940	1.3180	1.2214	1.0000

Degrees of Freedom for Denominator

2.5% Points of the F-Distribution

ν_2 \ ν_1	1	2	3	4	5	6	7	8	9
1	647.79	799.50	864.16	899.58	921.85	937.11	948.22	956.66	963.28
2	38.506	39.000	39.165	39.248	39.298	39.331	39.355	39.373	39.387
3	17.443	16.044	15.439	15.101	14.885	14.735	14.624	14.540	14.473
4	12.218	10.649	9.9792	9.6045	9.3645	9.1973	9.0741	8.9796	8.9047
5	10.007	8.4336	7.7636	7.3879	7.1464	6.9777	6.8531	6.7572	6.6810
6	8.8131	7.2598	6.5988	6.2272	5.9876	5.8197	5.6955	5.5996	5.5234
7	8.0727	6.5415	5.8898	5.5226	5.2852	5.1186	4.9949	4.8994	4.8232
8	7.5709	6.0595	5.4160	5.0526	4.8173	4.6517	4.5286	4.4332	4.3572
9	7.2093	5.7147	5.0781	4.7181	4.4844	4.3197	4.1971	4.1020	4.0260
10	6.9367	5.4564	4.8256	4.4683	4.2361	4.0721	3.9498	3.8549	3.7790
11	6.7241	5.2559	4.6300	4.2751	4.0440	3.8807	3.7586	3.6638	3.5879
12	6.5538	5.0959	4.4742	4.1212	3.8911	3.7283	3.6065	3.5118	3.4358
13	6.4143	4.9653	4.3472	3.9959	3.7667	3.6043	3.4827	3.3880	3.3120
14	6.2979	4.8567	4.2417	3.8919	3.6631	3.5014	3.3799	3.2853	3.2093
15	6.1995	4.7650	4.1528	3.8043	3.5764	3.4147	3.2934	3.1987	3.1227
16	6.1151	4.6867	4.0768	3.7294	3.5021	3.3406	3.2194	3.1248	3.0488
17	6.0420	4.6189	4.0112	3.6648	3.4379	3.2767	3.1556	3.0610	2.9849
18	5.9781	4.5597	3.9539	3.6083	3.3820	3.2209	3.0999	3.0053	2.9291
19	5.9216	4.5075	3.9034	3.5587	3.3327	3.1718	3.0509	2.9563	2.8800
20	5.8715	4.4613	3.8587	3.5147	3.2891	3.1283	3.0074	2.9128	2.8365
21	5.8266	4.4199	3.8188	3.4754	3.2501	3.0895	2.9686	2.8740	2.7977
22	5.7863	4.3828	3.7829	3.4401	3.2151	3.0546	2.9338	2.8392	2.7628
23	5.7498	4.3492	3.7505	3.4083	3.1835	3.0232	2.9024	2.8077	2.7313
24	5.7167	4.3187	3.7211	3.3794	3.1548	2.9946	2.8738	2.7791	2.7027
25	5.6864	4.2909	3.6943	3.3530	3.1287	2.9685	2.8478	2.7531	2.6766
26	5.6586	4.2655	3.6697	3.3289	3.1048	2.9447	2.8240	2.7293	2.6528
27	5.6331	4.2421	3.6472	3.3067	3.0828	2.9228	2.8021	2.7074	2.6309
28	5.6096	4.2205	3.6264	3.2863	3.0625	2.9027	2.7820	2.6872	2.6106
29	5.5878	4.2006	3.6072	3.2674	3.0438	2.8840	2.7633	2.6686	2.5919
30	5.5675	4.1821	3.5894	3.2499	3.0265	2.8667	2.7460	2.6513	2.5746
40	5.4239	4.0510	3.4633	3.1261	2.9037	2.7444	2.6238	2.5289	2.4519
60	5.2857	3.9253	3.3425	3.0077	2.7863	2.6274	2.5068	2.4117	2.3344
120	5.1524	3.8046	3.2270	2.8943	2.6740	2.5154	2.3948	2.2994	2.2217
∞	5.0239	3.6889	3.1161	2.7858	2.5665	2.4082	2.2875	2.1918	2.1136

Degrees of Freedom for Numerator

Degrees of Freedom for Denominator

TABLE C-3 (*continued*)

r_1 →				Degrees of Freedom for Numerator						
r_2	10	12	15	20	24	30	40	60	120	∞
1	968.63	976.71	984.87	993.10	997.25	1,001.4	1,005.6	1,009.8	1,014.0	1,018.3
2	39.398	39.415	39.431	39.448	39.456	39.465	39.473	39.481	39.490	39.498
3	14.419	14.337	14.253	14.167	14.124	14.081	14.037	13.992	13.947	13.902
4	8.8439	8.7512	8.6565	8.5599	8.5109	8.4613	8.4111	8.3604	8.3092	8.2573
5	6.6192	6.5246	6.4277	6.3285	6.2780	6.2269	6.1751	6.1225	6.0693	6.0153
6	5.4613	5.3662	5.2687	5.1684	5.1172	5.0652	5.0125	4.9589	4.9045	4.8491
7	4.7611	4.6658	4.5678	4.4667	4.4150	4.3624	4.3089	4.2544	4.1989	4.1423
8	4.2951	4.1997	4.1012	3.9995	3.9472	3.8940	3.8398	3.7844	3.7279	3.6702
9	3.9639	3.8682	3.7694	3.6669	3.6142	3.5604	3.5055	3.4493	3.3918	3.3329
10	3.7168	3.6209	3.5217	3.4186	3.3654	3.3110	3.2554	3.1984	3.1399	3.0798
11	3.5257	3.4296	3.3299	3.2261	3.1725	3.1176	3.0613	3.0035	2.9441	2.8828
12	3.3736	3.2773	3.1772	3.0728	3.0187	2.9633	2.9063	2.8478	2.7874	2.7249
13	3.2497	3.1532	3.0527	2.9477	2.8932	2.8373	2.7797	2.7204	2.6590	2.5955
14	3.1469	3.0501	2.9493	2.8437	2.7888	2.7324	2.6742	2.6142	2.5519	2.4872
15	3.0602	2.9633	2.8621	2.7559	2.7006	2.6437	2.5850	2.5242	2.4611	2.3953
16	2.9862	2.8890	2.7875	2.6808	2.6252	2.5678	2.5085	2.4471	2.3831	2.3163
17	2.9222	2.8249	2.7230	2.6158	2.5598	2.5021	2.4422	2.3801	2.3153	2.2474
18	2.8664	2.7689	2.6667	2.5590	2.5027	2.4445	2.3842	2.3214	2.2558	2.1869
19	2.8173	2.7196	2.6171	2.5089	2.4523	2.3937	2.3329	2.2695	2.2032	2.1333
20	2.7737	2.6758	2.5731	2.4645	2.4076	2.3486	2.2873	2.2234	2.1562	2.0853
21	2.7348	2.6368	2.5338	2.4247	2.3675	2.3082	2.2465	2.1819	2.1141	2.0422
22	2.6998	2.6017	2.4984	2.3890	2.3315	2.2718	2.2097	2.1446	2.0760	2.0032
23	2.6682	2.5699	2.4665	2.3567	2.2989	2.2389	2.1763	2.1107	2.0415	1.9677
24	2.6396	2.5412	2.4374	2.3273	2.2693	2.2090	2.1460	2.0799	2.0099	1.9353
25	2.6135	2.5149	2.4110	2.3005	2.2422	2.1816	2.1183	2.0517	1.9811	1.9055
26	2.5895	2.4909	2.3867	2.2759	2.2174	2.1565	2.0928	2.0257	1.9545	1.8781
27	2.5676	2.4688	2.3644	2.2533	2.1946	2.1334	2.0693	2.0018	1.9299	1.8527
28	2.5473	2.4484	2.3438	2.2324	2.1735	2.1121	2.0477	1.9796	1.9072	1.8291
29	2.5286	2.4295	2.3248	2.2131	2.1540	2.0923	2.0276	1.9591	1.8861	1.8072
30	2.5112	2.4120	2.3072	2.1952	2.1359	2.0739	2.0089	1.9400	1.8664	1.7867
40	2.3882	2.2882	2.1819	2.0677	2.0069	1.9429	1.8752	1.8028	1.7242	1.6371
60	2.2702	2.1692	2.0613	1.9445	1.8817	1.8152	1.7440	1.6668	1.5810	1.4822
120	2.1570	2.0548	1.9450	1.8249	1.7597	1.6899	1.6141	1.5299	1.4327	1.3104
∞	2.0483	1.9447	1.8326	1.7085	1.6402	1.5660	1.4835	1.3883	1.2684	1.0000

Degrees of Freedom for Denominator

1% Points of the F-Distribution

v_2 \ v_1	1	2	3	4	5	6	7	8	9
1	4,052.2	4,999.5	5,403.3	5,624.6	5,763.7	5,859.0	5,928.3	5,981.6	6,022.5
2	98.503	99.000	99.166	99.249	99.299	99.332	99.356	99.374	99.388
3	34.116	30.817	29.457	28.710	28.237	27.911	27.672	27.489	27.345
4	21.198	18.000	16.694	15.977	15.522	15.207	14.976	14.799	14.659
5	16.258	13.274	12.060	11.392	10.967	10.672	10.456	10.289	10.158
6	13.745	10.925	9.7795	9.1483	8.7459	8.4661	8.2600	8.1016	7.9761
7	12.246	9.5466	8.4513	7.8467	7.4604	7.1914	6.9928	6.8401	6.7188
8	11.259	8.6491	7.5910	7.0060	6.6318	6.3707	6.1776	6.0289	5.9106
9	10.561	8.0215	6.9919	6.4221	6.0569	5.8018	5.6129	5.4671	5.3511
10	10.044	7.5594	6.5523	5.9943	5.6363	5.3858	5.2001	5.0567	4.9424
11	9.6460	7.2057	6.2167	5.6683	5.3160	5.0692	4.8861	4.7445	4.6315
12	9.3302	6.9266	5.9526	5.4119	5.0643	4.8206	4.6395	4.4994	4.3675
13	9.0738	6.7010	5.7394	5.2053	4.8616	4.6204	4.4410	4.3021	4.1911
14	8.8616	6.5149	5.5639	5.0354	4.6950	4.4558	4.2779	4.1399	4.0297
15	8.6831	6.3589	5.4170	4.8932	4.5556	4.3183	4.1415	4.0045	3.8948
16	8.5310	6.2262	5.2922	4.7726	4.4374	4.2016	4.0259	3.8896	3.7804
17	8.3997	6.1121	5.1850	4.6690	4.3359	4.1015	3.9267	3.7910	3.6822
18	8.2854	6.0129	5.0919	4.5790	4.2479	4.0146	3.8406	3.7054	3.5971
19	8.1850	5.9259	5.0103	4.5003	4.1708	3.9386	3.7653	3.6305	3.5225
20	8.0960	5.8489	4.9382	4.4307	4.1027	3.8714	3.6987	3.5644	3.4567
21	8.0166	5.7804	4.8740	4.3688	4.0421	3.8117	3.6396	3.5056	3.3981
22	7.9454	5.7190	4.8166	4.3134	3.9880	3.7583	3.5867	3.4530	3.3458
23	7.8811	5.6637	4.7649	4.2635	3.9392	3.7102	3.5390	3.4057	3.2986
24	7.8229	5.6136	4.7181	4.2184	3.8951	3.6667	3.4959	3.3629	3.2560
25	7.7698	5.5680	4.6755	4.1774	3.8550	3.6272	3.4568	3.3239	3.2172
26	7.7213	5.5263	4.6366	4.1400	3.8183	3.5911	3.4210	3.2884	3.1818
27	7.6767	5.4881	4.6009	4.1056	3.7848	3.5580	3.3882	3.2558	3.1494
28	7.6356	5.4529	4.5681	4.0740	3.7539	3.5276	3.3581	3.2259	3.1195
29	7.5976	5.4205	4.5378	4.0449	3.7254	3.4995	3.3302	3.1982	3.0920
30	7.5625	5.3904	4.5097	4.0179	3.6990	3.4735	3.3045	3.1726	3.0665
40	7.3141	5.1785	4.3126	3.8283	3.5138	3.2910	3.1238	2.9930	2.8876
60	7.0771	4.9774	4.1259	3.6491	3.3389	3.1187	2.9530	2.8233	2.7185
120	6.8510	4.7865	3.9493	3.4796	3.1735	2.9559	2.7918	2.6629	2.5586
∞	6.6349	4.6052	3.7816	3.3192	3.0173	2.8020	2.6393	2.5113	2.4673

Degrees of Freedom for Denominator

TABLE C-3 (continued)

		Degrees of Freedom for Numerator								
r_2	r_1	10	12	15	20	24	30	40	60	120	∞
1		6,055.8	6,106.3	6,157.3	6,208.7	6,234.6	6,260.7	6,286.8	6,313.0	6,339.4	6,366.0
2		99.399	99.416	99.432	99.449	99.458	99.466	99.474	99.483	99.491	99.501
3		27.229	27.052	26.872	26.690	26.598	26.505	26.411	26.316	26.221	26.125
4		14.546	14.374	14.198	14.020	13.929	13.838	13.745	13.652	13.558	13.463
5		10.051	9.8883	9.7222	9.5527	9.4665	9.3793	9.2912	9.2020	9.1118	9.0204
6		7.8741	7.7183	7.5590	7.3958	7.3127	7.2285	7.1432	7.0568	6.9690	6.8801
7		6.6201	6.4691	6.3143	6.1554	6.0743	5.9921	5.9084	5.8236	5.7372	5.6495
8		5.8143	5.6668	5.5151	5.3591	5.2793	5.1981	5.1156	5.0316	4.9460	4.8588
9		5.2565	5.1114	4.9621	4.8080	4.7290	4.6486	4.5667	4.4831	4.3978	4.3105
10		4.8492	4.7059	4.5582	4.4054	4.3269	4.2169	4.1653	4.0819	3.9965	3.9090
11		4.5393	4.3974	4.2509	4.0990	4.0209	3.9411	3.8596	3.7761	3.6904	3.6025
12		4.2961	4.1553	4.0096	3.8584	3.7805	3.7008	3.6192	3.5355	3.4494	3.3608
13		4.1003	3.9603	3.8154	3.6646	3.5868	3.5070	3.4253	3.3413	3.2548	3.1654
14		3.9394	3.8001	3.6557	3.5052	3.4274	3.3476	3.2656	3.1813	3.0942	3.0040
15		3.8049	3.6662	3.5222	3.3719	3.2940	3.2141	3.1319	3.0471	2.9595	2.8684
16		3.6909	3.5527	3.4089	3.2588	3.1808	3.1007	3.0182	2.9330	2.8447	2.7528
17		3.5931	3.4552	3.3117	3.1615	3.0835	3.0032	2.9205	2.8348	2.7459	2.6530
18		3.5082	3.3706	3.2273	3.0771	2.9990	2.9185	2.8354	2.7493	2.6597	2.5660
19		3.4338	3.2965	3.1533	3.0031	2.9249	2.8442	2.7608	2.6742	2.5839	2.4893
20		3.3682	3.2311	3.0880	2.9377	2.8594	2.7785	2.6947	2.6077	2.5168	2.4212
21		3.3098	3.1729	3.0299	2.8796	2.8011	2.7200	2.6359	2.5484	2.4568	2.3603
22		3.2576	3.1209	2.9780	2.8274	2.7488	2.6675	2.5831	2.4951	2.4029	2.3055
23		3.2106	3.0740	2.9311	2.7805	2.7017	2.6202	2.5355	2.4471	2.3542	2.2559
24		3.1681	3.0316	2.8887	2.7380	2.6591	2.5773	2.4923	2.4035	2.3099	2.2107
25		3.1294	2.9931	2.8502	2.6993	2.6203	2.5383	2.4530	2.3637	2.2695	2.1694
26		3.0941	2.9579	2.8150	2.6640	2.5848	2.5026	2.4170	2.3273	2.2325	2.1315
27		3.0618	2.9256	2.7827	2.6316	2.5522	2.4699	2.3840	2.2938	2.1984	2.0965
28		3.0320	2.8959	2.7530	2.6017	2.5223	2.4397	2.3535	2.2629	2.1670	2.0642
29		3.0045	2.8685	2.7256	2.5742	2.4946	2.4118	2.3253	2.2344	2.1378	2.0342
30		2.9791	2.8431	2.7002	2.5487	2.4689	2.3860	2.2992	2.2079	2.1107	2.0062
40		2.8005	2.6648	2.5216	2.3689	2.2880	2.2034	2.1142	2.0194	1.9172	1.8047
60		2.6318	2.4961	2.3523	2.1978	2.1154	2.0285	1.9360	1.8363	1.7263	1.6006
120		2.4721	2.3363	2.1915	2.0346	1.9500	1.8600	1.7628	1.6557	1.5330	1.3805
∞		2.3209	2.1848	2.0385	1.8783	1.7908	1.6964	1.5923	1.4730	1.3246	1.0000

Degrees of Freedom for Denominator

394

0.5% POINTS OF THE F-DISTRIBUTION

Degrees of Freedom for Numerator

ν_2 \ ν_1	1	2	3	4	5	6	7	8	9
1	16,211	20,000	21,615	22,500	23,036	23,437	23,715	23,925	24,091
2	198.50	199.00	199.17	199.25	199.30	199.33	199.36	199.37	199.39
3	55.552	49.799	47.467	46.195	45.392	44.838	44.434	44.126	43.882
4	31.333	26.284	24.259	23.155	22.456	21.975	21.622	21.352	21.139
5	22.785	18.314	16.530	15.556	14.940	14.513	14.200	13.961	13.772
6	18.635	14.544	12.917	12.028	11.461	11.073	10.786	10.566	10.391
7	16.236	12.404	10.882	10.050	9.5221	9.1554	8.8854	8.6781	8.5138
8	14.688	11.042	9.5965	8.8051	8.3018	7.9520	7.6942	7.4960	7.3386
9	13.614	10.107	8.7171	7.9559	7.4711	7.1338	6.8849	6.6933	6.5411
10	12.826	9.4270	8.0807	7.3428	6.8723	6.5446	6.3025	6.1159	5.9676
11	12.226	8.9122	7.6004	6.8809	6.4217	6.1015	5.8648	5.6821	5.5368
12	11.754	8.5096	7.2258	6.5211	6.0711	5.7570	5.5245	5.3451	5.2021
13	11.374	8.1865	6.9257	6.2335	5.7910	5.4819	5.2529	5.0761	4.9351
14	11.060	7.9217	6.6803	5.9984	5.5623	5.2574	5.0313	4.8566	4.7173
15	10.798	7.7008	6.4760	5.8029	5.3721	5.0708	4.8473	4.6743	4.5364
16	10.575	7.5138	6.3034	5.6378	5.2117	4.9134	4.6920	4.5207	4.3838
17	10.384	7.3536	6.1556	5.4967	5.0746	4.7789	4.5594	4.3893	4.2535
18	10.218	7.2148	6.0277	5.3746	4.9560	4.6627	4.4448	4.2759	4.1410
19	10.073	7.0935	5.9161	5.2681	4.8526	4.5614	4.3448	4.1770	4.0428
20	9.9439	6.9865	5.8177	5.1743	4.7616	4.4721	4.2569	4.0900	3.9564
21	9.8295	6.8914	5.7304	5.0911	4.6808	4.3931	4.1789	4.0128	3.8799
22	9.7271	6.8064	5.6524	5.0168	4.6088	4.3225	4.1094	3.9440	3.8116
23	9.6348	6.7300	5.5823	4.9500	4.5141	4.2591	4.0469	3.8822	3.7502
24	9.5513	6.6610	5.5190	4.8898	4.4857	4.2019	3.9905	3.8264	3.6949
25	9.4753	6.5982	5.4615	4.8351	4.4327	4.1500	3.9394	3.7758	3.6447
26	9.4059	6.5409	5.4091	4.7852	4.3844	4.1027	3.8928	3.7297	3.5989
27	9.3423	6.4885	5.3611	4.7396	4.3402	4.0594	3.8501	3.6875	3.5571
28	9.2838	6.4403	5.3170	4.6977	4.2996	4.0197	3.8110	3.6487	3.5186
29	9.2297	6.3958	5.2764	4.6591	4.2622	3.9830	3.7749	3.6130	3.4832
30	9.1797	6.3547	5.2388	4.6233	4.2276	3.9492	3.7416	3.5801	3.4505
40	8.8278	6.0664	4.9759	4.3738	3.9860	3.7129	3.5088	3.3498	3.2220
60	8.4916	5.7950	4.7290	4.1399	3.7600	3.4918	3.2911	3.1344	3.0083
120	8.1790	5.5393	4.4973	3.9207	3.5482	3.2849	3.0874	2.9330	2.8083
∞	7.8794	5.2983	4.2794	3.7151	3.3499	3.0913	2.8968	2.7444	2.6210

Degrees of Freedom for Denominator

TABLE C-3 (continued)

r_2		Degrees of Freedom for Numerator								
r_1 →	10	12	15	20	24	30	40	60	120	∞
1	24,224	24,426	24,630	24,836	24,910	25,044	25,148	25,253	25,359	25,465
2	199.40	199.42	199.43	199.45	199.46	199.47	199.47	199.48	199.49	199.51
3	43.685	43.387	43.085	42.778	42.622	42.466	42.308	42.149	41.989	41.829
4	21.967	20.705	20.438	20.167	20.030	19.892	19.752	19.611	19.468	19.325
5	13.618	13.384	13.146	12.903	12.780	12.656	12.530	12.402	12.274	12.144
6	10.250	10.034	9.8140	9.5888	9.4741	9.3583	9.2408	9.1219	9.0015	8.8793
7	8.3803	8.1764	7.9678	7.7540	7.6450	7.5345	7.4225	7.3088	7.1933	7.0760
8	7.2107	7.0149	6.8143	6.6082	6.5029	6.3961	6.2875	6.1772	6.0649	5.9505
9	6.4171	6.2274	6.0325	5.8318	5.7292	5.6248	5.5186	5.4104	5.3001	5.1875
10	5.8467	5.6613	5.4707	5.2732	5.1732	5.0705	4.9659	4.8592	4.7501	4.6385
11	5.4182	5.2363	5.0489	4.8552	4.7557	4.6543	4.5508	4.4450	4.3367	4.2256
12	5.0855	4.9063	4.7214	4.5299	4.4315	4.3309	4.2282	4.1229	4.0149	3.9039
13	4.8199	4.6429	4.4600	4.2703	4.1726	4.0727	3.9701	3.8655	3.7577	3.6465
14	4.6034	4.4281	4.2468	4.0585	3.9614	3.8619	3.7600	3.6553	3.5473	3.4359
15	4.4236	4.2498	4.0698	3.8826	3.7859	3.6867	3.5850	3.4803	3.3722	3.2602
16	4.2719	4.0994	3.9205	3.7342	3.6378	3.5388	3.4372	3.3324	3.2240	3.1115
17	4.1423	3.9709	3.7929	3.6073	3.5112	3.4124	3.3107	3.2058	3.0971	2.9839
18	4.0305	3.8599	3.6827	3.4977	3.4017	3.3030	3.2014	3.0962	2.9871	2.8732
19	3.9329	3.7631	3.5866	3.4020	3.3062	3.2075	3.1058	3.0004	2.8908	2.7762
20	3.8470	3.6779	3.5020	3.3178	3.2220	3.1234	3.0215	2.9159	2.8058	2.6904
21	3.7709	3.6024	3.4270	3.2431	3.1474	3.0488	2.9467	2.8408	2.7302	2.6140
22	3.7030	3.5350	3.3600	3.1764	3.0807	2.9821	2.8799	2.7736	2.6625	2.5455
23	3.6420	3.4745	3.2999	3.1165	3.0208	2.9221	2.8198	2.7132	2.6015	2.4837
24	3.5870	3.4199	3.2456	3.0624	2.9667	2.8679	2.7654	2.6585	2.5463	2.4276
25	3.5370	3.3704	3.1963	3.0133	2.9176	2.8187	2.7160	2.6088	2.4960	2.3765
26	3.4916	3.3252	3.1515	2.9685	2.8728	2.7738	2.6709	2.5633	2.4501	2.3297
27	3.4499	3.2839	3.1104	2.9275	2.8318	2.7327	2.6296	2.5217	2.4078	2.2867
28	3.4117	3.2460	3.0727	2.8899	2.7941	2.6949	2.5916	2.4834	2.3689	2.2469
29	3.3765	3.2111	3.0379	2.8551	2.7594	2.6601	2.5565	2.4479	2.3330	2.2102
30	3.3440	3.1787	3.0057	2.8230	2.7272	2.6278	2.5241	2.4151	2.2997	2.1760
40	3.1167	2.9531	2.7811	2.5984	2.5020	2.4015	2.2958	2.1838	2.0635	1.9318
60	2.9042	2.7419	2.5705	2.3872	2.2898	2.1874	2.0789	1.9622	1.8341	1.6885
120	2.7052	2.5439	2.3727	2.1881	2.0890	1.9839	1.8709	1.7469	1.6055	1.4311
∞	2.5188	2.3583	2.1868	1.9998	1.8983	1.7891	1.6691	1.5325	1.3637	1.0000

Degrees of Freedom for Denominator

From "Tables of Percentage Points of the Inverted Beta (F) Distribution" by Maxine Merrington and Catherine M. Thompson, *Biometrika*, Vol. 33(1943), pp. 73–88. Reproduced by permission of the editor of *Biometrika*.

TABLE C-4   UPPER TAIL PROBABILITIES FOR THE NULL DISTRIBUTION OF THE KRUSKAL-WALLIS H STATISTIC: $k = 3$, $n_1 = 1(1)5$, $n_2 = n_1(1)5$, $2 \leq n_3 = n_2(1)5$

For $k = 3$ and sample sizes n_1, n_2, n_3, the tabled entry for the point x is $P_0\{H \geq x\}$. Thus if x is such that $P_0\{H \geq x\} = \alpha$, then $h(\alpha, 3, (n_1, n_2, n_3)) = x$.

$n_1 = 1, n_2 = 1, n_3 = 2$		$n_1 = 1, n_2 = 1, n_3 = 5$		$n_1 = 1, n_2 = 2, n_3 = 4$		$n_1 = 1, n_2 = 2, n_3 = 5$	
x	$P_0\{H \geq x\}$	x	$P_0\{H \geq x\}$	x	$P_0\{H \geq x\}$	x	$P_0\{H \geq x\}$
.300	1.000	2.314	.524	.000	1.000	.583	.821
1.800	.833	2.829	.333	.161	.971	.667	.798
2.700	.500	3.857	.143	.268	.933	.717	.774
				.321	.895	1.000	.750
				.536	.857	1.117	.738
$n_1 = 1, n_2 = 1, n_3 = 3$		$n_1 = 1, n_2 = 2, n_3 = 2$		.643	.819	1.200	.714
				.696	.800	1.250	.655
x	$P_0\{H \geq x\}$	x	$P_0\{H \geq x\}$	1.018	.781	1.383	.619
				1.071	.743	1.533	.583
.533	1.000	.000	1.000	1.125	.705	1.783	.560
.800	.800	.400	.933	1.286	.667	1.800	.536
2.133	.700	.600	.867	1.393	.629	1.917	.488
3.200	.300	1.400	.733	1.446	.590	2.050	.464
		2.000	.600	1.875	.533	2.333	.429
		2.400	.467	2.036	.495	2.450	.393
$n_1 = 1, n_2 = 1, n_3 = 4$		3.000	.333	2.143	.476	2.717	.298
		3.600	.200	2.250	.457	2.800	.286
x	$P_0\{H \geq x\}$			2.411	.400	2.867	.214
				2.571	.305	3.133	.202
.143	1.000	$n_1 = 1, n_2 = 2, n_3 = 3$		2.786	.286	3.333	.190
.786	.933			2.893	.267	3.383	.179
1.000	.800	x	$P_0\{H \geq x\}$	3.161	.190	3.783	.131
1.286	.667			3.696	.171	4.050	.119
2.143	.600	.095	1.000	3.750	.133	4.200	.095
2.500	.467	.238	.933	4.018	.114	4.450	.071
3.571	.200	.429	.900	4.500	.076	5.000	.048
		.810	.833	4.821	.057	5.250	.036
		.857	.800				
$n_1 = 1, n_2 = 1, n_3 = 5$		1.238	.700				
		1.381	.600	$n_1 = 1, n_2 = 2, n_3 = 5$		$n_1 = 1, n_2 = 3, n_3 = 3$	
x	$P_0\{H \geq x\}$	1.952	.567				
		2.143	.533	x	$P_0\{H \geq x\}$	x	$P_0\{H \geq x\}$
.257	1.000	2.381	.433				
.429	.905	3.095	.267	.050	1.000	.000	1.000
1.029	.857	3.524	.200	.133	.964	.143	.986
1.114	.762	3.857	.133	.200	.940	.286	.957
1.457	.667	4.286	.100	.450	.905	.571	.871
1.714	.571			.467	.845	1.000	.771

TABLE C-4 (continued)

$n_1 = 1, n_2 = 3, n_3 = 3$		$n_1 = 1, n_2 = 3, n_3 = 4$		$n_1 = 1, n_2 = 3, n_3 = 5$		$n_1 = 1, n_2 = 4, n_3 = 4$	
x	$P_0\{H \ge x\}$	x	$P_0\{H \ge x\}$	x	$P_0\{H \ge x\}$	x	$P_0\{H \ge x\}$
1.143	.743	3.764	.136	2.844	.258	2.267	.410
1.286	.600	3.889	.129	2.951	.218	2.400	.384
1.571	.571	4.056	.093	3.040	.210	2.467	.349
2.000	.514	4.097	.086	3.218	.190	2.667	.305
2.286	.486	4.208	.079	3.271	.183	2.700	.260
2.571	.329	4.764	.071	3.378	.143	2.967	.235
3.143	.243	5.000	.057	3.484	.135	3.000	.222
3.286	.157	5.208	.050	3.804	.131	3.267	.178
4.000	.129	5.389	.036	3.840	.123	3.367	.171
4.571	.100	5.833	.021	4.018	.095	3.467	.152
5.143	.043			4.284	.083	3.867	.121

$n_1 = 1, n_2 = 3, n_3 = 5$

continued table:

x	$P_0\{H \ge x\}$	x	$P_0\{H \ge x\}$
4.338	.079	3.900	.108
4.551	.075	4.067	.102
4.711	.056	4.167	.083
4.871	.052	4.267	.070
4.960	.048	4.800	.067
5.404	.044	4.867	.054
5.440	.036	4.967	.048
5.760	.028	5.100	.041
6.044	.020	5.667	.035
6.400	.012	6.000	.029
		6.167	.022
		6.667	.010

$n_1 = 1, n_2 = 3, n_3 = 4$	
x	$P_0\{H \ge x\}$
.056	1.000
.097	.971
.208	.950
.333	.921
.431	.900
.500	.871
.556	.843
.764	.786
.875	.743
1.097	.721
1.208	.707
1.222	.679
1.389	.629
1.431	.557
1.764	.536
1.833	.514
1.875	.471
2.097	.457
2.208	.443
2.333	.429
2.431	.371
2.722	.300
2.764	.229
3.000	.221
3.097	.214
3.208	.200
3.222	.157

$n_1 = 1, n_2 = 3, n_3 = 5$	
x	$P_0\{H \ge x\}$
.000	1.000
.071	.992
.160	.972
.178	.952
.284	.929
.338	.889
.551	.869
.604	.853
.640	.833
.711	.770
.818	.750
.960	.730
1.084	.694
1.138	.683
1.351	.651
1.404	.611
1.440	.591
1.511	.571
1.600	.560
1.671	.520
1.778	.488
1.884	.480
1.938	.468
2.044	.452
2.204	.437
2.400	.413
2.418	.405
2.560	.341

$n_1 = 1, n_2 = 4, n_3 = 4$	
x	$P_0\{H \ge x\}$
.000	1.000
.067	.987
.167	.968
.267	.930
.300	.911
.567	.873
.600	.835
.667	.803
.867	.759
.967	.721
1.067	.689
1.200	.676
1.367	.644
1.500	.600
1.667	.537
1.767	.498
2.167	.460

$n_1 = 1, n_2 = 4, n_3 = 5$	
x	$P_0\{H \ge x\}$
.033	1.000
.060	.983
.104	.968
.186	.952
.273	.938
.278	.922
.295	.906
.360	.890
.409	.875
.540	.848
.622	.821
.731	.806
.758	.794
.796	.778
.818	.762

$n_1 = 1, n_2 = 4, n_3 = 5$		$n_1 = 1, n_2 = 4, n_3 = 5$		$n_1 = 1, n_2 = 5, n_3 = 5$		$n_1 = 1, n_2 = 5, n_3 = 5$	
x	$P_0\{H > x\}$	x	$P_0\{H \geqslant x\}$	x	$P_0\{H \geqslant x\}$	x	$P_0\{H \geqslant x\}$
.906	.730	4.287	.071	1.309	.630	7.309	.009
.933	.719	4.549	.067	1.346	.605	7.527	.008
.976	.690	4.636	.063	1.600	.584	7.746	.005
1.151	.676	4.724	.060	1.636	.571	8.182	.002
1.167	.665	4.833	.059	1.709	.509		
1.195	.651	4.860	.056	1.746	.493	$n_1 = 2, n_2 = 2, n_3 = 2$	
1.233	.640	4.986	.044	1.782	.468		
1.342	.625	5.078	.041	1.927	.462	x	$P_0\{H > x\}$
1.369	.614	5.160	.038	2.000	.438		
1.495	.606	5.515	.037	2.146	.422	.000	1.000
1.500	.589	5.558	.035	2.182	.411	.286	.933
1.587	.562	5.596	.033	2.327	.379	.857	.800
1.604	.535	5.733	.027	2.436	.374	1.143	.667
1.669	.517	5.776	.025	2.509	.361	2.000	.533
1.778	.498	5.858	.024	2.582	.314	2.571	.400
1.806	.483	5.864	.022	2.727	.286	3.429	.333
1.849	.468	5.967	.021	2.909	.242	3.714	.200
1.931	.460	6.431	.019	2.946	.227	4.571	.067
2.040	.441	6.578	.016	3.236	.188		
2.067	.432	6.818	.013	3.346	.168		
2.106	.419	6.840	.011	3.382	.161	$n_1 = 2, n_2 = 2, n_3 = 3$	
2.242	.406	6.954	.008	3.527	.141		
2.286	.400	7.364	.005	3.600	.132	x	$P_0\{H \geqslant x\}$
2.455	.394			3.636	.116		
2.460	.354	$n_1 = 1, n_2 = 5, n_3 = 5$		3.927	.113	.000	1.000
2.504	.346			4.036	.105	.179	.971
2.591	.300	x	$P_0\{H \geqslant x\}$	4.109	.086	.214	.895
2.651	.286			4.182	.082	.500	.857
2.896	.251	.000	1.000	4.400	.076	.607	.800
2.913	.222	.036	.994	4.546	.074	.714	.743
2.940	.216	.109	.982	4.800	.056	.857	.686
3.000	.208	.146	.956	4.909	.053	1.179	.657
3.087	.194	.182	.944	5.127	.046	1.357	.619
3.158	.187	.327	.920	5.236	.039	1.464	.562
3.240	.183	.400	.885	5.636	.033	1.607	.524
3.349	.151	.436	.872	5.709	.030	1.929	.467
3.524	.146	.546	.847	5.782	.027	2.000	.438
3.595	.138	.582	.802	6.000	.022	2.214	.419
3.682	.132	.727	.792	6.146	.019	2.429	.381
3.813	.110	.836	.771	6.509	.018	2.464	.362
3.960	.102	.909	.752	6.546	.015	2.750	.324
3.987	.098	.982	.716	6.582	.014	2.857	.286
4.206	.095	1.127	.669	6.727	.012	3.179	.267
4.222	.087	1.200	.646	6.836	.011	3.429	.248

TABLE C-4 (*continued*)

$n_1 = 2, n_2 = 2, n_3 = 3$

x	$P_0\{H \geqslant x\}$
3.607	.238
3.750	.219
3.929	.181
4.464	.105
4.500	.067
4.714	.048
5.357	.029

$n_1 = 2, n_2 = 2, n_3 = 4$

x	$P_0\{H \geqslant x\}$
.000	1.000
.125	.971
.167	.914
.333	.890
.458	.862
.500	.814
.667	.757
.792	.733
1.000	.695
1.125	.657
1.333	.581
1.500	.552
1.792	.514
1.833	.486
2.000	.448
2.125	.410
2.458	.362
2.667	.333
2.792	.314
2.833	.295
3.000	.276
3.125	.248
3.167	.229
3.458	.210
3.667	.190
4.000	.181
4.125	.152
4.167	.105
4.458	.100
4.500	.090
5.125	.052

$n_1 = 2, n_2 = 2, n_3 = 4$

x	$P_0\{H \geqslant x\}$
5.333	.033
5.500	.024
6.000	.014

$n_1 = 2, n_2 = 2, n_3 = 5$

x	$P_0\{H \geqslant x\}$
.000	1.000
.093	.984
.133	.937
.240	.913
.360	.881
.373	.844
.533	.807
.573	.791
.773	.759
.840	.722
.893	.685
.960	.653
1.093	.638
1.200	.606
1.373	.590
1.440	.563
1.493	.542
1.533	.516
1.693	.495
1.800	.474
2.133	.452
2.160	.444
2.173	.402
2.293	.381
2.333	.365
2.373	.344
2.693	.317
2.760	.296
2.973	.275
3.093	.265
3.133	.254
3.240	.238
3.333	.206
3.360	.196
3.573	.185

$n_1 = 2, n_2 = 2, n_3 = 5$

x	$P_0\{H \geqslant x\}$
3.773	.175
3.840	.164
3.973	.159
4.093	.148
4.200	.138
4.293	.122
4.373	.090
4.573	.085
4.800	.063
4.893	.061
5.040	.056
5.160	.034
5.693	.029
6.000	.019
6.133	.013
6.533	.008

$n_1 = 2, n_2 = 3, n_3 = 3$

x	$P_0\{H \geqslant x\}$
.028	1.000
.111	.968
.222	.946
.250	.896
.472	.864
.556	.807
.694	.757
1.000	.686
1.111	.671
1.139	.600
1.361	.564
1.444	.539
1.806	.511
1.889	.446
2.000	.425
2.028	.396
2.250	.368
2.472	.357
2.694	.329
2.778	.307
2.889	.286
3.139	.243

$n_1 = 2, n_2 = 3, n_3 = 3$

x	$P_0\{H \geqslant x\}$
3.222	.221
3.361	.207
3.778	.200
3.806	.179
4.028	.164
4.111	.129
4.250	.121
4.556	.100
4.694	.093
5.000	.075
5.139	.061
5.361	.032
5.556	.025
6.250	.011

$n_1 = 2, n_2 = 3, n_3 = 4$

x	$P_0\{H \geqslant x\}$
.000	1.000
.078	.987
.100	.965
.111	.944
.244	.922
.278	.902
.311	.881
.344	.862
.400	.844
.444	.829
.544	.811
.600	.794
.611	.770
.700	.756
.778	.722
.811	.703
.900	.689
.978	.673
1.000	.660
1.078	.627
1.111	.614
1.178	.602
1.244	.586
1.344	.571

$n_1 = 2, n_2 = 3, n_3 = 4$		$n_1 = 2, n_2 = 3, n_3 = 4$		$n_1 = 2, n_2 = 3, n_3 = 5$		$n_1 = 2, n_2 = 3, n_3 = 5$	
x	$P_0\{H > x\}$	x	$P_0\{H > x\}$	x	$P_0\{H \geqslant x\}$	x	$P_0\{H > x\}$
1.378	.559	4.378	.105	.713	.743	3.069	.243
1.411	.548	4.444	.102	.724	.714	3.167	.237
1.500	.537	4.511	.098	.767	.703	3.186	.233
1.600	.511	4.544	.086	.887	.692	3.273	.222
1.611	.502	4.611	.083	.942	.680	3.331	.211
1.678	.478	4.711	.079	1.014	.659	3.342	.206
1.711	.468	4.811	.076	1.058	.648	3.386	.201
1.778	.457	4.878	.073	1.091	.638	3.414	.193
1.844	.448	4.900	.071	1.149	.616	3.506	.189
1.944	.437	4.978	.059	1.178	.593	3.546	.183
2.144	.417	5.078	.057	1.276	.579	3.604	.175
2.178	.406	5.144	.054	1.324	.569	3.676	.171
2.200	.398	5.378	.052	1.378	.537	3.767	.167
2.211	.376	5.400	.051	1.451	.529	3.778	.159
2.244	.368	5.444	.046	1.586	.519	3.822	.156
2.378	.357	5.500	.040	1.596	.510	3.909	.152
2.400	.346	5.611	.032	1.614	.502	3.942	.146
2.411	.338	5.800	.030	1.713	.483	3.996	.139
2.444	.329	6.000	.024	1.727	.474	4.058	.137
2.500	.321	6.111	.021	1.760	.459	4.069	.132
2.778	.294	6.144	.014	1.814	.451	4.204	.129
2.800	.284	6.300	.011	1.858	.444	4.214	.125
2.911	.271	6.444	.008	1.876	.429	4.233	.122
2.944	.262	7.000	.005	2.022	.420	4.258	.120
3.011	.256			2.033	.403	4.331	.117
3.100	.251	$n_1 = 2, n_2 = 3, n_3 = 5$		2.076	.396	4.378	.113
3.111	.238			2.106	.389	4.494	.101
3.244	.232	x	$P_0\{H > x\}$	2.196	.382	4.651	.091
3.278	.225			2.251	.375	4.694	.089
3.300	.216	.014	1.000	2.294	.368	4.724	.087
3.311	.203	.069	.981	2.367	.362	4.727	.085
3.444	.197	.113	.966	2.454	.356	4.814	.071
3.478	.190	.131	.951	2.458	.350	4.869	.067
3.544	.184	.142	.932	2.469	.336	4.913	.063
3.600	.175	.273	.917	2.487	.330	4.942	.062
3.811	.168	.276	.901	2.546	.321	5.076	.060
3.844	.163	.306	.886	2.633	.294	5.087	.053
3.911	.159	.331	.869	2.749	.287	5.106	.052
3.978	.156	.364	.855	2.818	.279	5.251	.049
4.000	.149	.451	.823	2.894	.269	5.349	.046
4.078	.140	.549	.807	2.924	.263	5.513	.044
4.200	.137	.567	.794	2.949	.257	5.524	.043
4.278	.124	.622	.781	2.978	.252	5.542	.041
4.311	.108	.636	.769	3.022	.248	5.727	.037

TABLE C-4 (continued)

$n_1 = 2, n_2 = 3, n_3 = 5$		$n_1 = 2, n_2 = 4, n_3 = 4$		$n_1 = 2, n_2 = 4, n_3 = 4$		$n_1 = 2, n_2 = 4, n_3 = 5$	
x	$P_0\{H > x\}$	x	$P_0\{H > x\}$	x	$P_0\{H > x\}$	x	$P_0\{H > x\}$
5.742	.034	1.636	.510	6.546	.020	1.050	.623
5.786	.033	1.718	.488	6.600	.017	1.091	.614
5.804	.033	1.827	.441	6.627	.016	1.200	.607
5.949	.026	1.964	.426	6.873	.011	1.204	.599
6.004	.025	2.046	.400	7.036	.006	1.268	.592
6.033	.024	2.236	.386	7.282	.004	1.291	.576
6.091	.021	2.264	.375	7.854	.002	1.314	.569
6.124	.020	2.373	.363			1.318	.562
6.294	.017	2.454	.338	$n_1 = 2, n_2 = 4, n_3 = 5$		1.391	.554
6.386	.016	2.509	.317	x	$P_0\{H > x\}$	1.414	.537
6.414	.015	2.673	.301			1.450	.529
6.818	.012	2.809	.281	.000	1.000	1.473	.521
6.822	.010	2.918	.272	.041	.992	1.518	.507
6.909	.009	2.946	.263	.064	.979	1.591	.499
6.949	.006	3.054	.239	.068	.965	1.618	.491
7.182	.004	3.136	.228	.141	.952	1.641	.485
7.636	.002	3.327	.220	.154	.939	1.664	.479
		3.354	.210	.164	.926	1.704	.472
		3.464	.192	.223	.913	1.750	.465
$n_1 = 2, n_2 = 4, n_3 = 4$		3.491	.185	.254	.902	1.754	.459
x	$P_0\{H > x\}$	3.682	.180	.273	.891	1.814	.452
		3.764	.166	.300	.880	1.823	.432
.000	1.000	3.818	.152	.323	.866	1.973	.427
.054	.988	4.009	.142	.368	.855	2.004	.420
.082	.970	4.364	.125	.404	.832	2.018	.403
.191	.940	4.418	.120	.504	.823	2.073	.398
.218	.910	4.446	.103	.518	.812	2.114	.392
.273	.893	4.554	.098	.541	.801	2.118	.387
.327	.879	4.582	.094	.564	.791	2.141	.381
.409	.848	4.691	.080	.573	.781	2.164	.375
.491	.820	4.773	.075	.614	.759	2.223	.371
.627	.779	4.854	.071	.618	.749	2.254	.366
.736	.757	4.991	.065	.654	.740	2.291	.361
.764	.712	5.127	.057	.723	.730	2.318	.351
.873	.685	5.236	.052	.791	.720	2.323	.346
.954	.671	5.454	.046	.841	.710	2.391	.335
1.091	.651	5.509	.044	.864	.701	2.454	.329
1.146	.638	5.536	.042	.891	.691	2.473	.324
1.173	.596	5.646	.039	.904	.683	2.504	.320
1.282	.577	5.727	.034	.914	.674	2.550	.315
1.309	.559	5.946	.028	.950	.657	2.618	.311
1.364	.537	6.082	.025	.954	.649	2.700	.306
1.582	.526	6.327	.024	1.018	.640	2.723	.301
		6.409	.022	1.023	.632	2.754	.296

$n_1 = 2, n_2 = 4, n_3 = 5$		$n_1 = 2, n_2 = 4, n_3 = 5$		$n_1 = 2, n_2 = 4, n_3 = 5$		$n_1 = 2, n_2 = 5, n_3 = 5$	
x	$P_0\{H > x\}$	x	$P_0\{H > x\}$	x	$P_0\{H > x\}$	x	$P_0\{H > x\}$
2.768	.285	4.404	.110	6.564	.016	.908	.674
2.773	.273	4.500	.104	6.654	.016	.931	.661
2.868	.267	4.518	.101	6.723	.015	1.115	.638
2.891	.262	4.541	.098	6.904	.014	1.154	.611
2.904	.258	4.614	.090	6.914	.013	1.185	.593
2.914	.249	4.664	.088	7.000	.013	1.277	.569
2.973	.246	4.768	.079	7.018	.012	1.300	.558
3.023	.237	4.791	.078	7.064	.012	1.362	.552
3.050	.234	4.800	.076	7.118	.010	1.431	.539
3.064	.231	4.818	.074	7.204	.009	1.485	.528
3.118	.226	4.841	.072	7.254	.009	1.523	.516
3.164	.221	4.868	.071	7.291	.008	1.554	.506
3.268	.217	4.950	.063	7.450	.007	1.646	.496
3.314	.214	5.073	.061	7.500	.007	1.669	.486
3.341	.208	5.154	.059	7.568	.006	1.731	.463
3.364	.200	5.164	.053	7.573	.005	1.854	.445
3.414	.197	5.254	.052	7.773	.004	1.915	.434
3.454	.193	5.268	.051	7.814	.003	1.923	.424
3.523	.190	5.273	.049	8.018	.002	2.015	.407
3.564	.187	5.300	.048	8.114	.001	2.038	.398
3.568	.184	5.314	.046	8.591	.001	2.223	.379
3.573	.181	5.414	.045			2.262	.374
3.618	.178	5.518	.043	$n_1 = 2, n_2 = 5, n_3 = 5$		2.285	.363
3.641	.175	5.523	.042			2.292	.353
3.654	.170	5.564	.038	x	$P_0\{H > x\}$	2.385	.345
3.700	.164	5.641	.037			2.408	.330
3.704	.160	5.664	.036	.008	1.000	2.469	.323
3.791	.157	5.754	.035	.046	.988	2.538	.315
3.800	.151	5.823	.034	.069	.978	2.592	.300
3.818	.148	5.891	.032	.077	.966	2.662	.292
3.823	.145	5.954	.030	.169	.947	2.754	.286
3.864	.143	5.973	.029	.192	.928	2.777	.279
4.041	.139	6.004	.026	.254	.896	2.908	.276
4.064	.135	6.041	.025	.323	.877	2.962	.270
4.073	.133	6.068	.025	.377	.859	3.023	.243
4.091	.130	6.118	.024	.415	.830	3.031	.234
4.141	.128	6.141	.023	.446	.822	3.123	.228
4.154	.126	6.223	.022	.538	.807	3.146	.218
4.200	.123	6.368	.021	.562	.775	3.331	.210
4.223	.121	6.391	.021	.623	.759	3.369	.203
4.250	.119	6.473	.020	.692	.749	3.392	.198
4.323	.116	6.504	.020	.746	.735	3.492	.190
4.364	.114	6.541	.017	.808	.719	3.515	.186
4.368	.112	6.550	.017	.815	.688	3.577	.181

TABLE C-4 (continued)

$n_1 = 2, n_2 = 5, n_3 = 5$

x	$P_0\{H \geqslant x\}$
3.646	.169
3.738	.165
3.769	.163
3.862	.150
3.885	.146
4.015	.136
4.069	.132
4.131	.130
4.138	.127
4.231	.124
4.254	.114
4.438	.106
4.477	.103
4.508	.100
4.623	.097
4.685	.092
4.754	.084
4.808	.081
4.846	.073
4.877	.068
4.992	.066
5.054	.060
5.177	.057
5.238	.054
5.246	.051
5.338	.047
5.546	.045
5.585	.041
5.608	.040
5.615	.039
5.708	.037
5.731	.036
5.792	.032
5.915	.030
5.985	.028
6.077	.027
6.231	.026
6.346	.025
6.354	.021
6.446	.020
6.469	.019
6.654	.017
6.692	.016
6.815	.015
6.838	.014

$n_1 = 2, n_2 = 5, n_3 = 5$

x	$P_0\{H \geqslant x\}$
6.969	.013
7.023	.013
7.185	.012
7.208	.011
7.269	.010
7.338	.010
7.392	.009
7.462	.008
7.577	.007
7.762	.007
7.923	.006
8.008	.006
8.077	.006
8.131	.005
8.169	.003
8.292	.003
8.377	.002
8.562	.002
8.685	.001
8.938	.001
9.423	.000

$n_1 = 3, n_2 = 3, n_3 = 3$

x	$P_0\{H \geqslant x\}$
.000	1.000
.089	.993
.267	.929
.356	.879
.622	.829
.800	.721
1.067	.664
1.156	.629
1.422	.543
1.689	.511
1.867	.439
2.222	.382
2.400	.361
2.489	.339
2.756	.296
3.200	.254
3.289	.232

$n_1 = 3, n_2 = 3, n_3 = 3$

x	$P_0\{H \geqslant x\}$
3.467	.196
3.822	.168
4.267	.139
4.356	.132
4.622	.100
5.067	.086
5.422	.071
5.600	.050
5.689	.029
5.956	.025
6.489	.011
7.200	.004

$n_1 = 3, n_2 = 3, n_3 = 4$

x	$P_0\{H \geqslant x\}$
.018	1.000
.046	.984
.118	.970
.164	.941
.200	.925
.336	.895
.346	.869
.409	.842
.454	.830
.482	.817
.636	.791
.700	.764
.746	.717
.891	.690
1.064	.656
1.073	.633
1.136	.611
1.182	.602
1.209	.582
1.427	.541
1.473	.523
1.573	.513
1.618	.497
1.654	.481
1.791	.447
1.800	.433

$n_1 = 3, n_2 = 3, n_3 = 4$

x	$P_0\{H \geqslant x\}$
1.864	.415
2.091	.402
2.200	.389
2.227	.368
2.300	.351
2.382	.326
2.518	.314
2.527	.303
2.664	.291
2.882	.281
2.927	.273
2.954	.253
3.027	.244
3.073	.234
3.109	.220
3.254	.212
3.364	.203
3.391	.196
3.609	.188
3.682	.180
3.754	.178
3.800	.165
3.836	.150
3.973	.143
4.046	.132
4.091	.126
4.273	.123
4.336	.117
4.382	.111
4.564	.106
4.700	.101
4.709	.092
4.818	.085
4.846	.081
5.000	.074
5.064	.070
5.109	.068
5.254	.064
5.436	.062
5.500	.056
5.573	.053
5.727	.050
5.791	.046
5.936	.036

$n_1 = 3, n_2 = 3, n_3 = 4$		$n_1 = 3, n_2 = 3, n_3 = 5$		$n_1 = 3, n_2 = 3, n_3 = 5$		$n_1 = 3, n_2 = 4, n_3 = 4$	
x	$P_0\{H \geqslant x\}$	x	$P_0\{H \geqslant x\}$	x	$P_0\{H \geqslant x\}$	x	$P_0\{H \geqslant x\}$
5.982	.034	1.515	.512	4.533	.097	.000	1.000
6.018	.027	1.527	.505	4.679	.094	.046	.993
6.154	.025	1.576	.491	4.776	.090	.053	.981
6.300	.023	1.648	.478	4.800	.087	.144	.959
6.564	.017	1.746	.450	4.848	.085	.167	.937
6.664	.014	1.770	.437	4.861	.082	.182	.925
6.709	.013	1.867	.425	4.909	.079	.212	.913
6.746	.010	2.012	.414	5.042	.077	.326	.890
7.000	.006	2.048	.403	5.079	.069	.348	.870
7.318	.004	2.061	.393	5.103	.067	.386	.850
7.436	.002	2.133	.382	5.212	.065	.409	.829
8.018	.001	2.170	.367	5.261	.062	.477	.819
		2.182	.358	5.346	.058	.576	.799
		2.194	.352	5.442	.055	.598	.779

$n_1 = 3, n_2 = 3, n_3 = 5$	
x	$P_0\{H \geqslant x\}$

Continuing columns:

x	$P_0\{H \geqslant x\}$ (col 1)	x	$P_0\{H \geqslant x\}$ (col 2)	x	$P_0\{H \geqslant x\}$ (col 3)	x	$P_0\{H \geqslant x\}$ (col 4)
		2.315	.342	5.503	.053	.659	.761
		2.376	.334	5.515	.051	.667	.742
.000	1.000	2.594	.315	5.648	.049	.712	.731
.048	.994	2.667	.306	5.770	.047	.727	.713
.061	.970	2.679	.298	5.867	.042	.848	.704
.133	.958	2.715	.291	6.012	.040	.894	.685
.170	.948	2.836	.267	6.061	.033	.932	.668
.194	.926	2.861	.258	6.109	.032	.962	.651
.242	.902	2.970	.242	6.194	.027	1.053	.635
.315	.890	3.079	.239	6.303	.026	1.076	.620
.376	.868	3.103	.232	6.315	.021	1.136	.604
.412	.847	3.333	.218	6.376	.020	1.144	.597
.436	.826	3.382	.215	6.533	.019	1.296	.582
.533	.804	3.394	.209	6.594	.019	1.303	.568
.546	.794	3.442	.196	6.715	.014	1.326	.553
.594	.783	3.467	.184	6.776	.013	1.394	.539
.679	.765	3.503	.179	6.861	.012	1.417	.524
.776	.725	3.576	.173	6.982	.011	1.500	.510
.848	.686	3.648	.167	7.079	.009	1.546	.503
.970	.668	3.709	.162	7.333	.008	1.598	.490
1.042	.641	3.879	.156	7.467	.008	1.636	.477
1.079	.624	3.927	.149	7.503	.006	1.682	.470
1.103	.609	4.012	.144	7.515	.005	1.750	.457
1.200	.594	4.048	.139	7.636	.004	1.803	.444
1.212	.587	4.170	.135	7.879	.003	1.909	.421
1.261	.571	4.194	.126	8.048	.002	1.962	.409
1.442	.539	4.242	.122	8.242	.001	2.053	.388
1.503	.526	4.303	.117	8.727	.001	2.144	.378
		4.315	.113			2.227	.368
		4.412	.109			2.296	.354

TABLE C-4 (*continued*)

$n_1 = 3, n_2 = 4, n_3 = 4$		$n_1 = 3, n_2 = 4, n_3 = 4$		$n_1 = 3, n_2 = 4, n_3 = 4$		$n_1 = 3, n_2 = 4, n_3 = 5$	
x	$P_0\{H \geqslant x\}$	x	$P_0\{H \geqslant x\}$	x	$P_0\{H \geqslant x\}$	x	$P_0\{H \geqslant x\}$
2.303	.344	5.053	.078	8.909	.001	1.062	.621
2.326	.334	5.144	.073			1.103	.615
2.394	.325	5.182	.068			1.106	.609
2.417	.315	5.212	.066	$n_1 = 3, n_2 = 4, n_3 = 5$		1.118	.602
2.598	.306	5.296	.063			1.137	.590
2.636	.290	5.303	.061	x	$P_0\{H \geqslant x\}$	1.164	.584
2.667	.281	5.326	.058			1.188	.578
2.712	.276	5.386	.054	.010	1.000	1.241	.572
2.848	.269	5.500	.052	.030	.990	1.246	.566
2.894	.261	5.576	.051	.060	.981	1.260	.553
2.909	.254	5.598	.049	.081	.972	1.349	.548
2.932	.250	5.667	.047	.092	.963	1.414	.542
2.962	.243	5.803	.045	.118	.953	1.445	.537
3.076	.230	5.932	.043	.138	.944	1.465	.522
3.136	.218	5.962	.041	.173	.935	1.472	.516
3.326	.212	6.000	.040	.180	.926	1.487	.506
3.386	.207	6.046	.039	.214	.917	1.506	.495
3.394	.201	6.053	.035	.241	.908	1.558	.490
3.417	.195	6.144	.032	.256	.900	1.568	.479
3.477	.190	6.167	.031	.265	.891	1.599	.475
3.576	.184	6.182	.030	.276	.882	1.615	.465
3.598	.178	6.348	.027	.323	.874	1.718	.460
3.659	.173	6.386	.026	.337	.865	1.733	.455
3.682	.162	6.394	.025	.368	.857	1.753	.450
3.727	.160	6.409	.023	.426	.841	1.780	.446
3.803	.154	6.417	.022	.430	.833	1.814	.441
3.848	.150	6.546	.021	.462	.825	1.856	.437
3.932	.145	6.659	.020	.491	.817	1.906	.427
3.962	.140	6.712	.019	.503	.809	1.927	.423
4.144	.135	6.727	.018	.542	.784	1.938	.418
4.167	.131	6.962	.017	.549	.777	1.964	.400
4.212	.129	7.000	.016	.626	.769	1.968	.396
4.296	.125	7.053	.014	.645	.754	1.985	.391
4.303	.121	7.076	.011	.692	.746	2.019	.387
4.326	.116	7.136	.011	.727	.738	2.030	.383
4.348	.113	7.144	.010	.737	.716	2.060	.379
4.409	.106	7.212	.009	.799	.709	2.103	.375
4.477	.102	7.477	.006	.830	.696	2.112	.366
4.546	.099	7.598	.004	.831	.689	2.169	.358
4.576	.097	7.636	.004	.856	.667	2.272	.354
4.598	.093	7.682	.003	.953	.660	2.308	.350
4.712	.090	7.848	.003	1.004	.654	2.337	.346
4.750	.087	8.227	.002	1.041	.641	2.349	.343
4.894	.084	8.326	.001	1.045	.628	2.368	.335

$n_1 = 3, n_2 = 5, n_3 = 5$		$n_1 = 4, n_2 = 4, n_3 = 4$		$n_1 = 4, n_2 = 4, n_3 = 4$		$n_1 = 4, n_2 = 4, n_3 = 5$	
x	$P_0\{H > x\}$	x	$P_0\{H > x\}$	x	$P_0\{H > x\}$	x	$P_0\{H > x\}$
6.998	.015	.000	1.000	4.654	.097	.119	.952
7.050	.015	.038	.994	4.769	.094	.132	.937
7.121	.014	.115	.968	4.885	.086	.201	.930
7.209	.014	.154	.941	4.962	.080	.218	.916
7.226	.012	.269	.913	5.115	.074	.228	.903
7.288	.012	.346	.864	5.346	.063	.267	.889
7.306	.012	.462	.840	5.538	.057	.297	.875
7.314	.011	.500	.815	6.654	.055	.343	.869
7.437	.011	.615	.770	5.692	.049	.376	.862
7.543	.010	.731	.746	5.808	.044	.382	.849
7.578	.010	.808	.706	6.000	.040	.399	.836
7.622	.009	.962	.667	6.038	.037	.425	.823
7.736	.009	1.038	.648	6.269	.033	.475	.811
7.763	.008	1.077	.630	6.500	.030	.528	.798
7.780	.008	1.192	.592	6.577	.026	.544	.792
7.859	.007	1.385	.557	6.615	.024	.597	.780
7.894	.007	1.423	.540	6.731	.021	.610	.768
7.912	.007	1.500	.510	6.962	.019	.613	.757
8.026	.006	1.654	.480	7.038	.018	.640	.745
8.079	.006	1.846	.452	7.269	.016	.689	.734
8.106	.006	1.885	.436	7.385	.015	.742	.723
8.237	.005	2.000	.397	7.423	.013	.771	.711
8.264	.005	2.192	.370	7.538	.011	.804	.706
8.316	.005	2.346	.348	7.654	.008	.824	.695
8.334	.005	2.423	.327	7.731	.007	.860	.690
8.545	.004	2.462	.307	8.000	.005	.870	.679
8.571	.004	2.577	.296	8.115	.003	.903	.668
8.580	.004	2.808	.277	8.346	.002	.910	.658
8.650	.003	2.885	.260	8.654	.001	.940	.647
8.659	.003	2.923	.252	8.769	.001	1.019	.637
8.791	.002	3.038	.234	9.269	.001	1.058	.627
8.809	.002	3.115	.219	9.846	.000	1.068	.617
8.950	.002	3.231	.212			1.124	.607
9.002	.002	3.500	.197			1.167	.598
9.055	.001	3.577	.173	$n_1 = 4, n_2 = 4, n_3 = 5$		1.187	.589
9.284	.001	3.731	.162			1.190	.584
9.336	.001	3.846	.151	x	$P_0\{H > x\}$	1.203	.574
9.398	.001	3.962	.145			1.256	.565
9.521	.000	4.154	.136	.000	1.000	1.272	.556
9.635	.000	4.192	.131	.030	.996	1.299	.548
9.916	.000	4.269	.122	.033	.981	1.371	.539
10.057	.000	4.308	.114	.086	.974	1.404	.534
10.550	.000	4.500	.104	.096	.967	1.414	.526

TABLE C-4 (*continued*)

$n_1 = 4, n_2 = 4, n_3 = 5$		$n_1 = 4, n_2 = 4, n_3 = 5$		$n_1 = 4, n_2 = 4, n_3 = 5$		$n_1 = 4, n_2 = 4, n_3 = 5$	
x	$P_0\{H \geqslant x\}$	x	$P_0\{H \geqslant x\}$	x	$P_0\{H \geqslant x\}$	x	$P_0\{H \geqslant x\}$
1.454	.518	3.013	.228	4.701	.094	6.214	.034
1.503	.509	3.086	.224	4.711	.092	6.228	.033
1.530	.501	3.119	.221	4.728	.091	6.267	.032
1.533	.493	3.129	.217	4.747	.089	6.310	.031
1.586	.485	3.168	.214	4.760	.088	6.343	.030
1.596	.477	3.218	.210	4.813	.086	6.382	.029
1.615	.469	3.260	.206	4.830	.084	6.399	.028
1.668	.465	3.297	.202	4.833	.082	6.462	.027
1.701	.458	3.330	.200	4.896	.081	6.544	.027
1.718	.450	3.382	.197	4.975	.077	6.547	.026
1.744	.443	3.432	.190	5.014	.076	6.597	.026
1.810	.436	3.442	.187	5.024	.074	6.672	.024
1.876	.429	3.481	.183	5.028	.073	6.676	.024
1.899	.422	3.511	.180	5.090	.071	6.804	.023
1.929	.414	3.590	.176	5.172	.069	6.860	.022
1.942	.408	3.613	.170	5.196	.068	6.870	.022
1.958	.401	3.630	.167	5.225	.066	6.887	.021
2.047	.388	3.640	.164	5.344	.065	6.890	.021
2.110	.375	3.656	.160	5.360	.063	6.943	.020
2.140	.371	3.696	.157	5.370	.062	6.953	.020
2.143	.365	3.758	.154	5.387	.061	6.976	.019
2.176	.362	3.828	.151	5.410	.060	7.058	.018
2.196	.356	3.910	.146	5.440	.059	7.075	.017
2.275	.344	3.986	.143	5.476	.057	7.101	.017
2.387	.338	3.989	.141	5.486	.056	7.124	.016
2.390	.332	4.025	.139	5.489	.056	7.190	.016
2.403	.327	4.042	.134	5.519	.054	7.203	.015
2.440	.316	4.068	.132	5.568	.052	7.233	.015
2.443	.310	4.075	.130	5.571	.051	7.240	.014
2.453	.305	4.118	.127	5.618	.050	7.256	.014
2.558	.299	4.170	.125	5.657	.049	7.418	.014
2.575	.293	4.200	.122	5.687	.048	7.467	.013
2.601	.288	4.233	.121	5.756	.047	7.470	.013
2.667	.283	4.253	.119	5.782	.046	7.497	.013
2.670	.279	4.272	.117	5.815	.045	7.503	.012
2.733	.271	4.289	.114	5.819	.043	7.586	.012
2.756	.267	4.332	.112	5.914	.042	7.596	.012
2.799	.262	4.381	.108	6.003	.042	7.714	.011
2.881	.257	4.447	.106	6.013	.041	7.744	.011
2.904	.253	4.497	.104	6.030	.040	7.760	.009
2.918	.249	4.553	.102	6.096	.039	7.767	.009
2.967	.245	4.619	.100	6.119	.038	7.797	.009
2.987	.240	4.668	.098	6.132	.037	7.810	.009
2.997	.236	4.685	.096	6.201	.036	7.833	.008

$n_1 = 4, n_2 = 4, n_3 = 5$		$n_1 = 4, n_2 = 5, n_3 = 5$		$n_1 = 4, n_2 = 5, n_3 = 5$		$n_1 = 4, n_2 = 5, n_3 = 5$	
x	$P_0\{H \geqslant x\}$	x	$P_0\{H \geqslant x\}$	x	$P_0\{H \geqslant x\}$	x	$P_0\{H \geqslant x\}$
7.942	.007	.111	.958	1.366	.525	2.783	.272
7.981	.007	.131	.946	1.411	.518	2.786	.268
8.047	.006	.143	.935	1.423	.512	2.831	.257
8.113	.006	.180	.929	1.483	.505	2.840	.254
8.130	.006	.203	.923	1.551	.498	2.886	.250
8.140	.005	.223	.912	1.560	.492	2.931	.246
8.156	.005	.226	.901	1.606	.485	2.946	.239
8.189	.005	.271	.890	1.620	.479	2.966	.236
8.403	.004	.280	.879	1.643	.470	2.991	.232
8.440	.004	.326	.874	1.651	.458	3.023	.229
8.456	.004	.360	.863	1.686	.455	3.083	.224
8.525	.003	.371	.852	1.711	.449	3.103	.221
8.558	.003	.386	.841	1.731	.443	3.160	.218
8.571	.003	.463	.821	1.743	.437	3.240	.215
8.575	.003	.500	.805	1.803	.431	3.243	.211
8.604	.003	.523	.800	1.826	.425	3.266	.209
8.703	.003	.543	.790	1.871	.420	3.286	.203
8.733	.002	.546	.781	1.963	.414	3.311	.200
8.782	.002	.591	.771	1.971	.409	3.343	.197
8.868	.002	.600	.752	1.986	.398	3.380	.188
8.997	.001	.691	.742	2.006	.393	3.403	.187
9.053	.001	.706	.738	2.031	.382	3.471	.184
9.099	.001	.726	.729	2.051	.377	3.540	.176
9.129	.001	.751	.720	2.063	.372	3.571	.174
9.168	.001	.771	.711	2.100	.369	3.586	.170
9.396	.001	.783	.693	2.143	.364	3.651	.167
9.528	.001	.843	.684	2.191	.354	3.743	.162
9.590	.001	.863	.675	2.246	.349	3.746	.160
9.613	.000	.866	.667	2.280	.344	3.791	.155
9.758	.000	.966	.658	2.306	.339	3.800	.153
10.118	.000	.980	.654	2.351	.335	3.846	.151
10.187	.000	1.000	.650	2.371	.330	3.883	.148
10.681	.000	1.003	.642	2.383	.326	3.891	.144
		1.011	.626	2.420	.322	3.906	.142
		1.046	.617	2.443	.319	3.926	.140
$n_1 = 4, n_2 = 5, n_3 = 5$		1.071	.610	2.463	.307	3.951	.137
		1.140	.594	2.466	.302	3.971	.135
x	$P_0\{H \geqslant x\}$	1.183	.587	2.511	.298	4.043	.133
		1.186	.579	2.520	.294	4.063	.131
.006	1.000	1.286	.572	2.566	.292	4.166	.127
.020	.994	1.300	.553	2.600	.288	4.200	.124
.043	.988	1.323	.547	2.626	.284	4.203	.122
.051	.976	1.331	.540	2.691	.280	4.246	.120
.086	.970	1.346	.532	2.740	.276	4.271	.118

TABLE C-4 (*continued*)

$n_1 = 4, n_2 = 5, n_3 = 5$		$n_1 = 4, n_2 = 5, n_3 = 5$		$n_1 = 4, n_2 = 5, n_3 = 5$		$n_1 = 4, n_2 = 5, n_3 = 5$	
x	$P_0\{H > x\}$	x	$P_0\{H > x\}$	x	$P_0\{H > x\}$	x	$P_0\{H > x\}$
4.291	.115	5.711	.048	7.183	.017	8.683	.004
4.303	.113	5.780	.048	7.220	.017	8.691	.004
4.363	.111	5.803	.047	7.243	.017	8.726	.004
4.383	.110	5.811	.046	7.266	.016	8.751	.004
4.386	.108	5.871	.045	7.311	.015	8.771	.004
4.486	.106	5.903	.043	7.320	.015	8.969	.003
4.500	.105	5.963	.042	7.426	.015	8.980	.003
4.520	.101	5.983	.042	7.446	.014	9.000	.003
4.523	.099	5.986	.041	7.471	.014	9.011	.003
4.531	.098	6.031	.040	7.491	.014	9.026	.003
4.591	.096	6.086	.040	7.503	.013	9.071	.002
4.611	.095	6.100	.038	7.563	.013	9.103	.002
4.660	.093	6.123	.037	7.586	.012	9.163	.002
4.706	.092	6.146	.037	7.631	.012	9.231	.002
4.806	.089	6.166	.035	7.640	.011	9.286	.002
4.843	.088	6.211	.035	7.686	.011	9.323	.001
4.851	.086	6.223	.034	7.720	.011	9.411	.001
4.866	.084	6.283	.034	7.766	.010	9.503	.001
4.886	.083	6.303	.033	7.791	.010	9.506	.001
4.911	.079	6.351	.032	7.823	.010	9.606	.001
4.943	.078	6.406	.031	7.860	.010	9.643	.001
4.980	.076	6.440	.030	7.903	.009	9.651	.001
5.023	.075	6.451	.029	7.906	.009	9.686	.001
5.071	.074	6.486	.029	8.006	.009	9.926	.001
5.126	.073	6.531	.028	8.043	.009	9.986	.000
5.163	.070	6.543	.028	8.051	.008	10.051	.000
5.171	.069	6.603	.027	8.066	.008	10.063	.000
5.186	.068	6.623	.026	8.086	.008	10.100	.000
5.206	.067	6.626	.026	8.131	.008	10.260	.000
5.231	.066	6.671	.025	8.143	.008	10.511	.000
5.263	.064	6.760	.025	8.223	.007	10.520	.000
5.323	.063	6.763	.024	8.226	.007	10.566	.000
5.400	.061	6.771	.024	8.271	.007	10.646	.000
5.446	.059	6.786	.023	8.280	.006	11.023	.000
5.460	.058	6.806	.022	8.340	.006	11.083	.000
5.483	.057	6.831	.022	8.363	.006	11.571	.000
5.491	.056	6.900	.021	8.371	.005		
5.526	.056	6.943	.020	8.386	.005	$n_1 = 5, n_2 = 5, n_3 = 5$	
5.571	.055	7.000	.019	8.431	.005		
5.583	.052	7.046	.019	8.463	.005	x	$P_0\{H > x\}$
5.620	.051	7.080	.018	8.523	.005		
5.643	.050	7.106	.018	8.543	.005	.000	1.000
5.666	.049	7.171	.018	8.546	.004	.020	.998

$n_1 = 5, n_2 = 5, n_3 = 5$		$n_1 = 5, n_2 = 5, n_3 = 5$		$n_1 = 5, n_2 = 5, n_3 = 5$		$n_1 = 5, n_2 = 5, n_3 = 5$	
x	$P_0\{H \geqslant x\}$	x	$P_0\{H \geqslant x\}$	x	$P_0\{H \geqslant x\}$	x	$P_0\{H \geqslant x\}$
.060	.983	2.340	.330	5.120	.072	8.000	.009
.080	.968	2.420	.319	5.180	.070	8.060	.009
.140	.954	2.480	.314	5.360	.065	8.180	.008
.180	.925	2.540	.304	5.420	.063	8.240	.008
.240	.911	2.580	.294	5.460	.060	8.340	.007
.260	.898	2.660	.284	5.540	.055	8.420	.007
.320	.871	2.780	.265	5.580	.053	8.540	.006
.380	.858	2.880	.256	5.660	.051	8.640	.006
.420	.832	2.940	.252	5.780	.049	8.660	.006
.500	.807	2.960	.239	5.820	.048	8.720	.005
.540	.794	3.020	.231	5.840	.046	8.780	.005
.560	.783	3.120	.223	6.000	.044	8.820	.005
.620	.759	3.140	.216	6.020	.043	8.880	.004
.720	.736	3.260	.208	6.080	.040	8.960	.004
.740	.725	3.380	.201	6.140	.038	9.060	.004
.780	.703	3.420	.190	6.180	.036	9.140	.003
.860	.681	3.440	.184	6.260	.035	9.260	.003
.960	.660	3.500	.177	6.320	.033	9.360	.003
.980	.650	3.620	.171	6.480	.032	9.380	.003
1.040	.620	3.660	.165	6.500	.031	9.420	.002
1.140	.601	3.780	.159	6.540	.030	9.500	.002
1.220	.582	3.840	.153	6.620	.028	9.620	.002
1.260	.564	3.860	.150	6.660	.027	9.680	.001
1.280	.547	3.920	.145	6.720	.026	9.740	.001
1.340	.538	3.980	.137	6.740	.025	9.780	.001
1.460	.521	4.020	.132	6.860	.024	9.920	.001
1.500	.505	4.160	.127	6.980	.021	9.980	.001
1.520	.497	4.220	.123	7.020	.020	10.140	.001
1.580	.481	4.340	.118	7.220	.019	10.220	.001
1.620	.466	4.380	.110	7.260	.018	10.260	.000
1.680	.459	4.460	.105	7.280	.018	10.500	.000
1.820	.444	4.500	.102	7.340	.016	10.580	.000
1.860	.416	4.560	.100	7.440	.015	10.640	.000
1.940	.403	4.580	.096	7.460	.015	10.820	.000
2.000	.390	4.740	.092	7.580	.014	11.060	.000
2.060	.383	4.820	.089	7.620	.013	11.180	.000
2.160	.371	4.860	.085	7.740	.012	11.520	.000
2.180	.365	4.880	.084	7.760	.012	11.580	.000
2.220	.353	4.940	.081	7.940	.011	12.020	.000
2.240	.342	5.040	.075	7.980	.011	12.500	.000

From Table F of *A Nonparametric Introduction to Statistics* by C. H. Kraft and C. van Eedan, Macmillan, New York, 1968. Reproduced by permission of the Macmillan Publishing Company.

TABLE C-5 SELECTED CRITICAL VALUES FOR ALL TREATMENTS MULTIPLE COMPARISONS BASED ON KRUSKAL-WALLIS RANK SUMS:

$$k = 3, n = 2(1)6; \qquad k = 6, 7, 8, n = 2, 3;$$
$$k = 4, 5, n = 2, 3, 4; \qquad k = 9(1)15, n = 2$$

For a given k and n, the entries in the table correspond to $P\{|R_i - R_j| < y(\alpha, k, n)$ $i = 1, \ldots, k - 1, j = u + 1, \ldots, k\} \approx 1 - \alpha$.

n

k	$y(\alpha, k, 2)$	α	$y(\alpha, k, 3)$	α	$y(\alpha, k, 4)$	α	$y(\alpha, k, 5)$	α	$y(\alpha, k, 6)$	α
			2		3		4		5		6	
3	8	.067	15*	.064	24*	.045	33*	.048	43*	.049
			16	.029	25	.031	35	.031	51*	.011
			17*	.011	27*	.011	39*	.009		
4	12	.029	22	.043	34	.049				
			23	.023	36	.026				
			24	.012	38	.012				
5	15	.048	28	.060	44	.056				
	16	.016	30	.023	46	.033				
			32	.007	50	.010				
6	19	.030	35	.055						
	20	.010	37	.024						
			39	.009						
7	22	.056	42	.054						
	23	.021	44	.026						
	24	.007	46	.012						
8	26	.041	49	.055						
	28	.005	51	.029						
			54	.010						
9	29	.063								
	30	.031								
	31	.012								
10	33	.050								
	34	.025								
	35	.009								
11	37	.040								
	38	.020								
	39	.008								
12	40	.062								
	41	.033								
	43	.006								
13	44	.052								
	45	.028								
	46	.014								
14	48	.044								
	49	.024								
	50	.012								
15	52	.038								
	54	.010								

From "Rank Sum Multiple Comparisons in One- and Two-Way Classification," by B. J. McDonald and W. A. Thompson, *Biometrika*, Vol. 54 (1967), pp. 487–497. Reproduced by permission of the editor of *Biometrika*. The starred values are from "Distribution-free Multiple Comparisons," Ph.D. Thesis (1963), P. Nemenyi, Princeton University, with the permission of the author.

TABLE C-6   SELECTED CRITICAL VALUES FOR THE RANGE OF k INDEPENDENT $N(0, 1)$ VARIABLES: $k = 2(1)20(2)40(10)100$

For a given k and α, the tabled entry is $q(\alpha, k, \infty)$.

					α				
k	.0001	.0005	.001	.005	.01	.025	.05	.10	.20
2	5.502	4.923	4.654	3.970	3.643	3.170	2.772	2.326	1.812
3	5.864	5.316	5.063	4.424	4.120	3.682	3.314	2.902	2.424
4	6.083	5.553	5.309	4.694	4.403	3.984	3.633	3.240	2.784
5	6.240	5.722	5.484	4.886	4.603	4.197	3.858	3.478	3.037
6	6.362	5.853	5.619	5.033	4.757	4.361	4.030	3.661	3.232
7	6.461	5.960	5.730	5.154	4.882	4.494	4.170	3.808	3.389
8	6.546	6.050	5.823	5.255	4.987	4.605	4.286	3.931	3.520
9	6.618	6.127	5.903	5.341	5.078	4.700	4.387	4.037	3.632
10	6.682	6.196	5.973	5.418	5.157	4.784	4.474	4.129	3.730
11	6.739	6.257	6.036	5.485	5.227	4.858	4.552	4.211	3.817
12	6.791	6.311	6.092	5.546	5.290	4.925	4.622	4.285	3.895
13	6.837	6.361	6.144	5.602	5.348	4.985	4.685	4.351	3.966
14	6.880	6.407	6.191	5.652	5.400	5.041	4.743	4.412	4.030
15	6.920	6.449	6.234	5.699	5.448	5.092	4.796	4.468	4.089
16	6.957	6.488	6.274	5.742	5.493	5.139	4.845	4.519	4.144
17	6.991	6.525	6.312	5.783	5.535	5.183	4.891	4.568	4.195
18	7.023	6.559	6.347	5.820	5.574	5.224	4.934	4.612	4.242
19	7.054	6.591	6.380	5.856	5.611	5.262	4.974	4.654	4.287
20	7.082	6.621	6.411	5.889	5.645	5.299	5.012	4.694	4.329
22	7.135	6.677	6.469	5.951	5.709	5.365	5.081	4.767	4.405
24	7.183	6.727	6.520	6.006	5.766	5.425	5.144	4.832	4.475
26	7.226	6.773	6.568	6.057	5.818	5.480	5.201	4.892	4.537
28	7.266	6.816	6.611	6.103	5.866	5.530	5.253	4.947	4.595
30	7.303	6.855	6.651	6.146	5.911	5.577	5.301	4.997	4.648
32	7.337	6.891	6.689	6.186	5.952	5.620	5.346	5.044	4.697
34	7.370	6.925	6.723	6.223	5.990	5.660	5.388	5.087	4.743
36	7.400	6.957	6.756	6.258	6.026	5.698	5.427	5.128	4.786
38	7.428	6.987	6.787	6.291	6.060	5.733	5.463	5.166	4.826
40	7.455	7.015	6.816	6.322	6.092	5.766	5.498	5.202	4.864
50	7.571	7.137	6.941	6.454	6.228	5.909	5.646	5.357	5.026
60	7.664	7.235	7.041	6.561	6.338	6.023	5.764	5.480	5.155
70	7.741	7.317	7.124	6.649	6.429	6.118	5.863	5.582	5.262
80	7.808	7.387	7.196	6.725	6.507	6.199	5.947	5.669	5.353
90	7.866	7.448	7.259	6.792	6.575	6.270	6.020	5.745	5.433
100	7.918	7.502	7.314	6.850	6.636	6.333	6.085	5.812	5.503

From "Table of Range and Studentized Range," by H. L. Harter, *Ann. Math. Statist.* (1960) Vol 31, pp. 1122–1147. Reproduced by permission of the editor of the *Annals of Mathematical Statistics*.

TABLE C-7 QUANTILES OF THE GAMMA DISTRIBUTION

PER CENT\γ	0.1	0.2	0.3	0.4	.= 0.5
0.1	6.0730398E-31	6.5254914E-16	6.9727015E-11	2.3449403E-08	7.8539858E-07
0.5	5.9307050E-24	2.0392159E-12	1.4903943E-08	1.3108620E-06	1.9635211E-05
0.7	3.4199421E-22	1.5485293E-11	5.7580040E-08	3.6123166E-06	4.4179955E-05
1.0	6.0730419E-21	6.5254908E-11	1.5022232E-07	7.4153871E-06	7.8543941E-05
1.5	3.5020208E-19	4.9552939E-10	5.8037043E-07	2.0434596E-05	1.7673543E-04
2.0	6.2187947E-18	2.0881566E-09	1.5141467E-06	4.1948807E-05	3.1422512E-04
2.5	5.7917027E-17	6.3725492E-09	3.1856788E-06	7.3283179E-05	4.9103468E-04
3.0	3.5860676E-16	1.5856933E-08	5.8497902E-06	1.1560341E-04	7.0719174E-04
4.0	6.3680429E-15	6.6821009F-08	1.5261804E-05	2.3733141E-04	1.2576912E-03
5.0	5.9307026E-14	2.0392152E-07	3.2110344E-05	4.1465349E-04	1.9660703E-03
7.5	3.4199416E-12	1.5485298E-06	1.2406423E-04	1.1432451E-03	4.4309273E-03
10.0	6.0730380E-11	6.5255202E-06	3.2372462E-04	2.3488758E-03	7.8953891E-03
15.0	3.5020194E-09	4.9554952E-05	1.2515738E-03	6.4918979E-03	1.7882892E-02
20.0	6.2187893E-08	2.0885191E-04	3.2703397E-03	1.3392227E-02	3.2092386E-02
25.0	5.7917008E-07	6.3759328E-04	6.8998047E-03	2.3564933E-02	5.0765538E-02
30.0	3.5860783E-06	1.5877920E-03	1.2726660E-02	3.7541870E-02	7.4235954E-02
40.0	6.3684073E-05	6.7195739E-03	3.3739803E-02	7.9361846E-02	1.3749799E-01
50.0	5.9339001E-04	2.0746364E+02	7.3131159E-02	1.4507806E-01	2.2746834E-01
60.0	3.6844445E-03	5.3010665E-02	1.4125254E-01	2.4475218E-01	3.5416332E-01
70.0	1.7427737E-02	1.2103773E-01	2.5656503E-01	3.9725703E-01	5.3709736E-01
75.0	3.5306306E-02	1.7885940E-01	3.4289970E-01	5.0480585E-01	6.6165223E-01
80.0	6.9389746E-02	2.6354398E-01	4.6007422E-01	6.4557067E-01	8.2118779E-01
85.0	1.3466307E-01	3.9239831E-01	6.2662798E-01	8.3910112E-01	1.0361264E 00
90.0	2.6615398E-01	6.0490358E-01	8.8481154E-01	1.1298418E 00	1.3527737E 00
92.5	3.8439176E-C1	7.7335898E-01	1.0811301E 00	1.3461192E 00	1.5850286E 00
95.0	5.8043370E-01	1.0305303E 00	1.3723524E 00	1.6619615E 00	1.9207334E 00
97.5	9.7790323E-01	1.5111122E 00	1.9002707E 00	2.2247126E 00	2.5119509E 00
98.0	1.1190291E 00	1.6744477E 00	2.0765711E 00	2.4107092E 00	2.7059585E 00
99.0	1.5884692E 00	2.2023303E 00	2.6394268E 00	3.0000888E 00	3.3174701E 00
99.5	2.0945469E 00	2.7547526E 00	3.2206054E 00	3.6035306E 00	3.9397731E 00
99.9	3.3636681E 00	4.1C23043E 00	4.6191710E 00	5.0425654E 00	5.4140809E 00

PER CENT\γ	0.6	0.8	1.0	1.2	1.4
0.1	8.2890664E-06	1.6272338E-04	1.0005003E-03	3.4337143E-03	8.4321789E-03
0.5	1.2119545E-04	1.2173541E-03	5.0125419E-03	1.3187497E-02	2.6824095E-02
0.7	2.3823386E-04	2.0217386E-03	7.5282661E-03	1.8533525E-02	3.5971068E-02
1.0	3.8483497E-04	2.8980758E-03	1.0050336E-02	2.3608727E-02	4.4330284E-02
1.5	7.5659043E-04	4.8159973E-03	1.5113637E-02	3.3243716E-02	5.9597079E-02
2.0	1.2224146E-03	6.9081841E-03	2.0202707E-02	4.2425602E-02	7.3617880E-02
2.5	1.7737191E-03	9.1419638E-03	2.5317809E-02	5.1301112E-02	8.6809725E-02
3.0	2.4044995E-03	1.1496962E-02	3.0459208E-02	5.9952244E-02	9.9398144E-02
4.0	3.8873902E-03	1.6518275E-02	4.0821993E-02	7.6773139E-02	1.2327700E-01
5.0	5.6448357E-03	2.1897531E-02	5.1293299E-02	9.3145212E-02	1.4592649E-01
7.5	1.1133291E-02	3.6647758E-02	7.7961543E-02	1.3293248E-01	1.9922959E-01
10.0	1.8060443E-02	5.2981821E-02	1.0536052E-01	1.7189840E-01	2.4973644E-01
15.0	3.5894110E-02	8.9739645E-02	1.6251895E-01	2.4937285E-01	3.4689076E-01
20.0	5.8803372E-02	1.3152850E-01	2.2314357E-01	3.2979601E-01	4.4246886E-01
25.0	8.6771754E-02	1.7827535E-01	2.8768212E-01	4.0903381E-01	5.3886871E-01
30.0	1.1998889E-01	2.3019176E-01	3.5667497E-01	4.9357830E-01	6.3769279E-01
40.0	2.0382268E-01	3.5143535E-01	5.1082570E-01	6.7712659E-01	8.4793257E-01
50.0	3.1570220E-01	5.0135124E-01	6.9314729E-01	8.8793657E-01	1.0843713E 00
60.0	4.6590956E-01	6.9127123E-01	9.1629100E-01	1.1400349E 00	1.3623938E 00
70.0	6.7474850E-01	9.4321526E-01	1.2039732E 00	1.4587471E 00	1.7088393E 00
75.0	8.1365435E-01	1.1058806E 00	1.3862949E 00	1.6581306E 00	1.9234758E 00
80.0	9.8899279E-01	1.3073711E 00	1.6094384E 00	1.9000906E 00	2.1822733E 00
85.0	1.2219616E 00	1.5702344E 00	1.8971206E 00	2.2093879E 00	2.5109470E 00
90.0	1.5605061E 00	1.9452591E 00	2.3025857E 00	2.6414636E 00	2.9669431E 00
92.5	1.8063390E 00	2.2138442E 00	2.5902694E 00	2.9459091E 00	3.2864993E 00
95.0	2.1590177E 00	2.5951444E 00	2.9957345E 00	3.3726729E 00	3.7325141E 00
97.5	2.7747879E 00	3.2526998E 00	3.6888885E 00	4.0973864E 00	4.4858795E 00
98.0	2.9757511E 00	3.4656007E 00	3.9120296E 00	4.3296577E 00	4.7264682E 00
99.0	3.6066221E 00	4.1298904E 00	4.6051941E 00	5.0486563E 00	5.4690573E 00
99.5	4.2455626E 00	4.7977945E 00	5.2983655E 00	5.7645890E 00	6.2058056E 00
99.9	5.7512864E 00	6.3586366E 00	6.9080574E 00	7.4187547E 00	7.9005137E 00

(NOTE. 6.07E-31 IS EQUIVALENT TO 6.07 X 10^{-31})

PER CENT \ γ	1.6	1.8	2.0	2.2	2.4
0.1	1.6780863E-02	2.9006379E-02	4.5402019E-02	6.6083795E-02	9.1045250E-02
0.5	4.6409061E-02	7.2C19480E-02	1.0349455E-01	1.4055196E-01	1.8285795E-01
0.7	6.0108387E-02	9.0817968E-02	1.2777440E-01	1.7057086E-01	2.1878191E-01
1.0	7.2283818E-02	1.0717552E-01	1.4855474E-01	1.9592441E-01	2.4879515E-01
1.5	9.39C2475E-02	1.3560738E-01	1.8407833E-01	2.3869266E-01	2.9887579E-01
2.0	1.1322857E-01	1.6050999E-01	2.1469912E-01	2.7508877E-01	3.4104964E-01
2.5	1.3105810E-01	1.8314475E-01	2.4220929E-01	3.0748374E-01	3.7830102E-01
3.0	1.4781163E-01	2.0416656E-01	2.6752685E-01	3.3707901E-01	4.1212894E-01
4.0	1.7902662E-01	2.4280678E-01	3.1357260E-01	3.9044885E-01	4.7270738E-01
5.0	2.0808401E-01	2.7826796E-01	3.5536154E-01	4.3845224E-01	5.2679405E-01
7.5	2.7489323E-01	3.5837546E-01	4.4846799E-01	5.4422134E-01	6.4488062E-01
10.0	3.3669067E-01	4.3112658E-01	5.3181166E-01	6.3780034E-01	7.4835367E-01
15.0	4.5272915E-01	5.6524168E-01	6.8323869E-01	8.0583392E-01	9.3234974E-01
20.0	5.6432575E-01	6.9199598E-01	8.2438848E-01	9.6070601E-01	1.1003482E 00
25.0	6.7504999E-01	8.1617533E-01	9.6127882E-01	1.1096652E 00	1.2608163E 00
30.0	7.8710858E-01	9.4C60314E-01	1.0973494E 00	1.2567599E 00	1.4184017E 00
40.0	1.0219104E CO	1.1982486E 00	1.3764216E 00	1.5560676E 00	1.7369269E 00
50.0	1.2817967E 00	1.4798565E CO	1.6783474E 00	1.8771417E 00	2.0761571E 00
60.0	1.5834684E 00	1.8C33957E 00	2.0223137E 00	2.2403417E 00	2.4575810E 00
70.0	1.9551916E 00	2.19848C7E 00	2.4392173E 00	2.6777864E 00	2.9144873E 00
75.0	2.1837358E 00	2.4398846E 00	2.6926356E 00	2.9425202E 00	3.1899446E 00
80.0	2.4579001E 00	2.7282669E 00	2.9943091E 00	3.2567191E 00	3.5160193E 00
85.0	2.8042824E 00	3.0910514E 00	3.3724436E 00	3.6493274E 00	3.9223586E 00
90.0	3.2821878E 00	3.5892844E 00	3.8897218E 00	4.1845905E 00	4.4746965E 00
92.5	3.6156165E 00	3.9356059E 00	4.2481450E 00	4.5544488E 00	4.8554263E 00
95.0	4.0793485E 00	4.4158303E 00	4.7438694E 00	5.0648451E 00	5.3797783E 00
97.5	4.8591827E 00	5.2203591E 00	5.5716554E 00	5.9146804E 00	6.2506155E 00
98.0	5.1074690E 00	5.4758451E 00	5.8339376E 00	6.1834139E 00	6.5255030E 00
99.0	5.8719870E 00	6.2608809E 00	6.6383883E 00	7.0063273E 00	7.3660060E 00
99.5	6.6282540E 00	7.0353775E 00	7.4301966E 00	7.8146312E 00	8.1900520E 00
99.9	8.3615569E 00	8.8C45225E 00	9.2336768E 00	9.6511943E 00	1.0057676E 01

PER CENT \ γ	2.6	2.8	3.0	3.2	3.4
0.1	1.2020084E-01	1.5341695E-01	1.9053338E-01	2.3137730E-01	2.7577227E-01
0.5	2.3006679E-01	2.8184123E-01	3.3786340E-01	3.9783849E-01	4.6149604E-01
0.7	2.7199522E-01	3.2982536E-01	3.9191958E-01	4.5795804E-01	5.2765206E-01
1.0	3.0670981E-01	3.6925137E-01	4.3604521E-01	5.0675595E-01	5.8108456E-01
1.5	3.6411288E-01	4.3394811E-01	5.0798078E-01	5.8585852E-01	6.6727104E-01
2.0	4.1203167E-01	4.8755657E-01	5.6720963E-01	6.5063010E-01	7.3750284E-01
2.5	4.5408805E-01	5.3435255E-01	6.1867216E-01	7.0668188E-01	7.9806521E-01
3.0	4.9208797E-01	5.7645527E-01	6.6480418E-01	7.5676816E-01	8.5203115E-01
4.0	5.5974186E-01	6.5104351E-01	7.4618340E-01	8.4479663E-01	9.4657009E-01
5.0	6.1977374E-01	7.1688152E-01	8.1769154E-01	9.2184297E-01	1.0290283E 00
7.5	7.4983816E-01	8.5859696E-01	9.7074668E-01	1.0859437E 00	1.2038973E 00
10.0	8.628849lE-01	9.8091840E-01	1.1020654E 00	1.2260021E 00	1.3524560E 00
15.0	1.0622576E 00	1.1951340E 00	1.3306367E 00	1.4684834E 00	1.6084377E 00
20.0	1.2428518E 00	1.3878487E 00	1.5350443E 00	1.6841965E 00	1.8351031E 00
25.0	1.4143367E 00	1.5699138E 00	1.7272996E 00	1.8862906E 00	2.0467182E 00
30.0	1.5819476E 00	1.7471400E 00	1.9137761E 00	2.0816902E 00	2.2507454E 00
40.0	1.9188084E 00	2.1015632E 00	2.2850772E 00	2.4692581E 00	2.6540295E 00
50.0	2.2753401E 00	2.4746498E 00	2.6740610E 00	2.8735522E 00	3.0731070E 00
60.0	2.6741217E 00	2.8900326E 00	3.1053791E 00	3.3202146E 00	3.5345823E 00
70.0	3.1495635E 00	3.3832035E 00	3.6155689E 00	3.8467910E 00	4.0769786E 00
75.0	3.4352389E 00	3.6786528E 00	3.9204034E 00	4.1606653E 00	4.3995793E 00
80.0	3.7726309E 00	4.0268778E 00	4.2790313E 00	4.5293148E 00	4.7779093E 00
85.0	4.1920675E 00	4.4588519E 00	4.7230534E 00	4.9849489E 00	5.2447592E 00
90.0	4.7607076E 00	5.0431069E 00	5.3223249E 00	5.5986977E 00	5.8724992E 00
92.5	5.1518242E 00	5.4441808E 00	5.7329780E 00	6.0185937E 00	6.3013325E 00
95.0	5.6895199E 00	5.9946764E 00	6.2958025E 00	6.5933195E 00	6.8875671E 00
97.5	6.5804864E 00	6.9049583E 00	7.2247059E 00	7.5402121E 00	7.8518687E 00
98.0	6.8612753E 00	7.1914166E 00	7.5166242E 00	7.8374205E 00	8.1542076E 00
99.0	7.7186885E 00	8.0650784E 00	8.4059950E 00	8.7419776E 00	9.0734380E 00
99.5	8.5579089E 00	8.9188460E 00	9.2738498E 00	9.6235664E 00	9.9682106E 00
99.9	1.0456185E C1	1.0846008E 01	1.1229337E 01	1.1606503E 01	1.1977438E 01

415

TABLE C-7 (*continued*)

(NOTE. 6.07E-31 IS EQUIVALENT TO 6.07 X 10^{-31})

PER CENT\γ	3.6	3.8	4.0	4.2	4.4
0.1	3.2354432E-C1	3.7452469E-01	4.2855244E-01	4.8547497E-01	5.4514865E-01
0.5	5.2858958E-01	5.9889438E-01	6.7220659E-01	7.4834049E-01	8.2712702E-01
0.7	6.0074238E-01	6.7699502E-01	7.5619978E-01	8.3816664E-01	9.2272364E-01
1.0	6.5876450E-01	7.3955699E-01	8.2324872E-01	9.0964804E-01	9.9858204E-01
1.5	7.5194496E-01	8.3963767E-01	9.3013401E-01	1.0232418E 00	1.1187889E 00
2.0	8.2755259E-01	9.2053593E-01	1.0162386E 00	1.1144695E 00	1.2150586E 00
2.5	8.9254687E-01	9.8988446E-01	1.0898654E 00	1.1923011E 00	1.2970235E 00
3.0	9.5031885E-01	1.0513911E 00	1.1550374E 00	1.2610717E 00	1.3693284E 00
4.0	1.0512345E 00	1.1585543E 00	1.2683247E 00	1.3803650E 00	1.4945144E 00
5.0	1.1389841E 00	1.2514805E 00	1.3663186E 00	1.4833227E 00	1.6023380E 00
7.5	1.3243603E 00	1.4471184E 00	1.5719868E 00	1.6988037E 00	1.8274260E 00
10.0	1.4811965E 00	1.6120247E 00	1.7447698E 00	1.8792817E 00	2.0154290E 00
15.0	1.7503018E 00	1.8939042E 00	2.0390997E 00	2.1857612E 00	2.3337768E 00
20.0	1.9875967E 00	2.1415323E 00	2.2967870E 00	2.4532543E 00	2.6108400E 00
25.0	2.2084427E 00	2.3713439E 00	2.5353205E 00	2.7002840E 00	2.8661566E 00
30.0	2.4208299E C0	2.5918456E 00	2.7637115E 00	2.9363569E 00	3.1097192E 00
40.0	2.8393317E 00	3.0251106E 00	3.2113231E 00	3.3979324E 00	3.5849041E 00
50.0	3.2727174E 00	3.4723707E '00	3.6720615E 00	3.8717843E 00	4.0715334E 00
60.0	3.7485254E 00	3.9620748E 00	4.1752641E 00	4.3881181E 00	4.6006591E 00
70.0	4.3062314E 00	4.5346230E 00	4.7622305E 00	4.9891121E 00	5.2153195E 00
75.0	4.6372802E 00	4.8738642E 00	5.1094297E 00	5.3440557E 00	5.5778067E 00
80.0	5.0249804E 00	5.2706529E 00	5.5150478E 00	5.7582676E 00	6.0003931E 00
85.0	5.5026938E 00	5.7589013E 00	6.0135390E 00	6.2667284E 00	6.5185704E 00
90.0	6.1439911E 00	6.4133576E 00	6.6807875E 00	6.9464370E 00	7.2104183E 00
92.5	6.5814911E 00	6.8592682E 00	7.1348786E 00	7.4084946E 00	7.6802465E 00
95.0	7.1788929E 00	7.4675052E 00	7.7536630E 00	8.0375611E 00	8.3193413E 00
97.5	8.1601093E 00	8.4651448E 00	8.7672890E 00	9.0667806E 00	9.3637615E 00
98.0	8.4674107E 00	8.7772766E 00	9.0841264E 00	9.3882200E 00	9.6896919E 00
99.0	9.4009742E 00	9.7247509E 00	1.0045182E 01	1.0362505E 01	1.0676853E 01
99.5	1.0308698E 01	1.0644946E 01	1.0977605E 01	1.1306876E 01	1.1632821E 01
99.9	1.2344066E 01	1.2705508E 01	1.3062754E 01	1.3416387E 01	1.3765564E 01

PER CENT\γ	4.6	4.8	5.0	5.5	6.0
0.1	6.0743883E-01	6.7221912E-01	7.3937174E-01	9.1692643E-01	1.1071047E 00
0.5	9.0841244E-01	9.9205569E-01	1.0779283E 00	1.3016111E 00	1.5369119E 00
0.7	1.0097155E 00	1.099C001E 00	1.1904491E 00	1.4277901E 00	1.6763566E 00
1.0	1.0898952E 00	1.1834459E 00	1.2791062E 00	1.5267421E 00	1.7852847E 00
1.5	1.2166211E 00	1.3165983E 00	1.4185948E 00	1.6816918E 00	1.9551829E 00
2.0	1.3178539E 00	1.4227173E 00	1.5295258E 00	1.8043437E 00	2.0891438E 00
2.5	1.4038829E 00	1.5127442E 00	1.6234866E 00	1.9078744E 00	2.2018944E 00
3.0	1.4796606E 00	1.5919354E 00	1.7060345E 00	1.9985808E 00	2.3004521E 00
4.0	1.6106316E 00	1.7285883E 00	1.8482710E 00	2.1543736E 00	2.4692747E 00
5.0	1.7232276E 00	1.8458680E 00	1.9701498E 00	2.2874067E 00	2.6130151E 00
7.5	1.9577285E 00	2.0895982E 00	2.2229352E 00	2.5621264E 00	2.9087574E 00
10.0	2.1530969E 00	2.2921812E 00	2.4325913E 00	2.7888928E 00	3.1518984E 00
15.0	2.4830503E 00	2.6334934E 00	2.7850300E 00	3.1682177E 00	3.5569183E 00
20.0	2.7694638E 00	2.9290513E 00	3.0895401E 00	3.4943373E 00	3.9036642E 00
25.0	3.0328729E C0	3.2003714E 00	3.3686008E 00	3.7920720E 00	4.2192098E 00
30.0	3.2837474E 00	3.4583909E 00	3.6336099E 00	4.0739347E 00	4.5171390E 00
40.0	3.7722123E 00	3.9598298E 00	4.1477370E 00	4.6186435E 00	5.0909867E 00
50.0	4.2713065E 00	4.4710989E 00	4.6709096E 00	5.1705011E 00	5.6701630E 00
60.0	4.8129125E 00	5.0248932E 00	5.2366202E 00	5.7649193E 00	6.2919204E 00
70.0	5.4409055E 00	5.6659053E 00	5.8903626E 00	6.4493365E 00	7.0055531E 00
75.0	5.8107550E 00	6.0429444E 00	6.2744328E 00	6.8503502E 00	7.4227052E 00
80.0	6.2415134E 00	6.4816867E 00	6.7209831E 00	7.3157148E 00	7.9059964E 00
85.0	6.7691775E 00	7.0186161E 00	7.2669716E 00	7.8835530E 00	8.4946588E 00
90.0	7.4728816E 00	7.7339032E 00	7.9935977E 00	8.6375158E 00	9.2746798E 00
92.5	7.9503019E 00	8.2187408E 00	8.4856923E 00	9.1471189E 00	9.8009979E 00
95.0	8.5991930E 00	8.8772030E 00	9.1535279E 00	9.8375876E 00	1.0513056E 01
97.5	9.6584998E 00	9.9510543E 00	1.0241620E 01	1.0960063E 01	1.1668369E 01
98.0	9.9888175E 00	1.0285654E 01	1.0580429E 01	1.1309022E 01	1.2027013E 01
99.0	1.0988642E 01	1.1297808E 01	1.1604693E 01	1.2362570E 01	1.3108568E 01
99.5	1.1956081E 01	1.2276368E 01	1.2594225E 01	1.3378652E 01	1.4150016E 01
99.9	1.4112402E 01	1.4454964E 01	1.4794938E 01	1.5632967E 01	1.6455929E 01

PER CENT \γ	6.5	7.0	7.5	3.0	8.5
0.1	1.3086092E 00	1.5203364E 00	1.7413424E 00	1.9708140E 00	2.2080464E 00
0.5	1.7825174E 00	2.0373377E 00	2.3004580E 00	2.5711030E 00	2.8486089E 00
0.7	1.9348533E 00	2.2022113E 00	2.4775370E 00	2.7600756E 00	3.0491823E 00
1.0	2.0534580E 00	2.3302127E 00	2.6146747E 00	2.9061064E 00	3.2038803E 00
1.5	2.2378305E 00	2.5286208E 00	2.8267128E 00	3.1313983E 00	3.4420770E 00
2.0	2.3827228E 00	2.6840991E 00	2.9924585E 00	3.3071189E 00	3.6275021E 00
2.5	2.5043755E 00	2.8143633E 00	3.1310691E 00	3.4538326E 00	3.7820935E 00
3.0	2.6105067E 00	2.9278158E 00	3.2516130E 00	3.5812568E 00	3.9162049E 00
4.0	2.7918850E 00	3.1213215E 00	3.4568573E 00	3.7978841E 00	4.1438895E 00
5.0	2.9459324E 00	3.2853161E 00	3.6304726E 00	3.9808231E 00	4.3358805E 00
7.5	3.2618875E 00	3.6207577E 00	3.9847463E 00	4.3533348E 00	4.7260877E 00
10.0	3.5207531E 00	3.8947672E 00	4.2733786E 00	4.6561186E 00	5.0425941E 00
15.0	3.9504190E 00	4.3481487E 00	4.7496417E 00	5.1545107E 00	5.5624310E 00
20.0	4.3169312E 00	4.7336645E 00	5.1534806E 00	5.5760590E 00	6.0011336E 00
25.0	4.6495332E 00	5.0826579E 00	5.5182703E 00	5.9561104E 00	6.3959645E 00
30.0	4.9628425E 00	5.4107397E 00	5.8605856E 00	6.3121755E 00	6.7653399E 00
40.0	5.5645716E 00	6.0392421E 00	6.5148761E 00	6.9913707E 00	7.4686382E 00
50.0	6.1698797E 00	6.6696393E 00	7.1694322E 00	7.6692510E 00	8.1690941E 00
60.0	6.8177888E 00	7.3426500E 00	7.8666144E 00	8.3897712E 00	8.9121968E 00
70.0	7.5593646E 00	8.1110528E 00	8.6608512E 00	9.2089498E 00	9.7555171E 00
75.0	7.9919587E 00	8.5584704E 00	9.1225476E 00	9.6844350E 00	1.0244344E 01
80.0	8.4924042E 00	9.0753887E 00	9.6553365E 00	1.0232545E 01	1.0807289E 01
85.0	9.1009954E 00	9.7031254E 00	1.0301515E 01	1.0896538E 01	1.1488524E 01
90.0	9.9059781E 00	1.0532086E 01	1.1153581E 01	1.1770923E 01	1.2384529E 01
92.5	1.0448295E 01	1.1089785E 01	1.1726100E 01	1.2357771E 01	1.2985264E 01
95.0	1.1181042E 01	1.1842418E 01	1.2497918E 01	1.3148141E 01	1.3793597E 01
97.5	1.2367854E 01	1.3059525E 01	1.3744270E 01	1.4422725E 01	1.5095555E 01
98.0	1.2735805E 01	1.3436437E 01	1.4129837E 01	1.4816652E 01	1.5497598E 01
99.0	1.3844243E 01	1.4570737E 01	1.5289087E 01	1.6000054E 01	1.6704526E 01
99.5	1.4909999E 01	1.5659867E 01	1.6401025E 01	1.7133823E 01	1.7859475E 01
99.9	1.7265614E 01	1.8062641E 01	1.8849757E 01	1.9627376E 01	2.0397113E 01

PER CENT \γ	9.0	9.5	10.0	11.0	12.0
0.1	2.4524246E 00	2.7034082E 00	2.9605206E 00	3.4914847E 00	4.0424412E 00
0.5	3.1324024E 00	3.4219864E 00	3.7169226E 00	4.3213587E 00	4.9431173E 00
0.7	3.3443029E 00	3.6449553E 00	3.9507180E 00	4.5761350E 00	5.2180602E 00
1.0	3.5074561E 00	3.8163654E 00	4.1301997E 00	4.7712465E 00	5.4281815E 00
1.5	3.7582328E 00	4.0794198E 00	4.4052488E 00	5.0695062E 00	5.7487117E 00
2.0	3.9531113E 00	4.2835185E 00	4.6183496E 00	5.3000156E 00	5.9959114E 00
2.5	4.1153736E 00	4.4532587E 00	4.7953894E 00	5.4911615E 00	6.2005758E 00
3.0	4.2559932E 00	4.6002209E 00	4.9485406E 00	5.6562685E 00	6.3771362E 00
4.0	4.4944343E 00	4.8491418E 00	5.2076828E 00	5.9351506E 00	6.6749172E 00
5.0	4.6952280E 00	5.0585075E 00	5.4254066E 00	6.1690078E 00	6.9242137E 00
7.5	5.1026338E 00	5.4826530E 00	5.8658682E 00	6.6409487E 00	7.4262587E 00
10.0	5.4324691E 00	5.8254560E 00	6.2213050E 00	7.0207483E 00	7.8293430E 00
15.0	5.9731265E 00	6.3863619E 00	6.8019311E 00	7.6393782E 00	8.4842802E 00
20.0	6.4284776E 00	6.8578969E 00	7.2892214E 00	8.1570214E 00	9.0309031E 00
25.0	6.8376465E 00	7.2810003E 00	7.7258881E 00	8.6198114E 00	9.5186276E 00
30.0	7.2199330E 00	7.6758316E 00	8.1329297E 00	9.0503625E 00	9.9716172E 00
40.0	7.9466078E 00	8.4252195E 00	8.9044168E 00	9.8643986E 00	1.0826247E 01
50.0	8.6689538E 00	9.1688287E 00	9.6687176E 00	1.0668527E 01	1.1668365E 01
60.0	9.4339553E 00	9.9551048E 00	1.0475689E 01	1.1515335E 01	1.2553178E 01
70.0	1.0300684E 01	1.0844570E 01	1.1387277E 01	1.2469514E 01	1.3547990E 01
75.0	1.080245E 01	1.1358912E 01	1.1913852E 01	1.3019639E 01	1.4120585E 01
80.0	1.1379781E 01	1.1950216E 01	1.2518760E 01	1.3650738E 01	1.4776670E 01
85.0	1.2077742E 01	1.2664436E 01	1.3248803E 01	1.4411246E 01	1.5566245E 01
90.0	1.2994726E 01	1.3601808E 01	1.4206015E 01	1.5406661E 01	1.6598145E 01
92.5	1.3608924E 01	1.4229087E 01	1.4846038E 01	1.6071234E 01	1.7286167E 01
95.0	1.4414687E 01	1.5071797E 01	1.5705250E 01	1.6962263E 01	1.8207551E 01
97.5	1.5763251E 01	1.6426253E 01	1.7084891E 01	1.8390440E 01	1.9682139E 01
98.0	1.6173158E 01	1.6843826E 01	1.7509872E 01	1.8829848E 01	2.0135297E 01
99.0	1.7402804E 01	1.8095583E 01	1.8783283E 01	2.0144920E 01	2.1490142E 01
99.5	1.8578442E 01	1.9291487E 01	1.9998780E 01	2.1398395E 01	2.2779675E 01
99.9	2.1158045E 01	2.1912130E 01	2.2658903E 01	2.4136056E 01	2.5590844E 01

TABLE C-7 (*continued*)

(NOTE. 6.07E-31 IS EQUIVALENT TO 6.07 x 10^{-31})

PER CENT\γ	13.0	14.0	15.0	16.0	17.0
0.1	4.6110638E 00	5.1954403E 00	5.7934761E 00	6.4053278E 00	7.0283506E 00
0.5	5.5801195E 00	6.2306684E 00	6.8933615E 00	7.5670171E 00	8.2506377E 00
0.7	5.8744963E 00	6.5438208E 00	7.2246892E 00	7.9159749E 00	8.6167239E 00
1.0	6.0990738E 00	6.7823563E 00	7.4767294E 00	8.1811088E 00	8.8945751E 00
1.5	6.4410392E 00	7.1450050E 00	7.8593836E 00	8.5831507E 00	9.3154399E 00
2.0	6.7042936E 00	7.4237416E 00	8.1530889E 00	8.8913561E 00	9.6377208E 00
2.5	6.9219536E 00	7.6539314E 00	8.3953875E 00	9.1453834E 00	9.9031279E 00
3.0	7.1095212E 00	7.8521072E 00	8.6038142E 00	9.3637351E 00	1.0131110E 01
4.0	7.4254577E 00	8.1855372E 00	8.9541411E 00	9.7304205E 00	1.0513662E 01
5.0	7.6895794E 00	8.4639391E 00	9.2463311E 00	1.0035958E 01	1.0832143E 01
7.5	8.2205200E 00	9.0226990E 00	9.8319496E 00	1.0647562E 01	1.1468939E 01
10.0	8.6459442E 00	9.4696220E 00	1.0299619E 01	1.1135299E 01	1.1976129E 01
15.0	9.3356943E 00	1.0192867E 01	1.1055175E 01	1.1922099E 01	1.2793207E 01
20.0	9.9100994E 00	1.079398BE 01	1.1682060E 01	1.2573894E 01	1.3469138E 01
25.0	1.0421718E 01	1.1328581E 01	1.2238806E 01	1.3152057E 01	1.4068045E 01
30.0	1.0896202E 01	1.1823737C 01	1.2753882E 01	1.3686390E 01	1.4621032E 01
40.0	1.1789720E 01	1.2754630E 01	1.3720816C 01	1.4688148C 01	1.5656520E 01
50.0	1.2668234E 01	1.3668121L 01	1.4668023E 01	1.5667934E 01	1.6667863E 01
60.0	1.3589448E 01	1.4624319E 01	1.5657936E 01	1.6690443E 01	1.7721924E 01
70.0	1.4623175E 01	1.5695444E 01	1.6765128E 01	1.7832464E 01	1.8897699E 01
75.0	1.5217292E 01	1.6310262E 01	1.7399882E 01	1.8486499E 01	1.9570403E 01
80.0	1.5897320E 01	1.7013296E 01	1.8125106E 01	1.9233168E 01	2.0337850E 01
85.0	1.6714752E 01	1.7857509E 01	1.8995142E 01	2.0128173E 01	2.1257023E 01
90.0	1.7781609E 01	1.8957984E 01	2.0128054E 01	2.1292423E 01	2.2451631E 01
92.5	1.8492109E 01	1.969010BE 01	2.0881007L 01	2.2065500E 01	2.3244234E 01
95.0	1.9442629E 01	2.0668632E 01	2.1886569E 01	2.3097180E 01	2.4301234E 01
97.5	2.0961681E 01	2.2230526E 01	2.3489723E 01	2.4740289E 01	2.5983167E 01
98.0	2.1428066E 01	2.2709562E 01	2.3981039E 01	2.5243490E 01	2.6497842E 01
99.0	2.2821100E 01	2.4139324E 01	2.5446398E 01	2.6743183E 01	2.8030844E 01
99.5	2.4145321E 01	2.5497245E 01	2.6836654E 01	2.8166678C 01	2.9482640E 01
99.9	2.7028923E 01	2.8449401E 01	2.9854549E 01	3.1246408E 01	3.2626712E 01

PER CENT\γ	18.0	19.0	20.0	21.0	22.0
0.1	7.6620569E 00	8.3055943E 00	8.9582144E 00	9.6192597E 00	1.0288149E 01
0.5	8.9433638E 00	9.6444577E 00	1.0353269E 01	1.1069233E 01	1.1791848E 01
0.7	9.3261169E 00	1.0043450E 01	1.0768106E 01	1.1499544E 01	1.2237289E 01
1.0	9.6163391E 00	1.0345723E 01	1.1092133E 01	1.1825048E 01	1.2574014E 01
1.5	1.0055510E 01	1.0802721E 01	1.1556513E 01	1.2316403E 01	1.3081958E 01
2.0	1.0391474E 01	1.1152007E 01	1.1918788E 01	1.2691355E 01	1.3469300E 01
2.5	1.0667943E 01	1.1439243E 01	1.2216521E 01	1.2999333E 01	1.3787786E 01
3.0	1.0905281E 01	1.1685690E 01	1.2471847E 01	1.3263322E 01	1.4059740E 01
4.0	1.1303252E 01	1.2098665C 01	1.2899443E 01	1.3705187E 01	1.4515540E 01
5.0	1.1634307E 01	1.2441954E 01	1.3254654E 01	1.4072029E 01	1.4893742L 01
7.5	1.2295574E 01	1.3127027E 01	1.3962919E 01	1.4802916E 01	1.5646724E 01
10.0	1.2821652E 01	1.3671478E 01	1.4525265E 01	1.5382713E 01	1.6243567E 01
15.0	1.3668129E 01	1.4546540C 01	1.5428164C 01	1.6312768E 01	1.7200125E 01
20.0	1.4367484E 01	1.5268673E 01	1.6172481E 01	1.7078708E 01	1.7987179E 01
25.0	1.4986524E 01	1.5907287E 01	1.6830154E 01	1.7754958E 01	1.8681570E 01
30.0	1.5557613E 01	1.6495978E 01	1.7435974E 01	1.8377486E 01	1.9320403E 01
40.0	1.6625837E 01	1.7596011E 01	1.8566986E 01	1.9538694E 01	2.0511088E 01
50.0	1.7667794E 01	1.8667731E 01	1.9667683E 01	2.0667638E 01	2.1667592E 01
60.0	1.8752478E 01	1.9782188E 01	2.0811108E 01	2.1839308E 01	2.2866829E 01
70.0	1.9961006E 01	2.1022545E 01	2.2082456E 01	2.3140855E 01	2.4197863E 01
75.0	2.0651832E 01	2.1730978E 01	2.2808022E 01	2.3883145E 01	2.4956471E 01
80.0	2.1439427E 01	2.2538165E 01	2.3634289E 01	2.4728014E 01	2.5819494E 01
85.0	2.2382071E 01	2.3503613E 01	2.4621950E 01	2.5737329L 01	2.6849966E 01
90.0	2.3606131E 01	2.4756340E 01	2.5902573E 01	2.7045162E 01	2.8184337E 01
92.5	2.4417700E 01	2.5586345E 01	2.6750551E 01	2.7910633E 01	2.9066888E 01
95.0	2.5499299C 01	2.6691864C 01	2.7879356E 01	2.9062158E 01	3.0240542E 01
97.5	2.7218828C 01	2.8447969C 01	2.9671091E 01	3.0888593E 01	3.2100914E 01
98.0	2.7744701E 01	2.8984689E 01	3.0218336E 01	3.1446061E 01	3.2668638E 01
99.0	2.9310026E 01	3.0581402E 01	3.1845937C 01	3.3103699C 01	3.4355160E 01
99.5	3.0791157L 01	3.2091351E 01	3.3383925E 01	3.4668762E 01	3.5946929E 01
99.9	3.3995731E 01	3.5355368E 01	3.6703910E 01	3.8046123E 01	3.9379733E 01

From "Probability Plots for the Gamma Distribution," by M. B. Wilk, R. Gnanadesikan, and M. J. Huyett, *Technometrics* (1962), Vol. 4, pp. 1–20. Reproduced by permission of the editor of *Technometrics*.

TABLE C-8 PERCENTAGE POINTS OF
THE WE_1 TEST STATISTIC

	95% Range		90% Range	
n	Lower Point	Upper Point	Lower Point	Upper Point
7	0.025	0.260	0.033	0.225
8	0.025	0.230	0.032	0.200
9	0.025	0.205	0.031	0.177
10	0.025	0.184	0.030	0.159
11	0.025	0.166	0.030	0.145
12	0.025	0.153	0.029	0.134
13	0.025	0.140	0.028	0.124
14	0.024	0.128	0.027	0.115
15	0.024	0.119	0.026	0.106
16	0.023	0.113	0.025	0.098
17	0.023	0.107	0.024	0.093
18	0.022	0.101	0.024	0.087
19	0.022	0.096	0.023	0.083
20	0.021	0.090	0.023	0.077
21	0.020	0.085	0.022	0.074
22	0.020	0.080	0.022	0.069
23	0.019	0.075	0.021	0.065
24	0.019	0.069	0.021	0.062
25	0.018	0.065	0.020	0.058
26	0.018	0.062	0.020	0.056
27	0.017	0.058	0.020	0.054
28	0.017	0.056	0.019	0.052
29	0.016	0.054	0.019	0.050
30	0.016	0.053	0.019	0.048
31	0.016	0.051	0.018	0.047
32	0.015	0.050	0.018	0.045
33	0.015	0.048	0.018	0.044
34	0.014	0.046	0.017	0.043
35	0.014	0.045	0.017	0.041

From *Statistical Methods in Engineering* by Gerald J. Hahn and Samuel S. Shapiro, John Wiley & Sons, 1967. Reproduced by permission of John Wiley & Sons.

TABLE C-9 PERCENTAGE POINTS OF THE WE_2 TEST STATISTIC

n	95% Range Lower Point	95% Range Upper Point	90% Range Lower Point	90% Range Upper Point
7	0.062	0.404	0.071	0.358
8	0.054	0.342	0.062	0.301
9	0.050	0.301	0.058	0.261
10	0.049	0.261	0.056	0.231
11	0.046	0.234	0.052	0.208
12	0.044	0.215	0.050	0.191
13	0.040	0.195	0.046	0.173
14	0.038	0.178	0.043	0.159
15	0.036	0.163	0.040	0.145
16	0.034	0.150	0.038	0.134
17	0.030	0.135	0.034	0.120
18	0.028	0.123	0.031	0.109
19	0.026	0.114	0.029	0.102
20	0.025	0.106	0.028	0.095
21	0.024	0.101	0.027	0.091
22	0.023	0.094	0.026	0.084
23	0.022	0.087	0.025	0.078
24	0.021	0.082	0.024	0.074
25	0.021	0.078	0.023	0.070
26	0.020	0.073	0.022	0.066
27	0.020	0.070	0.022	0.063
28	0.019	0.067	0.021	0.061
29	0.019	0.064	0.021	0.058
30	0.018	0.060	0.020	0.054
31	0.017	0.057	0.019	0.052
32	0.017	0.055	0.019	0.050
33	0.017	0.053	0.018	0.048
34	0.017	0.051	0.018	0.047
35	0.016	0.049	0.018	0.045

From *Statistical Methods in Engineering* by Gerald J. Hahn and Samuel S. Shapiro, John Wiley & Sons, 1967. Reproduced by permission of John Wiley & Sons.

i \ n	3	4	5	6	7	8	9	10	11	12	13	14	15	16	17	18
1	0.7071	0.6872	0.6646	0.6431	0.6233	0.6052	0.5888	0.5739	0.5601	0.5475	0.5359	0.5251	0.5150	0.5056	0.4968	0.4886
2		0.1677	0.2413	0.2806	0.3031	0.3164	0.3244	0.3291	0.3315	0.3325	0.3325	0.3318	0.3306	0.3290	0.3273	0.3253
3				0.0875	0.1401	0.1743	0.1976	0.2141	0.2260	0.2347	0.2412	0.2460	0.2495	0.2521	0.2540	0.2553
4						0.0561	0.0947	0.1224	0.1429	0.1586	0.1707	0.1802	0.1878	0.1939	0.1988	0.2027
5								0.0399	0.0695	0.0922	0.1099	0.1240	0.1353	0.1447	0.1524	0.1587
6										0.0303	0.0539	0.0727	0.0880	0.1005	0.1109	0.1197
7												0.0240	0.0433	0.0593	0.0725	0.0837
8														0.0196	0.0359	0.0496
9																0.0163

i \ n	19	20	21	22	23	24	25	26	27	28	29	30	31	32	33	34
1	0.4808	0.4734	0.4643	0.4590	0.4542	0.4493	0.4450	0.4407	0.4366	0.4328	0.4291	0.4254	0.4220	0.4188	0.4156	0.4127
2	0.3232	0.3211	0.3185	0.3156	0.3126	0.3098	0.3069	0.3043	0.3018	0.2992	0.2968	0.2944	0.2921	0.2898	0.2876	0.2854
3	0.2561	0.2565	0.2578	0.2571	0.2563	0.2554	0.2543	0.2533	0.2522	0.2510	0.2499	0.2487	0.2475	0.2463	0.2451	0.2439
4	0.2059	0.2085	0.2119	0.2131	0.2139	0.2145	0.2148	0.2151	0.2152	0.2151	0.2150	0.2148	0.2145	0.2141	0.2137	0.2132
5	0.1641	0.1686	0.1736	0.1764	0.1787	0.1807	0.1822	0.1836	0.1848	0.1857	0.1864	0.1870	0.1874	0.1878	0.1880	0.1882
6	0.1271	0.1334	0.1399	0.1443	0.1480	0.1512	0.1539	0.1563	0.1584	0.1601	0.1616	0.1630	0.1641	0.1651	0.1660	0.1667
7	0.0932	0.1013	0.1092	0.1150	0.1201	0.1245	0.1283	0.1316	0.1346	0.1372	0.1395	0.1415	0.1433	0.1449	0.1463	0.1475
8	0.0612	0.0711	0.0804	0.0878	0.0941	0.0997	0.1046	0.1089	0.1128	0.1162	0.1192	0.1219	0.1243	0.1265	0.1284	0.1301
9	0.0303	0.0422	0.0530	0.0618	0.0696	0.0764	0.0823	0.0876	0.0923	0.0965	0.1002	0.1036	0.1066	0.1093	0.1118	0.1140
10		0.0140	0.0263	0.0368	0.0459	0.0539	0.0610	0.0672	0.0728	0.0778	0.0822	0.0862	0.0899	0.0931	0.0961	0.0988
11				0.0122	0.0228	0.0321	0.0403	0.0476	0.0540	0.0598	0.0650	0.0697	0.0739	0.0777	0.0812	0.0844
12						0.0107	0.0200	0.0284	0.0358	0.0424	0.0483	0.0537	0.0585	0.0629	0.0669	0.0706
13								0.0094	0.0178	0.0253	0.0320	0.0381	0.0435	0.0485	0.0530	0.0572
14										0.0084	0.0159	0.0227	0.0289	0.0344	0.0395	0.0441
15												0.0076	0.0144	0.0206	0.0262	0.0314
16														0.0068	0.0131	0.0187
17																0.0062

TABLE C-10 (*continued*)

i \ n	35	36	37	38	39	40	41	42	43	44	45	46	47	48	49	50
1	0.4096	0.4068	0.4040	0.4015	0.3989	0.3964	0.3940	0.3917	0.3894	0.3872	0.3850	0.3830	0.3808	0.3789	0.3770	0.3751
2	0.2834	0.2813	0.2794	0.2774	0.2755	0.2737	0.2719	0.2701	0.2684	0.2667	0.2651	0.2635	0.2620	0.2604	0.2589	0.2574
3	0.2427	0.2415	0.2403	0.2391	0.2380	0.2368	0.2357	0.2345	0.2334	0.2323	0.2313	0.2302	0.2291	0.2281	0.2271	0.2260
4	0.2127	0.2121	0.2116	0.2110	0.2104	0.2098	0.2091	0.2085	0.2078	0.2072	0.2065	0.2058	0.2052	0.2045	0.2038	0.2032
5	0.1883	0.1883	0.1883	0.1881	0.1880	0.1878	0.1876	0.1874	0.1871	0.1868	0.1865	0.1862	0.1859	0.1855	0.1851	0.1847
6	0.1673	0.1678	0.1683	0.1686	0.1689	0.1691	0.1693	0.1694	0.1695	0.1695	0.1695	0.1695	0.1695	0.1693	0.1692	0.1691
7	0.1487	0.1496	0.1505	0.1513	0.1520	0.1526	0.1531	0.1535	0.1539	0.1542	0.1545	0.1548	0.1550	0.1551	0.1553	0.1554
8	0.1317	0.1331	0.1344	0.1356	0.1366	0.1376	0.1384	0.1392	0.1398	0.1405	0.1410	0.1415	0.1420	0.1423	0.1427	0.1430
9	0.1160	0.1179	0.1196	0.1211	0.1225	0.1237	0.1249	0.1259	0.1269	0.1278	0.1286	0.1293	0.1300	0.1306	0.1312	0.1317
10	0.1013	0.1036	0.1056	0.1075	0.1092	0.1108	0.1123	0.1136	0.1149	0.1160	0.1170	0.1180	0.1189	0.1197	0.1205	0.1212
11	0.0873	0.0900	0.0924	0.0947	0.0967	0.0986	0.1004	0.1020	0.1035	0.1049	0.1062	0.1073	0.1085	0.1095	0.1105	0.1113
12	0.0739	0.0770	0.0798	0.0824	0.0848	0.0870	0.0891	0.0909	0.0927	0.0943	0.0959	0.0972	0.0986	0.0998	0.1010	0.1020
13	0.0610	0.0645	0.0677	0.0706	0.0733	0.0759	0.0782	0.0804	0.0824	0.0842	0.0860	0.0876	0.0892	0.0906	0.0919	0.0932
14	0.0484	0.0523	0.0559	0.0592	0.0622	0.0651	0.0677	0.0701	0.0724	0.0745	0.0765	0.0783	0.0801	0.0817	0.0832	0.0846
15	0.0361	0.0404	0.0444	0.0481	0.0515	0.0546	0.0575	0.0602	0.0628	0.0651	0.0673	0.0694	0.0713	0.0731	0.0748	0.0764
16	0.0239	0.0287	0.0331	0.0372	0.0409	0.0444	0.0476	0.0506	0.0534	0.0560	0.0584	0.0607	0.0628	0.0648	0.0667	0.0685
17	0.0119	0.0172	0.0220	0.0264	0.0305	0.0343	0.0379	0.0411	0.0442	0.0471	0.0497	0.0522	0.0546	0.0568	0.0588	0.0608
18		0.0057	0.0110	0.0158	0.0203	0.0244	0.0283	0.0318	0.0352	0.0383	0.0412	0.0439	0.0465	0.0489	0.0511	0.0532
19				0.0053	0.0101	0.0146	0.0188	0.0227	0.0263	0.0296	0.0328	0.0357	0.0385	0.0411	0.0436	0.0459
20						0.0049	0.0094	0.0136	0.0175	0.0211	0.0245	0.0277	0.0307	0.0335	0.0361	0.0386
21								0.0045	0.0087	0.0126	0.0163	0.0197	0.0229	0.0259	0.0288	0.0314
22										0.0042	0.0081	0.0118	0.0153	0.0185	0.0215	0.0244
23												0.0039	0.0076	0.0111	0.0143	0.0174
24														0.0037	0.0071	0.0104
25																0.0035

From *Statistical Methods in Engineering* by Gerald J. Hahn and Samuel S. Shapiro, John Wiley & Sons, 1967. Reproduced by permission of John Wiley & Sons.

TABLE C-11 PERCENTAGE POINTS OF THE W TEST STATISTIC FOR $n = 3(1)50$

n	1	2	5	10	50
3	0.753	0.756	0.767	0.789	0.959
4	0.687	0.707	0.748	0.792	0.935
5	0.686	0.715	0.762	0.806	0.927
6	0.713	0.743	0.788	0.826	0.927
7	0.730	0.760	0.803	0.838	0.928
8	0.749	0.778	0.818	0.851	0.932
9	0.764	0.791	0.829	0.859	0.935
10	0.781	0.806	0.342	0.869	0.938
11	0.792	0.817	0.850	0.876	0.940
12	0.805	0.828	0.859	0.883	0.943
13	0.814	0.837	0.866	0.889	0.945
14	0.325	0.846	0.874	0.895	0.947
15	0.835	0.855	0.881	0.901	0.950
16	0.844	0.863	0.887	0.906	0.952
17	0.851	0.869	0.892	0.910	0.954
18	0.858	0.874	0.897	0.914	0.956
19	0.863	0.879	0.901	0.917	0.957
20	0.868	0.884	0.905	0.920	0.959
21	0.873	0.888	0.908	0.923	0.960
22	0.878	0.892	0.911	0.926	0.961
23	0.881	0.895	0.914	0.928	0.962
24	0.884	0.898	0.916	0.930	0.963
25	0.888	0.901	0.918	0.931	0.964
26	0.891	0.904	0.920	0.933	0.965
27	0.894	0.906	0.923	0.935	0.965
28	0.896	0.908	0.924	0.936	0.966
29	0.898	0.910	0.926	0.937	0.966
30	0.900	0.912	0.927	0.939	0.967
31	0.902	0.914	0.929	0.940	0.967
32	0.904	0.915	0.930	0.941	0.968
33	0.906	0.917	0.931	0.942	0.968
34	0.908	0.919	0.933	0.943	0.969
35	0.910	0.920	0.934	0.944	0.969
36	0.912	0.922	0.935	0.945	0.970
37	0.914	0.924	0.936	0.946	0.970
38	0.916	0.925	0.938	0.947	0.971
39	0.917	0.927	0.939	0.948	0.971
40	0.919	0.928	0.940	0.949	0.972
41	0.920	0.929	0.941	0.950	0.972
42	0.922	0.930	0.942	0.951	0.972
43	0.923	0.932	0.943	0.951	0.973
44	0.924	0.933	0.944	0.952	0.973
45	0.926	0.934	0.945	0.953	0.973
46	0.927	0.935	0.945	0.953	0.974
47	0.928	0.936	0.946	0.954	0.974
48	0.929	0.937	0.947	0.954	0.974
49	0.929	0.937	0.947	0.955	0.974
50	0.930	0.938	0.947	0.955	0.974

From *Statistical Methods in Engineering* by Gerald J. Hahn and Samuel S. Shapiro, John Wiley & Sons, 1967. Reproduced by permission of John Wiley & Sons.

TABLE C-12   CONSTANTS USED IN OBTAINING
PROBABILITY OF COMPUTED W IN THE W TEST

n	γ	η	ε	n	γ	η	ε
3	−0.625	0.386	0.7500	27	−5.905	1.905	0.1980
4	−1.107	0.714	0.6297	28	−5.988	1.919	0.1943
5	−1.530	0.935	0.5521	29	−6.074	1.934	0.1907
6	−2.010	1.138	0.4963	30	−6.160	1.949	0.1872
7	−2.356	1.245	0.4533	31	−6.248	1.965	0.1840
8	−2.696	1.333	0.4186	32	−6.324	1.976	0.1811
9	−2.968	1.400	0.3900	33	−6.402	1.988	0.1781
10	−3.262	1.471	0.3660	34	−6.480	2.000	0.1755
11	−3.485	1.515	0.3451	35	−6.559	2.012	0.1727
12	−3.731	1.571	0.3270	36	−6.640	2.024	0.1702
13	−3.936	1.613	0.3111	37	−6.721	2.037	0.1677
14	−4.155	1.655	0.2969	38	−6.803	2.049	0.1656
15	−4.373	1.695	0.2842	39	−6.887	2.062	0.1633
16	−4.567	1.724	0.2727	40	−6.961	2.075	0.1612
17	−4.713	1.739	0.2622	41	−7.035	2.088	0.1591
18	−4.885	1.770	0.2528	42	−7.111	2.101	0.1572
19	−5.018	1.786	0.2440	43	−7.188	2.114	0.1552
20	−5.153	1.802	0.2359	44	−7.266	2.128	0.1534
21	−5.291	1.818	0.2264	45	−7.345	2.141	0.1516
22	−5.413	1.835	0.2207	46	−7.414	2.155	0.1499
23	−5.508	1.848	0.2157	47	−7.484	2.169	0.1482
24	−5.605	1.862	0.2106	48	−7.555	2.183	0.1466
25	−5.704	1.876	0.2063	49	−7.615	2.198	0.1451
26	−5.803	1.890	0.2020	50	−7.677	2.212	0.1436

From *Statistical Methods in Engineering* by Gerald J. Hahn and Samuel S. Shapiro, John Wiley & Sons, 1967. Reproduced by permission of John Wiley & Sons.

TABLE C-13 PERCENTAGE POINTS OF THE
t-DISTRIBUTION

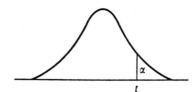

d.f.	$t_{.100}$	$t_{.050}$	$t_{.025}$	$t_{.010}$	$t_{.005}$	d.f.
1	3.078	6.314	12.706	31.821	63.657	1
2	1.886	2.920	4.303	6.965	9.925	2
3	1.638	2.353	3.182	4.541	5.841	3
4	1.533	2.132	2.776	3.747	4.604	4
5	1.476	2.015	2.571	3.365	4.032	5
6	1.440	1.943	2.447	3.143	3.707	6
7	1.415	1.895	2.365	2.998	3.499	7
8	1.397	1.860	2.306	2.896	3.355	8
9	1.383	1.833	2.262	2.821	3.250	9
10	1.372	1.812	2.228	2.764	3.169	10
11	1.363	1.796	2.201	2.718	3.106	11
12	1.356	1.782	2.179	2.681	3.055	12
13	1.350	1.771	2.160	2.650	3.012	13
14	1.345	1.761	2.145	2.624	2.977	14
15	1.341	1.753	2.131	2.602	2.947	15
16	1.337	1.746	2.120	2.583	2.921	16
17	1.333	1.740	2.110	2.567	2.898	17
18	1.330	1.734	2.101	2.552	2.878	18
19	1.328	1.729	2.093	2.539	2.861	19
20	1.325	1.725	2.086	2.528	2.845	20
21	1.323	1.721	2.080	2.518	2.831	21
22	1.321	1.717	2.074	2.508	2.819	22
23	1.319	1.714	2.069	2.500	2.807	23
24	1.318	1.711	2.064	2.492	2.797	24
25	1.316	1.708	2.060	2.485	2.787	25
26	1.315	1.706	2.056	2.479	2.779	26
27	1.314	1.703	2.052	2.473	2.771	27
28	1.313	1.701	2.048	2.467	2.763	28
29	1.311	1.699	2.045	2.462	2.756	29
∞	1.282	1.645	1.960	2.326	2.576	∞

From "Table of Percentage Points of the t-Distribution" by Maxine Merrington, *Biometrika*, Vol. 32, p. 300 (1941). Reproduced by permission of the editor of *Biometrika*.

TABLE C-14   COEFFICIENTS (a_i AND b_i) OF THE BEST ESTIMATES OF THE MEAN (μ) AND STANDARD DEVIATION (σ) IN CENSORED SAMPLES UP TO $n = 20$ FROM A NORMAL POPULATION

$n = 2$

$n-r$		$t_{(1)}$	$t_{(2)}$	$t_{(3)}$
0	μ	.5000	.5000	
	σ	−.8862	.8862	

$n = 3$

$n-r$		$t_{(1)}$	$t_{(2)}$	$t_{(3)}$
0	μ	.3333	.3333	.3333
	σ	−.5908	.0000	.5908
1	μ	.0000	1.0000	
	σ	−1.1816	1.1816	

$n = 4$

$n-r$		$t_{(1)}$	$t_{(2)}$	$t_{(3)}$	$t_{(4)}$
0	μ	.2500	.2500	.2500	.2500
	σ	−.4539	−.1102	.1102	.4539
1	μ	.1161	.2408	.6431	
	σ	−.6971	−.1268	.8239	
2	μ	−.4056	1.4056		
	σ	−1.3654	1.3654		

$n = 5$

$n-r$		$t_{(1)}$	$t_{(2)}$	$t_{(3)}$	$t_{(4)}$	$t_{(5)}$
0	μ	.2000	.2000	.2000	.2000	.2000
	σ	−.3724	−.1352	.0000	.1352	.3724
1	μ	.1252	.1830	.2147	.4771	
	σ	−.5117	−.1668	.0274	.6511	
2	μ	−.0638	.1498	.9139		
	σ	−.7696	−.2121	.9817		
3	μ	−.7411	1.7411			
	σ	−1.4971	1.4971			

$n = 6$

$n-r$		$t_{(1)}$	$t_{(2)}$	$t_{(3)}$	$t_{(4)}$	$t_{(5)}$	$t_{(6)}$
0	μ	.1667	.1667	.1667	.1667	.1667	.1667
	σ	−.3175	−.1386	−.0432	.0432	.1386	.3175
1	μ	.1183	.1510	.1680	.1828	.3799	
	σ	−.4097	−.1685	−.0406	.0740	.5448	
2	μ	.0185	.1226	.1761	.6828		
	σ	−.5528	−.2091	−.0290	.7909		
3	μ	−.2159	.0649	1.1511			
	σ	−.8244	−.2760	1.1004			
4	μ	−1.0261	2.0261				
	σ	−1.5988	1.5988				

$n = 7$

$n-r$		$t_{(1)}$	$t_{(2)}$	$t_{(3)}$	$t_{(4)}$	$t_{(5)}$	$t_{(6)}$	$t_{(7)}$
0	μ	.1429	.1429	.1429	.1429	.1429	.1429	.1429
	σ	−.2778	−.1351	−.0625	.0000	.0625	.1351	.2778
1	μ	.1088	.1295	.1400	.1487	.1571	.3159	
	σ	−.3440	−.1610	−.0681	.0114	.0901	.4716	
2	μ	.0465	.1072	.1375	.1626	.5462		
	σ	−.4370	−.1943	−.0718	.0321	.6709		
3	μ	−.0738	.0677	.1375	.8686			
	σ	−.5848	−.2428	−.0717	.8994			
4	μ	−.3474	−.0135	1.3609				
	σ	−.8682	−.3269	1.1951				
5	μ	−1.2733	2.2733					
	σ	−1.6812	1.6812					

n = 8

n − r		$t_{(1)}$	$t_{(2)}$	$t_{(3)}$	$t_{(4)}$	$t_{(5)}$	$t_{(6)}$	$t_{(7)}$	$t_{(8)}$
0	μ	.1250	.1250	.1250	.1250	.1250	.1250	.1250	.1250
	σ	−.2476	−.1294	−.0713	−.0230	.0230	.0713	.1294	.2476
1	μ	.0997	.1139	.1208	.1265	.1318	.1370	.2704	
	σ	−.2978	−.1515	−.0796	−.0200	.0364	.0951	.4175	
2	μ	.0569	.0962	.1153	.1309	.1451	.4555		
	σ	−.3638	−.1788	−.0881	−.0132	.0570	.5868		
3	μ	−.0167	.0677	.1084	.1413	.6993			
	σ	−.4586	−.2156	−.0970	.0002	.7709			
4	μ	−.1549	.0176	.1001	1.0372				
	σ	−.6110	−.2707	−.1061	.9878				
5	μ	−.4632	−.0855	1.5487					
	σ	−.9045	−.3690	1.2735					
6	μ	−1.4915	2.4915						
	σ	−1.7502	1.7502						

n = 9

n − r		$t_{(1)}$	$t_{(2)}$	$t_{(3)}$	$t_{(4)}$	$t_{(5)}$	$t_{(6)}$	$t_{(7)}$	$t_{(8)}$	$t_{(9)}$
0	μ	.1111	.1111	.1111	.1111	.1111	.1111	.1111	.1111	.1111
	σ	−.2237	−.1233	−.0751	−.0360	.0000	.0360	.0751	.1233	.2237
1	μ	.0915	.1018	.1067	.1106	.1142	.1177	.1212	.2365	
	σ	−.2633	−.1421	−.0841	−.0370	.0062	.0492	.0954	.3757	
2	μ	.0602	.0876	.1006	.1110	.1204	.1294	.3909		
	σ	−.3129	−.1647	−.0938	−.0364	.0160	.0678	.5239		
3	μ	.0104	.0660	.0923	.1133	.1320	.5860			
	σ	−.3797	−.1936	−.1048	−.0333	.0317	.6797			
4	μ	−.0731	.0316	.0809	.1199	.8408				
	σ	−.4766	−.2335	−.1181	−.0256	.8537				
5	μ	−.2272	−.0284	.0644	1.1912					
	σ	−.6330	−.2944	−.1348	1.0622					
6	μ	−.5664	−.1521	1.7185						
	σ	−.9355	−.4047	1.3402						
7	μ	−1.6868	2.6868							
	σ	−1.8092	1.8092							

n = 10

n − r		$t_{(1)}$	$t_{(2)}$	$t_{(3)}$	$t_{(4)}$	$t_{(5)}$	$t_{(6)}$	$t_{(7)}$	$t_{(8)}$	$t_{(9)}$	$t_{(10)}$
0	μ	.1000	.1000	.1000	.1000	.1000	.1000	.1000	.1000	.1000	.1000
	σ	−.2044	−.1172	−.0763	−.0436	−.0142	.0142	.0436	.0763	.1172	.2044
1	μ	.0843	.0921	.0957	.0986	.1011	.1036	.1060	.1085	.2101	
	σ	−.2364	−.1334	−.0851	−.0465	−.0119	.0215	.0559	.0937	.3423	
2	μ	.0605	.0804	.0898	.0972	.1037	.1099	.1161	.3424		
	σ	−.2753	−.1523	−.0947	−.0488	−.0077	.0319	.0722	.4746		
3	μ	.0244	.0636	.0818	.0962	.1089	.1207	.5045			
	σ	−.3252	−.1758	−.1058	−.0502	−.0006	.0469	.6107			
4	μ	−.0316	.0383	.0707	.0962	.1185	.7078				
	σ	−.3930	−.2063	−.1192	−.0501	.0111	.7576				
5	μ	−.1240	−.0016	.0549	.0990	.9718					
	σ	−.4919	−.2491	−.1362	−.0472	.9243					
6	μ	−.2923	−.0709	.0305	1.3327						
	σ	−.6520	−.3150	−.1593	1.1263						
7	μ	−.6596	−.2138	1.8734							
	σ	−.9625	−.4357	1.3981							
8	μ	−1.8634	2.8634								
	σ	−1.8608	1.8608								

TABLE C-14 (*continued*)

$n = 11$

$n - r$		$t_{(1)}$	$t_{(2)}$	$t_{(3)}$	$t_{(4)}$	$t_{(5)}$	$t_{(6)}$	$t_{(7)}$	$t_{(8)}$	$t_{(9)}$	$t_{(10)}$	$t_{(11)}$
0	μ	.0909	.0909	.0909	.0909	.0909	.0909	.0909	.0909	.0909	.0909	.0909
	σ	−.1883	−.1115	−.0760	−.0481	−.0234	.0000	.0234	.0481	.0760	.1115	.1883
1	μ	.0781	.0841	.0869	.0891	.0910	.0928	.0945	.0963	.0982	.1891	
	σ	−.2149	−.1256	−.0843	−.0519	−.0233	.0038	.0309	.0593	.0911	.3149	
2	μ	.0592	.0744	.0814	.0869	.0917	.0962	.1005	.1049	.3047		
	σ	−.2463	−.1417	−.0934	−.0555	−.0220	.0095	.0409	.0736	.4349		
3	μ	.0320	.0609	.0741	.0845	.0935	.1020	.1101	.4430			
	σ	−.2852	−.1610	−.1038	−.0589	−.0194	.0178	.0545	.5562			
4	μ	−.0082	.0415	.0642	.0820	.0974	.1116	.6116				
	σ	−.3357	−.1854	−.1163	−.0621	−.0146	.0299	.6842				
5	μ	−.0698	.0128	.0504	.0797	.1049	.8220					
	σ	−.4045	−.2175	−.1317	−.0647	−.0061	.8246					
6	μ	−.1702	−.0323	.0303	.0786	1.0937						
	σ	−.5053	−.2627	−.1519	−.0657	.9857						
7	μ	−.3516	−.1104	−.0016	1.4636							
	σ	−.6687	−.3331	−.1807	1.1825							
8	μ	−.7445	−.2712	2.0157								
	σ	−.9862	−.4630	1.4402								
9	μ	−2.0245	3.0245									
	σ	−1.9065	1.9065									

$n = 12$

$n - r$		$t_{(1)}$	$t_{(2)}$	$t_{(3)}$	$t_{(4)}$	$t_{(5)}$	$t_{(6)}$	$t_{(7)}$	$t_{(8)}$	$t_{(9)}$	$t_{(10)}$	$t_{(11)}$	$t_{(12)}$
0	μ	.0833	.0833	.0833	.0833	.0833	.0833	.0833	.0833	.0833	.0833	.0833	.0833
	σ	−.1748	−.1061	−.0749	−.0506	−.0294	−.0097	.0097	.0294	.0506	.0749	.1061	.1748
1	μ	.0726	.0775	.0796	.0813	.0828	.0842	.0855	.0868	.0882	.0896	.1719	
	σ	−.1972	−.1185	−.0827	−.0548	−.0305	−.0079	.0142	.0367	.0608	.0881	.2919	
2	μ	.0574	.0693	.0747	.0789	.0825	.0859	.0891	.0923	.0956	.2745		
	σ	−.2232	−.1324	−.0911	−.0590	−.0310	−.0050	.0203	.0461	.0733	.4020		
3	μ	.0360	.0581	.0682	.0759	.0827	.0888	.0948	.1006	.3950			
	σ	−.2545	−.1487	−.1007	−.0633	−.0308	−.0007	.0286	.0582	.5119			
4	μ	.0057	.0428	.0595	.0724	.0836	.0938	.1036	.5386				
	σ	−.2937	−.1686	−.1119	−.0678	−.0296	.0058	.0400	.6259				
5	u	−.0382	.0210	.0477	.0684	.0861	.1022	.7128					
	σ	−.3448	−.1939	−.1255	−.0726	−.0267	.0155	.7479					
6	μ	−.1048	−.0109	.0313	.0637	.0915	.9292						
	σ	−.4146	−.2274	−.1428	−.0774	−.0210	.8833						
7	μ	−.2125	−.0609	.0070	.0589	1.2075							
	σ	−.5171	−.2749	−.1659	−.0820	1.0399							
8	μ	−.4059	−.1472	−.0321	1.5852								
	σ	−.6836	−.3493	−.1996	1.2324								
9	μ	−.8225	−.3249	2.1474									
	σ	−1.0075	−.4874	1.4948									
10	μ	−2.1728	3.1728										
	σ	−1.9474	1.9474										

$n = 13$

$n - r$		$t_{(1)}$	$t_{(2)}$	$t_{(3)}$	$t_{(4)}$	$t_{(5)}$	$t_{(6)}$	$t_{(7)}$	$t_{(8)}$	$t_{(9)}$	$t_{(10)}$	$t_{(11)}$	$t_{(12)}$	$t_{(13)}$
0	μ	.0769	.0769	.0769	.0769	.0769	.0769	.0769	.0769	.0769	.0769	.0769	.0769	.0769
	σ	-.1632	-.1013	-.0735	-.0520	-.0335	-.0164	.0000	.0164	.0335	.0520	.0735	.1013	.1632
1	μ	-.0679	.0718	.0735	.0749	.0761	.0771	.0781	.0792	.0802	.0813	.0824	.1576	
	σ	-.1824	-.1122	-.0806	-.0563	-.0353	-.0160	.0026	.0212	.0404	.0612	.0850	.2724	
2	μ	.0552	.0648	.0691	.0724	.0752	.0778	.0803	.0827	.0852	.0877	.2497		
	σ	-.2043	-.1243	-.0884	-.0607	-.0368	-.0148	.0063	.0273	.0490	.0723	.3743		
3	μ	-.0380	.0555	.0633	.0693	.0745	.0792	.0836	.0880	.0924	.3564			
	σ	-.2301	-.1382	-.0970	-.0653	-.0379	-.0128	.0113	.0352	.0598	.4750			
4	μ	.0144	.0430	.0557	.0655	.0739	.0816	.0888	.0958	.4813				
	σ	-.2616	-.1549	-.1071	-.0703	-.0386	-.0095	.0182	.0456	.5781				
5	μ	-.0185	.0259	.0457	.0610	.0740	.0857	.0968	.6294					
	σ	-.3011	-.1754	-.1191	-.0758	-.0386	-.0046	.0278	.6867					
6	μ	-.0659	.0020	.0322	.0553	.0750	.0928	.8085						
	σ	-.3528	-.2015	-.1339	-.0819	-.0374	.0032	.8042						
7	μ	-.1371	-.0330	.0132	.0484	.0784	1.0101							
	σ	-.4236	-.2363	-.1528	-.0888	-.0341	.9355							
8	μ	-.2516	-.0876	-.0151	.0400	1.3143								
	σ	-.5276	-.2859	-.1785	-.0964	1.0884								
9	μ	-.4561	-.1817	-.0610	1.6988									
	σ	-.6969	-.3638	-.2165	1.2773									
10	μ	-.8946	-.3753	2.2699										
	σ	-1.0266	-.5004	1.5360										
11	μ	-2.3101	3.3101											
	σ	-1.9845	1.9845											

TABLE C-14 (*continued*)

n = 14

n − r		$t_{(1)}$	$t_{(2)}$	$t_{(3)}$	$t_{(4)}$	$t_{(5)}$	$t_{(6)}$	$t_{(7)}$	$t_{(8)}$	$t_{(9)}$	$t_{(10)}$	$t_{(11)}$	$t_{(12)}$	$t_{(13)}$	$t_{(14)}$
0	μ	.0714	.0714	.0714	.0714	.0714	.0714	.0714	.0714	.0714	.0714	.0714	.0714	.0714	.0714
	σ	−.1532	−.0968	−.0717	−.0526	−.0362	−.0212	−.0070	.0070	.0212	.0362	.0526	.0717	.0968	.1532
1	μ	.0637	.0669	.0683	.0694	.0704	.0712	.0721	.0728	.0736	.0745	.0753	.0762	.1455	
	σ	−.1698	−.1065	−.0784	−.0568	−.0384	−.0216	−.0056	.0100	.0259	.0426	.0609	.0820	.2556	
2	μ	.0530	.0609	.0643	.0670	.0692	.0713	.0732	.0751	.0770	.0789	.0809	.2291		
	σ	−.1885	−.1171	−.0854	−.0612	−.0404	−.0215	−.0036	.0140	.0319	.0505	.0707	.3506		
3	μ	.0388	.0529	.0592	.0639	.0680	.0717	.0752	.0785	.0819	.0852	.3247			
	σ	−.2102	−.1292	−.0933	−.0658	−.0423	−.0209	−.0006	.0192	.0393	.0601	.4438			
4	μ	.0199	.0426	.0526	.0602	.0667	.0726	.0782	.0835	.0887	.4350				
	σ	−.2361	−.1434	−.1023	−.0709	−.0440	−.0196	.0035	.0260	.0487	.5382				
5	μ	−.0057	.0288	.0440	.0557	.0655	.0744	.0828	.0908	.5637					
	σ	−.2678	−.1604	−.1129	−.0765	−.0455	−.0174	.0092	.0350	.6363					
6	μ	−.0411	.0102	.0328	.0500	.0646	.0777	.0899	.7159						
	σ	−.3077	−.1815	−.1256	−.0829	−.0466	−.0137	.0172	.7407						
7	μ	−.0915	−.0158	.0175	.0429	.0643	.0835	.8992							
	σ	−.3599	−.2084	−.1414	−.0903	−.0469	−.0077	.8546							
8	μ	−.1670	−.0537	−.0040	.0338	.0655	1.1255								
	σ	−.4317	−.2444	−.1618	−.0990	−.0457	.9825								
9	μ	−.2879	−.1127	−.0360	.0218	1.4148									
	σ	−.5372	−.2959	−.1898	−.1094	1.1322									
10	μ	−.5027	−.2142	−.0886	1.8054										
	σ	−.7091	−.3771	−.2318	1.3180										
11	μ	−.9616	−.4228	2.3843											
	σ	−1.0441	−.5293	1.5734											
12	μ	−2.4378	3.4378												
	σ	−2.0182	2.0182												

$n = 15$

$n - r$		$t_{(1)}$	$t_{(2)}$	$t_{(3)}$	$t_{(4)}$	$t_{(5)}$	$t_{(6)}$	$t_{(7)}$	$t_{(8)}$	$t_{(9)}$	$t_{(10)}$	$t_{(11)}$	$t_{(12)}$	$t_{(13)}$	$t_{(14)}$	$t_{(15)}$
0	μ	.0667	.0667	.0667	.0667	.0667	.0667	.0667	.0667	.0667	.0667	.0667	.0667	.0667	.0667	.0667
	σ	-.1444	-.0927	-.0699	-.0526	-.0379	-.0247	-.0122	.0000	.0122	.0247	.0379	.0526	.0699	.0927	.1444
1	μ	.0599	.0627	.0639	.0648	.0655	.0662	.0669	.0675	.0682	.0688	.0695	.0702	.0709	.1351	
	σ	-.1590	-.1013	-.0760	-.0568	-.0404	-.0256	-.0116	.0019	.0154	.0293	.0440	.0602	.0791	.2409	
2	μ	.0508	.0574	.0602	.0624	.0642	.0659	.0675	.0690	.0704	.0719	.0735	.0751	.2116		
	σ	-.1752	-.1108	-.0825	-.0610	-.0427	-.0262	-.0106	.0044	.0195	.0349	.0512	.0690	.3300		
3	μ	.0390	.0506	.0556	.0595	.0628	.0657	.0685	.0711	.0737	.0763	.0790	.2982			
	σ	-.1937	-.1214	-.0897	-.0655	-.0450	-.0265	-.0091	.0078	.0246	.0417	.0598	.4169			
4	μ	.0234	.0418	.0498	.0560	.0611	.0658	.0701	.0743	.0784	.0824	.3969				
	σ	-.2154	-.1336	-.0977	-.0705	-.0473	-.0264	-.0068	.0122	.0310	.0502	.5042				
5	μ	.0030	.0305	.0425	.0516	.0593	.0663	.0727	.0789	.0849	.5104					
	σ	-.2414	-.1481	-.1071	-.0760	-.0496	-.0258	-.0035	.0180	.0393	.5940					
6	μ	-.0244	.0155	.0330	.0462	.0574	.0674	.0767	.0856	.6425						
	σ	-.2733	-.1654	-.1181	-.0822	-.0518	-.0244	.0012	.0258	.6882						
7	μ	-.0621	-.0046	.0205	.0395	.0555	.0698	.0830	.7983							
	σ	-.3136	-.1870	-.1315	-.0894	-.0538	-.0219	.0079	.7892							
8	μ	-.1155	-.0326	.0036	.0309	.0539	.0743	.9854								
	σ	-.3664	-.2146	-.1482	-.0979	-.0555	-.0174	.9001								
9	μ	-.1950	-.0732	-.0203	.0196	.0531	1.2157									
	σ	-.4390	-.2518	-.1700	-.1082	-.0562	1.0252									
10	μ	-.3217	-.1364	-.0560	.0043	1.5097										
	σ	-.5459	-.3050	-.2002	-.1211	1.1722										
11	μ	-.5462	-.2448	-.1148	1.9058											
	σ	-.7201	-.3892	-.2458	1.3552											
12	μ	-1.0242	-.4676	2.4918												
	σ	-1.0601	-.5477	1.6077												
13	μ	-2.5574	3.5574													
	σ	-2.0493	2.0493													

TABLE C-14 (continued)

n = 16

n − r		t(1)	t(2)	t(3)	t(4)	t(5)	t(6)	t(7)	t(8)	t(9)	t(10)	t(11)	t(12)	t(13)	t(14)	t(15)	t(16)
0	μ	.0625	.0625	.0625	.0625	.0625	.0625	.0625	.0625	.0625	.0625	.0625	.0625	.0625	.0625	.0625	.0625
	σ	−.1366	−.0889	−.0681	−.0524	−.0391	−.0272	−.0160	−.0053	.0053	.0160	.0272	.0391	.0524	.0681	.0889	.1366
1	μ	.0566	.0589	.0599	.0607	.0613	.0619	.0625	.0630	.0635	.0640	.0645	.0651	.0657	.0663	.1261	
	σ	−.1495	−.0967	−.0737	−.0563	−.0416	−.0284	−.0161	−.0042	.0075	.0193	.0316	.0447	.0593	.0763	.2279	
2	μ	.0487	.0543	.0566	.0585	.0600	.0614	.0626	.0638	.0650	.0662	.0674	.0687	.0700	.1967		
	σ	−.1637	−.1051	−.0797	−.0604	−.0441	−.0294	−.0158	−.0027	.0103	.0233	.0369	.0513	.0671	.3120		
3	μ	.0386	.0483	.0525	.0557	.0584	.0608	.0630	.0652	.0673	.0693	.0714	.0736	.2757			
	σ	−.1797	−.1145	−.0862	−.0647	−.0466	−.0303	−.0151	−.0006	.0138	.0282	.0432	.0590	.3935			
4	μ	.0257	.0408	.0474	.0524	.0566	.0604	.0638	.0671	.0704	.0736	.0768	.3649				
	σ	−.1982	−.1252	−.0935	−.0694	−.0491	−.0310	−.0140	.0022	.0182	.0343	.0508	.4748				
5	μ	.0090	.0313	.0410	.0484	.0545	.0601	.0652	.0700	.0747	.0794	.4664					
	σ	−.2200	−.1376	−.1018	−.0747	−.0518	−.0313	−.0123	.0060	.0239	.0419	.5577					
6	μ	−.0129	.0191	.0329	.0434	.0522	.0601	.0673	.0742	.0808	.5829						
	σ	−.2461	−.1523	−.1114	−.0806	−.0546	−.0313	−.0097	.0110	.0312	.6439						
7	μ	−.0420	.0030	.0225	.0373	.0496	.0606	.0707	.0803	.7180							
	σ	−.2782	−.1700	−.1229	−.0874	−.0575	−.0307	−.0059	.0177	.7350							
8	μ	−.0817	−.0185	.0089	.0295	.0467	.0621	.0762	.8769								
	σ	−.3189	−.1920	−.1369	−.0954	−.0604	−.0292	−.0004	.8333								
9	μ	−.1379	−.0484	−.0096	.0194	.0438	.0653	1.0674									
	σ	−.3723	−.2204	−.1545	−.1049	−.0632	−.0262	.9415									
10	μ	−.2211	−.0916	−.0358	.0061	.0410	1.3015										
	σ	−.4457	−.2586	−.1776	−.1167	−.0657	1.0642										
11	μ	−.3534	−.1587	−.0750	−.0125	1.5996											
	σ	−.5538	−.3134	−.2097	−.1319	1.2088											
12	μ	−.5869	−.2739	−.1398	2.0006												
	σ	−.7303	−.4004	−.2586	1.3894												
13	μ	−1.0829	−.5101	2.5931													
	σ	−1.0748	−.5645	1.6394													
14	μ	−2.6696	3.6696														
	σ	−2.0779	2.0779														

$n = 17$

$n-r$		$t_{(1)}$	$t_{(2)}$	$t_{(3)}$	$t_{(4)}$	$t_{(5)}$	$t_{(6)}$	$t_{(7)}$	$t_{(8)}$	$t_{(9)}$	$t_{(10)}$	$t_{(11)}$	$t_{(12)}$	$t_{(13)}$	$t_{(14)}$	$t_{(15)}$	$t_{(16)}$	$t_{(17)}$
0	μ	.0588	.0588	.0588	.0588	.0588	.0588	.0588	.0588	.0588	.0588	.0588	.0588	.0588	.0588	.0588	.0588	.0588
	σ	−.1297	−.0854	−.0663	−.0519	−.0398	−.0290	−.0189	−.0094	.0000	.0094	.0189	.0290	.0398	.0519	.0663	.0854	.1297
1	μ	.0536	.0556	.0565	.0571	.0577	.0582	.0586	.0590	.0595	.0599	.0603	.0607	.0612	.0617	.0622	.1183	
	σ	−.1412	−.0925	−.0715	−.0556	−.0423	−.0304	−.0194	−.0089	.0014	.0117	.0222	.0332	.0450	.0582	.0737	.2164	
2	μ	.0468	.0515	.0535	.0550	.0563	.0574	.0585	.0595	.0605	.0615	.0624	.0634	.0645	.0656	.1837		
	σ	−.1537	−.1001	−.0769	−.0595	−.0448	−.0317	−.0196	−.0080	.0033	.0146	.0261	.0381	.0510	.0653	.2960		
3	μ	.0381	.0463	.0498	.0525	.0547	.0567	.0585	.0603	.0620	.0636	.0653	.0670	.0688	.2564			
	σ	−.1677	−.1085	−.0829	−.0636	−.0474	−.0329	−.0196	−.0068	.0057	.0181	.0308	.0439	.0580	.3728			
4	μ	.0271	.0398	.0453	.0494	.0528	.0559	.0588	.0615	.0641	.0666	.0692	.0718	.3378				
	σ	−.1837	−.1179	−.0895	−.0681	−.0501	−.0341	−.0192	−.0051	.0087	.0225	.0364	.0509	.4491				
5	μ	.0131	.0317	.0397	.0457	.0507	.0552	.0593	.0632	.0670	.0707	.0744	.4294					
	σ	−.2022	−.1287	−.0969	−.0730	−.0529	−.0351	−.0185	−.0028	.0126	.0278	.0433	.5263					
6	μ	−.0047	.0214	.0327	.0411	.0482	.0540	.0603	.0658	.0710	.0762	.5334						
	σ	−.2241	−.1412	−.1055	−.0786	−.0560	−.0359	−.0173	.0004	.0176	.0346	.6059						
7	μ	−.0278	.0083	.0239	.0356	.0454	.0540	.0620	.0695	.0767	.6525							
	σ	−.2504	−.1561	−.1154	−.0849	−.0592	−.0364	−.0154	.0046	.0240	.6892							
8	μ	−.0585	−.0088	.0126	.0286	.0421	.0539	.0648	.0750	.7903								
	σ	−.2828	−.1742	−.1274	−.0922	−.0627	−.0365	−.0124	.0105	.7777								
9	μ	−.1002	−.0317	−.0022	.0199	.0383	.0545	.0694	.9521									
	σ	−.3238	−.1967	−.1419	−.1009	−.0665	−.0359	−.0079	.8736									
10	μ	−.1589	−.0633	−.0223	.0084	.0340	.0565	1.1456										
	σ	−.3777	−.2257	−.1603	−.1113	−.0704	−.0341	.9796										
11	μ	−.2457	−.1091	−.0506	−.0070	.0293	1.3831											
	σ	−.4519	−.2649	−.1846	−.1245	−.0744	1.1002											
12	μ	−.3832	−.1799	−.0932	−.0287	1.6850												
	σ	−.5612	−.3212	−.2185	−.1418	1.3427												
13	μ	−.6253	−.3014	−.1637	2.0904													
	σ	−.7398	−.4108	−.2705	1.4211													
14	μ	−1.1383	−.5506	2.6888														
	σ	−1.0885	−.5802	1.6687														
15	μ	−2.7754	3.7754															
	σ	−2.1046	2.1046															

TABLE C-14 (*continued*)

$n = 18$

$n-r$		$t_{(1)}$	$t_{(2)}$	$t_{(3)}$	$t_{(4)}$	$t_{(5)}$	$t_{(6)}$	$t_{(7)}$	$t_{(8)}$	$t_{(9)}$	$t_{(10)}$	$t_{(11)}$	$t_{(12)}$	$t_{(13)}$	$t_{(14)}$	$t_{(15)}$	$t_{(16)}$	$t_{(17)}$	$t_{(18)}$
0	μ	.0556	.0556	.0556	.0556	.0556	.0556	.0556	.0556	.0556	.0556	.0556	.0556	.0556	.0556	.0556	.0556	.0556	.0556
	σ	−.1235	−.0822	−.0645	−.0512	−.0401	−.0302	−.0211	−.0125	−.0041	.0041	.0125	.0211	.0302	.0401	.0512	.0645	.0822	.1235
1	μ	.0509	.0526	.0534	.0540	.0544	.0548	.0552	.0556	.0559	.0563	.0566	.0570	.0574	.0577	.0582	.0586	.1113	
	σ	−.1338	−.0887	−.0693	−.0548	−.0426	−.0318	−.0219	−.0124	−.0033	.0058	.0149	.0243	.0342	.0450	.0570	.0712	.2061	
2	μ	.0449	.0489	.0507	.0520	.0531	.0540	.0549	.0558	.0566	.0574	.0582	.0590	.0598	.0607	.0616	.1723		
	σ	−.1449	−.0955	−.0743	−.0584	−.0451	−.0333	−.0224	−.0121	−.0021	.0078	.0178	.0281	.0389	.0505	.0634	.2818		
3	μ	.0373	.0443	.0474	.0496	.0515	.0532	.0547	.0562	.0576	.0589	.0603	.0617	.0631	.0646	.2396			
	σ	−.1573	−.1030	−.0798	−.0623	−.0477	−.0347	−.0228	−.0115	−.0005	.0103	.0213	.0325	.0442	.0568	.3545			
4	μ	.0279	.0387	.0433	.0468	.0497	.0522	.0546	.0568	.0589	.0610	.0631	.0652	.0674	.3144				
	σ	−.1713	−.1114	−.0858	−.0665	−.0504	−.0361	−.0230	−.0105	.0015	.0135	.0254	.0377	.0505	.4264				
5	μ	.0161	.0317	.0384	.0434	.0475	.0512	.0546	.0578	.0609	.0639	.0669	.0698	.3979					
	σ	−.1872	−.1209	−.0925	−.0711	−.0533	−.0375	−.0230	−.0092	.0042	.0173	.0305	.0440	.4988					
6	μ	−.0013	.0230	.0323	.0392	.0450	.0502	.0549	.0593	.0636	.0677	.0718	.4918						
	σ	−.2058	−.1318	−.1001	−.0763	−.0564	−.0388	−.0226	−.0073	.0075	.0221	.0367	.5728						
7	μ	−.0176	.0121	.0247	.0342	.0421	.0491	.0555	.0615	.0673	.0729	.5980							
	σ	−.2278	−.1445	−.1089	−.0821	−.0598	−.0400	−.0219	−.0047	.0119	.0282	.6497							
8	μ	−.0419	−.0019	.0153	.0281	.0387	.0482	.0568	.0648	.0726	.7194								
	σ	−.2543	−.1597	−.1191	−.0888	−.0635	−.0411	−.0205	−.0011	.0176	.7305								
9	μ	−.0741	−.0200	.0031	.0204	.0348	.0474	.0590	.0698	.8597									
	σ	−.2869	−.1781	−.1315	−.0966	−.0676	−.0418	−.0183	.0039	.8168									
10	μ	−.1176	−.0441	−.0128	.0106	.0300	.0471	.0627	1.0241										
	σ	−.3283	−.2010	−.1466	−.1059	−.0720	−.0421	−.0147	.9107										
11	μ	−.1788	−.0775	−.0343	−.0022	.0245	.0479	1.2204											
	σ	−.3827	−.2307	−.1657	−.1173	−.0770	−.0414	1.0148											
12	μ	−.2689	−.1257	−.0648	−.0195	.0180	1.4609												
	σ	−.4576	−.2707	−.1911	−.1317	−.0824	1.1335												
13	μ	−.4113	−.2001	−.1106	−.0442	1.7662													
	σ	−.5681	−.3285	−.2266	−.1509	1.2741													
14	μ	−.6615	−.3276	−.1867	2.1758														
	σ	−.7486	−.4205	−.2815	1.4505														
15	μ	−1.1905	−.5891	2.7796															
	σ	−1.1013	−.5948	1.6960															
16	μ	−2.8756	3.8756																
	σ	−2.1294	2.1294																

n = 19

n − r		t(1)	t(2)	t(3)	t(4)	t(5)	t(6)	t(7)	t(8)	t(9)	t(10)	t(11)	t(12)	t(13)	t(14)	t(15)	t(16)	t(17)	t(18)	t(19)
0	μ	.0526	.0526	.0526	.0526	.0526	.0526	.0526	.0526	.0526	.0526	.0526	.0526	.0526	.0526	.0526	.0526	.0526	.0526	.0526
	σ	−.1178	−.0792	−.0628	−.0505	−.0402	−.0312	−.0228	−.0150	−.0074	.0000	.0074	.0150	.0228	.0312	.0402	.0505	.0628	.0792	.1178
1	μ	.0485	.0500	.0506	.0511	.0515	.0519	.0522	.0525	.0528	.0531	.0534	.0537	.0540	.0543	.0547	.0550	.0554	.1052	
	σ	−.1272	−.0851	−.0672	−.0538	−.0427	−.0328	−.0238	−.0152	−.0070	.0011	.0092	.0174	.0259	.0349	.0447	.0558	.0689	.1969	
2	μ	.0431	.0467	.0482	.0493	.0502	.0511	.0518	.0525	.0532	.0539	.0546	.0552	.0559	.0566	.0574	.0581	.1623		
	σ	−.1372	−.0914	−.0719	−.0573	−.0451	−.0344	−.0245	−.0152	−.0063	.0025	.0113	.0202	.0295	.0392	.0498	.0616	.2690		
3	μ	.0365	.0426	.0451	.0471	.0487	.0501	.0514	.0526	.0538	.0550	.0561	.0572	.0584	.0596	.0608	.2249			
	σ	−.1482	−.0982	−.0769	−.0609	−.0477	−.0359	−.0252	−.0150	−.0053	.0043	.0139	.0236	.0336	.0442	.0556	.3381			
4	μ	.0283	.0376	.0415	.0445	.0469	.0491	.0510	.0529	.0547	.0564	.0582	.0599	.0616	.0634	.2940				
	σ	−.1605	−.1057	−.0823	−.0649	−.0504	−.0375	−.0257	−.0146	−.0040	.0065	.0170	.0276	.0385	.0499	.4062				
5	μ	.0182	.0315	.0371	.0413	.0448	.0479	.0507	.0534	.0559	.0584	.0609	.0631	.0658	.3707					
	σ	−.1744	−.1141	−.0884	−.0692	−.0532	−.0391	−.0262	−.0140	−.0022	.0093	.0207	.0323	.0442	.4744					
6	μ	.0057	.0240	.0318	.0376	.0424	.0467	.0506	.0542	.0577	.0611	.0644	.0678	.4561						
	σ	−.1905	−.1237	−.0952	−.0740	−.0563	−.0407	−.0264	−.0129	.0000	.0127	.0253	.0380	.5437						
7	μ	−.0100	.0147	.0253	.0331	.0396	.0453	.0506	.0554	.0601	.0647	.0691	.5520							
	σ	−.2091	−.1347	−.1030	−.0794	−.0597	−.0423	−.0264	−.0114	.0029	.0170	.0309	.6152							
8	μ	−.0298	−.0032	.0172	.0276	.0363	.0439	.0509	.0574	.0636	.0696	.6603								
	σ	−.2312	−.1476	−.1120	−.0855	−.0634	−.0439	−.0261	−.0093	.0067	.0224	.6898								
9	μ	−.0553	−.0115	.0070	.0209	.0324	.0425	.0516	.0602	.0684	.7839									
	σ	−.2578	−.1630	−.1226	−.0925	−.0675	−.0454	−.0252	−.0063	.0118	.7685									
10	μ	−.0888	−.0306	−.0059	.0125	.0277	.0411	.0532	.0646	.9264										
	σ	−.2907	−.1817	−.1353	−.1007	−.0720	−.0468	−.0237	−.0021	.8530										
11	μ	−.1341	−.0560	−.0229	.0017	.0221	.0399	.0561	1.0931											
	σ	−.3324	−.2051	−.1509	−.1106	−.0772	−.0478	−.0210	.9450											
12	μ	−.1976	−.0910	−.0459	−.0124	.0153	.0395	1.2920												
	σ	−.3873	−.2353	−.1708	−.1228	−.0831	−.0482	1.0474												
13	μ	−.2909	−.1415	−.0783	−.0316	.0070	1.5353													
	σ	−.4629	−.2762	−.1971	−.1384	−.0898	1.1645													
14	μ	−.4379	−.2194	−.1273	−.0593	1.8438														
	σ	−.5744	−.3353	−.2342	−.1594	1.3014														
15	μ	−.6958	−.3527	−.2086	2.2572															
	σ	−.7568	−.4295	−.2918	1.4781															
16	μ	−1.2401	−.6258	2.8659																
	σ	−1.1132	−.6084	1.7216																
17	μ	−2.9705	3.9705																	
	σ	−2.1527	2.1527																	

TABLE C-14 (continued)

n = 20

n − r		$t_{(1)}$	$t_{(2)}$	$t_{(3)}$	$t_{(4)}$	$t_{(5)}$	$t_{(6)}$	$t_{(7)}$	$t_{(8)}$	$t_{(9)}$	$t_{(10)}$	$t_{(11)}$	$t_{(12)}$	$t_{(13)}$	$t_{(14)}$	$t_{(15)}$	$t_{(16)}$	$t_{(17)}$	$t_{(18)}$	$t_{(19)}$	$t_{(20)}$
0	μ	.0500	.0500	.0500	.0500	.0500	.0500	.0500	.5000	.0100	.0500	.0500	.0500	.0500	.0500	.0500	.0500	.0500	.0500	.0500	.0500
	σ	−.1128	−.0765	−.0611	−.0497	−.0402	−.0318	−.0241	−.0169	−.0101	−.0033	.0033	.0101	.0169	.0241	.0318	.0402	.0497	.0611	.0765	.1128
1	μ	.0462	.0476	.0482	.0486	.0489	.0493	.0495	.0498	.0501	.0503	.0506	.0508	.0511	.0513	.0516	.0519	.0522	.0525	.0996	
	σ	−.1212	−.0819	−.0652	−.0528	−.0425	−.0335	−.0252	−.0174	−.0099	−.0026	.0046	.0119	.0193	.0271	.0354	.0444	.0546	.0667	.1884	
2	μ	.0415	.0446	.0459	.0469	.0477	.0484	.0491	.0497	.0502	.0508	.0514	.0519	.0525	.0531	.0537	.0544	.0550	.1533		
	σ	−.1303	−.0876	−.0695	−.0561	−.0449	−.0351	−.0261	−.0177	−.0096	−.0017	.0061	.0140	.0221	.0305	.0394	.0491	.0599	.2574		
3	μ	.0356	.0409	.0431	.0448	.0462	.0474	.0485	.0496	.0506	.0516	.0525	.0535	.0544	.0554	.0564	.0575	.2119			
	σ	−.1401	−.0938	−.0741	−.0595	−.0474	−.0367	−.0269	−.0178	−.0090	−.0004	.0080	.0166	.0253	.0344	.0440	.0543	.3333			
4	μ	.0284	.0365	.0399	.0424	.0445	.0463	.0480	.0496	.0511	.0526	.0540	.0554	.0569	.0584	.0599	.2762				
	σ	−.1511	−.1006	−.0792	−.0632	−.0500	−.0384	−.0277	−.0178	−.0082	.0011	.0103	.0196	.0291	.0389	.0492	.3880				
5	μ	.0197	.0311	.0359	.0395	.0425	.0451	.0475	.0497	.0519	.0539	.0560	.0580	.0600	.0621	.3470					
	σ	−.1634	−.1081	−.0847	−.0673	−.0528	−.0401	−.0285	−.0176	−.0071	.0030	.0131	.0232	.0335	.0441	.4526					
6	μ	.0090	.0246	.0312	.0361	.0402	.0438	.0470	.0501	.0530	.0558	.0586	.0613	.0640	.4254						
	σ	−.1773	−.1166	−.0908	−.0717	−.0558	−.0418	−.0291	−.0171	−.0057	.0055	.0165	.0275	.0387	.5179						
7	μ	−.0042	.0167	.0255	.0321	.0375	.0423	.0466	.0507	.0545	.0583	.0619	.0656	.5126							
	σ	−.1934	−.1262	−.0978	−.0766	−.0591	−.0437	−.0296	−.0164	−.0038	.0085	.0206	.0327	.5848							
8	μ	−.0206	.0069	.0185	.0272	.0343	.0406	.0463	.0517	.0567	.0616	.0664	.6103								
	σ	−.2121	−.1374	−.1057	−.0822	−.0627	−.0455	−.0299	−.0153	−.0013	.0123	.0257	.6541								
9	μ	−.0414	−.0053	.0099	.0213	.0306	.0388	.0463	.0532	.0598	.0662	.7205									
	σ	−.2343	−.1504	−.1149	−.0886	−.0667	−.0475	−.0300	−.0136	.0020	.0172	.7268									
10	μ	−.0680	−.0203	−.0008	.0140	.0262	.0369	.0466	.0557	.0642	.8460										
	σ	−.2612	−.1660	−.1258	−.0959	−.0712	−.0494	−.0296	−.0111	.0065	.8038										
11	μ	−.1028	−.0408	−.0146	.0048	.0209	.0349	.0476	.0594	.9905											
	σ	−.2943	−.1850	−.1389	−.1046	−.0762	−.0513	−.0287	−.0076	.8866											
12	μ	−.1498	−.0673	−.0326	−.0068	.0144	.0329	.0497	1.1595												
	σ	−.3363	−.2088	−.1550	−.1151	−.0820	−.0531	−.0268	.9771												
13	μ	−.2154	−.1039	−.0569	−.0222	.0064	.0313	1.3607													
	σ	−.3916	−.2396	−.1755	−.1280	−.0888	−.0544	1.0779													
14	μ	−.3117	−.1565	−.0914	−.0433	−.0037	1.6066														
	σ	−.4679	−.2811	−.2028	−.1447	−.0967	1.1914														
15	μ	−.4632	−.2378	−.1433	−.0718	1.9180															
	σ	−.5804	−.3417	−.2414	−.1673	1.3308															
16	μ	−.7284	−.3766	−.2298	2.3148																
	σ	−.7645	−.4380	−.3014	1.5039																
17	μ	−1.2872	−.6610	2.9482																	
	σ	−1.1244	−.6212	1.7456																	
18	μ	−3.0609	4.0609																		
	σ	−2.1745	2.1745																		

From "Estimation of Location and Scale Parameters by Order Statistics from Singly and Doubly Censored Samples, Parts I and II" by A. E. Sarhan and B. G. Greenberg, *Ann. Math. Statistic*, Vol. 27 (1956). Reproduced by permission of the editor of the *Annals of Mathematical Statistics*.

TABLE C-15 VARIANCES AND COVARIANCES OF THE BEST LINEAR ESTIMATES OF THE MEAN ($\hat{\mu}$) AND STANDARD DEVIATION ($\hat{\sigma}$) FOR CENSORED SAMPLES UP TO SIZE 20 FROM A NORMAL POPULATION

n		\multicolumn{13}{c}{$n-r$}
		0	1	2	3	4	5	6	7	8	9	10	11	12
2	$V(\hat{\mu})$	.5000												
	$V(\hat{\sigma})$	.5708												
	Cov $(\hat{\mu}, \hat{\sigma})$	0												
3	$V(\hat{\mu})$	.3333	.4487											
	$V(\hat{\sigma})$	.2755	.6378											
	Cov $(\hat{\mu}, \hat{\sigma})$	0	.2044											
4	$V(\hat{\mu})$	.2500	.2870	.5130										
	$V(\hat{\sigma})$	.1801	.3021	.6730										
	Cov $(\hat{\mu}, \hat{\sigma})$	0	.0672	.3567										
5	$V(\hat{\mu})$	.2000	.2177	.2839	.6112									
	$V(\hat{\sigma})$	.1333	.1948	.3181	.6957									
	Cov $(\hat{\mu}, \hat{\sigma})$	0	.0330	.1234	.4749									
6	$V(\hat{\mu})$	.1667	.1769	.2068	.2999	.7186								
	$V(\hat{\sigma})$	.1057	.1428	.2044	.3292	.7119								
	Cov $(\hat{\mu}, \hat{\sigma})$	0	.0195	.0624	.1702	.5705								
7	$V(\hat{\mu})$	.1429	.1494	.1660	.2071	.3248	.8264							
	$V(\hat{\sigma})$	.0875	.1123	.1493	.2114	.3375	.7243							
	Cov $(\hat{\mu}, \hat{\sigma})$	0	.0128	.0375	.0881	.2099	.6503							
8	$V(\hat{\mu})$	.1250	.1295	.1399	.1623	.2138	.3541	.9310						
	$V(\hat{\sigma})$	.0746	.0924	.1171	.1542	.2168	.3441	.7342						
	Cov $(\hat{\mu}, \hat{\sigma})$	0	.0090	.0250	.0538	.1106	.2442	.7186						
9	$V(\hat{\mu})$	.1111	.1144	.1214	.1352	.1629	.2241	.3854	1.0313					
	$V(\hat{\sigma})$	.0650	.0784	.0960	.1207	.1581	.2212	.3494	.7423					
	Cov $(\hat{\mu}, \hat{\sigma})$	0	.0067	.0178	.0362	.0684	.1305	.2743	.7781					
10	$V(\hat{\mu})$	.1000	.1025	.1075	.1167	.1336	.1664	.2366	.4174	1.1269				
	$V(\hat{\sigma})$	.0576	.0681	.0813	.0989	.1237	.1613	.2248	.3539	.7491				
	Cov $(\hat{\mu}, \hat{\sigma})$	0	.0051	.0132	.0260	.0465	.0816	.1483	.3011	.8306				
11	$V(\hat{\mu})$	.0909	.0929	.0966	.1031	.1143	.1342	.1718	.2504	.4493	1.2179			
	$V(\hat{\sigma})$	.0517	.0601	.0704	.0836	.1013	.1262	.1640	.2279	.3577	.7550			
	Cov $(\hat{\mu}, \hat{\sigma})$	0	.0041	.0102	.0195	.0336	.0559	.0936	.1644	.3251	.8777			
12	$V(\hat{\mu})$	.0833	.0849	.0878	.0926	.1004	.1136	.1363	.1784	.2650	.4809	1.3044		
	$V(\hat{\sigma})$	.0469	.0538	.0620	.0723	.0855	.1033	.1283	.1663	.2305	.3610	.7601		
	Cov $(\hat{\mu}, \hat{\sigma})$	0	.0033	.0081	.0152	.0254	.0406	.0645	.1045	.1791	.3469	.9202		

TABLE C-15 (continued)

First block — columns are $n-r$

n	Statistic	0	1	2	3	4	5	6	7	8	9	10	11	12	13	14	15	16
13	$V(\hat{\mu})$	.0769	.0782	.0805	.0841	.0899	.0991	.1141	.1395	.1858	.2799	.5118	1.5404					
	$V(\hat{\sigma})$	.0429	.0486	.0554	.0636	.0739	.0872	.1050	.1302	.1688	.2329	.3640	.7723					
	$\mathrm{Cov}(\hat{\mu},\hat{\sigma})$	0	.0027	.0066	.0121	.0198	.0309	.0472	.0725	.1146	.1925	.3668	1.0273					
14	$V(\hat{\mu})$	.0714	.0725	.0743	.0772	.0816	.0883	.0988	.1155	.1435	.1938	.2950	.5419	1.4654				
	$V(\hat{\sigma})$	.0395	.0444	.0500	.0568	.0650	.0753	.0886	.1065	.1318	.1702	.2350	.3666	.7687				
	$\mathrm{Cov}(\hat{\mu},\hat{\sigma})$	0	.0023	.0055	.0099	.0159	.0242	.0360	.0533	.0799	.1239	.2048	.3851	.9945				
15	$V(\hat{\mu})$	.0667	.0676	.0691	.0714	.0748	.0799	.0875	.0992	.1176	.1480	.2022	.3101	.5713	1.3868			
	$V(\hat{\sigma})$	.0366	.0408	.0456	.0512	.0580	.0662	.0765	.0899	.1078	.1332	.1718	.2368	.3689	.7646			
	$\mathrm{Cov}(\hat{\mu},\hat{\sigma})$	0	.0019	.0046	.0082	.0130	.0195	.0284	.0409	.0590	.0868	.1325	.2163	.4020	.9589			
16	$V(\hat{\mu})$	.0625	.0633	.0645	.0664	.0692	.0731	.0789	.0874	.1002	.1202	.1529	.2107	.3250	.5998	1.6121		
	$V(\hat{\sigma})$	.0341	.0378	.0419	.0467	.0523	.0590	.0672	.0776	.0910	.1090	.1345	.1732	.2385	.3710	.7756		
	$\mathrm{Cov}(\hat{\mu},\hat{\sigma})$	0	.0017	.0039	.0069	.0109	.0160	.0229	.0323	.0454	.0644	.0933	.1406	.2269	.4177	1.0578		
17	$V(\hat{\mu})$	.0588	.0595	.0605	.0621	.0644	.0675	.0720	.0784	.0877	.1016	.1231	.1581	.2194	.3398	.6274	1.6808	
	$V(\hat{\sigma})$	.0320	.0352	.0387	.0428	.0476	.0532	.0599	.0682	.0786	.0920	.1101	.1357	.1745	.2400	.3729	.7786	
	$\mathrm{Cov}(\hat{\mu},\hat{\sigma})$	0	.0015	.0034	.0059	.0092	.0134	.0189	.0262	.0360	.0497	.0694	.0993	.1481	.2369	.4324	1.0862	
18	$V(\hat{\mu})$	.0556	.0561	.0570	.0584	.0602	.0628	.0664	.0713	.0784	.0885	.1034	.1264	.1635	.2282	.3544	.6543	1.7467
	$V(\hat{\sigma})$	.0301	.0329	.0360	.0395	.0436	.0483	.0540	.0607	.0690	.0794	.0929	.1111	.1368	.1757	.2414	.3746	.7814
	$\mathrm{Cov}(\hat{\mu},\hat{\sigma})$	0	.0013	.0030	.0051	.0079	.0114	.0158	.0216	.0293	.0395	.0538	.0742	.1050	.1552	.2463	.4462	1.1127

Second block — columns are $n-r$

n	Statistic	0	1	2	3	4	5	6	7	8	9	10	11	12	13	14	15	16	17	18
19	$V(\hat{\mu})$	.0526	.0531	.0539	.0550	.0566	.0587	.0616	.0656	.0710	.0787	.0895	.1055	.1299	.1691	.2369	.3687	.6804	1.8101	
	$V(\hat{\sigma})$	.0284	.0309	.0336	.0367	.0403	.0443	.0491	.0547	.0614	.0698	.0802	.0938	.1120	.1377	.1768	.2426	.3763	.7839	
	$\mathrm{Cov}(\hat{\mu},\hat{\sigma})$	0	.0011	.0026	.0045	.0068	.0098	.0135	.0182	.0242	.0322	.0429	.0576	.0787	.1104	.1619	.2551	.4591	1.1377	
20	$V(\hat{\mu})$	.0500	.0504	.0511	.0521	.0534	.0552	.0576	.0608	.0651	.0710	.0792	.0909	.1078	.1335	.1747	.2456	.3828	.7056	1.8710
	$V(\hat{\sigma})$	.0268	.0291	.0316	.0343	.0374	.0409	.0450	.0497	.0553	.0621	.0705	.0810	.0945	.1128	.1386	.1778	.2438	.3778	.7863
	$\mathrm{Cov}(\hat{\mu},\hat{\sigma})$	0	.0010	.0023	.0039	.0060	.0085	.0116	.0155	.0204	.0267	.0350	.0460	.0612	.0829	.1155	.1682	.2634	.4713	1.1613

Up to $n = 15$ of this table is reproduced from A. E. Sarhan and B. G. Greenberg, "Estimation of location and scale parameters by order statistics from singly and censored samples, Parts I and II," *Ann. Math. Statist.*, Vol. 27 (1956), pp. 427–451, and Vol. 29 (1958), pp. 79–105, with permission of the editor of the *Annals of Mathematical Statistics*. The rest of the table is produced from A. E. Sarhan and B. G. Greenberg, "Estimation of location and scale parameters by order statistics from singly and doubly censored samples. Part III," Tech. Rep. 4-OOR, Project 1597, U.S. Army Research Office.

TABLE C-16 $1/(1-R)$ AND $\hat{\gamma}$ FOR THE ESTIMATION OF THE PARAMETERS OF THE GAMMA DISTRIBUTION WHEN THERE ARE NO CENSORED OBSERVATIONS

$1/(1-R)$	$\hat{\gamma}$	$1/(1-R)$	$\hat{\gamma}$	$1/(1-R)$	$\hat{\gamma}$
1·000	0·00000	1·11	0·29854	2·4	1·06335
1·001	0·11539	1·12	0·30749	2·5	1·11596
1·002	0·12663	1·13	0·31616	2·6	1·16833
1·003	0·13434	1·14	0·32461	2·7	1·22050
1·004	0·14043	1·15	0·33285	2·8	1·27248
1·005	0·14556	1·16	0·34090	2·9	1·32430
1·006	0·15005	1·18	0·35654	3·0	1·37599
1·007	0·15408	1·20	0·37165	3·2	1·47899
1·008	0·15775	1·22	0·38631	3·4	1·58158
1·009	0·16115	1·24	0·40060	3·6	1·68385
1·010	0·16432	1·26	0·41457	3·8	1·78585
1·012	0·17011	1·28	0·42825	4·0	1·88763
1·014	0·17535	1·30	0·44170	4·2	1·98921
1·016	0·18017	1·32	0·45492	4·4	2·09063
1·018	0·18464	1·34	0·46795	4·6	2·19191
1·020	0·18884	1·36	0·48081	4·8	2·29308
1·022	0·19282	1·38	0·49351	5·0	2·39414
1·024	0·19659	1·40	0·50608	5·5	2·64643
1·026	0·20020	1·45	0·53694	6·0	2·89830
1·028	0·20366	1·50	0·56715	6·5	3·14984
1·030	0·20700	1·55	0·59682	7·0	3·40115
1·035	0·21486	1·60	0·62604	8·0	3·90325
1·040	0·22217	1·65	0·65488	9·0	4·40482
1·045	0·22904	1·70	0·68340	10·0	4·90608
1·050	0·23554	1·75	0·71163	12·0	5·90790
1·055	0·24175	1·80	0·73961	14·0	6·90914
1·060	0·24771	1·85	0·76737	16·0	7·91005
1·065	0·25345	1·90	0·79494	18·0	8·91073
1·070	0·25899	1·95	0·82233	20·0	9·91125
1·075	0·26437	2·00	0·84957	30·0	14·91301
1·080	0·26959	2·10	0·90364	40·0	19·91401
1·090	0·27966	2·20	0·95724	50·0	24·91457
1·100	0·28928	2·30	1·01046		

From "Estimation of Parameters of the Gamma Distribution Using Order Statistics" by M. B. Wilk, R. Gnanadesikan and Marilyn J. Huyett, *Biometrika*, Vol. 49, pp. 525–545 (1962). Reproduced by permission of the editor of *Biometrika*.

TABLE C-17 $\hat{\gamma}(P, S)$ AND $\hat{\mu}(P, S)$ FOR:

$n/r = 1.0$

S		0·96	0·92	0·88	0·84	0·80	0·76	0·72	0·68	0·64	0·60	0·56			P/S
0·08	γ	**0·850**		12·412	6·158	4·070	3·024	2·394	1·972	1·669	1·440	1·261	1·116	0·996	1·00
	μ	**0·080**		1·000	1·000	1·000	1·000	1·000	1·000	1·000	1·000	1·000	1·000	1·000	
0·12	γ	**0·567**	**1·376**		11·912	5·908	3·903	2·898	2·293	1·888	1·596	1·376	1·203	1·063	0·96
	μ	**0·120**	**0·120**		0·960	0·960	0·960	0·960	0·960	0·960	0·960	0·960	0·960	0·960	
0·16	γ	**0·464**	**0·850**	**1·887**		11·412	5·657	3·736	2·772	2·192	1·803	1·523	1·311	1·145	0·92
	μ	**0·160**	**0·160**	**0·160**		0·920	0·920	0·920	0·920	0·920	0·920	0·920	0·920	0·920	
0·20	γ	**0·408**	**0·664**	**1·116**	**2·394**		10·912	5·407	3·569	2·646	2·091	1·718	1·450	1·247	0·88
	μ	**0·200**	**0·200**	**0·200**	**0·200**		0·880	0·880	0·880	0·880	0·880	0·880	0·880	0·880	
0·24	γ	**0·372**	**0·567**	**0·850**	**1·376**	**2·898**		10·411	5·157	3·401	2·520	1·989	1·633	1·376	0·84
	μ	**0·240**	**0·240**	**0·240**	**0·240**	**0·240**		0·840	0·840	0·840	0·840	0·840	0·840	0·840	
0·28	γ	**0·346**	**0·506**	**0·712**	**1·028**	**1·633**	**3·401**		9·911	4·906	3·234	2·394	1·888	1·547	0·80
	μ	**0·280**	**0·280**	**0·280**	**0·280**	**0·280**	**0·280**		0·800	0·800	0·800	0·800	0·800	0·800	
0·32	γ	**0·327**	**0·464**	**0·626**	**0·850**	**1·203**	**1·888**	**3·903**		9·411	4·655	3·066	2·268	1·786	0·76
	μ	**0·320**	**0·320**	**0·320**	**0·320**	**0·320**	**0·320**	**0·320**		0·760	0·760	0·760	0·760	0·760	
0·36	γ	**0·312**	**0·432**	**0·567**	**0·740**	**0·984**	**1·376**	**2·141**	**4·405**		8·911	4·405	2·898	2·141	0·72
	μ	**0·360**	**0·360**	**0·360**	**0·360**	**0·360**	**0·360**	**0·360**	**0·360**		0·720	0·720	0·720	0·720	
0·40	γ	**0·300**	**0·408**	**0·524**	**0·664**	**0·850**	**1·116**	**1·547**	**2·394**	**4·906**		***8·410***	4·154	2·730	0·68
	μ	**0·400**	**0·400**	**0·400**	**0·400**	**0·400**	**0·400**	**0·400**	**0·400**	**0·400**		***0·680***	0·680	0·680	
0·44	γ	**0·289**	**0·388**	**0·490**	**0·609**	**0·758**	**0·957**	**1·247**	**1·718**	**2·646**	**5·407**		***7·910***	3·903	0·64
	μ	**0·440**	**0·440**	**0·440**	**0·440**	**0·440**	**0·440**	**0·440**	**0·440**	**0·440**	**0·440**		***0·640***	0·640	
0·48	γ	**0·281**	**0·372**	**0·464**	**0·567**	**0·691**	**0·850**	**1·063**	**1·376**	**1·888**	**2·898**	**5·908**		***7·409***	0·60
	μ	**0·480**	**0·480**	**0·480**	**0·480**	**0·480**	**0·480**	**0·480**	**0·480**	**0·480**	**0·480**	**0·480**		***0·600***	
0·52	γ	**0·273**	**0·358**	**0·442**	**0·534**	**0·641**	**0·771**	**0·939**	**1·168**	**1·505**	***2·057***	***3·150***	***6·408***		0·56
	μ	**0·520**	**0·520**	**0·520**	**0·520**	**0·520**	**0·520**	**0·520**	**0·520**	**0·520**	***0·520***	***0·520***	***0·520***		
0·56	γ	0·266	0·346	0·423	0·506	0·600	0·712	0·850	1·028	1·272	1·633	***2·226***	***3·401***	***6·909***	
	μ	0·560	0·560	0·560	0·560	0·560	0·560	0·560	0·560	0·560	0·560	***0·560***	***0·560***	***0·560***	
0·60	γ	0·261	0·336	0·408	0·483	0·567	0·664	0·781	0·927	1·116	1·376	***1·760***	***2·394***	***3·652***	
	μ	0·600	0·600	0·600	0·600	0·600	0·600	0·600	0·600	0·600	0·600	***0·600***	***0·600***	***0·600***	
0·64	γ	0·255	0·327	0·394	0·464	0·540	0·626	0·727	0·850	1·003	1·203	1·479	***1·888***	***2·562***	
	μ	0·640	0·640	0·640	0·640	0·640	0·640	0·640	0·640	0·640	0·640	0·640	0·640	0·640	
0·68	γ	0·251	0·319	0·382	0·447	0·516	0·594	0·683	0·789	0·917	1·078	1·290	***1·582***	***2·015***	
	μ	0·680	0·680	0·680	0·680	0·680	0·680	0·680	0·680	0·680	0·680	0·680	0·680	0·680	
0·72	γ	0·246	0·312	0·372	0·432	0·496	0·567	0·647	0·740	0·850	0·984	1·153	1·376	***1·684***	
	μ	0·720	0·720	0·720	0·720	0·720	0·720	0·720	0·720	0·720	0·720	0·720	0·720	0·720	
0·76	γ	0·242	0·305	0·362	0·419	0·479	0·544	0·616	0·699	0·795	0·910	1·050	1·228	1·462	
	μ	0·760	0·760	0·760	0·760	0·760	0·760	0·760	0·760	0·760	0·760	0·760	0·760	0·760	
0·80	γ	0·239	0·300	0·354	0·408	0·464	0·524	0·590	0·664	0·750	0·850	0·969	1·116	1·302	
	μ	0·800	0·800	0·800	0·800	0·800	0·800	0·800	0·800	0·800	0·800	0·800	0·800	0·800	
0·84	γ	0·236	0·294	0·346	0·397	0·450	0·506	0·567	0·635	0·712	0·800	0·904	1·028	1·181	
	μ	0·840	0·840	0·840	0·840	0·840	0·840	0·840	0·840	0·840	0·840	0·840	0·840	0·840	
0·88	γ	0·232	0·289	0·339	0·388	0·438	0·490	0·547	0·609	0·679	0·758	0·850	0·957	1·087	
	μ	0·880	0·880	0·880	0·880	0·880	0·880	0·880	0·880	0·880	0·880	0·880	0·880	0·880	
0·92	γ	0·230	0·285	0·333	0·379	0·427	0·476	0·529	0·587	0·651	0·722	0·804	0·899	1·010	
	μ	0·920	0·920	0·920	0·920	0·920	0·920	0·920	0·920	0·920	0·920	0·920	0·920	0·920	
0·96	γ	0·227	0·281	0·327	0·372	0·417	0·464	0·513	0·567	0·626	0·691	0·765	0·850	0·948	
	μ	0·960	0·960	0·960	0·960	0·960	0·960	0·960	0·960	0·960	0·960	0·960	0·960	0·960	
1·00	γ	0·225	0·277	0·322	0·365	0·408	0·452	0·499	0·549	0·604	0·664	0·732	0·808	0·895	
	μ	1·000	1·000	1·000	1·000	1·000	1·000	1·000	1·000	1·000	1·000	1·000	1·000	1·000	
S/P		0·04	0·08	0·12	0·16	0·20	0·24	0·28	0·32	0·36	0·40	0·44	0·48	0·52	

For $P \leqslant 0.52$ read S from the left-hand margin, and for $P \geqslant 0.56$ read S from the right-hand margin. Note that the figures in region 2 are printed in bold roman type and those in region 3 in bold italic type; the remainder of the table (outside of regions 2 and 3) is region 1.

$$n/r = 1.10$$

S		1·00	0·96	0·92	0·88	0·84	0·80	0·76	0·72	0·68	0·64	0·60	0·56		P/S /S
0·08	γ	0·501		12·505	6·224	4·124	3·070	2·435	2·009	1·703	1·471	1·290	1·143	1·021	1·00
	μ	0·195		1·023	1·034	1·043	1·051	1·058	1·065	1·072	1·079	1·085	1·092	1·099	
0·12	γ	0·438	0·759		11·866	5·927	3·932	2·929	2·323	1·916	1·624	1·402	1·228	1·087	0·96
	μ	0·243	0·226		0·986	0·996	1·005	1·013	1·021	1·028	1·035	1·041	1·048	1·055	
0·16	γ	0·398	0·634	1·029		11·212	5·628	3·740	2·787	2·211	1·824	1·544	1·332	1·166	0·92
	μ	0·291	0·272	0·259		0·948	0·959	0·968	0·976	0·983	0·990	0·997	1·004	1·011	
0·20	γ	0·370	0·558	0·827	1·321		10·546	5·326	3·547	2·646	2·099	1·731	1·465	1·263	0·88
	μ	0·338	0·318	0·305	0·294		0·911	0·922	0·931	0·939	0·946	0·953	0·960	0·967	
0·24	γ	0·349	0·507	0·711	1·025	1·639		9·872	5·023	3·354	2·504	1·987	1·638	1·385	0·84
	μ	0·386	0·364	0·351	0·339	0·329		0·874	0·885	0·893	0·901	0·909	0·916	0·923	
0·28	γ	0·333	0·470	0·634	0·863	1·232	1·986		9·193	4·721	3·161	2·363	1·875	1·545	0·80
	μ	0·433	0·411	0·396	0·385	0·374	0·365		0·837	0·847	0·856	0·864	0·872	0·879	
0·32	γ	0·319	0·442	0·580	0·759	1·018	1·449	2·365		8·516	4·419	2·968	2·222	1·764	0·76
	μ	0·480	0·457	0·442	0·430	0·419	0·410	0·400		0·800	0·810	0·819	0·827	0·835	
0·36	γ	0·308	0·419	0·539	0·687	0·884	1·177	1·677	2·776		7·846	4·119	2·777	2·082	0·72
	μ	0·528	0·503	0·487	0·475	0·464	0·455	0·446	0·436		0·763	0·774	0·782	0·790	
0·40	γ	0·299	0·401	0·507	0·633	0·792	1·010	1·340	1·914	3·220		7·186	3·821	2·588	0·68
	μ	0·575	0·549	0·533	0·520	0·509	0·500	0·491	0·482	0·472		0·727	0·737	0·745	
0·44	γ	0·291	0·385	0·481	0·591	0·724	0·897	1·138	1·507	2·162	3·698		6·543	3·529	0·64
	μ	0·622	0·595	0·578	0·565	0·554	0·544	0·535	0·527	0·518	0·508		0·690	0·700	
0·48	γ	0·284	0·372	0·460	0·557	0·672	0·815	1·002	1·267	1·678	2·419	4·208		5·920	0·60
	μ	0·669	0·641	0·623	0·610	0·599	0·589	0·580	0·571	0·563	0·554	0·544		0·653	
0·52	γ	0·277	0·361	0·442	0·529	0·630	0·752	0·905	1·109	1·399	1·854	2·685	4·750		
	μ	0·716	0·687	0·669	0·655	0·644	0·634	0·624	0·616	0·607	0·599	0·590	0·581		
0·56	γ	0·272	0·351	0·426	0·506	0·596	0·702	0·831	0·996	1·216	1·532	2·033	2·959	5·321	
	μ	0·763	0·733	0·714	0·700	0·688	0·678	0·669	0·660	0·652	0·644	0·635	0·627	0·617	
0·60	γ	0·267	0·342	0·412	0·486	0·568	0·662	0·773	0·910	1·086	1·324	1·668	2·215	3·241	
	μ	0·810	0·779	0·759	0·745	0·733	0·723	0·713	0·705	0·696	0·688	0·680	0·672	0·663	
0·64	γ	0·262	0·334	0·401	0·469	0·544	0·628	0·726	0·844	0·989	1·178	1·433	1·804	2·400	
	μ	0·856	0·824	0·805	0·790	0·778	0·767	0·758	0·749	0·741	0·733	0·725	0·717	0·709	
0·68	γ	0·258	0·327	0·390	0·454	0·523	0·600	0·687	0·790	0·914	1·068	1·269	1·543	1·942	
	μ	0·903	0·870	0·850	0·835	0·822	0·811	0·802	0·793	0·785	0·777	0·769	0·761	0·754	
0·72	γ	0·254	0·320	0·380	0·441	0·505	0·575	0·654	0·745	0·853	0·984	1·147	1·361	1·653	
	μ	0·950	0·916	0·895	0·879	0·867	0·856	0·846	0·837	0·829	0·821	0·813	0·806	0·798	
0·76	γ	0·251	0·314	0·372	0·429	0·489	0·554	0·626	0·708	0·803	0·916	1·054	1·227	1·453	
	μ	0·996	0·962	0·940	0·924	0·911	0·900	0·890	0·881	0·873	0·865	0·857	0·850	0·842	
0·80	γ	0·247	0·309	0·364	0·418	0·475	0·535	0·602	0·676	0·761	0·860	0·979	1·123	1·306	
	μ	1·043	1·007	0·985	0·969	0·956	0·944	0·934	0·925	0·917	0·909	0·901	0·894	0·887	
0 84	γ	0·244	0·304	0·357	0·409	0·462	0·519	0·580	0·648	0·725	0·814	0·917	1·041	1·193	
	μ	1·089	1·053	1·030	1·014	1·000	0·989	0·978	0·969	0·961	0·953	0·945	0·938	0·931	
0·88	γ	0·241	0·300	0·350	0·400	0·451	0·504	0·561	0·624	0·695	0·774	0·866	0·974	1·103	
	μ	1·136	1·098	1·075	1·058	1·045	1·033	1·023	1·013	1·005	0·997	0·989	0·982	0·974	
0·92	γ	0·239	0·295	0·344	0·392	0·440	0·491	0·544	0·603	0·668	0·740	0·822	0·918	1·030	
	μ	1·182	1·144	1·120	1·103	1·089	1·077	1·067	1·057	1·049	1·040	1·033	1·025	1·018	
0·96	γ	0·236	0·291	0·339	0·385	0·431	0·479	0·529	0·584	0·644	0·710	0·785	0·870	0·970	
	μ	1·229	1·189	1·165	1·148	1·134	1·121	1·111	1·101	1·092	1·084	1·076	1·069	1·062	
1·00	γ	0·234	0·288	0·334	0·378	0·422	0·467	0·515	0·567	0·623	0·684	0·752	0·830	0·918	
	μ	1·275	1·235	1·210	1·192	1·178	1·166	1·155	1·145	1·136	1·128	1·120	1·113	1·106	
S/P		0·04	0·08	0·12	0·16	0·20	0·24	0·28	0·32	0·36	0·40	0·44	0·48	0·52	

TABLE C-17 (*continued*)

$$n/r = 1.20$$

P		0.96	0.92	0.88	0.84	0.80	0.76	0.72	0.68	0.64	0.60	0.56			P/S
0.08	γ	0.423		12.587	6.283	4.172	3.111	2.471	2.042	1.733	1.499	1.315	1.167	1.043	1.00
	μ	0.343		1.044	1.065	1.082	1.098	1.112	1.126	1.140	1.153	1.167	1.180	1.194	
0.12	γ	0.393	0.616		11.827	5.945	3.958	2.956	2.350	1.942	1.648	1.425	1.249	1.107	0.96
	μ	0.400	0.352		1.009	1.030	1.047	1.063	1.078	1.092	1.106	1.119	1.133	1.147	
0.16	γ	0.370	0.554	0.816		11.045	5.603	3.744	2.801	2.228	1.842	1.563	1.351	1.184	0.92
	μ	0.456	0.405	0.373		0.974	0.995	1.013	1.028	1.043	1.058	1.072	1.086	1.100	
0.20	γ	0.352	0.510	0.713	1.033		10.248	5.259	3.529	2.646	2.107	1.742	1.478	1.277	0.88
	μ	0.512	0.458	0.424	0.398		0.940	0.961	0.978	0.994	1.009	1.023	1.038	1.052	
0.24	γ	0.337	0.476	0.643	0.879	1.274		9.447	4.916	3.315	2.491	1.986	1.643	1.394	0.84
	μ	0.567	0.511	0.476	0.449	0.426		0.906	0.926	0.944	0.960	0.975	0.990	1.004	
0.28	γ	0.325	0.450	0.592	0.779	1.056	1.542		8.651	4.574	3.103	2.338	1.866	1.545	0.80
	μ	0.623	0.564	0.527	0.500	0.476	0.455		0.872	0.892	0.910	0.926	0.941	0.956	
0.32	γ	0.315	0.429	0.553	0.708	0.919	1.245	1.841		7.871	4.237	2.893	2.186	1.747	0.76
	μ	0.678	0.616	0.579	0.550	0.526	0.505	0.485		0.838	0.858	0.876	0.892	0.908	
0.36	γ	0.307	0.412	0.522	0.654	0.825	1.066	1.447	2.175		7.116	3.906	2.685	2.036	0.72
	μ	0.733	0.669	0.630	0.601	0.577	0.555	0.535	0.515		0.805	0.825	0.842	0.859	
0.40	γ	0.299	0.397	0.497	0.612	0.755	0.945	1.219	1.664	2.547		6.393	3.583	2.482	0.68
	μ	0.788	0.721	0.681	0.651	0.626	0.605	0.585	0.566	0.546		0.772	0.791	0.809	
0.44	γ	0.292	0.384	0.476	0.579	0.702	0.858	1.069	1.379	1.895	2.961		5.708	3.270	0.64
	μ	0.842	0.774	0.732	0.701	0.676	0.654	0.634	0.616	0.597	0.578		0.739	0.758	
0.48	γ	0.286	0.373	0.458	0.551	0.659	0.791	0.962	1.196	1.546	2.142	3.418		5.066	0.60
	μ	0.897	0.826	0.783	0.751	0.726	0.704	0.684	0.665	0.647	0.629	0.609		0.706	
0.52	γ	0.281	0.363	0.442	0.527	0.624	0.739	0.881	1.068	1.327	1.721	2.403	3.920		
	μ	0.951	0.878	0.834	0.802	0.775	0.753	0.733	0.714	0.696	0.678	0.661	0.641		
0.56	γ	0.276	0.355	0.429	0.507	0.594	0.696	0.819	0.973	1.176	1.463	1.902	2.679	4.470	
	μ	1.005	0.930	0.885	0.852	0.825	0.802	0.782	0.763	0.745	0.728	0.711	0.693	0.673	
0.60	γ	0.272	0.347	0.417	0.490	0.570	0.661	0.768	0.899	1.065	1.287	1.601	2.090	2.968	
	μ	1.059	0.982	0.935	0.901	0.874	0.851	0.830	0.811	0.794	0.777	0.760	0.743	0.725	
0.64	γ	0.268	0.340	0.406	0.474	0.548	0.631	0.726	0.840	0.980	1.159	1.399	1.743	2.283	
	μ	1.113	1.033	0.986	0.951	0.923	0.900	0.879	0.860	0.842	0.825	0.809	0.793	0.776	
0.68	γ	0.264	0.333	0.397	0.461	0.529	0.605	0.691	0.791	0.912	1.061	1.254	1.514	1.888	
	μ	1.167	1.085	1.037	1.001	0.973	0.949	0.927	0.908	0.890	0.874	0.857	0.841	0.825	
0.72	γ	0.261	0.328	0.388	0.449	0.513	0.583	0.661	0.751	0.856	0.984	1.144	1.350	1.630	
	μ	1.221	1.137	1.087	1.051	1.022	0.997	0.976	0.957	0.939	0.922	0.906	0.890	0.874	
0.76	γ	0.257	0.322	0.380	0.438	0.498	0.563	0.635	0.716	0.810	0.922	1.057	1.226	1.447	
	μ	1.274	1.188	1.137	1.100	1.071	1.046	1.024	1.005	0.987	0.970	0.954	0.938	0.923	
0.80	γ	0.254	0.317	0.373	0.428	0.485	0.545	0.612	0.686	0.771	0.870	0.987	1.130	1.310	
	μ	1.328	1.240	1.188	1.150	1.120	1.095	1.073	1.053	1.035	1.018	1.002	0.986	0.971	
0.84	γ	0.252	0.313	0.366	0.419	0.473	0.530	0.592	0.660	0.737	0.826	0.929	1.052	1.203	
	μ	1.381	1.291	1.238	1.200	1.169	1.143	1.121	1.101	1.082	1.065	1.049	1.034	1.019	
0.88	γ	0.249	0.308	0.360	0.411	0.462	0.516	0.574	0.637	0.708	0.788	0.880	0.988	1.118	
	μ	1.434	1.343	1.288	1.249	1.218	1.192	1.169	1.149	1.130	1.113	1.097	1.081	1.067	
0.92	γ	0.247	0.304	0.355	0.403	0.452	0.503	0.558	0.617	0.682	0.755	0.839	0.935	1.048	
	μ	1.487	1.394	1.338	1.298	1.267	1.240	1.217	1.197	1.178	1.161	1.144	1.129	1.114	
0.96	γ	0.244	0.301	0.349	0.396	0.443	0.492	0.543	0.599	0.659	0.727	0.802	0.889	0.989	
	μ	1.540	1.445	1.389	1.348	1.316	1.289	1.265	1.245	1.226	1.208	1.192	1.176	1.161	
1.00	γ	0.242	0.297	0.344	0.389	0.434	0.481	0.530	0.582	0.639	0.701	0.771	0.849	0.939	
	μ	1.593	1.496	1.439	1.397	1.364	1.337	1.313	1.292	1.273	1.256	1.239	1.224	1.209	
S/P		0.04	0.08	0.12	0.16	0.20	0.24	0.28	0.32	0.36	0.40	0.44	0.48	0.52	

$n/r = 1.40$

		1.00	0.96	0.92	0.88	0.84	0.80	0.76	0.72	0.68	0.64	0.60	0.56	P/S
0.08	γ	0.368		12.728	6.382	4.253	3.181	2.533	2.097	1.783	1.545	1.358	1.207	1.081	1.00
	μ	0.760		1.081	1.120	1.153	1.183	1.211	1.239	1.266	1.293	1.321	1.349	1.378	
0.12	γ	0.355	0.514		11.765	5.975	4.002	3.002	2.395	1.985	1.689	1.463	1.286	1.142	0.96
	μ	0.835	0.671		1.051	1.090	1.123	1.154	1.182	1.211	1.238	1.266	1.295	1.324	
0.16	γ	0.343	0.486	0.661		10.779	5.565	3.752	2.824	2.257	1.873	1.594	1.382	1.214	0.92
	μ	0.909	0.739	0.643		1.021	1.061	1.094	1.125	1.154	1.183	1.211	1.240	1.269	
0.20	γ	0.334	0.463	0.614	0.820		9.789	5.156	3.503	2.647	2.121	1.762	1.501	1.301	0.88
	μ	0.982	0.806	0.708	0.637		0.992	1.032	1.065	1.097	1.126	1.156	1.185	1.214	
0.24	γ	0.325	0.445	0.578	0.749	0.996		8.818	4.753	3.258	2.473	1.986	1.653	1.409	0.84
	μ	1.055	0.873	0.772	0.699	0.641		0.964	1.003	1.037	1.069	1.099	1.129	1.159	
0.28	γ	0.318	0.429	0.548	0.695	0.893	1.194		7.884	4.358	3.017	2.302	1.854	1.545	0.80
	μ	1.128	0.940	0.836	0.761	0.702	0.651		0.936	0.976	1.010	1.042	1.073	1.103	
0.32	γ	0 311	0.415	0.524	0.652	0.817	1.049	1.417		7.000	3.977	2.782	2.134	1.724	0.76
	μ	1.200	1.007	0.899	0.823	0.763	0.711	0.664		0.909	0.948	0.983	1.015	1.047	
0.36	γ	0.306	0.403	0.503	0.617	0.759	0.947	1.220	1.671		6.177	3.611	2.554	1.971	0.72
	μ	1.271	1.073	0.962	0.884	0.823	0.771	0.725	0.681		0.883	0.922	0.957	0.990	
0.40	γ	0.300	0.393	0.486	0.589	0.713	0.870	1.085	1.406	1.960		*5.423*	3.263	2.335	0.68
	μ	1.342	1.138	1.025	0.946	0.883	0.831	0.784	0.741	0.699		*0.857*	0.896	0.931	
0.44	γ	0.296	0.384	0.470	0.565	0.675	0.810	0.987	1.233	1.611	*2.290*		*4.738*	2.935	0.64
	μ	1.413	1.204	1.088	1.007	0.943	0.890	0.843	0.800	0.759	*0.719*		*0.832*	0.871	
0.48	γ	0.291	0.376	0.457	0.544	0.643	0.762	0.911	1.109	1.390	*1.834*	2.665		*4.124*	0.60
	μ	1.483	1.269	1.150	1.067	1.002	0.948	0.901	0.859	0.818	*0.779*	0.740		*0.808*	
0.52	γ	0.287	0.368	0.445	0.526	0.616	0.722	0.851	1.016	1.237	*1.558*	2.077	3.092		
	μ	1.553	1.334	1.213	1.128	1.062	1.007	0.959	0.916	0.877	*0.839*	0.801	0.761		
0.56	γ	0.283	0.361	0.434	0.510	0.593	0.689	0.803	0.943	1.124	1.372	*1.737*	*2.342*	3.577	
	μ	1.623	1.399	1.275	1.188	1.121	1.065	1.017	0.974	0.934	0.897	*0.860*	*0.823*	0.784	
0.60	γ	0.280	0.355	0.424	0.496	0.573	0.660	0.762	0.884	1.037	1.237	1.513	*1.926*	2.627	
	μ	1.692	1.463	1.336	1.248	1.180	1.123	1.075	1.031	0.991	0.954	0.918	*0.883*	0.847	
0.64	γ	0.277	0.349	0.416	0.483	0.555	0.636	0.727	0.836	0.968	1.134	1.353	1.660	2.126	
	μ	1.762	1.528	1.398	1.308	1.238	1.181	1.132	1.088	1.048	1.011	0.976	0.941	0.907	
0.68	γ	0.274	0.344	0.408	0.472	0.539	0.614	0.698	0.795	0.911	1.053	1.233	1.473	1.813	
	μ	1.830	1.592	1.460	1.368	1.297	1.239	1.189	1.145	1.105	1.067	1.032	0.998	0.965	
0.72	γ	0.271	0.339	0.400	0.461	0.525	0.595	0.672	0.760	0.863	0.987	1.140	1.335	1.597	
	μ	1.899	1.656	1.521	1.428	1.355	1.296	1.246	1.201	1.161	1.123	1.088	1.055	1.022	
0.76	γ	0.268	0.335	0.394	0.452	0.512	0.578	0.649	0.730	0.823	0.932	1.064	1.228	1.439	
	μ	1.967	1.719	1.582	1.487	1.414	1.354	1.302	1.257	1.217	1.179	1.144	1.111	1.079	
0.80	γ	0.266	0.330	0.387	0.443	0.501	0.562	0.629	0.703	0.788	0.886	1.002	1.142	1.318	
	μ	2.036	1.783	1.643	1.546	1.472	1.411	1.359	1.313	1.272	1.235	1.200	1.166	1.135	
0.84	γ	0.263	0.326	0.382	0.435	0.490	0.548	0.611	0.680	0.757	0.846	0.949	1.072	1.221	
	μ	2.104	1.846	1.704	1.606	1.530	1.468	1.415	1.369	1.328	1.290	1.255	1.222	1.190	
0.88	γ	0.261	0.323	0.376	0.428	0.480	0.535	0.595	0.659	0.731	0.812	0.904	1.013	1.143	
	μ	2.171	1.910	1.765	1.665	1.588	1.525	1.472	1.425	1.383	1.345	1.310	1.276	1.245	
0.92	γ	0.259	0.319	0.371	0.421	0.471	0.524	0.580	0.640	0.707	0.781	0.866	0.963	1.077	
	μ	2.239	1.973	1.826	1.724	1.646	1.582	1.528	1.481	1.438	1.400	1.364	1.331	1.299	
0.96	γ	0.257	0.316	0.366	0.415	0.463	0.513	0.566	0.623	0.685	0.754	0.832	0.920	1.022	
	μ	2.306	2.036	1.886	1.783	1.704	1.639	1.584	1.536	1.494	1.455	1.419	1.385	1.354	
1.00	γ	0.255	0.313	0.362	0.409	0.455	0.503	0.554	0.608	0.666	0.730	0.802	0.882	0.974	
	μ	2.374	2.099	1.947	1.842	1.761	1.696	1.640	1.592	1.548	1.509	1.473	1.440	1.408	
S/P		0.04	0.08	0.12	0.16	0.20	0.24	0.28	0.32	0.36	0.40	0.44	0.48	0.52	

TABLE C-17 (*continued*)

$$n/r = 1.60$$

		1·00	0·96	0·92	0·88	0·84	0·80	0·76	0·72	0·68	0·64	0·60	0·56	P/S
0·08	γ	0·346		12·845	6·465	4·320	3·238	2·583	2·143	1·825	1·584	1·394	1·239	1·111	1·00
	μ	1·366		1·113	1·168	1·215	1·259	1·300	1·341	1·381	1·422	1·464	1·507	1·552	
0·12	γ	0·338	0·473		11·716	6·000	4·039	3·040	2·432	2·020	1·722	1·495	1·316	1·170	0·96
	μ	1·460	1·089		1·087	1·143	1·191	1·235	1·277	1·319	1·361	1·403	1·446	1·491	
0·16	γ	0·331	0·456	0·598		10·575	5·536	3·760	2·844	2·281	1·899	1·620	1·408	1·239	0·92
	μ	1·553	1·172	0·975		1·062	1·119	1·167	1·212	1·255	1·298	1·341	1·385	1·430	
0·20	γ	0·325	0·441	0·570	0·732		9·451	5·079	3·485	2·650	2·133	1·779	1·520	1·321	0·88
	μ	1·645	1·255	1·053	0·916		1·038	1·095	1·144	1·190	1·234	1·278	1·323	1·368	
0·24	γ	0·320	0·429	0·546	0·688	0·880		8·373	4·633	3·216	2·461	1·988	1·662	1·421	0·84
	μ	1·736	1·336	1·130	0·991	0·884		1·016	1·073	1·123	1·169	1·215	1·260	1·306	
0·28	γ	0·315	0·418	0·526	0·652	0·816	1·046		7·363	4·205	2·956	2·277	1·846	1·547	0·80
	μ	1·827	1·418	1·206	1·064	0·956	0·866		0·995	1·052	1·102	1·150	1·196	1·243	
0·32	γ	0·310	0·408	0·508	0·623	0·764	0·954	1·234		6·435	3·798	2·705	2·098	1·709	0·76
	μ	1·916	1·498	1·282	1·137	1·027	0·936	0·857		0·975	1·032	1·083	1·131	1·180	
0·36	γ	0·306	0·400	0·493	0·597	0·723	0·884	1·107	1·448		5·597	3·414	2·466	1·927	0·72
	μ	2·006	1·578	1·358	1·210	1·098	1·006	0·927	0·854		0·956	1·013	1·065	1·115	
0·40	γ	0·302	0·392	0·480	0·576	0·689	0·828	1·012	1·275	1·694		4·850	3·057	2·239	0·68
	μ	2·094	1·658	1·433	1·282	1·168	1·075	0·996	0·924	0·856		0·938	0·995	1·048	
0·44	γ	0·298	0·385	0·468	0·557	0·660	0·783	0·940	1·151	1·461	1·976		4·192	2·727	0·64
	μ	2·182	1·737	1·507	1·354	1·238	1·144	1·064	0·992	0·926	0·862		0·921	0·979	
0·48	γ	0·295	0·378	0·457	0·541	0·635	0·745	0·881	1·058	1·301	1·667	2·302		3·615	0·60
	μ	2·269	1·816	1·582	1·425	1·307	1·212	1·131	1·060	0·994	0·932	0·870		0·906	
0·52	γ	0·292	0·372	0·447	0·526	0·613	0·713	0·834	0·984	1·183	1·462	1·895	2·679		
	μ	2·356	1·894	1·656	1·496	1·376	1·280	1·198	1·126	1·061	1·000	0·940	0·880		
0·56	γ	0·289	0·367	0·439	0·513	0·594	0·685	0·793	0·925	1·092	1·316	1·637	2·146	3·114	
	μ	2·442	1·972	1·729	1·567	1·445	1·347	1·265	1·192	1·127	1·066	1·008	0·951	0·893	
0·60	γ	0·286	0·361	0·430	0·501	0·577	0·661	0·759	0·876	1·020	1·205	1·457	1·824	2·423	
	μ	2·528	2·050	1·802	1·637	1·514	1·414	1·331	1·258	1·192	1·132	1·075	1·019	0·964	
0·64	γ	0·284	0·357	0·423	0·490	0·561	0·640	0·730	0·835	0·961	1·118	1·324	1·606	2·025	
	μ	2·613	2·127	1·876	1·708	1·582	1·481	1·396	1·323	1·257	1·196	1·140	1·086	1·033	
0·68	γ	0·281	0·352	0·416	0·480	0·548	0·621	0·704	0·799	0·911	1·048	1·221	1·447	1·762	
	μ	2·699	2·204	1·948	1·778	1·650	1·548	1·462	1·387	1·321	1·261	1·204	1·151	1·099	
0·72	γ	0·279	0·348	0·410	0·471	0·535	0·604	0·681	0·768	0·869	0·990	1·138	1·326	1·576	
	μ	2·783	2·281	2·021	1·847	1·718	1·614	1·527	1·452	1·385	1·324	1·268	1·215	1·164	
0·76	γ	0·277	0·344	0·404	0·463	0·524	0·589	0·661	0·741	0·833	0·941	1·070	1·230	1·435	
	μ	2·868	2·358	2·093	1·917	1·785	1·680	1·592	1·516	1·449	1·388	1·331	1·279	1·228	
0·80	γ	0·274	0·341	0·399	0·455	0·514	0·575	0·643	0·717	0·802	0·899	1·014	1·152	1·325	
	μ	2·951	2·434	2·165	1·986	1·852	1·746	1·657	1·580	1·512	1·450	1·394	1·341	1·291	
0·84	γ	0·272	0·337	0·393	0·448	0·504	0·563	0·626	0·696	0·774	0·862	0·966	1·088	1·236	
	μ	3·035	2·510	2·237	2·055	1·920	1·811	1·721	1·643	1·575	1·513	1·456	1·404	1·354	
0·88	γ	0·271	0·334	0·389	0·441	0·495	0·551	0·611	0·676	0·749	0·830	0·924	1·033	1·163	
	μ	3·118	2·585	2·308	2·124	1·987	1·877	1·785	1·707	1·638	1·575	1·518	1·465	1·416	
0·92	γ	0·269	0·331	0·384	0·435	0·487	0·540	0·597	0·659	0·727	0·802	0·888	0·986	1·101	
	μ	3·201	2·661	2·380	2·193	2·053	1·942	1·850	1·770	1·700	1·637	1·580	1·527	1·477	
0·96	γ	0·267	0·328	0·380	0·429	0·479	0·531	0·585	0·643	0·707	0·777	0·856	0·945	1·048	
	μ	3·284	2·736	2·451	2·261	2·120	2·007	1·914	1·833	1·762	1·699	1·641	1·588	1·538	
1·00	γ	0·265	0·325	0·376	0·424	0·472	0·521	0·573	0·629	0·688	0·754	0·827	0·909	1·003	
	μ	3·367	2·811	2·522	2·330	2·186	2·072	1·977	1·896	1·825	1·760	1·702	1·649	1·599	
S/P		0·04	0·08	0·12	0·16	0·20	0·24	0·28	0·32	0·36	0·40	0·44	0·48	0·52	

$$n/r = 1.80$$

S		1·00	0·96	0·92	0·88	0·84	0·80	0·76	0·72	0·68	0·64	0·60	0·56	P/S
0·08	γ	0·335	12·945	6·535	4·377	3·287	2·626	2·181	1·860	1·616	1·423	1·267	1·137	1·00
	μ	2·196	1·140	1·211	1·271	1·327	1·381	1·434	1·487	1·542	1·598	1·656	1·717	
0·12	γ	0·330	0·451	11·677	6·023	4·070	3·072	2·463	2·050	1·750	1·522	1·341	1·194	0·96
	μ	2·310	1·613	1·119	1·190	1·251	1·309	1·364	1·419	1·474	1·531	1·589	1·650	
0·16	γ	0·325	0·439	0·564	10·412	5·515	3·767	2·861	2·302	1·920	1·642	1·429	1·260	0·92
	μ	2·422	1·712	1·370	1·099	1·171	1·233	1·292	1·349	1·405	1·463	1·522	1·583	
0·20	γ	0·321	0·429	0·544	0·685	9·188	5·019	3·471	2·654	2·144	1·794	1·536	1·338	0·88
	μ	2·534	1·810	1·461	1·237	1·080	1·153	1·217	1·276	1·335	1·394	1·453	1·515	
0·24	γ	0·317	0·420	0·527	0·654	0·817	8·039	4·542	3·185	2·453	1·990	1·670	1·432	0·84
	μ	2·644	1·907	1·551	1·323	1·153	1·064	1·137	1·201	1·263	1·323	1·384	1·446	
0·28	γ	0·313	0·412	0·512	0·627	0·771	0·965	6·984	4·090	2·910	2·259	1·842	1·549	0·80
	μ	2·753	2·003	1·641	1·408	1·236	1·098	1·049	1·122	1·188	1·251	1·313	1·377	
0·32	γ	0·310	0·404	0·499	0·605	0·733	0·899	1·132	6·038	3·667	2·648	2·072	1·698	0·76
	μ	2·861	2·098	1·729	1·492	1·318	1·179	1·061	1·036	1·109	1·176	1·241	1·306	
0·36	γ	0·306	0·398	0·487	0·585	0·701	0·845	1·039	1·322	5·201	3·274	2·401	1·895	0·72
	μ	2·968	2·192	1·817	1·576	1·399	1·259	1·140	1·035	1·024	1·098	1·167	1·234	
0·40	γ	0·303	0·391	0·476	0·568	0·674	0·802	0·967	1·195	*1·541*	4·471	2·913	2·170	0·68
	μ	3·075	2·286	1·904	1·659	1·480	1·337	1·218	1·114	1·018	*1·015*	1·089	1·160	
0·44	γ	0·301	0·386	0·467	0·553	0·650	0·765	0·909	1·099	*1·369*	*1·795*	3·839	2·585	0·64
	μ	3·181	2·380	1·991	1·742	1·559	1·416	1·295	1·191	1·096	*1·007*	*1·007*	1·083	
0·48	γ	0·298	0·380	0·458	0·539	0·629	0·734	0·862	1·024	1·243	*1·562*	*2·088*	*3·296*	0·60
	μ	3·286	2·472	2·077	1·824	1·639	1·493	1·372	1·267	1·173	*1·085*	*1·000*	*1·001*	
0·52	γ	0·295	0·375	0·450	0·527	0·611	0·707	0·822	0·963	1·147	*1·399*	*1·778*	*2·431*	0·56
	μ	3·390	2·564	2·163	1·905	1·718	1·570	1·447	1·342	1·248	*1·161*	*1·079*	*0·998*	
0·56	γ	0·293	0·371	0·442	0·515	0·595	0·684	0·788	0·913	1·071	1·278	*1·569*	*2·019*	*2·830*
	μ	3·494	2·656	2·248	1·987	1·796	1·646	1·522	1·416	1·322	1·236	*1·155*	*1·077*	*0·998*
0·60	γ	0·291	0·366	0·435	0·505	0·580	0·663	0·758	0·871	1·009	1·184	1·418	*1·754*	*2·287*
	μ	3·597	2·747	2·333	2·067	1·874	1·722	1·597	1·490	1·395	1·309	1·229	*1·153*	*1·078*
0·64	γ	0·289	0·362	0·429	0·496	0·567	0·644	0·732	0·834	0·957	1·108	1·303	1·568	1·954
	μ	3·699	2·838	2·418	2·148	1·952	1·798	1·671	1·563	1·468	1·382	1·302	1·227	1·155
0·68	γ	0·287	0·359	0·423	0·487	0·555	0·628	0·709	0·803	0·913	1·046	1·212	1·429	1·727
	μ	3·802	2·928	2·502	2·228	2·029	1·873	1·745	1·635	1·539	1·453	1·374	1·300	1·229
0·72	γ	0·285	0·355	0·418	0·479	0·544	0·613	0·689	0·775	0·875	0·993	1·138	1·320	1·560
	μ	3·903	3·018	2·586	2·308	2·106	1·948	1·818	1·708	1·611	1·524	1·445	1·371	1·301
0·76	γ	0·283	0·352	0·412	0·472	0·533	0·599	0·670	0·751	0·842	0·949	1·076	1·233	1·433
	μ	4·004	3·108	2·669	2·387	2·183	2·022	1·891	1·779	1·682	1·595	1·515	1·442	1·372
0·80	γ	0·281	0·348	0·408	0·465	0·524	0·586	0·654	0·729	0·813	0·910	1·024	1·161	1·331
	μ	4·105	3·197	2·752	2·467	2·259	2·097	1·964	1·851	1·752	1·665	1·585	1·511	1·442
0·84	γ	0·279	0·345	0·403	0·459	0·515	0·575	0·639	0·709	0·787	0·876	0·979	1·101	1·249
	μ	4·205	3·286	2·835	2·545	2·335	2·171	2·036	1·922	1·823	1·734	1·654	1·580	1·512
0·88	γ	0·278	0·343	0·399	0·452	0·507	0·564	0·625	0·691	0·764	0·846	0·940	1·050	1·180
	μ	4·305	3·374	2·918	2·624	2·411	2·245	2·108	1·993	1·892	1·803	1·723	1·649	1·580
0·92	γ	0·276	0·340	0·394	0·447	0·499	0·554	0·612	0·674	0·743	0·819	0·906	1·005	1·121
	μ	4·404	3·462	3·000	2·703	2·487	2·318	2·180	2·063	1·962	1·872	1·791	1·717	1·648
0·96	γ	0·275	0·337	0·391	0·441	0·492	0·545	0·600	0·660	0·724	0·795	0·875	0·966	1·071
	μ	4·503	3·550	3·082	2·781	2·562	2·391	2·252	2·134	2·031	1·941	1·859	1·784	1·715
1·00	γ	0·273	0·335	0·387	0·436	0·486	0·536	0·589	0·646	0·707	0·774	0·848	0·932	1·027
	μ	4·601	3·638	3·164	2·859	2·637	2·465	2·323	2·204	2·100	2·009	1·926	1·851	1·782
S/P		0·04	0·08	0·12	0·16	0·20	0·24	0·28	0·32	0·36	0·40	0·44	0·48	0·52

TABLE C-17 (continued)

$$n/r = 2.0$$

		1.00	0.96	0.92	0.88	0.84	0.80	0.76	0.72	0.68	0.64	0.60	0.56	P/S
0.08	γ	0.328	13.029	6.596	4.426	3.329	2.663	2.215	1.890	1.644	1.449	1.291	1.159	1.00
	μ	3.287	1.165	1.249	1.321	1.389	1.455	1.520	1.586	1.654	1.724	1.798	1.875	
0.12	γ	0.325	0.437	11.643	6.042	4.097	3.100	2.490	2.076	1.774	1.545	1.362	1.214	0.96
	μ	3.420	2.252	1.147	1.233	1.307	1.376	1.444	1.512	1.581	1.651	1.725	1.803	
0.16	γ	0.321	0.429	0.543	10.278	5.497	3.774	2.875	2.319	1.939	1.661	1.447	1.277	0.92
	μ	3.552	2.366	1.830	1.132	1.218	1.294	1.365	1.435	1.506	1.577	1.652	1.730	
0.20	γ	0.318	0.421	0.528	0.655	8.977	4.971	3.461	2.657	2.154	1.806	1.550	1.352	0.88
	μ	3.683	2.479	1.934	1.596	1.119	1.206	1.283	1.357	1.429	1.502	1.577	1.656	
0.24	γ	0.315	0.414	0.515	0.631	0.777	7.776	4.470	3.160	2.446	1.993	1.677	1.442	0.84
	μ	3.812	2.591	2.037	1.694	1.449	1.108	1.196	1.275	1.350	1.426	1.502	1.581	
0.28	γ	0.312	0.408	0.503	0.611	0.742	0.913	6.695	4.000	2.874	2.245	1.838	1.552	0.80
	μ	3.940	2.702	2.139	1.791	1.542	1.348	1.099	1.188	1.269	1.347	1.425	1.505	
0.32	γ	0.309	0.402	0.493	0.593	0.712	0.862	1.067	5.741	3.566	2.604	2.053	1.690	0.76
	μ	4.067	2.812	2.240	1.886	1.634	1.438	1.276	1.093	1.182	1.265	1.346	1.428	
0.36	γ	0.307	0.396	0.483	0.577	0.686	0.820	0.994	1.242	4.913	3.168	2.352	1.871	0.72
	μ	4.193	2.921	2.340	1.981	1.726	1.527	1.364	1.223	1.089	1.180	1.265	1.349	
0.40	γ	0.305	0.391	0.474	0.563	0.663	0.784	0.936	1.142	1.443	4.201	2.807	2.118	0.68
	μ	4.318	3.029	2.440	2.075	1.816	1.616	1.451	1.310	1.183	1.088	1.180	1.267	
0.44	γ	0.302	0.387	0.466	0.550	0.643	0.753	0.889	1.064	1.306	1.676	3.593	2.481	0.64
	μ	4.442	3.137	2.539	2.169	1.906	1.703	1.537	1.395	1.269	1.153	1.089	1.183	
0.48	γ	0.300	0.382	0.459	0.538	0.626	0.727	0.848	1.001	1.202	1.490	1.947	3.076	0.60
	μ	4.566	3.244	2.637	2.261	1.995	1.789	1.622	1.480	1.354	1.239	1.130	1.093	
0.52	γ	0.298	0.378	0.452	0.528	0.610	0.704	0.814	0.949	1.121	1.355	1.697	2.265	
	μ	4.688	3.350	2.735	2.354	2.083	1.875	1.706	1.563	1.437	1.323	1.217	1.113	
0.56	γ	0.296	0.374	0.445	0.518	0.596	0.683	0.784	0.905	1.055	1.251	1.521	1.929	2.638
	μ	4.810	3.455	2.832	2.445	2.171	1.961	1.790	1.645	1.519	1.405	1.301	1.201	1.101
0.60	γ	0.294	0.371	0.439	0.509	0.583	0.665	0.758	0.867	1.000	1.168	1.390	1.703	2.189
	μ	4.931	3.560	2.928	2.537	2.259	2.045	1.872	1.727	1.600	1.486	1.382	1.284	1.190
0.64	γ	0.293	0.367	0.434	0.501	0.571	0.648	0.735	0.835	0.954	1.101	1.288	1.540	1.902
	μ	5.051	3.664	3.024	2.627	2.346	2.130	1.955	1.807	1.680	1.566	1.462	1.366	1.274
0.68	γ	0.291	0.364	0.429	0.493	0.560	0.633	0.714	0.806	0.914	1.044	1.206	1.415	1.700
	μ	5.171	3.768	3.120	2.718	2.432	2.213	2.036	1.888	1.759	1.645	1.541	1.445	1.355
0.72	γ	0.289	0.361	0.424	0.486	0.550	0.619	0.695	0.781	0.880	0.997	1.138	1.316	1.549
	μ	5.290	3.872	3.215	2.808	2.518	2.297	2.118	1.967	1.838	1.723	1.619	1.524	1.434
0.76	γ	0.288	0.358	0.419	0.479	0.541	0.607	0.679	0.758	0.850	0.955	1.082	1.236	1.432
	μ	5.408	3.974	3.310	2.897	2.604	2.380	2.199	2.047	1.916	1.800	1.696	1.601	1.512
0.80	γ	0.286	0.355	0.415	0.473	0.532	0.595	0.663	0.738	0.823	0.919	1.033	1.169	1.337
	μ	5.526	4.077	3.404	2.986	2.690	2.462	2.279	2.126	1.993	1.877	1.772	1.677	1.588
0.84	γ	0.285	0.352	0.411	0.467	0.524	0.585	0.649	0.720	0.798	0.888	0.991	1.112	1.259
	μ	5.644	4.179	3.498	3.075	2.775	2.545	2.359	2.204	2.071	1.953	1.848	1.752	1.664
0.88	γ	0.283	0.350	0.407	0.462	0.517	0.575	0.636	0.703	0.777	0.859	0.954	1.064	1.194
	μ	5.760	4.280	3.592	3.163	2.859	2.627	2.439	2.282	2.147	2.029	1.923	1.827	1.738
0.92	γ	0.282	0.347	0.403	0.456	0.510	0.565	0.624	0.688	0.757	0.834	0.921	1.021	1.138
	μ	5.877	4.382	3.685	3.252	2.944	2.708	2.519	2.360	2.224	2.105	1.998	1.901	1.812
0.96	γ	0.281	0.345	0.399	0.451	0.503	0.557	0.613	0.673	0.739	0.811	0.892	0.984	1.090
	μ	5.993	4.482	3.778	3.339	3.028	2.790	2.598	2.437	2.300	2.180	2.072	1.975	1.886
1.00	γ	0.279	0.342	0.396	0.447	0.497	0.549	0.603	0.660	0.722	0.791	0.866	0.951	1.047
	μ	6.108	4.583	3.871	3.427	3.112	2.871	2.677	2.515	2.376	2.254	2.146	2.048	1.958
S/P		0.04	0.08	0.12	0.16	0.20	0.24	0.28	0.32	0.36	0.40	0.44	0.48	0.52

$$n/r = 2.30$$

		1·00	0·96	0·92	0·88	0·84	0·80	0·76	0·72	0·68	0·64	0·60	0·56		P/S
0·08	γ	**0·322**		13·143	6·674	4·489	3·383	2·711	2·257	1·929	1·679	1·482	1·321	1·187	1·00
	μ	**5·488**		1·197	1·299	1·389	1·473	1·556	1·638	1·723	1·810	1·902	1·998	2·101	
0·12	γ	**0·320**	0·425		11·606	6·067	4·131	3·135	2·524	2·108	1·805	1·573	1·390	1·239	0·96
	μ	**5·651**	3·438		1·185	1·290	1·381	1·468	1·554	1·640	1·729	1·821	1·917	2·020	
0·16	γ	**0·318**	0·419	**0·524**		10·116	5·477	3·783	2·895	2·342	1·963	1·685	1·471	1·300	0·92
	μ	**5·812**	3·575	2·641		1·176	1·283	1·377	1·467	1·555	1·646	1·739	1·836	1·939	
0·20	γ	**0·315**	0·414	**0·513**	**0·627**		8·727	4·913	3·449	2·662	2·166	1·822	1·568	1·370	0·88
	μ	**5·972**	3·711	2·764	2·209		1·170	1·278	1·375	1·468	1·560	1·655	1·753	1·856	
0·24	γ	**0·313**	0·409	**0·504**	**0·610**	**0·740**		7·473	4·385	3·132	2·440	1·997	1·686	1·454	0·84
	μ	**6·130**	3·845	2·886	2·324	1·937		1·168	1·277	1·376	1·473	1·569	1·669	1·773	
0·28	γ	**0·311**	0·404	**0·495**	**0·595**	**0·714**	**0·864**		6·368	3·897	2·833	2·229	1·836	1·556	0·80
	μ	**6·287**	3·978	**3·007**	2·437	2·046	1·751		1·168	1·279	1·381	1·481	1·583	1·688	
0·32	γ	**0·309**	0·400	**0·487**	**0·581**	**0·691**	**0·826**	**1·004**		5·414	3·452	2·555	2·031	1·682	0·76
	μ	**6·443**	4·110	3·127	2·549	2·153	1·855	1·616		1·172	1·285	1·390	1·495	1·601	
0·36	γ	**0·308**	0·396	**0·479**	**0·569**	**0·671**	**0·794**	**0·950**	**1·164**		4·601	3·050	2·297	1·844	0·72
	μ	**6·598**	4·241	3·246	2·661	2·260	1·958	1·717	1·514		1·180	1·294	1·403	1·513	
0·40	γ	**0·306**	0·392	**0·472**	**0·558**	**0·653**	**0·766**	**0·905**	**1·088**	**1·347**		*3·914*	2·690	2·061	0·68
	μ	**6·751**	4·371	3·364	2·771	2·365	2·060	1·817	1·614	1·435		*1·191*	1·308	1·422	
0·44	γ	**0·304**	0·388	**0·466**	**0·547**	**0·637**	**0·741**	**0·867**	**1·027**	**1·243**	*1·559*		*3·336*	2·369	0·64
	μ	**6·903**	4·501	3·481	2·881	2·469	2·161	1·916	1·711	1·534	*1·373*		*1·207*	1·327	
0·48	γ	**0·303**	0·385	**0·460**	**0·538**	**0·623**	**0·720**	**0·834**	**0·976**	**1·160**	*1·416*	*1·807*		*2·851*	0·60
	μ	**7·055**	4·629	3·598	2·989	2·573	2·261	2·013	1·808	1·630	*1·471*	*8·324*		*1·227*	
0·52	γ	**0·301**	0·381	**0·455**	**0·529**	**0·610**	**0·700**	**0·806**	**0·933**	**1·094**	*1·308*	*1·612*	*2·099*		
	μ	**7·205**	4·756	3·713	3·097	2·675	2·360	2·110	1·903	1·724	*1·566*	*1·422*	*1·284*		
0·56	γ	**0·300**	0·378	**0·449**	**0·521**	**0·598**	**0·683**	**0·781**	**0·896**	**1·039**	1·222	*1·470*	*1·834*	*2·443*	
	μ	**7·354**	4·882	3·828	3·205	2·778	2·458	2·206	1·996	1·817	1·660	*1·516*	*1·382*	*1·252*	
0·60	γ	**0·298**	0·375	**0·444**	**0·514**	**0·587**	**0·667**	**0·758**	**0·864**	0·992	1·152	1·360	*1·648*	*2·035*	
	μ	**7·503**	5·008	3·942	3·311	2·879	2·556	2·300	2·089	1·909	1·751	1·609	*1·477*	*1·351*	
0·64	γ	**0·297**	0·372	**0·440**	**0·507**	**0·577**	**0·653**	**0·738**	**0·836**	0·952	1·093	1·272	1·509	1·845	
	μ	**7·651**	5·134	4·055	3·417	2·980	2·653	2·395	2·181	2·000	1·841	1·699	1·568	1·446	
0·68	γ	**0·296**	0·370	**0·435**	**0·500**	**0·567**	**0·640**	**0·720**	**0·811**	0·917	1·044	1·200	1·400	1·670	
	μ	**7·798**	5·258	4·168	3·523	3·080	2·749	2·488	2·273	2·090	1·930	1·788	1·658	1·537	
0·72	γ	**0·294**	0·367	**0·431**	**0·494**	**0·559**	**0·628**	**0·704**	**0·789**	0·886	1·001	1·140	1·313	1·537	
	μ	**7·944**	5·382	4·281	3·628	3·180	2·845	2·581	2·364	2·179	2·019	1·876	1·746	1·626	
0·76	γ	**0·293**	0·365	**0·427**	**0·488**	**0·551**	**0·617**	**0·689**	**0·768**	0·859	0·964	1·089	1·240	1·431	
	μ	**8·089**	5·505	4·392	3·732	3·279	2·940	2·673	2·454	2·268	2·106	1·962	1·833	1·713	
0·80	γ	**0·292**	0·362	**0·423**	**0·483**	**0·543**	**0·606**	**0·675**	**0·750**	0·835	0·931	1·044	1·179	1·345	
	μ	**8·234**	5·627	4·503	3·836	3·378	3·035	2·765	2·543	2·355	2·193	2·048	1·918	1·799	
0·84	γ	**0·291**	0·360	**0·420**	**0·477**	**0·536**	**0·597**	**0·662**	**0·733**	0·813	0·902	1·005	1·127	1·273	
	μ	**8·378**	5·749	4·614	3·939	3·476	3·130	2·857	2·632	2·443	2·278	2·133	2·003	1·883	
0·88	γ	**0·290**	0·358	**0·416**	**0·473**	**0·529**	**0·588**	**0·650**	**0·718**	0·793	0·876	0·971	1·082	1·212	
	μ	**8·521**	5·871	4·724	4·042	3·574	3·224	2·948	2·721	2·530	2·364	2·218	2·086	1·967	
0·92	γ	**0·288**	0·356	**0·413**	**0·468**	**0·523**	0·579	**0·639**	**0·704**	0·774	0·853	0·941	1·042	1·160	
	μ	**8·664**	5·992	4·834	4·145	3·671	3·317	3·038	2·809	2·616	2·449	2·301	2·169	2·049	
0·96	γ	**0·287**	0·354	**0·410**	**0·463**	**0·517**	0·572	**0·629**	**0·691**	0·757	0·831	0·913	1·007	1·114	
	μ	**8·806**	6·112	4·943	4·247	3·768	3·411	3·129	2·897	2·702	2·533	2·385	2·252	2·131	
1·00	γ	**0·286**	0·352	**0·407**	**0·459**	**0·511**	0·564	**0·620**	**0·678**	0·742	0·812	0·889	0·975	1·073	
	μ	**8·947**	6·232	5·052	4·349	3·865	3·504	3·218	2·985	2·787	2·617	2·467	2·333	2·212	
S/P		0·04	0·08	0·12	0·16	0·20	0·24	0·28	0·32	0·36	0·40	0·44	0·48	0·52	

TABLE C-17 (*continued*)

$$n/r = 2.60$$

S		1·00	0·96	0·92	0·88	0·84	0·80	0·76	0·72	0·68	0·64	0·60	0·56		P/S
0·08	γ	0·319	13·237	6·740	4·542	3·428	2·751	2·293	1·961	1·708	1·509	1·346	1·210		1·00
	μ	8·478	1·226	1·344	1·449	1·548	1·647	1·746	1·848	1·955	2·067	2·186	2·315		
0·12	γ	0·317	0·418	11·575	6·089	4·160	3·165	2·553	2·135	1·831	1·598	1·412	1·261		0·96
	μ	8·670	4·915	1·219	1·340	1·448	1·552	1·654	1·758	1·866	1·978	2·098	2·226		
0·16	γ	0·315	0·413	0·512	9·987	5·461	3·792	2·911	2·361	1·983	1·704	1·490	1·318		0·92
	μ	8·861	5·075	3·600	1·216	1·340	1·451	1·559	1·666	1·775	1·889	2·009	2·137		
0·20	γ	0·314	0·409	0·504	0·610	8·532	4·869	3·441	2·667	2·177	1·836	1·582	1·386		0·88
	μ	9·049	5·233	3·743	2·906	1·217	1·344	1·458	1·570	1·682	1·797	1·918	2·047		
0·24	γ	0·312	0·405	0·496	0·596	0·716	7·241	4·320	3·110	2·436	2·001	1·694	1·465		0·84
	μ	9·237	5·390	3·884	3·038	2·477	1·222	1·351	1·470	1·586	1·704	1·826	1·955		
0·28	γ	0·311	0·402	0·489	0·584	0·695	0·833	6·124	3·818	2·802	2·218	1·834	1·560		0·80
	μ	9·423	5·545	4·023	3·168	2·601	2·185	1·232	1·363	1·486	1·607	1·731	1·862		
0·32	γ	0·310	0·398	0·483	0·574	0·677	0·803	0·964	5·174	3·366	2·518	2·014	1·677		0·76
	μ	9·607	5·699	4·162	3·297	2·723	2·303	1·975	1·246	1·380	1·507	1·635	1·767		
0·36	γ	0·308	0·395	0·477	0·564	0·661	0·776	0·921	1·113	4·377	2·963	2·257	1·825		0·72
	μ	9·790	5·852	4·299	3·424	2·844	2·420	2·089	1·816	1·265	1·402	1·535	1·669		
0·40	γ	0·307	0·392	0·471	0·554	0·646	0·753	0·884	1·052	1·284	*3·711*	2·605	2·019		0·68
	μ	9·972	6·004	4·435	3·551	2·964	2·535	2·201	1·928	1·693	*1·289*	1·430	1·569		
0·44	γ	0·306	0·389	0·466	0·546	0·633	0·733	0·852	1·002	1·200	*1·483*	*3·157*	2·289		0·64
	μ	10·153	6·154	4·570	3·676	3·082	2·649	2·312	2·037	1·803	*1·595*	*1·318*	1·465		
0·48	γ	0·304	0·386	0·461	0·538	0·621	0·715	0·825	0·959	1·132	*1·366*	*1·714*	*2·696*		0·60
	μ	10·333	6·304	4·705	3·800	3·200	2·762	2·422	2·145	1·910	*1·704*	*1·516*	*1·355*		
0·52	γ	0·303	0·384	0·457	0·531	0·610	0·698	0·800	0·923	1·075	*1·275*	*1·554*	*1·987*		
	μ	10·512	6·453	4·838	3·924	3·317	2·874	2·530	2·251	2·015	*1·810*	*1·625*	*1·452*		
0·56	γ	0·302	0·381	0·452	0·524	0·600	0·683	0·779	0·891	1·028	1·201	*1·433*	*1·768*	*2·311*	
	μ	10·689	6·600	4·970	4·046	3·433	2·985	2·638	2·356	2·119	1·913	*1·729*	*1·560*	*1·399*	
0·60	γ	0·301	0·379	0·448	0·517	0·590	0·670	0·759	0·863	0·987	1·140	1·338	*1·609*	*2·012*	
	μ	10·866	6·747	5·102	4·168	3·548	3·095	2·744	2·460	2·221	2·015	1·831	*1·664*	*1·508*	
0·64	γ	0·300	0·376	0·444	0·511	0·581	0·657	0·741	0·838	0·951	1·088	1·261	1·487	1·804	
	μ	11·041	6·893	5·233	4·289	3·662	3·204	2·850	2·563	2·322	2·115	1·932	1·766	1·612	
0·68	γ	0·299	0·374	0·440	0·506	0·573	0·645	0·725	0·815	0·920	1·044	1·196	1·390	1·649	
	μ	11·216	7·038	5·363	4·410	3·776	3·313	2·955	2·665	2·422	2·214	2·030	1·865	1·712	
0·72	γ	0·298	0·372	0·437	0·500	0·565	0·635	0·710	0·795	0·892	1·005	1·142	1·311	1·529	
	μ	11·390	7·183	5·492	4·529	3·888	3·421	3·059	2·766	2·522	2·312	2·127	1·962	1·810	
0·76	γ	0·297	0·370	0·433	0·495	0·558	0·625	0·697	0·777	0·867	0·971	1·095	1·244	1·432	
	μ	11·563	7·326	5·621	4·649	4·001	3·528	3·163	2·867	2·620	2·408	2·223	2·057	1·906	
0·80	γ	0·296	0·368	0·430	0·490	0·551	0·615	0·684	0·760	0·845	0·941	1·054	1·188	1·352	
	μ	11·735	7·469	5·749	4·767	4·113	3·635	3·265	2·967	2·718	2·504	2·318	2·151	2·001	
0·84	γ	0·295	0·366	0·427	0·486	0·545	0·607	0·673	0·744	0·824	0·914	1·018	1·139	1·285	
	μ	11·906	7·611	5·876	4·885	4·224	3·741	3·368	3·066	2·815	2·599	2·412	2·244	2·093	
0·88	γ	0·294	0·364	0·424	0·481	0·539	0·598	0·662	0·730	0·805	0·890	0·986	1·096	1·227	
	μ	12·076	7·753	6·003	5·002	4·335	3·847	3·470	3·165	2·911	2·694	2·505	2·337	2·185	
0·92	γ	0·293	0·362	0·421	0·477	0·533	0·591	0·652	0·717	0·788	0·868	0·957	1·059	1·177	
	μ	12·246	7·894	6·129	5·119	4·445	3·952	3·571	3·263	3·007	2·788	2·597	2·428	2·276	
0·96	γ	0·292	0·361	0·418	0·473	0·528	0·583	0·642	0·705	0·773	0·847	0·931	1·025	1·134	
	μ	12·415	8·034	6·255	5·236	4·555	4·057	3·672	3·361	3·102	2·881	2·689	2·518	2·366	
1·00	γ	0·291	0·359	0·416	0·469	0·522	0·577	0·633	0·693	0·758	0·829	0·907	0·995	1·095	
	μ	12·583	8·174	6·380	5·352	4·664	4·161	3·772	3·458	3·197	2·974	2·780	2·608	2·455	
S/P		0·04	0·08	0·12	0·16	0·20	0·24	0·28	0·32	0·36	0·40	0·44	0·48	0·52	

$$n/r = 3.00$$

		1.00	0.96	0.92	0.88	0.84	0.80	0.76	0.72	0.68	0.64	0.60	0.56		P/S
0.08	γ	**0.316**	13.343	6.815	4.602	3.479	2.795	2.333	1.997	1.742	1.539	1.374	1.236		1.00
	μ	13.912	1.258	1.396	1.520	1.638	1.756	1.876	2.001	2.132	2.271	2.421	2.584		
0.12	γ	**0.315**	0.411	11.540	6.113	4.193	3.199	2.585	2.166	1.860	1.625	1.438	1.284		0.96
	μ	14.142	7.370	1.258	1.400	1.528	1.651	1.775	1.902	2.034	2.174	2.324	2.486		
0.16	γ	**0.314**	0.408	0.502	9.850	5.445	3.802	2.929	2.383	2.005	1.727	1.511	1.339		0.92
	μ	14.371	7.560	5.113	1.263	1.409	1.541	1.670	1.800	1.934	2.075	2.225	2.387		
0.20	γ	**0.313**	0.405	0.496	0.595	8.328	4.822	3.433	2.674	2.189	1.851	1.599	1.402		0.88
	μ	14.598	7.747	5.280	3.966	1.273	1.422	1.560	1.694	1.831	1.974	2.124	2.286		
0.24	γ	**0.312**	0.403	0.490	0.585	0.695	7.003	4.252	3.088	2.433	2.006	1.704	1.477		0.84
	μ	14.823	7.933	5.446	4.119	3.273	1.288	1.441	1.584	1.725	1.870	2.022	2.185		
0.28	γ	**0.311**	0.400	0.485	0.575	0.679	0.805	5.878	3.738	2.771	2.207	1.834	1.565		0.80
	μ	15.047	8.118	5.610	4.270	3.416	2.810	1.309	1.466	1.615	1.764	1.918	2.081		
0.32	γ	**0.310**	0.397	0.480	0.567	0.665	0.782	0.928	4.938	3.281	2.480	1.999	1.672		0.76
	μ	15.270	8.301	5.773	4.420	3.557	2.946	2.480	1.336	1.498	1.653	1.811	1.976		
0.36	γ	**0.309**	0.395	0.475	0.559	0.652	0.761	0.894	1.068	4.160	2.877	2.217	1.806		0.72
	μ	15.491	8.483	5.934	4.569	3.697	3.080	2.611	2.234	1.370	1.536	1.699	1.868		
0.40	γ	**0.308**	0.393	0.471	0.552	0.640	0.742	0.865	1.020	1.228	*3.518*	2.522	1.978		0.68
	μ	15.711	8.663	6.094	4.716	3.835	3.212	2.739	2.360	2.044	*1.411*	1.583	1.756		
0.44	γ	**0.307**	0.390	0.467	0.545	0.630	0.726	0.839	0.979	1.161	*1.413*	*2.989*	2.212		0.64
	μ	15.929	8.842	6.253	4.862	3.973	3.343	2.866	2.485	2.167	*1.893*	*2.460*	1.640		
0.48	γ	**0.306**	0.388	0.463	0.538	0.620	0.711	0.816	0.944	1.105	*1.320*	*1.630*	*2.552*		0.60
	μ	16.147	9.020	6.410	5.007	4.108	3.472	2.991	2.607	2.289	*2.015*	*1.771*	*1.518*		
0.52	γ	**0.305**	0.386	0.459	0.532	0.610	0.697	0.796	0.913	1.058	*1.244*	*1.500*	*1.886*		
	μ	16.363	9.197	6.567	5.150	4.243	3.601	3.114	2.727	2.407	*2.134*	*1.892*	*1.671*		
0.56	γ	**0.304**	0.384	0.455	0.527	0.602	0.684	0.777	0.886	1.017	1.182	*1.399*	*1.706*	*2.189*	
	μ	16.578	9.372	6.722	5.293	4.377	3.728	3.237	2.847	2.525	2.250	*2.009*	*1.792*	*1.588*	
0.60	γ	**0.304**	0.382	0.452	0.521	0.594	0.672	0.760	0.862	0.982	1.129	1.318	*1.572*	*1.942*	
	μ	16.792	9.547	6.877	5.435	4.510	3.854	3.358	2.964	2.640	2.364	2.124	*1.908*	*1.709*	
0.64	γ	**0.303**	0.380	0.449	0.516	0.586	0.661	0.745	0.840	0.951	1.084	1.250	1.466	1.764	
	μ	17.005	9.721	7.031	5.575	4.641	3.979	3.479	3.081	2.754	2.477	2.236	2.021	1.825	
0.68	γ	**0.302**	0.379	0.446	0.511	0.579	0.651	0.731	0.820	0.923	1.045	1.193	1.380	1.628	
	μ	17.217	9.893	7.183	5.715	4.772	4.104	3.598	3.197	2.867	2.588	2.346	2.132	1.937	
0.72	γ	**0.301**	0.377	0.443	0.507	0.572	0.642	0.718	0.802	0.898	1.010	1.144	1.309	1.521	
	μ	17.428	10.065	7.335	5.855	4.903	4.228	3.717	3.312	2.979	2.698	2.455	2.240	2.046	
0.76	γ	**0.300**	0.375	0.440	0.502	0.566	0.633	0.706	0.786	0.876	0.979	1.102	1.249	1.433	
	μ	17.638	10.236	7.486	5.993	5.032	4.350	3.834	3.425	3.090	2.807	2.562	2.347	2.153	
0.80	γ	**0.300**	0.373	0.437	0.498	0.560	0.625	0.694	0.770	0.855	0.952	1.064	1.197	1.360	
	μ	17.847	10.406	7.636	6.131	5.161	4.473	3.951	3.539	3.200	2.914	2.668	2.452	2.258	
0.84	γ	**0.299**	0.372	0.434	0.494	0.554	0.617	0.684	0.756	0.837	0.927	1.031	1.152	1.298	
	μ	18.055	10.575	7.786	6.268	5.289	4.594	4.068	3.651	3.309	3.021	2.773	2.556	2.361	
0.88	γ	**0.298**	0.370	0.432	0.490	0.549	0.610	0.674	0.743	0.820	0.905	1.001	1.113	1.244	
	μ	18.262	10.744	7.935	6.404	5.417	4.715	4.184	3.763	3.418	3.127	2.877	2.658	2.463	
0.92	γ	**0.298**	0.369	0.429	0.487	0.544	0.603	0.665	0.731	0.804	0.884	0.974	1.077	1.197	
	μ	18.468	10.912	8.083	6.540	5.543	4.835	4.299	3.874	3.525	3.232	2.981	2.760	2.564	
0.96	γ	**0.297**	0.367	0.427	0.483	0.539	0.596	0.656	0.720	0.789	0.865	0.950	1.016	1.156	
	μ	18.674	11.078	8.230	6.675	5.670	4.955	4.413	3.984	3.633	3.337	3.083	2.861	2.663	
1.00	γ	**0.296**	0.366	0.425	0.480	0.534	0.590	0.648	0.709	0.776	0.848	0.928	1.017	1.119	
	μ	18.878	11.245	8.377	6.809	5.795	5.074	4.527	4.094	3.739	3.441	3.185	2.961	2.762	
S/P		0.04	0.08	0.12	0.16	0.20	0.24	0.28	0.32	0.36	0.40	0.44	0.48	0.52	

From "Estimation of Parameters of the Gamma Distribution Using Order Statistics" by M. B. Wilk, R. Gnanadesikan and Marilyn J. Huyett, *Biometrika*, Vol. 49, pp. 525–545 (1962). Reproduced by permission of the editor of *Biometrika*.

TABLE C-18   PERCENTAGE POINTS, l_α, SUCH THAT $P[\hat{\gamma}_1/\hat{\gamma}_2 < l_\alpha] = 1 - \alpha$

$n \backslash 1 - \alpha$	.60	.70	.75	.80	.85	.90	.95	.98
5	1.158	1.351	1.478	1.636	1.848	2.152	2.725	3.550
6	1.135	1.318	1.418	1.573	1.727	1.987	2.465	3.146
7	1.127	1.283	1.370	1.502	1.638	1.869	2.246	2.755
8	1.119	1.256	1.338	1.450	1.573	1.780	2.093	2.509
9	1.111	1.236	1.311	1.410	1.524	1.711	1.982	2.339
10	1.104	1.220	1.290	1.380	1.486	1.655	1.897	2.213
11	1.098	1.206	1.273	1.355	1.454	1.609	1.829	2.115
12	1.093	1.195	1.258	1.334	1.428	1.571	1.774	2.036
13	1.088	1.186	1.245	1.317	1.406	1.538	1.727	1.922
14	1.084	1.177	1.233	1.301	1.386	1.509	1.688	1.917
15	1.081	1.170	1.224	1.288	1.369	1.485	1.654	1.870
16	1.077	1.164	1.215	1.277	1.355	1.463	1.624	1.829
17	1.075	1.158	1.207	1.266	1.341	1.444	1.598	1.792
18	1.072	1.153	1.200	1.257	1.329	1.426	1.574	1.762
19	1.070	1.148	1.194	1.249	1.318	1.411	1.553	1.733
20	1.068	1.144	1.188	1.241	1.308	1.396	1.534	1.708
22	1.064	1.136	1.178	1.227	1.291	1.372	1.501	1.663
24	1.061	1.129	1.169	1.216	1.276	1.351	1.473	1.625
26	1.058	1.124	1.162	1.206	1.263	1.333	1.449	1.593
28	1.055	1.119	1.155	1.197	1.252	1.318	1.428	1.566
30	1.053	1.114	1.149	1.190	1.242	1.304	1.409	1.541
32	1.051	1.110	1.144	1.183	1.233	1.292	1.393	1.520
34	1.049	1.107	1.139	1.176	1.224	1.281	1.378	1.500
36	1.047	1.103	1.135	1.171	1.217	1.272	1.365	1.483
38	1.046	1.100	1.131	1.166	1.210	1.263	1.353	1.467
40	1.045	1.098	1.127	1.161	1.204	1.255	1.342	1.453
42	1.043	1.095	1.124	1.156	1.198	1.248	1.332	1.439
44	1.042	1.093	1.121	1.152	1.193	1.241	1.323	1.427
46	1.041	1.091	1.118	1.149	1.188	1.235	1.314	1.416
48	1.040	1.088	1.115	1.145	1.184	1.229	1.306	1.405
50	1.039	1.087	1.113	1.142	1.179	1.224	1.299	1.396
52	1.038	1.085	1.111	1.139	1.175	1.219	1.292	1.387
54	1.037	1.083	1.108	1.136	1.172	1.215	1.285	1.378
56	1.036	1.081	1.106	1.133	1.168	1.210	1.279	1.370
58	1.036	1.080	1.104	1.131	1.165	1.206	1.274	1.363
60	1.035	1.078	1.102	1.128	1.162	1.203	1.268	1.355
62	1.034	1.077	1.101	1.126	1.159	1.199	1.263	1.349
64	1.034	1.076	1.099	1.124	1.156	1.196	1.258	1.342
66	1.033	1.075	1.097	1.122	1.153	1.192	1.253	1.336
68	1.032	1.073	1.096	1.120	1.151	1.189	1.249	1.331
70	1.032	1.072	1.094	1.118	1.148	1.186	1.245	1.325
72	1.031	1.071	1.093	1.116	1.146	1.184	1.241	1.320
74	1.031	1.070	1.091	1.114	1.143	1.181	1.237	1.315
76	1.030	1.069	1.090	1.112	1.141	1.179	1.233	1.310
78	1.030	1.068	1.089	1.111	1.139	1.176	1.230	1.306
80	1.030	1.067	1.088	1.109	1.137	1.174	1.227	1.301
90	1.028	1.063	1.082	1.102	1.128	1.164	1.212	1.282
100	1.026	1.060	1.078	1.097	1.121	1.155	1.199	1.266
120	1.023	1.054	1.071	1.087	1.109	1.142	1.180	1.240

From "Two Sample Test in the Weibull Distribution," by D. R. Thoman and L. J. Bain, *Technometrics*, Vol. 11, pp. 805–815 (1969). Reproduced by permission of the editor of *Technometrics*.

TABLE C-19   PERCENTAGE POINTS, z_α, SUCH THAT $P(G < z_\alpha) = 1 - \alpha$

n	$1-\alpha$ .60	.70	.75	.80	.85	.90	.95	.98
5	.228	.476	.608	.777	.960	1.226	1.670	2.242
6	.190	.397	.522	.642	.821	1.050	1.404	1.840
7	.164	.351	.461	.573	.726	.918	1.215	1.592
8	.148	.320	.415	.521	.658	.825	1.086	1.421
9	.136	.296	.383	.481	.605	.757	.992	1.294
10	.127	.277	.356	.449	.563	.704	.918	1.195
11	.120	.261	.336	.423	.528	.661	.860	1.115
12	.115	.248	.318	.401	.499	.625	.811	1.049
13	.110	.237	.303	.383	.474	.594	.770	.993
14	.106	.227	.290	.366	.453	.567	.734	.945
15	.103	.218	.279	.352	.434	.544	.704	.904
16	.099	.210	.269	.339	.417	.523	.676	.867
17	.096	.203	.260	.328	.403	.505	.654	.834
18	.094	.197	.251	.317	.389	.488	.631	.805
19	.091	.191	.244	.308	.377	.473	.611	.779
20	.089	.186	.237	.299	.366	.459	.593	.755
22	.085	.176	.225	.284	.347	.435	.561	.712
24	.082	.168	.215	.271	.330	.414	.534	.677
26	.079	.161	.206	.259	.316	.396	.510	.646
28	.076	.154	.198	.249	.303	.380	.490	.619
30	.073	.149	.191	.240	.292	.366	.472	.595
32	.071	.144	.185	.232	.282	.354	.455	.574
34	.069	.139	.179	.225	.273	.342	.441	.555
36	.067	.135	.174	.218	.265	.332	.427	.537
38	.065	.131	.169	.212	.258	.323	.415	.522
40	-.064	.127	.165	.206	.251	.314	.404	.507
42	.062	.124	.160	.201	.245	.306	.394	.494
44	.061	.121	.157	.196	.239	.298	.384	.482
46	.059	.118	.153	.192	.234	.292	.376	.470
48	.058	.115	.150	.188	.229	.285	.367	.460
50	.057	.113	.147	.184	.224	.279	.360	.450
52	.056	.110	.144	.180	.220	.273	.353	.440
54	.055	.108	.141	.176	.215	.268	.346	.432
56	.054	.106	.138	.173	.212	.263	.340	.423
58	.053	.104	.136	.170	.208	.258	.334	.416
60	.052	.102	.134	.167	.204	.254	.328	.408
62	.051	.100	.131	.164	.201	.250	.323	.402
64	.050	.099	.129	.162	.198	.246	.317	.395
66	.049	.097	.127	.159	.195	.242	.313	.389
68	.049	.095	.125	.157	.192	.238	.308	.383
70	.048	.094	.123	.154	.190	.235	.304	.377
72	.047	.092	.122	.152	.187	.231	.299	.372
74	.046	.091	.120	.150	.184	.228	.295	.366
76	.046	.090	.118	.148	.182	.225	.291	.361
78	.045	.089	.117	.146	.180	.222	.288	.357
80	.045	.087	.115	.144	.178	.219	.284	.352
90	.042	.082	.109	.136	.168	.207	.268	.332
100	.040	.077	.103	.128	.160	.196	.255	.315
120	.036	.070	.094	.117	.147	.179	.233	.287

From "Two Sample Test in the Weibull Distribution," by D. R. Thoman and L. J. Bain, *Technometrics*, Vol. 11, pp. 805–815 (1969). Reproduced by permission of the editor of *Technometrics*.

TABLE C-20 RANDOM DIGITS

```
12 67  73 29  44 54  12 73  97 48  79 91  20 20  17 31  83 20  85 66     43 83  39 24  50 74  10 05  38 11  25 80  44 14  98 31  87 41  02 74
06 24  89 57  11 27  43 03  14 29  84 52  86 13  51 70  65 88  60 88     63 19  91 27  08 59  02 28  47 13  05 53  02 28  81 96  46 90  95 52
29 15  84 77  17 86  64 87  06 55  36 44  92 58  64 91  94 48  64 65     23 87  60 31  98 97  76 57  82 47  64 87  50 45  73 54  26 47  62 10
49 56  97 93  91 59  41 21  98 03  70 95  31 99  74 45  67 94  47 79     07 04  47 34  36 03  87 67  03 28  72 19  98 99  32 98  78 76  85 40
50 77  60 28  58 75  70 96  60 66  05 95  58 39  20 25  96 89            98 61  67 62  09 89  73 50  06 81  29 09  43 43  30 21  32 69  82 19

00 31  32 48  23 12  31 08  51 06  23 44  26 43  56 34  78 65  50 80     36 86  50 21  42 18  20 55  00 90  01 96  42 12  68 18  45 93  52 99
01 67  45 57  55 98  93 69  07 81  62 35  22 03  89 22  54 94  83 31     70 64  92 95  09 09  79 63  09 29  69 99  98 26  19 83  94 88  95 37
24 00  48 34  15 45  34 50  02 37  43 57  36 13  76 71  95 40  34 10     41 71  91 61  31 86  38 01  71 79  44 75  67 69  35 31  69 47  81 64
77 52  60 27  64 16  06 83  38 73  51 32  62 85  24 58  95 29  64 56     23 48  32 36  88 50  29 07  27 32  21 28  73 41  77 39  00 78  92 65
36 29  93 93  10 00  51 34  81 26  13 53  26 29  16 94  19 01  40 45     13 32  99 81  00 28  87 13  00 86  56 16  81 20  63 29  37 45  08 91

94 82  03 96  49 78  32 61  17 78  70 12  91 69  99 62  75 16  50 69     70 55  85 27  24 96  91 83  89 17  89 98  51 31  17 29  05 77  62 95
23 12  21 19  67 27  86 47  43 25  25 05  76 17  50 55  70 32  83 36     12 50  84 01  63 40  74 86  88 90  63 76  97 74  08 70  88 88  98 96
77 58  90 38  66 53  45 85  13 93  00 65  30 59  39 44  86 75  90 73     97 00  24 63  47 63  47 66  21 79  28 66  67 24  33 20  01 52  09 59
92 37  51 97  83 78  12 70  41 42  01 72  10 48  88 95  05 24  44 21     16 99  63 29  67 89  14 55  70 31  45 56  05 71  84 30  48 32  90 94
28 93  48 44  13 02  49 32  07 95  26 47  67 70  72 71  08 47  26 18     57 95  93 54  30 74  11 18  31 26  75 39  81 28  63 34  31 23  77 67

09 68  01 98  80 27  49 78  56 67  49 22  13 66  61 33  53 18  36 03     01 32  91 11  23 65  44 58  69 77  58 86  35 20  92 12  48 15  56 67
61 73  92 33  89 48  20 42  32 33  79 37  68 88  44 59  35 17  97 61     00 30  26 68  89 38  13 99  47 38  06 82  49 47  40 33  23 72  01 50
82 35  37 33  53 42  52 04  16 54  08 25  48 89  57 87  59 89  96 76     48 15  27 13  97 70  18 48  14 28  26 30  74 16  13 07  36 21  94 84
39 20  77 72  55 19  66 58  57 91  38 43  67 97  52 66  45 29  74 67     58 86  65 76  67 05  99 53  33 56  92 61  63 98  55 39  15 77  61 67
51 90  71 05  82 38  37 40  94 52  24 09  35 44  37 33  35 20  65 89     75 07  14 81  41 16  12 21  79 82  16 42  70 43  73 33  78 22  63 25

97 49  53 79  17 25  02 65  77 70  88 45  53 51  63 30  89 66  42 03     86 19  97 09  64 04  21 26  65 11  20 32  82 38  52 94  79 21  85 07
73 18  91 38  25 82  29 71  56 89  86 74  68 58  75 36  93 13  33 31     66 17  52 10  35 14  21 89  54 32  61 49  63 06  36 25  63 84  78 24
17 79  34 97  25 89  01 17  67 92  62 25  54 70  52 88  28 05  61 17     56 70  95 77  25 19  21 15  29 88  57 75  51 19  31 06  48 50  09 65
97 27  26 86  17 67  59 56  95 07  49 05  70 06  70 35  21 35  26 18     14 43  67 32  81 78  19 72  32 70  34 86  11 90  37 02  54 39  45 87
56 06  63 00  07 40  65 87  09 49  70 34  67 02  33 39  04 40  01 51     04 17  91 71  96 90  85 68  32 35  77 20  71 43  55 95  28 90  51 69
```

References

Aaby, P., Bukh, J., Kronborg, D., Lisse, I. M., and Da Silva, M. C. (1990). Delayed Excess Mortality after Exposure to Measles during the First Six Months of Life. *American Journal of Epidemiology*, **132**, 211–219.

Abdi, E. A., Hanson, J., and McPherson, T. A. (1987). Adjuvant Chemoimmunotherapy after Regional Lymphadenectomy for Malignant Melanoma. *American Journal of Clinical Oncology*, **10**, 117–122.

Abramowitz, M., and Stegun, I. A. (1964). *Handbook of Mathematical Functions with Formulas, Graphs, and Mathematical Tables*. National Bureau of Standards, Applied Mathematics Series 55.

Aitchison, J. (1970). Statistical Problems of Treatment Allocation. *Journal of the Royal Statistical Society, Series A*, **133**, 206–238.

Aitchison, J., and Brown, J. A. C. (1957). *The Lognormal Distribution*. Cambridge University Press, Cambridge.

Aitkin, M., Laird, N., and Francis, B. (1983). A Reanalysis of the Stanford Heart Transplant Data, with discussion. *Journal of the American Statistical Association*, **78**, 264–292.

Altshuler, B. (1970). Theory for Measurement of Competing Risks in Animal Experiments. *Mathematical Biosciences*, **6**, 1–11.

Anderson, J. A. (1972). Separate Sample Logistic Discrimination. *Biometrika*, **59**, 19–35.

Anderson, P. K. (1982). Testing Goodness of Fit of Cox's Regression Model. *Biometrics*, **38**, 67–77.

ARIC Investigators. (1989). The Atherosclerosis Risk in Communities (ARIC) Study: Design and Objectives. *American Journal of Epidemiology*, **129**, 687–702.

Arjas, E. (1988). A Graphical Method for Assessing Goodness of Fit in Cox's Proportional Hazards Model. *Journal of the American Statistical Association*, **83**, 204–212.

Armitage, P. (1958). Sequential Methods in Clinical Trials. *American Journal of Public Health*, **48**, 1395–1402.

Armitage, P. (1959). The Comparison of Survival Curves. *Journal of the Royal Statistical Society, Series A*, **122**, 279–300.

453

Armitage, P. (1971). *Statistical Methods in Medical Research*. Blackwell Scientific Publications, Oxford.

Armitage, P. (1975). *Sequential Medical Trials*, 2nd ed. Blackwell Scientific Publications, Oxford.

Armitage, P. (1981). Importance of Prognostic Factors in the Analysis of Data from Clinical Trials. *Controlled Clinical Trials*, **1**, 347–353.

Armitage, P., and Gehan, E. A. (1974). Statistical Methods for the Identification and Use of Prognostic Factors. *International Journal of Cancer*, **13**, 16–35.

Armitage, P., McPherson, K., and Rowe, B. C. (1969). Repeated Significance Tests on Accumulating Data. *Journal of The Royal Statistical Society*, *Series A*, **132**, 235–244.

Asal, N. R., Geyer, J. R., Risser, D. R., Lee, E. T., Kadamani, S., and Cherng, N. (1988a). Risk Factors in Renal Cell Carcinoma I. Methodology, Demographics, Tobacco, Beverage and Obesity. *Cancer Detection and Prevention*, **11**, 359–377.

Asal, N. R., Geyer, J. R., Risser, D. R., Lee, E. T., Kadamani, S., and Cherng, N. (1988b). Risk Factors in Renal Cell Carcinoma II. Medical History, Occupation, Multivariate Analysis and Conclusion. *Cancer Detection and Prevention*, **13**, 263–279.

Bain, L. J. (1972). Inferences Based on Censored Sampling from the Weibull or Extreme-Value Distribution. *Technometrics*, **14**, 693–702.

Barnard, G. A. (1963). Some aspects of the Fiducial Argument. *Journal of the Royal Statistical Society*, *Series B*, **34**, 216–217.

Bartholomew, D. J. (1957). A Problem in Life Testing. *Journal of the American Statistical Association*, **52**, 350–355.

Bartholomew, D. J. (1963). The Sampling Distribution of an Estimate Arising in Life Testing. *Technometrics*, **5**, 361–374.

Baumgartner, R. N., Roche, A. F., et al. (1987). Fatness and Fat Patterns: Associations with Plasma Lipids and Blood Pressure in Adults, 18 to 57 Years of Age. *American Journal of Epidemiology*, **126**, 614–628.

Beale, E. M. L., Kendall, M. G., and Mann, D. W. (1976). The Discarding of Variable in Multivariate Analysis. *Biometrika*, **54**, 357–366.

Bentzen, S. M., Poulsen, H. S., Kaae, S., Jensen, O. M., Johansen, H., Mouridsen, H. T., Daugaard, S., and Arnoldi, C. (1988). Prognostic Factors in Osteosarcomas. *Cancer*, **62**, 194–202.

Berkson, J. (1942). The Calculation of Survival Rates, in *Carcinoma and Other Malignant Lesions of the Stomach*, edited by W. Walters, H. K. Gray and J. T. Priestley. W. B. Saunders, Philadelphia.

Berkson, J., and Gage, R. R. (1950). Calculation of Survival Rates for Cancer. *Proceedings of Staff Meetings*, Mayo Clinic, **25**, 250.

Billmann, B. R., and Antle, C. E. (1972). Statistical Inference from Censored Weibull Samples. *Technometrics*, **14**, 831–840.

Birnbaum, Z. W., and Saunders, S. C. (1958). A Statistical Model for Life-Length of Materials. *Journal of American Statistical Association*, **53**, 151–160.

Boag, J. W. (1949). Maximum Likelihood Estimates of Proportion of Patients Cured by Cancer Therapy. *Journal of the Royal Statistical Society*, *Series B*, **11**, 15.

Bodey, G. P., Gehan, E. A., Feireich, E. J., et al. (1971). Protected Environment—Prophylactic Antibiotic Program in the Chemotherapy of Acute Leukemia. *American Journal of Medical Science*, **262**, 138–151.

Bolin, R. B., and Greene, J. R. (1986). Stored Platelet Survival Data Analysis by a Gamma Model. *Transfusion*, **26**, 28–30.

Bonadonna, G., et al. (1976). Combination Chemotherapy as an Adjuvant Treatment in Operable Breast Cancer. *The New England Journal of Medicine*, **294**, 405–410.

Breslow, N. (1970). A Generalized Kruskal–Wallis Test for Comparing *K* Samples Subject to Unequal Pattern of Censorship. *Biometrika*, **57**, 579–594.

Breslow, N. (1974). Covariance Analysis of Survival Data under the Proportional Hazards Model. *International Statistics Review*, **43**, 43–54.

Breslow, N. E. (1975). Analysis of Survival Data Under the Proportional Hazards Model. *International Statistical Review*, **43**, 45–48.

Breslow, N. E., and Crowley, J. (1974). A Large Sample Study of the Life Table and Product Limit Estimates under Random Censoring. *Annals of Statistics*, **2**, 437–453.

Breslow, N. E., and Day, N. E. (1980). Statistical Methods in Cancer Research, Vol 1—The Analysis of Case–Control Studies. International Agency for Research on Cancer, Lyon.

Breslow, N., and Powers, W. (1978). Are There Two Logistic Regressions for Retrospective Studies? *Biometrics*, **34**, 100–105.

Breslow, N. E., Day, N. E., Halvorsen, K. T., Prentice, R. L., and Sabai, C. (1978). Estimation of Multiple Relative Risk Functions in Matched Case–Control Studies. *American Journal of Epidemiology*, **108**, 299–307.

Bristol, D. R. (1988). A One-Sided Interim Analysis with Binary Outcomes. *Controlled Clinical Trials*, **9**, 206–211.

Broadbent, S. (1958). Simple Mortality Rates. *Journal of Applied Statistics*, **7**, 86.

Broderick, A., Mori, M., Nettleman, M. D., Streed, S. A., and Wenzel, R. P. (1990). Nosocomial Infections: Validation of Surveillance and Computer Modeling to Identify Patients at Risk. *American Journal of Epidemiology*, **131**, 734–742.

Brookmeyer, R., and Goedert, J. J. (1989). Censoring in an Epidemic with an Application to Hemophilia-associated AIDS. *Biometrics*, **45**, 325–335.

Brown, B. W. Jr. (1970). Controls in Protocol Design. *Proceedings of Symposium on Statistical Aspects of Protocol Design*. National Cancer Institute, pp. 161 180.

Brown, B. W. Jr. (1980a). Statistical Controversies in the Design of Clinical Trials—Some Personal Views. *Control Clinical Trials*, **1**, 13–27.

Brown, B. W. Jr. (1980b). The Cross-Over Experiment for Clinical Trials. *Biometrics*, **36**, 69–80.

Brown, B. W., and Hollander, M. (1977). *Statistics, A Biomedical Introduction*. Wiley, New York.

Brown, C. C. (1982). On a goodness-of-Fit Test for the Logistic Model Based on Score Statistics. *Communications in Statistics*, **11**, 1087–1105.

Brown, G. W., and Flood, M. M. (1947). Tumbler Mortality, *Journal of the American Statistical Association*, **42**, 562–574.

Bull, J. P. (1959). The Historical Development of Clinical Therapeutic Trials. *Journal of Chronic Diseases*, **10**, 214–248.

Burdette, W. J., and Gehan, E. A. (1970). *Planning and Analysis of Clinical Studies*. Charles C. Thomas, Springfield, IL.

Buyse, M. E., Staquet, M. J., and Sylvester, R. J., Ed. (1984). *Cancer Clinical Trials: Methods and Practices*. Oxford University Press, Oxford.

Buzdar, A. U., Gutterman, J. U., Blumehscein, G. R., Hortobagiji, G. H., Tashima, C. K., Smith, T. L., Hersh, E. M., Freiriech, E. J., and Gehan, E. A. (1978). Intensive Postoperative Chemoimmunotherapy for Patients with Stage II and Stage III Breast Cancer. *Cancer*, **41**, 1064–1075.

Byar, D. P. (1974). Selecting Optimum Treatment in Clinical Trials Using Covariate Information. Presented at the 1974 Annual Meeting of the American Statistical Association, August 28, 1974.

Byar, D. P., Huse, R., and Bailar, J. C. III, and the Veterans Administration Cooperative Urological Research Group (1974). An exponential Model Relating Censored Survival Data and Concomitant Information for Prostatic Cancer Patients. *Journal of the National Cancer Institute*, **52**, 321–326.

Carbone, P., Kellerhouse, L., and Gehan, E. (1967). Plasmacytic Myeloma: A Study of the Relationship of Survival to Various Clinical Manifestations and Anomalous Protein Type in 112 Patients. *American Journal of Medicine*, **42**, 937–948.

Carbone, P., Spurr, C., Schneiderman, M., et al. (1968). Management of Patient with Malignant Lymphoma: A Comparative Study with Cyclophosphamide and Vinca Alkoids. *Cancer Research*, **28**, 811–822.

Carlo, W. A., Siner, B., and Chatburn, R. L. (1990). Early Randomized Intervention with High-Frequency Jet Ventilation in Respiratory Distress Syndrome, *Journal of Pediatrics*, **177**, 765–770.

Carnahan, B. Luther, H. A., and Wilkes, J. O. (1969). *Applied Numerical Methods*. Wiley, New York.

Carter, S. K., Oleg, S., and Slavik, M. (1977). Phase I Clinical Trials. In *Methods of Development of New Anticancer Drugs*, National Cancer Institute Monograph 45, Department of Health, Education, and Welfare Publication No. (NIH) 76-1037, National Cancer Institute, Bethesda, Maryland.

Chalmers, T. C. (1975). Symposium on Diseases of the Liver: Randomization of the First Patient. *Medical Clinics of North America*, **59**, 1035.

Chalmers, T. C., Block, J. B., and Lee, S. (1972). Controlled Studies in Clinical Cancer Research. *New England Journal of Medicine*, **287**(2) 75–78.

Chernoff, H., and Leiberman, G. J. (1954). Use of Normal Probability Paper. *Journal of the American Statistical Association*, **49**, 778–785.

Chiang, C. L. (1961). Standard Error of the Age-adjusted Death Rate. *Vital Statistics—Special Reports, Selected Studies*, 47, 9. U.S. Department of HEW, Washington, DC.

Chiang, C. L. (1968). *Introduction to Stochastic Processes in Biostatistics*. Wiley, New York.

Chiasson, M. A., Stoneburner, R. L., et al. (1990). Risk Factors for Human Immunodeficiency Virus Type 1 (HIV-1) Infection in Patients at a Sexually Transmitted Disease Clinic in New York City. *American Journal of Epidemiology*, **131**, 208–220.

Claus, E. B., Risch, N. J., and Thompson, W. D. (1990). Age at Onset as an Indicator of Familial Risk of Breast Cancer. *American Journal of Epidemiology*, **131**, 961–972.

Cochran, W. G., and Cox, G. M. (1957). *Experimental Designs*, 2nd. Wiley, New York.

Cohen, A. C., Jr. (1951). Estimating Parameters of Logarithmic–Normal Distributions by Maximum Likelihood. *Journal of the American Statistical Association*, **46**, 206–212.

Cohen, A. C., Jr. (1959). Simplified Estimators for the Normal Distribution When Samples are Singly Censored or Truncated. *Technometrics*, **1**(3) 217–237.

Cohen, A. C., Jr. (1961). Table for Maximum Likelihood Estimates: Singly Truncated and Singly Censored Samples. *Technometrics*, **3**, 535–541.

Cohen, A. C., Jr. (1963). Progressively Censored Sample in Life Testing. *Technometrics*, **5**, 327–339.

Cohen, A. C. Jr. (1965). Maximum Likelihood Estimation in the Weibull Distribution Based on Complete and on Censored Samples. *Technometrics*, **7**, 579–588.

Cohen, A. C. Jr. (1976). Progressively Censored Sampling in the Three Parameter Log-Normal Distribution. *Technometrics*, **18**.

Cohen, J., and Cohen, P. (1975). *Applied Multiple Regression/Correlation Analysis for the Behavioral Sciences*. Lawrence Erlbaum Associates, Hillsdale, NJ.

Collins, J. A., Garner, J. B., Wilson, E. H., Wrixon, W., and Casper, R. F. (1984). A Proportional Hazards Analysis of the Clinical Characteristics of Infertile Couples. *American Journal of Obstetrics and Gynecology*, **148**, 527–532.

Connelly, R. R., Cutler, S. J., and Baylis, P. (1966). End Result in Cancer of the Lung: Comparison of Male and Female Patients. *Journal of the National Cancer Institute*, **36**, 277–287.

Cornfield, J. (1951). A Method of Estimating Comparative Rates from Clinical Data. Applications to Cancer of the Lung, Breast and Cervix. *Journal of the National Cancer Institute*, **11**, 1269–1275.

Cornfield, J. (1956). A Statistical Problem Arising From Retrospective Studies. In J. Neyman (Ed.), *Proceedings of the Third Berkeley Symposium on Mathematical Statistics and Probability*, Vol. 4, 135–148, University of California Press, Berkeley, California.

Cornfield, J. (1962). Joint Dependence of Risk of Coronary Heart Disease in Serum Cholesterol and Systolic Blood Pressure: A Discriminant Function Analysis. *Federation Proceeding*, **21**, 58–61.

Correa, P., Pickle, L. W., Fortham, E., et al. (1983). Passive Smoking and Lung Cancer. *Lancet*, **2**, 595–597.

Cox, D. R. (1953). Some Simple Tests for Poisson Variates. *Biometrika*, **40**, 354–360.

Cox, D. R. (1958). *Planning of Experiments*. Wiley, New York.

Cox, D. R. (1959). The Analysis of Exponentially Distributed Life-times with Two Types of Failures. *Journal of the Royal Statistical Society, Series B*, **21**, 411–421.

Cox, D. R. (1962). *Renewal Theory*. Methuen, London.

Cox, D. R. (1964). Some Applications of Exponentially Distributed Life-times with

Two Types of Failures. *Journal of the Royal Statistical Society, Series, B,* **26,** 103–110.

Cox, D. R. (1970). *Analysis of Binary Data.* Methuen, London.

Cox, D. R. (1972). Regression Models and Life Tables. *Journal of the Royal Statistical Society,* **34,** 187–220.

Cox, D. R., Oakes, D. (1984). *Analysis of Survival Data.* Chapman and Hall, New York.

Cox, D. R., Snell, E. J. (1968). A General Definition of Residuals. *Journal of the Royal Statistical Society, Series B,* **30,** 248–275.

Crist, W., Boyett, J., and Jackson, J., et al. (1989). Prognostic Importance of the Pre-B-Cell Immunophenotype and Other Presenting Features in B-Lineage Childhood Acute Lymphoblastic Leukemia: A Pediatric Oncology Group Study. *Blood,* **74,** 1252–1259.

Crowley, J., and Hu, M. (1977). Covariance Analysis of Heart Transplant Survival Data. *Journal of the American Statistical Association,* **72,** 27–36.

Crowley, J., and Thomas D. R. (1975). Large Sample Theory for the Log Rank Test. Technical Report No. 415. Department of Statistics, University of Wisconsin.

Cutler, S. J., Griswold, M. H., and Eisenberg, H. (1957). An Interpretation of Survival Rates: Cancer of the Breast. *Journal of the National Cancer Institute,* **19,** 1107–1117.

Cutler, S. J., and Ederer, F. (1958). Maximum Utilization of the Life Table Method in Analyzing Survival. *Journal of Chronic Diseases,* **8,** 699–712.

Cutler, S. J., Ederer, F., Griswold, M. H., and Greenberg, R. A. (1959). Survival of Breast-Cancer Patients in Connecticut, 1935–54. *Journal of the National Cancer Institute,* **23,** 1137–1156.

Cutler, S. J., Ederer, F., Griswold, M. H., and Greenberg, R. A. (1960a). Survival of Patients with Uterine Cancer, Connecticut, 1935–54. *Journal of the National Cancer Institute,* **24,** 519–539.

Cutler, S. J., Ederer, F., Griswold, M. H., and Greenberg, R. A. (1960a). Survival of Patients with Ovarian Cancer, Connecticut, 1935–54. *Journal of the National Cancer Institute,* **24,** 541–549.

Cutler, S. J., Axtell, L., and Heise, H. (1967). Ten Thousand Cases of Leukemia: 1940–62. *Journal of the National Cancer Institute,* **39,** 993–1026.

Daniel, C. (1959). Use of Half-Normal Plots in Interpreting Factorial Two-Level Experiments. *Technometrics,* **1,** 311–341.

Daniel, W. W. (1987). *Biostatistics: A Foundation for Analysis in the Health Sciences.* Wiley, New York.

Daniels, M., and Hill, A. B. (1952). Chemotherapy of Pulmonary Tuberculosis in Young Adults. *British Medical Journal,* **1,** 1162–1168.

Davis, B. R., Blaufox, M. D., Hawkins, C. M., et al. (1989). Trial of Antihypertensive Interventions and Management. *Controlled Clinical Trials,* **10,** 11–30.

Davis, D. J. (1952). An Analysis of Some Failure Date. *Journal of the American Statistical Association,* **47,** 113–150.

Davis, H. T., and Feldstein, M. L. (1979). The Generalized Pareto Law as a Model for Progressively Censored Survival Data. *Biometrika*, **66**, 299–306.

Dawber, T. R. (1980). *The Framingham Study*. Harvard University Press, Cambridge, MA.

Dawber, T. R., Meadors, G. F., and Moore, F. E. Jr. (1951). Epidemiological Approaches to Heart Disease: The Framingham Study. *American Journal of Public Health*, **41**, 279–286.

Dempster, A. P., Selwyn, M. R., and Weeks, B. J. (1983). Combining Historical and Randomized Controls for Assessing Trends in Proportions. *Journal of American Statistical Association*, **78**, 221–227.

DeMets, D. L., and Lan, K. K. G. (1984). An Overview of Sequential Methods and Their Application in Clinical Trials. *Communications in Statistics: Theory and Methods*, **13**, 2315–2338.

DeMets, D. L., and Ware, J. H. (1980). Group Sequential Methods for Clinical Trials with a One-sided Hypothesis. *Biometrika*, **67**, 651–660.

DeWals, P., and Bouckaert, A. (1985). Methods for Estimating the Duration of Bacterial Carriage. *International Journal of Epidemiology*, **14**, 628–634.

Dixon, W. J., Brown, M. B., Engelman, L., Hill, M. A., and Jennrich, R. I. (1988). *BMDP Statistical Software Manual*, Vol. 2. University of California Press, Berkeley, Los Angeles, and London.

Donadio, J. V., and Offord, K. P. (1989). Reassessment of Treatment Results in Membranoproliferative Glomerulonephritis, with Emphasis on Life-Table Analysis. *American Journal of Kidney Diseases*, **XIV**, 445–451.

Draper, N. R., and Smith, H. (1966). *Applied Regression Analysis*. Wiley, New York.

Drenick, R. F. (1960). The Failure Law of Complex Equipment. *Journal of Social and Industrial Applied Mathematics*, **8**, 680.

Droz, J. P., Kramar, A., and Ghosn, M., et al. (1988). Prognostic Factors in Advanced Nonseminomatous Testicular Cancer. *Cancer*, **62**, 564–568.

Dunn, O. J. (1964). New Table for Multiple Comparisons with a Control. *Biometrics*, **20**, 482–491.

Ederer, F., Axtell, L. M., and Cutler, S. J. (1961). The Relative Survival Rate—A Statistical Methodology. *National Cancer Institute Monographs*, **6**, 101–121.

Efron, B. (1975). The Efficiency of Logistic Regression Compared to Normal Discriminant Analysis. *Journal of the American Statistical Association*, **70**, 892–898.

Efron, B. (1977). The Efficiency of Cox's Likelihood Function for Censored Data. *Journal of the American Statistical Association*, **72**, 557–565.

Eggermont, J. J. (1988). On the Rate of Maturation of Sensory Evoked Potentials. *Electroencephalography and Clinical Neurophysiology*, **70**, 293–305.

Eisenberger, M., Krasnow, S., Ellenberg, S., et al. (1989). A Comparison of Carboplatin Plus Methotrexate Versus Methotrexate Alone in Patients with Recurrent and Metastatic Head and Neck Cancer. *Journal of Clinical Oncology*, **7**, 1341–1345.

Elaad, E., and Ben-Shakhar, G. (1989). Effects of Motivation and Verbal Response

Type on Psychophysiological Detection of Information. *Psychophysiology*, **26**, 442–451.

Elandt-Johnson, R. C., and Johnson, N. L. (1980). *Survival Models and Data Analysis*. Wiley, New York.

Enas, G. G., Dornseit, B. E., Sampson, C. B., Rockhold, F. W., and Wuu, J. (1989). Monitoring Versus Interim Analysis of Clinical Trials: A Perspective from the Pharmaceutical Industry. *Controlled Clinical Trials*, **10**, 57–70.

Engelhardt, M. (1975). On Simple Estimation of Parameters of the Weibull or Extreme-value Distribution. *Technometrics*, **17**, 369–374.

Engelman, L., (1978). A Computer Program for Model Building with Stepwise Logistic Regression. Technical Report No. 37, Health Sciences Computing Facility, University of California, Los Angeles.

Epstein, B. (1958). The Exponential Distribution and Its Role in Life Testing. *Industrial Quality Control*, **15**, 2–7.

Epstein, B. (1960a). Estimation of the Parameters of Two Parameter Exponential Distribution from Censored Samples. *Technometrics*, **2**, 403–406.

Epstein, B. (1960b). Estimation from Life Test Data. *Technometrics*, **2**, 447–454.

Epstein, B., and Sobel, M. (1953). Life Testing. Journal of the American Statistical Association, **48**, 486–502.

Farchi, G., Menotti, A., and Conti, S. (1987). Coronary Risk Factors and Survival probability from Coronary and other Causes of Death. *American Journal of Epidemiology*, **126**, 400–408.

Feigl, P., and Zelen M. (1965). Estimation of Exponential Survival Probabilities with Concomitant Information. *Biometrics*, **21**, 826–838.

Feinleib, M. (1960). A Method of Analyzing Log-Normally Distributed Survival Data with Incomplete Follow-up. *Journal of the American Statistical Association*, **55**, 534–545.

Feinleib, M., and MacMahon, B. (1960). Variation in the Duration of Survival of Patients with Chronic Leukemias. *Blood*, **17**, 332–349.

Feinstein, A. R. (1977). *Clinical Biostatistics*. C. V. Mosby, St. Louis.

Feinstein, A. R., and Landis, J. R. (1976). The Role of Prognostic Stratification in Preventing the Bias Permitted by Random Allocation of Treatment. *Journal of Chronic Diseases*, **29**, 277–284.

Fisher, E. R., Redmond, C., Fisher, B., Bass, G., and Contributing NSABP Investigators. (1990). Pathologic Findings from the National Survival Adjuvant Breast and Bowel Projects (NSABP)—Prognostic Discriminants for 8-Year Survival for Node-Negative Invasive Breast Cancer Patients. *Cancer*, **65**, 2121–2128.

Fisher, R. A. (1922). On the Mathematical Foundation of Theoretical Statistics. *Philosophical Transactions of the Royal Society of London, Series A*, **222**.

Fisher, R. A. (1936). The Use of Multiple Measurements in Toxonomic Problems. *Annals of Eugenics*, **7**, 312–330.

Fleiss, J. L. (1979). Confidence Intervals for the Odds Ratio in Case–Control Studies: The State of the Art. *Journal of Chronic Diseases*, **32**, 69–82.

Fleiss, J. L. (1981). Statistical Methods for Rates and Proportions. Wiley, New York.

Fleming, T. R., and Harrington, D. P. (1979). Non-parametric Estimation of the Survival Distribution in Censored Data. Unpublished manuscript.

Fleming, T. R., Harrington, D. P., and O'Brien, P. C. (1984). Designs for Group Sequential Tests. *Controlled Clinical Trials*, **5**, 348–361.

Fleming, T. R., O'Fallon, J. R., O'Brian, P. C., and Harrington, D. P. (1980). Modified Kolmogorov-Smirnov Test Procedures with Application to Arbitrarily Right Censored Data. *Biometrics*, **36**, 607–626.

Fraser, D. A. S. (1968). *The Structure of Inference*. Wiley, New York.

Freedman, L. S. (1982). Tables of the Number of Patients Required in Clinical Trials Using the Logrank Test. *Statistics in Medicine*, **1**, 121–129.

Frei, E., et al. (1961). Studies of Sequential and Combination Antimetabolite Therapy in Acute Leukemia: 6-Mercaptopurine and Methotrexate. *Blood*, **18**, 431–454.

Frei, E. III, Luce, J. K., Gamble, J. F., et al. (1973). Combination Chemotherapy in Advanced Hodgkin's Disease: Induction and Maintenance of Remission. *Annals of Internal Medicine*, **79**, 376–382.

Freireich E. J., and Gehan, E. A. (1974). Historical Controls Reincarnated: An Examination of Techniques for Clinical Therapeutic Research. *Cancer Chemotherapy Reports, Part 1*, **58**, 623–626.

Freireich, E. J., and Gehan, E. A. (1979). The Limitations of the Randomized Clinical Trials. *Methods in Cancer Research*, **17**, 227–310.

Freireich, E. J., Gehan, E. A., Frei, E., et al. (1963). The Effect of 6-Mercaptopurine on the Duration of Steroid-Induced Remissions in Acute Leukemia: A Model for Evaluation of Other Potential useful therapy. *Blood*, **21**(6), 699–716.

Freireich, E. J., Gehan, E. A., Rall, D. P., Schmidt, L. H., and Skipper, H. E. (1966). Quantitative Comparison of Toxicity of Anticancer Agents in Mouse, Rat, Hamster, Dog, Monkey and Man. *Cancer Chemotherapy Report*, **50**, 4.

Freireich, E. J., Gehan, E. A., Bodey, G. P., Hersh, E. M., Hart, J. S., Gutterman, J. U., and McCredie, K. B. (1974). New Prognostic Factors Affecting Response and Survival in Adult Leukemia. *Transactions of the Association of American Physicians*, **87**, 298–305.

Friedman, L. M., Furberg, C. D., and DeMets, D. L. (1985). *Fundamentals of Clinical Trials*, 2nd edition. PSG Publishing Company, Inc., Littleton, Massachusetts.

Gaddum, J. H. (1945a). Log Normal Distributions. *Nature, London*, **156**, 463.

Gaddum, J. H. (1945b). Log Normal Distributions, *Nature, London*, **156**, 747.

Gail, M., and Gart, J. J. (1973). The Determination of Sample Sizes for Use with the Exact Conditional Test in 2×2 Comparative Trials. *Biometrics*, **29**, 441–448.

Gajjar, A. V., and Khatri, C. G. (1969). Progressively Censored Samples from Log-Normal and Logistic Distributions. *Technometrics*, **11**, 793–803.

Galli, G., Maini, C. L., Salvatori, M., and Andreasi, F. (1983). A Practical Approach to the Hepatobiliary Kinetics of $^{99M}T_c$—HIDA: Clinical Validation of the Method and a Preliminary Report on Its Use for Parametric Imaging. *European Journal of Nuclear Medicine*, **8**, 292–298.

Garside, M. J. (1965). The Best Sub-set in Multiple Regression Analysis. *Applied Statistics*, **14**, 196–200.

Gehan, E. A. (1961). The Determination of the Number of Patients Required in a Prelininary and a Follow-up Trial of a New Chemotherapeutic Agent. *Journal of Chronic Diseases*, **13**, 346–353.

Gehan, E. A. (1965a). A Generalized Wilcoxon Test for Comparing Arbitrarily Singly-Censored Samples. *Biometrika*, **52**, 203–223.

Gehan E. A. (1965b). A Generalized Two-Sample Wilcoxon Test for Doubly-Censored Data. *Biometrika*, **52**, 650–653.

Gehan, E. A. (1969). Estimating Survival Function from the Life Table. *Journal of Chronic Diseases*, **21**, 629–644.

Gehan, E. A. (1970). Unpublished notes on survival time studies. The University of Texas M.D. Anderson Cancer Center, Houston, Texas.

Gehan, E. A. (1982a). Progress of Therapy in Acute Leukemia 1948–1981: Randomized versus Nonrandomized Clinical Trials. *Controlled Clinical Trials*, **3**, 199–207.

Gehan, E. A. (1982b). Design of Controlled Clinical Trials: Use of Historical Controls. *Cancer Treatment Report*, **66**, 1089–1093.

Gehan, E. A., and Freireich, E. J. (1974). Non-Randomized Controls in Cancer Clinical Trials. *New England Journal of Medicine*, **290**, 198–203.

Gehan, E. A., and Schneiderman, M. A. (1973). Experimental Design of Clinical Trials, Chapter VIII in *Cancer Medicine*, edited by J. F. Holland and E. Frei, III. Lea & Febiger, Philadelphia.

Gehan, E. A., and Siddiqui, M. M. (1973). Simple Regression Methods for Survival Times Studies. *Journal of the American Statistical Association*, **68**, 848–856.

Gehan, E. A., and Thomas, D. G. (1969). The Performance of Some Two-Sample Tests in Small Samples with and without Censoring. *Biometrika*, **56**, 127–132.

Gelenberg, A. J., Kane, J. M., Keller, M. B., et al. (1989). Comparison of Standard and Low Serum Levels of Lithium for Maintenance Treatment of Bipolar Disorder. *New England Journal of Medicine*, **321**, 1489–1493.

Geller, N., Kemeny, N., Yagoda, A., Cheng, E., Sordillo, P., and Hollander, P. (1984). Randomized Clinical Trial for Colorectal Carcinoma with Planned Interim Data Analysis: A First Experience and Implications for Future Design. *American Association for Cancer Research Abstracts*, **C-628**, 159.

Geller, N. L. (1987). Planned Interim Analysis and Its Role in Cancer Clinical Trials. *Journal of Clinical Oncology*, **5**, 1485–1490.

Geller, N. L., and Pocock, S. J. (1987). Interim Analyses in Randomized Clinical Trials: Ramifications and Guidelines for Practitioners. *Biometrics*, **43**, 213–223.

Geller, N. L., Bosl, G. J., and Chan, E. Y. W. (1989). Prognostic Factors for Relapse after Complete Response in Patients with Metastatic Germ Cell Tumors. *Cancer*, **63**, 440–445.

George, S. L., and Desu, M. M. (1974). Planning the Size and Duration of a Clinical Trial Studying the Time to Some Critical Event. *Journal of Chronic Diseases*, **27**, 15–24.

George, S. L., Fernback, D. J., et al. (1973). Factors Influencing Survival in Pediatric Acute Leukemia, The SWCCSG Experience, 1959–1970. *Cancer*, **32**, 1542–1553.

Gertsbakh, I. B. (1989). *Statistical Reliability Theory*. Marcel Dekker, New York.

Gill, R., and Schumacher, M. (1987). A Simple Test of the Proportional Hazards Assumption. *Biometrika*, **74**, 289–300.

Gillum, R. F., Fortmann, S. P., Prineas, R. J., and Kottke, T. E. (1984). International Diagnostic Criteria for Acute Myocardial Infarction and Acute Stroke. *American Heart Journal*, **108**, 150–158.

Glasser, M. (1967). Exponential Survival with Covariance. *Journal of the American Statistical Association*, **62**, 561–568.

Gompertz, B. (1825). On the Nature of the Function Expressive of the Law of Human Mortality and on the New Mode of Determining the Value of Life Contingencies. *Philosophical Transactions*, **513**.

Gray, R. J. (1990). Some Diagnostic Methods for Cox Regression Models through Hazard Smoothing. *Biometrics*, **46**, 93–102.

Greenwood, J. A., and Durand, D. (1960). Aids for Fitting the Gamma Distribution by Maximum Likelihood. *Technometrics*, **2**, 55–65.

Greenwood, M. (1926). The Natural Duration of Cancer. *Reports on Public Health and Medical Subjects*, Her Majesty's Stationery Office, London, **33**, 1–26.

Griswold, M. H., and Cutler, S. J. (1956). The Connecticut Cancer Register, Seventeen Years of Experience. *Connecticut Medical Journal*, **20**, 366–372.

Griswold, M. H., Wilder, C. S., Cutler, S. J., and Pollack, E. S. (1955). *Cancer in Connecticut 1935–1951*, Monograph. Connecticut State Department of Health, Hartford, CT.

Grizzle, J. E. (1967). Continuity Correction in the χ^2-Test for 2×2 Tables. *The American Statistician*, **21**, 28–32.

Grizzle, J. E. (1982). A Note on Stratifying versus Complete Random Assignment in Clinical Trials. *Controlled Clincal Trials*, **3**, 365–368.

Gross, A. J., and Clark, V. A. (1975). *Survival Distributions: Reliability Applications in the Biomedical Sciences*. Wiley, New York.

Grove, R. D., and Hetzel, A. M. (1963). *Vital Statistics Rates in the United States, 1940–1960*. National Center for Health Statistics, Washington DC.

Gupta, A. K. (1952). Estimation of the Mean and Standard Deviation of a Normal Population from a Censored Sample. *Biometrika*, **39**, 260–273.

Gupta, S. S. (1960). Order Statistics form the Gamma Distribution. *Technometrics*, **2**, 243–262.

Gupta, S. S., and Groll, P. A. (1961). Gamma Distribution in Acceptance Sampling Based on Life Tests. *Journal of the American Statistical Association*, **56**, 942–970.

Hahn, G. J., and Shapiro, S. S. (1967). *Statistical Models in Engineering*. Wiley, New York.

Haldane, J. B. S. (1955). The Estimation and Significance of the Logarithm of a Ratio of Frequencies. *Annals of Human Genetics*, **20**, 309–311.

Halperin, M. (1952). Maximum Likelihood Estimation in Truncated Samples. *Annals of Mathematical Statistics*, **23**, 226–238.

Halperin, M., Blackwelder, W. C., and Verter, J. I. (1971). Estimation of the Multivariate Logistic Risk Function: A Comparison of the Discriminant Function and Maximum Likelihood Approaches. *Journal of Chronic Diseases*, **24**, 125–158.

Hammond, I. W., Lee, E. T., Davis, A. W., and Booze, C. F. (1984). Prognostic Factors Related to Survival and Complication-Free Times in Airmen Medically Certified after Coronary Bypass Surgery. *Aviation, Space, and Environmental Medicine*, April, 321–331.

Hanson, B. S., Isacsson, S-O., Janzon, L., and Lindell, S. E. (1989). Social Network and Social Support Influence Mortality in Elderly Men. *American Journal of Epidemiology*, **130**, 100–111.

Haldane, J. B. S. (1956). The Estimation and Significance of the Logarithm of a Ratio of Frequencies. *Annals of Human Genetics*, **20**, 309–311.

Harrell, F. (1980a). The PHGLM Procedure. SAS Technical Report S-109. SAS Institute, Raleigh, NC.

Harrell, F. (1980b). The LOGIST Procedure. SAS Technical Report S-110. SAS Institute, Raleigh, NC.

Harrison, J. D., Jones, J. A., and Morris, D. L. (1990). The Effect of the Gastrin Receptor Antagonist Proglumide on Survival in Gastric Carcinoma. *Cancer*, **66**, 1449–1452.

Hart, J. S., George, S. L., Frei, E., Bodey, G. P., Nickerson, R. C., and Freireich, E. J. (1977). Prognostic Significance of Pretreatment Proliferative Activity in Adult Acute Leukemia. *Cancer*, **39**, 1603–1617.

Harter, H. L., and Moore, A. H. (1965). Maximum Likelihood Estimation of the Parameters of Gamma and Weibull Populations from Complete and from Censored Samples. *Technometrics*, **7**, 639–643.

Harter, H. L., and Moore, A. H. (1966). Local Maximum Likelihood Estimation of the Parameters of Three-Parameter Log-Normal Population from Complete and Censored Sample. *Journal of the American Statistical Association*, **61**, 842–851.

Harter, H. L., and Moore, A. H. (1967). Asymptotic Variance and Covariances of Maximum Likelihood Estimators, from Censored Samples, of the Parameters of Weibull and Gamma Parameters. *Annals of Mathematical Statistics*, **38**, 557–570.

Hastings, N. A. J., and Peacock, J. B. (1974). *Statistical Distributions*. Butterworth, London.

Hauck, W. W. Jr., and Doner, A. (1977). Wald's Test as Applied to Hypotheses in Logit Analysis. *Journal of the American Statistical Association*, **72**, 851–853.

Hersey, P., Edwards, A., Coates, A., Shaw, H., McCarthy, W., and Milton, G. (1987). Evidence That Treatment with Vaccinia Melanoma Cell Lysates (VMCL) May Improve Survival of Patients with Stage II Melanoma. *Cancer Immunology and Immunotherapy*, **25**, 257–265.

Hertz-Picciotto, I., Swan, S. H., Neutra, R. R., and Samuels, S. J. (1989). Spontaneous Abortions in Relation to Consumption of Tap Water: An Application of Methods from Survival Analysis to a Pregnancy Follow-Up Study. *American Journal of Epidemiology*, **130**, 79–83.

Hill, A. B. (1960a). *Controlled Clinical Trials*. Blackwell Scientific, Oxford, England.

Hill, A. B. (1960b). *Statistical Methods in Clinical and Preventive Medicine*. Oxford University Press, Oxford.

Hill, A. B. (1971). Principles of Medical Statistics. Oxford University Press, New York.

Hirayama, T. (1981). Non-smoking Wives of Heavy Smokers Have a Higher Risk of Lung Cancer: A Study from Japan. British Medical Journal, 282, 183–185.

Hoel, D. G., Sobel, M., and Weiss, G. H. (1975). A Survey of Adaptive Sampling for Clinical Trials. Perspectives in Biometrics, 1, 29–61.

Hogan, T. F., Koss, W., Murgo, A. J., Amato, S., Fontana, J. A., and VanScoy, F. L. (1987). Acute Lymphoblastic Leukemia with Chromosomal 5; 14 Translocation and Hypereosinophilia: Case Report and Literature Review. Journal of Clinical Oncology, 5, 382–390.

Hokanson, J. A., Brown, B. W., Thompson, J. R., Drewinko, B., and Alexanian, R. (1977). Tumor Growth Patterns in Multiple Myeloma. Cancer, 39, 1077–1084.

Holford, T. R., White, C., and Kelsey, J. L. (1978). Multivariate Analysis for Matched Case–Control Studies. American Journal of Epidemiology, 107, 245–256.

Hollander, M., and Proschan, F. (1979). Testing to Determine the Underlying Distribution Using Randomly Censored Data. Biometrics, 35, 393–401.

Hollander, M., and Wolfe, D. A. (1973). Nonparametric Statistical Methods. Wiley, New York.

Horner, R. D. (1987). Age at Onset of Alzheimer's Disease: Clue to the Relative Importance of Etiologic Factors? American Journal of Epidemiology, 126, 409–414.

Hosmer, D. W., and Lemeshow, S. (1980). A Goodness-of-Fit Test for the Multiple Logistic Regression Model. Communications in Statistics, A10, 1043–1069.

Hosmer, D. W., and Lemeshow, S. (1989). Applied Logistic Regression. Wiley, New York.

Howell, D. W. (1987). Statistical Methods for Psychology. Duxbury, Boston.

Hung, C. T., Lim, J. K. C., and Zoest, A. R. (1988). Optimization of High-Performance Liquid Chromatographic Analysis for Isoxazolye Penicillins Using Factorial Design. Journal of Chromatography, 425, 331–341.

Hyde, J. (1977). Testing Survival Under Right Censoring and Left Truncation. Biometrika, 64, 225–230.

Ingelfinger, F. J. (1970). Editorial on the Randomized Cervical Trial. New England Journal of Medicine, 287, 100–101.

Ingram, D. D., and Kleinman, J. C. (1989). Empirical Comparisons of Proportional Hazards and Logistic Regression Models. Statistics in Medicine, 8, 525–538.

Irwin, J. O. (1949). The Standard Error of an Estimate of Expectational Life. Journal of Hygiene, 47, 188–189.

Jennings, D. E. (1986). Judging Inference Adequacy in Logistic Regression. Journal of the American Statistical Association, 81, 471–476.

Johnson, L. G. (1964). The Statistical Treatment of Fatigue Experiments. Elsevier, New York.

Johnson, P., and Pearce, J. M. (1990). Recurrent Spontaneous Abortion and Polycystic Ovarian Disease: Comparison of Two Regimens to Induce Ovulation. British Medical Journal, 300, 154–156.

Jones, B., and Kenward, M. G. (1989). *Design and Analysis of Crossover Trials*. Chapman & Hall, London.

Juckett, D. A., and Rosenberg, B. (1990). Periodic Clustering of Human Disease— Specific Mortality Distributions by Shape and Time Position, and a New Integer-based Law of Mortality. *Mechanisms of Ageing and Development*, **55**, 255–291.

Khan H. A. (1983). *An Introduction to Epidemiologic Methods*. Oxford University Press, New York.

Kalbfleisch, J. D. (1974). Some Extensions and Applications of Cox's Regression and Life Model. Presented at the joint meeting of the Biometrick Society and the American Statistical Association, Tallahassee, FL, March 20–22, 1974.

Kalbfleisch, J. D., and Prentice, R. L. (1973). Marginal Likelihoods Based on Cox's Regression and Life Table Model. *Biometrika*, **60**, 267–278.

Kalbfleisch, J. D., and Prentice, R. L. (1980). *The Statistical Analysis of Failure Time Data*. Wiley, New York.

Kao, J. H. K. (1958). Computer Methods for Estimating Weibull Parameters in Reliability Studies. *I.R.E. Transactions on Reliability and Quality Control*, **PGRQC 13**, 15–22.

Kao, J. H. K. (1959). A Graphical Estimation of Mixed Weibull Parameters in Life-Testing of Electron Tubes. *Technometrics*, **1**, 389–407.

Kaplan, E. L., and Meier, P. (1958). Nonparametric Estimation from Incomplete Observations. *Journal of the American Statistical Association*, **53**, 457–481.

Kaplan, L. D., Abrams, D. I., Feigal, E., et al. (1989). AIDS-associated Non-Hodgkin's Lymphoma in San Francisco. *Journal of American Medical Association*, **261**, 719–724.

Kashiwagi, S., et al. (1985). Prevalence of Immunologic Markers of Hepatitis A and B Infection in Hospital Personnel in Miyazaki Prefecture, Japan. *American Journal of Epidemiology*, **122**, 960–969.

Kay, R. (1979). Proportional Hazard Regression Models and the Analysis of Censored Survival Data. *Applied Statistics*, **26**, 227–237.

Kay, R. (1984). Goodness of Fit Methods for the Proportional Hazards Model: A Review. *Revue Epidemiologie et de Santé Publique*, **32**, 185–198.

Kaye, F. J., Bunn, P. A., Steinberg, S. M., et al. (1989). A Randomized Trial Comparing Combination Electron-Beam Radiation and Chemotherapy with Topical Therapy in the Initial Treatment of Mycosis Fungoides. *New England Journal of Medicine*, **321**, 1784–1790.

Kelsey, J. L., Thompson, W. D., and Evans, A. S. (1986). *Methods in Observational Epidemiology*. Oxford University Press, New York.

Kemeny, N., Daly, P., et al. (1984). Randomized Study of Intrahepatic vs System Infusion of Fluorodeoxyuridine in Patients with Liver Metastases from Colorectal Carcinoma. *American Society for Clinical Oncology Abstracts*, **C-551**, 141.

Kemeny, N., Reichman, B., Geller, N., and Hollander, P. (1988). Implementation of the Group Sequential Methodology in a Randomized Trial in Metastatic Colorectal Carcinoma. *American Journal of Clinical Oncology*, **11**, 66–72.

Kennedy, A. D., and Gehan, E. A. (1971). Computerized Simple Regression

Methods for Survival Time Studies. *Computer Programs in Biomedicine*, **1**, 235–244.

Kimball, B. F. (1960). On the Choice of Plotting Position on Probabiity Paper. *Journal of the American Stastical Association*, **55**, 546–560.

King, J. R. (1971). *Probability Charts for Decision Making*. Industrial Press, New York.

King, M., Bailey, D. M., Gibson, D. G., Pitha, J. V., and McCay, P. B. (1979). Incidence and Growth of Mammary Tumors Induced by 7,12-Dimethylbenz(α)antheacene as Related to the Dietary Content of Fat and Antioxidant. *Journal of the National Cancer Institute*, **63**, 656–664.

Kirk, A. P., et al. (1980). Late Results of the Royal Free Hospital Prospective Controlled trial of Predimione Therapy in Hepatitis B Surface Antigen Negative Chronic Active Hepatitis. *Gut*, **21**, 78–83.

Kitagawa, E. M. (1964). Standardized Comparisons in Population Research. *Demography*, **1**, 296–315.

Kleinbaum, D. G., Kupper, L. L., and Muller, K. E. (1988). *Applied Regression Analysis and Other Multivariate Methods (2nd edition)*, PWS-KENT Publishing Company, Boston.

Kodlin, D. (1967). A New Response Time Distribution. *Biometrics*, **23**, 227–239.

Koziol, J., and Green, S. (1976). A Cramer-von Mises Statistic for Randomly Censored Data. *Biometrika*, **63**, 465–474.

Krishna, I. P. V. (1951). A Non-parametric Method of Testing k Samples. *Nature*, **167**, 33.

Kruskal, W. H., and Wallis, W. A. (1952). Use of Ranks in One-Criterion Variance Analysis. *Journal of the American Statistical Association*, **47**, 583–621.

Kuzma, J. W. (1967). A Comparison of Two Life Table Methods. *Biometrics*, **23**, 51–64.

Lachin, J. M. (1981). Introduction to Sample Size Determination and Power Analysis for Clinical Trials. *Controlled Clinical Trials*, **2**, 93–113.

Lachin, J. M., and Foulkes, M. A. (1986). Evaluation of Sample Size and Power for Analysis of Survival and Allowance for Nonuniform Patient Entry, Losses-to-Follow-up, Noncompliance, and Stratification. *Biometrics*, **42**, 507–519.

Lagakos, S. W. (1980). The Graphical Evaluation of Explanatory Variables in Proportional Hazard Regression Models. *Biometrika*, **68**, 93–98.

Lan, K. K. G., and DeMets, D. L. (1983). Discrete Sequential Boundaries for Clinical Trials. *Biometrika*, **70**, 659–663.

Lan, K. K. G., and DeMets, D. L. (1989). Changing Frequency of Interim Analysis in Sequential Monitoring. *Biometrics*, **45**, 1017–1020.

Lawless, J. F. (1982). *Statistical Methods and Model for Lifetime Data*. Wiley, New York.

Lawless, J. F. (1983). Statistical Methods in Reliability. *Technometrics*, **25**, 305–316.

Lee, E. T. (1980). *Statistical Methods for Survival Data Analysis*. Lifetime Learning Publications, Belmont, California.

Lee, E. T., and Thomas, D. R. (1980). Confidence Interval for Comparing Two Life Distributions. *IEEE Transactions on Reliability*, **R-29**, 51–56.

Lee, E. T., Ishmael, D. R., Bottomley, R. H., and Murray, J. L. (1982). An Analysis of Skin Tests and Their Relationship to Recurrence and Survival in Stage III and Stage IV Melanoma patients. *Cancer*, **49**, 1336–2341.

Lee, E. T., Desu, M. M., and Gehan, E. A. (1975). A Monte-Carlo Study of the Power of Some Two-Sample Tests. *Biometrika*, **62**, 425–432.

Lee, E. T., Ishmael, D. R., Bottomley, R. H., and Murray, J. L. (1982). An Analysis of Skin Tests and Their Relationship to Recurrence and Survival in Stage III and Stage IV Melanoma Patients, *Cancer*, **49**, 2336–2341.

Lee, E. T., Yeh, J. L., Cleves, M. A., and Shafer, D. (1988). Vascular Complications in Noninsulin Dependent Diabetic Oklahoma, Indians. *Diabetes*, **37** (Suppl. 1).

Lee, E. T., Lee, V. S., Lu, M., et al. (1992). The Development of Diabetic Retinopathy in NIDDM: A Twelve-Year Follow-Up Study in Oklahoma Indians. Unpublished manuscript.

Lee, E. T., Russell, D., Jorge, N., Kenny, S., and Yu, M. (in press). A Follow-up Study of Diabetic Oklahoma Indians: Mortality and Caues of Death. *Diabetes Care*.

Leenen, F. H. H., Balfe, J. A., Pelech, A. N., et al. (1987). Postoperative Hypertension after Repair of Coarctation of Aorta in Children: Protective Effect of Propranolol, *American Heart Journal*, **113**, 1164–1173.

Lehmann, E. L. (1953). The Power of Rank Tests. *Annals of Mathematical Statistics*, **24**, 23–43.

Lemeshow, S., and Hosmer, D. W. (1982). A Review of Goodness-of-Fit Statistics for Use in the Development of Logistic Regression Models. *American Journal of Epidemiology*, **115**, 92–106.

Leyland-Jones, B., Donnelly, H., Groshen, S., Myskowski, P., Donner, A. L., Fanucchi, M., Fox, J., and the Memorial Sloan-Kettering Antiviral Working Group. (1986). 2'-Fluror-5-Iodoarabinosylcytosine, A New Potent Antiviral Agent: Efficacy in Immunosuppressed Individuals with Herpes Zoster. *Journal of Infectious Diseases*, **154**, 430–436.

Liang, K. Y., Self, S. G., and Liu , X. (1990). The Cox Proportional Hazards Model with Change Point: An Epidemiologic Application. *Biometrics*, **46**, 783–793.

Lieblein, J., and Zelen, M. (1956). Statistical Investigation of the Fatigue Life of Deep-Grove Ball Bearings. *Journal of Research*, National Bureau of Standards, **57**, 273–316.

Lilliefors, H. W. (1971). *Reducing the Bias of Estimators of Parameters for the Erlang and Gamma Distribution*. Unpublished manuscript.

Lindley, D. V. (1968). The Choice of Variables in Multiple Regression. *Journal of the Royal Stastical Society (B)*, **30**, 31–53.

Lippman, S. M., Alberts, D. S., Slymen, D. J., et al. (1988). Second-Look Laparotomy in Epithelial Ovarian Carcinoma—Prognostic Factors Associated with Survival Duration. *Cancer*, **61**, 2571–2577.

Liu, P. Y., and Crowely, J. (1978). Large Sample Theory of the MLE Based on Cox's Regression Model for Survival Data. Technial Report No. 1. Wisconsin Clinical Cancer Center (Biostatistics), University of Wisconsin, Madison.

Lubin, J. H. (1981). A Computer Program for the Analysis of Matched Case–Control Studies. *Computers and Biomedical Research*, **14**, 138–143.

McAlister, D. (1879). The Law of the Geometric Mean. *Proceedings of the Royal Society*, **29**, 367.

McCracken, D. D., and Dorn, W. S. (1964). *Numerical Methods and Fortran Programming*. Wiley, New York.

McFadden, D. (1976). A Comment on Discriminant Analysis "Versus" Logit Analysis. *Annals of Economic and Social Measurement*, **5**, 511–523.

McHugh, R., and Matts, J. (1983). Post-Stratification in the Randomized Clinical Trial. *Biometrics*, **39**, 217–225.

McPherson, C. K., and Armitage, P. (1971). Repeated Significance Tests on Accumulating Data When the Null Hypothesis Is not True. *Journal of the Statistical Society, Series A*, **134**, 15–25.

McPherson, K. (1974). Statistics: The Problem of Examining Accumulating Data More Than Once. *New England Journal of Medicine*, **290**, 501–502.

McPherson, K. (1984). Interim Analysis and Stopping Rules. *Cancer Clinical Trials: Methods and Practice*. Oxford University Press, New York.

Mainland, D. (1960). The Clinical Trial—Some Difficulties and Suggestions. *Journal of Chronic Diseases*, **11**, 484–496.

Makuch, R. W., and Simon, R. M. (1982). Sample Size Requirements for Comparing Time-to-Failure among k Treatment Groups. *Journal of Chronic Diseases*, **35**, 861–867.

Mann, H. B., and Whitney, D. R. (1947). On a Test of Whether One of Two Random Variables Is Stochastically Larger Than the Other. *Annals of Mathematical Statistics*, **18**, 50–60.

Mann, N. R. (1968). Point and Interval Estimation Procedures for the Two-Parameter Weibull and Extreme-Value Distributions. *Technometrics*, **10**, 231–256.

Mann, N. R. (1970). Estimators and Exact Confidence Bounds for Weibull Parameters Based on a Few Ordered Observations. *Technometrics*, **12**, 345–361.

Mann, N. R., Schafer, R. E., and Singpurwalla, N. D. (1974). *Methods for Statistical Analysis of Reliability and Life Data*. Wiley, New York.

Manninen, O. (1988). Changes in Hearing, Cardiovascular Functions, Haemodynamics, Upright Body Sway, Urinary Catecholamines and Their Correlates after Prolonged Successive Exposure to Complex Environmental Conditions. *International Archives of Occupational and Environmental Health*, **60**, 249–272.

Mantel, N. (1966). Evaluation of Survival Data and Two New Rank Order Statistics Arising in Its Consideration. *Cancer Chemotherapy Reports*, **50**, 163–170.

Mantel, N. (1967). Ranking Procedures for Arbitrarily Restricted Observations. *Biometrics*, **23**, 65–78.

Mantel, N. (1970). Why Stepdown Procedures in Variable Selection. *Technometrics*, **12**, 621–625.

Mantel, N. (1977). Test and Limits for the Common Odds Ratio of Several 2×2 Contingency Tables: Methods in Analogy with the Mantel-Haenszel Procedure. *Journal of Statistical Planning Information*, **1**, 179–189.

Mantel, N., and Haenszel, W. (1959). Statistical Aspects of the Analysis of Data from Retrospective Studies of Disease. *Journal of the National Cancer Institute*, **22**, 719–748.

Mantel, N., and Hankey, B. F. (1978). A Logistic Regression Analysis of Response– Time Data Where the Hazard Function Is Time Dependent. *Communications in Statistics Theory and Methods*, **A7**, 333–347.

Mantel, N., and Myers, M. (1971). Problems of Converegence of Maximum Likelihood Iterative Procedures in Multiparameter Situation. *Journal of the American Statistical Association*, **66**, 484–491.

Mantel, N., and Stark, C. R. (1968). Computation of Indirect Adjusted Rates in the Presence of Confounding. *Biometrics*, **24**, 997–1005.

Marascuilo, L. A., and McSweeney, M. (1977). *Nonparametric and Distribution- Free Methods for the Social Sciences*. Brooks/Cole, Monterey, CA.

Matthews, D. E., and Farewell, V. (1985). *Using and Understanding Medical Statistics*. Karger, New York.

Matthews, J. N. S. (1987). Optimal Crossover Designs for the Comparison of Two Treatments in the Presence of Carryover Efects and Autocorrelated Errors. *Biometrika*, **74**, 311–320; Correction, **75**, 396.

Mausner, J. S., and Kramer, S., (1985). *Epidemiology, An Introductory Text*. W. B. Saunders Company, Philadelphia.

Meier, P. (1975a). Statistics and Medical Experimentation. *Biometrics*, **31**, 511–529.

Meier, P. (1975b). Estimation of a Distribution Function from Incomplete Observa- tions, in *Perspectives in Probability and Statistics*, edited by J. Gaui. Applied Probability Trust, Sheffield, England.

Meier, P. (1981). Stratification in the Design of a Clinical Trial. *Controlled Clinical Trials*, **1**, 355–361.

Meinert, C. L. (1986). *Clinical Trials: Design, Conduct, and Analysis*. Oxford University Press, New York.

Menon, M. V. (1963). Estimation of the Shape and Scale Parameter of the Weibull Distribution. *Technometrics*, **5**, 175–182.

Micciolo, R., Valagussa, P., and Marubini, E. (1985). The Use of Hisotrical Controls in Breast Cancer. *Controlled Clinical Trials*, **6**, 259–270.

Miettinen, O. S. (1979). Comments on "Confidence Intervals for the Odds Ratio in Case-Control Studies: the State of the Art" by J. L. Fleiss. *Journal of Chronic Diseases*, **32**, 80–82.

Miller, R. G., Jr. (1966). *Simultaneous Statistical Inference*. McGraw-Hill, New York.

Miller, R. G. (1981). *Survival Analysis*. Wiley, New York.

Minow, R. A., Benjamin, R. S., Lee, E. T., and Gottlieb, J. A. (1977). Adriamycin Cardiomyopathy—Risk Factors. *Cancer*, **39**, 1397–1402.

Molloy, D. W., Guyatt, G. H., Wilson, D. B., et al. (1991). Effect of Tetrahydro- aminoacridine on Cognition, Function and Behaviour in Alzheimer's Disease. *Canadian Medical Association Journal*, **144**, 29–34.

Montaner, J. S. G., Lawson, L. M., Levitt, N., et al. (1990). Costicorsteroids Prevent Early Deterioration in Patients with Moderately Severe Pneumocystis

Carinii Pneumonia and the Acquired Immunodeficiency Syndrome (AIDS). *Annals of Internal Medicine*, **113**, 14–20.

Moolgavkar, S., Lustbader, E., and Venzon, D. J. (1985). Assessing the Adequacy of the Logistic Regression Model for Matched Case–Control Studies. *Statistics in Medicine*, **4**, 425–435.

Morabito, A., and Marubini, E. (1976). A Computer Suitable for Fitting Linear Models When the Dependent Variable is Dichotomous, Polichotomous or Censored Survival and Non-Linear Models When the Dependent Variable Is Quantative. *Computer Programs in Biomedicine*, **5**, 283–295.

Moreau, T., O'Quigley, J., and Mesbah, M. (1985). A Global Goodness-of-Fit Statistic for the Proportional Hazards Model. *Applied Statistics*, **34**, 212–218.

Morrison, D. F. (1967). *Multivariate Statistical Methods*, McGraw-Hill, New York.

Moussa, M. A. (1988). Planning the Size of Survival Time Clinical Trials with Allowance for Stratification. *Statistics in Medicine*, 7, 5, 559–569.

Mulders, P. F. A., Dijkman, G. A., et al. (1990). Analysis of Prognostic Factors in Disseminated Prostatic Cancer. *Cancer*, **65**, 2758–2761.

Myers, M., Hankey, B. F., and Mantel, N. (1973). A Logistic–Exponential Model for Use with the Response–Time Data Involving Regressor Variables. *Biometrics*, **29**, 257–269.

Myers, M. H. (1969). A Computing Procedure for a Significance Test of the Difference between Two Survival Curves, Methodological Note No. 18 in *Methodological Notes*. End Results Sections, National Cancer Institute, National Institutes of Health, Bethesda, MD.

Nadas, A. (1970). On Proportional Hazard Functions. *Technometrics*, **12**, 413–416.

National Cancer Institute (1970). *Proceedings of the Symposium on Statistical Aspects of Protocol Design*, San Juan, Puerto Rico, December 9–10, 1970.

Natrella, M. G. (1963). *Experimental Statistics*, National Bureau of Standards Handbook 91. Government Printing Office, Washington DC, Tables A-25, A-26.

Nelson, W. (1972). Theory and Applications of Hazard Plotting for Censored Failure Data. *Technometrics*, **14**, 945–966.

Nelson, W. (1982). *Applied Life Data Analysis*. Wiley, New York.

Nemenyi P. (1963). Distribution-Free Multiple Comparisons. Ph.D. Thesis, Princeton University.

Neter, J., and Wasserman, W. (1974). *Applied Linear Statistical Models*. Richard D. Irwin, Homewood, IL.

Nie, N. H., Hull, C. H., Jenkins, J. G., Steinbrenner, K., and Bent, D. H. (1975). *SPSS—Statistical Package for the Social Scienices*. McGraw-Hill, New York.

Niederjohn, R. J., and Haworth, D. J. (1986). Speech Amplitude Statistics and the Gamma Density Function as a Theoretical Approximation. *Journal of the Acoustic Society of America*, **80**, 1583–1588.

Numerical Algorithms Group (1987). *The Generalized Linear Interactive Modeling System: The GLIM System*. Royal Statistical Society, London.

O'Brien, P. C., and Fleming, T. R. (1979). A Multiple Testing Procedure for Clinical Trials. *Biometrics*, **35**, 549–556.

Oliver, I. N., et al. (1988). Nitrogen Mustard, Vincristine, Procarbazine, and for Relapse after Radiation in Hodgkin's Disease. *Cancer*, **62**, 233–239.

Osgood, E. W. (1958). Methods for Analyzing Survival Data, Illustrated by Hodgkin's Disease. *American Journal of Medicine*, **24**, 40–47.

Ozer, H., Golomb, H. M., Zimmerman, H., and Spiegel, R. J. (1989). Cost–Benefit Analysis of Interferon Alfa-2b in Treatment of Hairy Cell Leukemia. *Journal of the National Cancer Institute*, **81**, 594–602.

Packard, F. R. (1925). *The Life and Times of Ambroise Paré*. Paul B, Hoeber, New York.

Parker, R. L., Dry, T. J., Willius, F. A., and Gage, R. P. (1946). Life Expectancy in Angina Pectoris. *Journal of the American Medical Association*, **131**, 95–100.

Pearson, E. S., and Hartely, N. O. (1958). *Biometrika Tables for Statiscians*, Vol. 1. Cambridge University Press, Cambridge.

Pearson, K. (1922, 1957). *Tables of the Incomplete Γ-function*. Cambridge University Press, Cambridge.

Pershagen, G. (1986). Review of Epidemiology in Relation to Passive Smoking. *Archive of Toxicology*, **9**, (Suppl.) 63–73.

Pershagen, G., Hrubec, Z., and Svensson, C. (1987). Passive Smoking and Lung Cancer in Swedish Women. *American Journal of Epidemiology*, **125**, 17–24.

Peto, R., and Lee, P. N. (1973). Weibull Distributions for Continuous Carcinogenesis Experiments. *Biometrics*, **29**, 457–470.

Peto, R., and Peto, J. (1972). Asymptotically Efficient Rank Invariant Procedures. *Journal of the Royal Statistical Society*, Series A, **135**, 185–207.

Peto, R., Lee, P. N., and Paige, W. S. (1972). Statistical Analysis of the Bioassay of Continuous Carcinogens. *British Journal of Cancer*, **26**, 258–261.

Peto, R., Pike, M. C., Armitage, P., Breslow, N. E., Cox, D. R., Howard, S. V., Mantel, N., McPherson, K., Peto, J., and Smith, P. G. (1976, 1977). Design and Analysis of Randomized Clinical Trials Requiring Prolonged Observation of Each Patient. *British Journal of Cancer*, Part I, **34**, 585–612, 1976; Part II, **35**, 1–39, 1977.

Pierce, M., Borges, W. H., Heyn, R., Wolfe, J., and Gilbert, E. S. (1969). Epidemiological Factors and Survival Experience in 1770 Children with Acute Leukemia. *Cancer*, **23**, 1296–1304.

Pike, M. C. (1966). A Method of Analysis of a Certain Class of Experiments in Carcinogenesis. *Biometrics*, **22**, 142–161.

Piper, J. M., Matanoski, G. M., and Tonascia, J. (1986). Bladder Cancer in Young Women. *American Journal of Epidemiology*, **123**, 1033–1042.

Pocock, S. J. (1976a). The Combination of Randomized and Historical Controls in Clinical Trials. *Journal of Chronic Diseases*, **29**, 175–188.

Pocock, S. J. (1976b). Randomized versus Historical Control—A Compromise Solution. *Proceedings of the 9th International Biometric Conference*, **I**,-245–260.

Pocock, S. J. (1977). Group Sequential Methods in the Design and Analysis of Clinical Trials. *Biometrika*, **64**, 191–199.

Pocock, S. J. (1982). Interim Analyses for Randomized Clinical Trials: The Group Sequential Approach. *Biometrics*, **38**, 153–162.

Pocock, S. J. (1983). *Clinical Trials, A Practical Approach.* Wiley, New York.

Pocock, S. J., and Simon, R. (1975). Sequential Treatment Assignment with Balancing for Prognostic Factors in the Control Clinical Trials. *Biometrics*, **31**, 103–115.

Pregibon, D. (1984). Data Analytic Methods for Matched Case–Control Studies. *Biometrics*, **40**, 639–651.

Prentice, R. L. (1973). Exponential Survivals with Censoring and Explanatory Variables. *Biometrika*, **60**, 279–288.

Prentice, R. L. (1976). Use of the Logistic Model in Retrospective Studies. *Biometrics*, **32**, 599–606.

Prentice, R. L., and Gloeckler, L. A. (1978). Regression Analysis of Grouped Survival Data with Application to Breast Cancer Data. *Biometrics*, **34**, 57–67.

Prentice, R. L., and Kalbfleisch, J. D. (1979). Hazard Rate Models with Covariates. *Biometrics*, **35**, 25–39.

Prentice, R. L., and Marek, P. (1979). A Quantitative Discrepancy between Censored Data Rank Tests. *Biometrics*, **35**, 861–867.

Press, S. J. (1972). *Applied Multivariate Analysis.* Holt, Rinehart & Winston, New York.

Press, S. J., and Wilson, S. (1978). Choosing between Logistic Regression and Discriminant Analysis. *Journal of the American Statistical Association*, **73**, 699–705.

Ragland, D. R., and Brand, R. J. (1988). Coronary Heart Disease Mortality in the Western Collaborative Group Study. *American Journal of Epidemiology*, **127**, 462–475.

Ralston, A., and Wilf, H. (1967). *Mathematical Methods for Digital Computers.* Wiley, New York.

Rank, F., Dombernowsky, P., Jespersen, N. C. B., Pedersen, B. V., and Keiding, N. (1987). Histologic Malignancy Grading of Invasive Ductal Breast Carcinoma. *Cancer*, **60**, 1299–1305.

Rao, C. R. (1952). *Advanced Statistical Methods in Biometric Research.* Wiley, New York.

Rao, C. R. (1973). *Linear Statistical Inference and Its Application*, 2nd ed. Wiley, New York.

Reaman, G. H., Steinherz, P. G., Gaynon, P. S., Bleyer, W. A., et al. (1987). Improved Survival of Infants Less Than 1 Year of Age with Acute Lymphoblastic Leukemia Treated with Intensive Multiagent Chemotherapy[1,2,3]. *Cancer Treatment Reports*, **71**, 1033–1038.

Rosner, G. L., and Tsiatis, A. A. (1989). The Impact That Group Sequential Tests Would Have Made on ECOG Clinical Trials. *Statistics in Medicine*, **8**, 505–516.

Rowe-Jones, D. C., Peel, A. L. G., Kingston, R. D., Shaw, J. F. L., Teasdale, C., and Cole, D. S. (1990). Single Dose Cefotaxime Plus Metronidazole versus Three Dose Cefuroxime Plus Metronidazole as Prophylaxis against Wound Infection in Colorectal Surgery: Multicentre Prospective Randomised Study. *British Medical Journal*, **300**, 18–22.

Roy, S. N., Gnadesikan, R., and Srivastava, J. N. (1971). *Analysis and Design of Certain Quantitative Multiresponse Experiments.* Pergamon, New York.

Rümke, C. L. (1963). Clinical Application of Sequential Analysis: An Introductory Review. *Clinical Pharmacological Therapeutics*, **34**, 531.

Sacher, G. A. (1956). On the Statistical Nature of Mortality, with Special Reference to Chronic Radiation Mortality. *Radiology*, **67**, 250–257.

Sacks, H., Chalmers, T. C., and Smith, H. (1982). Randomized versus Historical Control for Clinical Trials. *American Journal of Medicine*, **72**, 233–240.

Sarhan, A. A., and Greenberg, B. G. (1956). Estimation of Location and Scale Parameters by Order Statistics from Singly and Doubly Censored Samples, Part I, The Normal Distribution up to Samples of Size 10. *Annals of Mathematical Statistics*, **27**, 427–451.

Sarhan, A. E., and Greenberg, B. G. (1957). Estimation of Location and Scale Parameters by Order Statistics from Singly and Doubly Censored Samples, Part III. Technical Report No. 4-OOR Project 1597, U.S. Army Research Office.

Sarhan, A. E., and Greenberg, B. G. (1958). Estimation of Location and Scale Parameters by Order Statistics from Singly and Doubly Censored Samples, Part II. *Annals of Mathematical Statistics*, **29**, 79–105.

Sarhan, A. E., and Greenberg, B. G. (1962). *Contribution to Order Statistics*, Wiley, New York.

SAS Institute. (1990). *SAS/STAT User's Guide*, Vols. 1 and 2, Version 6 4th ed. SAS Institute, Gary, NC.

Savage, I. R. (1956). Contributions to the Theory of Rank Order Statistics—The Two Sample Case. *Annals of Mathematical Statistics*, **27**, 590–615.

Saw, J. G. (1959). Estimation of the Normal Population Parameters Given A Singly Censored Sample. *Biometrika*, **46**, 150–159.

Schade, D. S., Mitchell, W. J., and Griego, G. (1987). Addition of Sulfonylurea to Insulin Treatment in Poorly Controlled Type II Diabetes. *Journal of the American Medical Association*, **257**, 2441–2445.

Schenker, M. B., Samuels, S. J., Green, R. S. and Wiggins, P. (1990). Adverse Reproductive Outcomes among Female Veterinarians, *American Journal of Epidemiology*, **132**, 96–106.

Schlesselman, J. J. (1982). *Case–Control Studies*, Oxford University Press, New York.

Schoenfeld, D. A., and Richter, J. R. (1982). Nomograms for Calculating the Number of Patients Needed for a Clinical Trial with Survival as an Endpoint. *Biometrics*, **38**, 163–170.

Schwartzbaum, J. A., Hulka, B. S., Fowler, W. C., Jr. Kaufman, D. G., and Hoberman, D. (1987). The Influence of Exogenous Estrogen Use on Survival after Diagnosis of Endometrial Cancer. *American Journal of Epidemiology*, **126**, 851–860.

Scott, B. R., and Hahn, F. F. (1980). A Model that Leads to the Weibull Distribution Function to Characterize Early Radiation Response Probabilities. *Health Physics*, **39**(3), 521–530.

Segal, M. R., and Bloch, D. A. (1989). A Comparison of Estimated Proportional Hazards Models and Regression Trees. *Statistics in Medicine*, **8**, 539–550.

Sellke, T., and Siegmund, D. (1983). Sequential Analysis of the Proportional Hazards Model. *Biometrika*, **70**, 315–326.

Shapiro, S. S., and Wilk, M. B. (1965). An Analysis of Variance Test for Normality (Complete Samples). *Biometrika*, **52**, 591.

Shapiro, S. S., and Wilk, M. B. (1965). Testing for Distributional Assumptions— Exponential and Uniform Distributions, unpublished.

Shryock, H. S., Sigel, J. S., and Associates (1971). *The Methods and Materials of Demography*, Vols. I and II. U.S. Dept. of Commerce, Bureau of the Census, U.S. Government Printing Office, Washington DC.

Shulz, D., Chernichovsky, D., and Allweis, C. (1986). A Novel Method for Quantifying Passive–Avoidance Behavior Based on the Exponential Distribution of Step-Through Latencies. *Pharmacology, Biochemistry and Behavior*, **25**, 979–983.

Sichieri, R., Everhart, J. E., and Roth, H. P. (1990). Low Incidence of Hospitalization with Gallbladder Disease among Blacks in the United States. *American Journal of Epidemiology*, **131**, 826–835.

Sillitto, G. P. (1949). Note on Approximations to the Power Function of the "2×2 Comparative Trial." *Biometrika*, **36**, 347–352.

Simon, R. (1979). Restricted Randomization Design in Clinical Trials. *Biometrics*, **35**, 503–512.

Slud, E. V., and Wei, L. J. (1982). Two-Sample Repeated Significance Tests Based on the Modified Wilcoxon Statistic. *Journal of the American Statistical Society*, **77**, 862–868.

Smith, T. L., Putman, J. E., and Gehan, E. A. (1970). A Computer Program for Estimating Survival Functions from the Life Table. *Computer Programs in Biomedicine*, **1**, 59–64.

Snedecor, G. W., and Cochran, W. G. (1967). *Statistical Methods*. The Iowa State University Press, Ames, IA.

SPSS. (1988). *SPSS-S User's Gudie*, 3rd ed. SPSS, Chicago, IL.

Stampfer, M. J., Buring, J. E., Willett, W., Rosner, B., Eberlein, K. Y., and Hennekens, C. H. (1985). The 2×2 Factorial Design: Its Application to a Randomized Trial of Aspirin and Carotene in U.S. Physicians. *Statistics in Medicine*, **4**, 111–116.

Staquet, M. J., editor (1978). *Radomized Trials in Cancer: A Critical Review by Sites*. Raven Press, New York.

Statistics and Epidemiology Research Corporation (SERC) (1988). *EGRET Statistical Software*. SERC, Seattle, WA.

Steering Committee on the Physicians Health Study Research Group. (1989). Final Report on the Aspirin Component of the Ongoing Physicians' Health Study. *New England Journal of Medicine*, **321**, 129–135.

Sullivan, M. P., Humphrey, G. B., Vietti, T. J., Haggard, M. E., and Lee, E. (1975). Superiority of Conventional Intrathecal Methotrexate Therapy with Maintenance over Intensive Intrathecal Methotraxate Therapy, Unmaintained, or Radiotherapy (2000–2500 Rads Tumor Dose) in Treatment for Meningeal Leukemia. *Cancer*, **35**, 1066–1073.

Tarone, R. E. (1982). The Use of Historical Control Information in Testing a Trend in Poisson Means. *Biometrics*, **38**(2), 457–462.

Tarone, R.E., and Ware, J. (1977). On Distribution-free Tests for Equality of Survival Distribution. *Biometrics*, **64**, 156–160.

Thoman, D. R., and Bain, L. J. (1969). Two Sample Tests in the Weibull Distribution. *Technometrics*, **11**, 805–815.

Thoman, D. R., Bain, L. J., and Antle, C. E. (1969). Inferences on the Parameters of the Weibull Distribution. *Technometrics*, **11**, 445–460.

Thoman, D. R., Bain, L. J., and Antle, C. E. (1970). Maximum Likelihood Estimation, Exact Confidence Intervals for Reliability and Tolerance Limits in the Weibull Distribution, *Technometrics*, **12**, 363–373.

Teitelman, A. M., Welch, L. S., Hellenbrand, K. G., and Bracken, M. B. (1990). Effect of Maternal Work Activity on Preterm Birth and Low Birth Weight. *American Journal of Epidemiology*, **131**, 104–113.

Tsiatis, A. A. (1980). A Note of a Goodness-of-Fit Test for the Logistic Regression Model. *Biometrika*, **67**, 250–251.

Tsiatis, A. A. (1981). A Large Sample Study of Cox's Regression Model. *Annals of Staitsitics*, **9**, 93–108.

Tsiatis, A. A. (1982). Repeated Significance Testing for a General Class of Statistics Used in Censored Survival Analysis. *Journal of the American Statistical Association*, **77**, 855–861.

Truett, J., Cornfield, J., and Kannel, W. B. (1967). A Multivariate Analysis of the Risk of Coronary Heart Disease in Framingham. *Journal of Chronic Diseases*, **20**, 511–524.

Tukey, J. W. (1977). Some Thoughts on Clinical Trials, Especially Problems of Multiplicity. *Science*, **198**, 4318.

Vega, G. L., and Grundy, S. M. (1989). Comparison of Lovastatin and Gemfibrozil in Normolipidemic Patients with Hypoalphalipoproteinemia. *Journal of the American Medical Association*, **262**, 3148–3153.

Wald, A. (1947). *Sequential Analysis*. Wiley, New York.

Walle, A. J., Al-Katib, A., Wong, G. Y., Jhanwar, S. C., Chaganti, R. S. K., and Koziner, B. (1987). Multiparameter Characterization of L3 Leukemia Cell Populations. *Leukemia Research*, **11**, 73–83.

Watson, G. S., and Wells, W. T. (1961). On the Possibility of Improving the Mean Useful Life of Items by Eliminating Those with Short Lives. *Technometrics*, **3**, 281–298.

Wei, L. J. (1984). Testing Goodness of Fit for Proportional Hazards Model with Censored Observations. *Journal of the American Statistical Association*, **79**, 649–652.

Weibull, W. (1939). A Statistical Theory of the Strength of Materials. *Ingeniors Vetenskaps Akakemien Handlingar*, **151**, The Phenomenon of Rupture in Solids, 293–297.

Weibull, W. (1951). A Statistical Distribution of Wide Applicability. *Journal of Applied Mechanics*, **18**, 293–297.

Weiss, H. (1963). A Survey of Some Mathematical Methods in the Theory of Reliability, in *Statistical Theory of Reliability*, edited by M. Zelen. University of Wisconsin Press, Madison, Wisconsin.

Whayne, T. F., Alaupovic, P., Curry, M. D., Lee, E. T., Anderson, P. S., and Schechter, E. (1981). Plasma Apolipoprotein B and VLDL-, LDL-, and HDL-

Cholesterol as Risk Factors in the Development of Coronary Artery Disease in Male Patients Examined by Angiography. *Atherosclerosis*, **39**, 411–424.

Whitehead, J., and Stratton, I. (1983). Group Sequential Clinical Trials with Triangular Continuation Regions. *Biometrics*, **39**, 227–236.

Wilcoxon, F. (1945). Individual Comparison by Ranking Methods. *Biometrics*, **1**, 80–83.

Wilk, M. B., Gnanadesikan, R., and Huyett, M. J. (1962a). Estimation of Parameters of the Gamma Distribution Using Order Statistics. *Biometrika*, **49**, 525–545.

Wilk, M. B., Gnanadesikan, R., and Huyett, M. J. (1962b). Probability Plots for the Gamma Distribution. *Technometrics*, **4**, 1–20.

Wilkinson, L. (1987). *SYSTAT: The System for Statistics*. Systat Inc., Evanston, Illinois.

Wilks, S. S. (1948). Order Statistics. *Bulletin of the American Mathematical Society*, **54**, 6–50.

Wilks, S. S. (1950). *Mathematical Statistics*. Princeton University Press, Princeton, NJ.

Williams, C. A. Jr. (1950). On the Choice of the Number and Width of Classes for the Chi-Square Test of Goodness of Fit. *Journal of the American Statistical Association*, **45**, 77–86.

Williams, J. S. (1978). Efficient Analysis of Weibull Survival Data from Experiments on Heterogenous Patient Populations. *Biometrics*, **34**, 209–222.

Winkleby, M. A., Ragland, D. R., and Syme, L. (1988). Self-Reported Stressors and Hypertension: Evidence of an Inverse Association. *American Journal of Epidemiology*, **127**, 124–134.

Winter, F. D., Snell, P. G., and Stray-Gundersen J. (1989). Effects of 100% Oxygen on Performance of Professional Soccer Players. *Journal of the American Medical Association*, **262**, 227–229.

Witts, L. J. (1964). *Medical Surveys and Clinical Trials*. Oxford University Press, London.

Woolf, B. (1955). On Estimating the Relation between Blood Group and Disease. *Annals of Human Genetics*, **19**, 251–253.

Wu, M., Fisher, M., and Demets, D. (1980). Sample Sizes for Long-Term Medical Trials with Time-dependent Dropout and Event Rates. *Controlled Clinical Trials*, **1**, 109–121.

Zelen, M. (1966). Applications of Exponential Models to Problems in Cancer Research. *Journal of the Royal Statistical Society*, Series A, **129**, 368–398.

Zelen, M. (1974). The Randomization and Stratification of Patients to Clinical Trials. *Journal of Chronic Diseases*, **27**, 365–375.

Zelen, M. (1979). A New Design for Randomized Clinical Trials. *New England Journal of Medicine*, **300**, 1242–1245.

Zippin, C., and Armitage, P. (1966). Use of Concomitant Variables and Incomplete Survival Information in the Estimation of an Exponential Survival Parameter. *Biometrics*, **22**, 665–672.

Index

*Now available in a lower priced paperback edition in the Wiley Classics Library.